ORNITHOLOGY

in Laboratory and Field

A reconstruction of *Archaeopteryx*. Painting by Rudolf Freund, courtesy of Carnegie Museum.

Fourth Edition

ORNITHOLOGY
in Laboratory and Field

OLIN SEWALL PETTINGILL, JR.

Laboratory of Ornithology
Cornell University
Ithaca, New York

Illustrated by
WALTER J. BRECKENRIDGE
University of Minnesota
Museum of Natural History

Burgess Publishing Company

Minneapolis, Minn.

Library of Congress Catalog Card Number 79-119563
ISBN 8087-1609-3

10   11   12   13   14   15   16   17   18   19   20

To the memory
of
ALFRED O. GROSS
(1883-1970)
Professor of Biology at Bowdoin College,
pre-eminent ornithologist and inspiring teacher.

PREFACE

The great disadvantage in writing a text in one's early years of teaching is that he will see it go out of date and become a source of embarrassment. Since 1937 I have brought this book through five stages of development, the first four under the title *A Laboratory and Field Manual of Ornithology*. The first stage of the book was a series of typewritten study directions, mimeographed, for class work in laboratory and field. Its second stage, reached in 1939, consisted of more extensive directions and was published by Burgess Publishing Company in mimeoprint, spiral bound. Its third stage, designated the second edition, with some explanatory text for the first time, was published in 1946, in photo offset, also spiral bound.

Encouraged as well as complimented in the years following by the ever-wider use of the book in colleges and universities, I became at the same time increasingly and unhappily conscious of its almost purely descriptive and insufficiently ecological approach to ornithology. Concepts of avian evolution, the functional aspects of anatomy, and the more comprehensive studies of birds in the field were coming to the fore as the principal objectives of ornithologists. Thus in 1956 I brought out the third edition, extensively revised and appropriately expanded, including more text, and this time in hard covers. My contentedness was short-lived, however, for in the next few years came a veritable explosion of new information and concepts, generated partly through the use of more sophisticated instrumentation in investigations but in a still greater measure from the increased concentration on studies of behavior. "So far behind the times" had the book become that I was frankly embarrassed to have my own classes use it.

With renewed contentment, temporary though it may be, I offer this edition, newly titled *Ornithology in Laboratory and Field*. Like its predecessors, it is intended as an aid to ornithological study at the college or university level. If a student lacks the background knowledge usually acquired during a course in general zoology or biology, he should keep handy for ready reference a standard elementary text on the subject.

In this book I make no pretense of covering the entire ornithological field. Somewhat arbitrarily I have selected those important aspects of ornithology that can be studied during a semester course or summer session of the academic year. There is much more subject matter than can be included in any one course; the instructor must make his own choice of topics. I assume that he will wish to supplement this book with other topics that he deems suited to the objectives of his course.

This edition contains extensive material for purely informational reading, possibly enough to supplant the need of an additional textbook. Nevertheless its principal purpose still complies with the title of its predecessors for it is essentially a *manual* to guide and assist the student in direct observations. All twenty sections, except the last ("The Origin, Evolution, and Decrease of Birds"), suggest methods and provide instructions for studies; and all conclude with an extensive list of references, frequently annotated, for further information.

The twenty sections of this book are more or less independent units. While they are presented in a fairly logical sequence, they can be taken up in almost any order and some may be omitted without affecting the instructional value of the others. (For this reason I have avoided the designation "chapters" which connotes a sequential series of steps in the development of a subject.) Since ornithology courses differ in accordance with time of the year given, number of class hours per week, and emphasis or objectives, each instructor can adapt the book to the particular needs of his course by selecting the sections in the order and number that seem the most workable.

A feature of this new edition is an introduction to birds and ornithology, intended for reading at the beginning of a course. The purpose is twofold: to show the significance of birds for study and to give an overall preview of ornithology, *the subject*, with emphasis on its wide scope, how it is studied, and some of the continuing and exciting opportunities that it offers for investigation.

In order to use this book effectively, certain equipment and materials are necessary. The student should provide himself with:

Binocular.

Field guide.

Key to the bird species found in the region where the study is being made.

Daily field check-lists (as many as there will be field trips).

Loose-leaf pocket notebook, preferably with aluminum cover.

A set of colored crayons.

The institution should make available to each student the following:

The American Ornithologists' Union's *Check-list of North American Birds*, latest edition.

Annotated check-list of birds found in the region of the study.

An atlas of the world.

Telescope (preferably the so-called "spotting scope") with either interchangeable eyepieces or a "zoom" lens and a tripod.

Portable tape recorder.

Compound microscope.

Meter stick.

A Common Pigeon and a House Sparrow, properly preserved but not plucked, for external study and dissection.

Several nestling House Sparrows preserved in spirit for study of feather tracts.

Pigeon skeleton, mounted.

Pigeon skeleton, completely disarticulated and mounted on a plaque.

Human skeleton, mounted; or a detailed chart of a human skeleton.

Human cervical vertebra (fifth).

Hyoid apparatus of a woodpecker.

A series of contour feathers illustrating specialized feather types. Some of the more exotic types may be obtained from zoos which usually save such material for educational institutions.

A semiplume, an adult down feather, a filoplume, a bristle, and a section of the vane of a contour feather mounted on slides for microscopic study.

Parts of feathers with different colors—red, orange, yellow, black, gray, brown, tawny, green (one from turaco, one not), white, blue, and iridescent—mounted on slides for microscopic study of color-producing elements.

A series of wings of a passerine species (e.g., the House Sparrow or the Starling) to illustrate the progress of molt.

A collection of bird skins representing all the orders and families of North American birds and all species found in the region of the study. (If possible, the collection should be sufficiently comprehensive to show sex, age, and other constant differences in plumage; abnormal plumage coloration; color phases; eclipse plumage; and plumage changes by wear and fading.) A transparent plastic tube (capped) for each of the smaller skins that will be handled often is recommended in order to prevent damage. (These can be ordered from any biological supply company.)

A record player with records of songs by species occurring regularly in the region of the study. (A selected list of records available can be obtained from the Cornell Laboratory of Ornithology, 159 Sapsucker Woods Road, Ithaca, New York 14850.)

In preparing all sections of this book I have consulted numerous authoritative books and papers for sources of basic information. The main sources for sections of the earlier editions repeated here I have acknowledged in the preface of those editions. The additional sources for this edition I have acknowledged or cited in the text where the information is first drawn upon.

Certain parts of the introduction to birds and ornithology I have taken, with modifications, from my foreword to *The Audubon Illustrated Handbook of American Birds* (McGraw-Hill Book Company, 1968) and from my contribution, "Masters of the Air," to *Birds in Our Lives* (U.S. Department of the Interior, 1966.)

The classification and nomenclature follow Alexander Wetmore's *A Classification for the Birds of the World* (1960) and the *A.O.U. Check-list of North American Birds* (1957 edition and corrections in *The Auk* for 1962, pages 493-494.)

I finished my preparation of the text in August, 1969, and in only a few instances have I altered it, or cited and listed publications appearing since that time.

The cover design is by Walter J. Breckenridge. All the illustrations are also his, except: the frontispiece, by Rudolf Freund (first reproduced in *The Living Bird* for 1966 and reproduced here, courtesy of the Carnegie Museum); Plates I, II, III, V, VI, and XV, by William Montagna; Plates IV and XXVII, by Ray S. Pierce; Plate VII (adapted from Jones, *Elements of Practical Aero-dynamics*, John Wiley and Sons, 1942): the drawing on page 49 (modified from Gower in *The Auk* for 1936) and the map on page 273, by Helen S. Chapman; Plates VIII, IX, X, XI, XII, XIII, XIV, XVI, XVII, XVIII, XIX, XXIV, and XXV, the drawings on pages 35, 36, 39, 40, 41, and the drawings on page 83 (modified from Häcker, *Der Gesang der Vogel*, Gustav Fischer, 1900), by Robert B. Ewing; the map on page 210 (first published in *The American Midland Naturalist* for 1941), by Frank A. Pitelka and reproduced here through his courtesy; the diagram on page 221 (adapted from Woodbury in *Ecology* for 1947), reproduced here as redrawn in *Fundamentals of Ecology* by E. P. Odum (W. B. Saunders Company, 1953); Plates XXIII and XXVI (from Portmann, *Traite de Zoologie*, Volume 15, Masson et Cie, 1950), reproduced by courtesy of the publisher; Plates XX, XXI (Figure 1), and XXII, by Barbara Downs; Figure 2 of Plate XXI (from George and Berger, *Avian Myology*, Academic Press, 1960), reproduced by courtesy of the publisher; Plate XXIX, by Alfred J. Hyslop; the spectrograms on page 325 (from Thorpe, *Bird-song*, Cambridge University Press, 1961), reproduced by courtesy of the author and publisher.

I wish to express my great indebtedness to many friends and associates working in ornithology who helped with the preceding editions. In the prefaces I have given their names and stated their contributions. Without their assistance I would not have had the substantial framework and much of the substance for this edition.

For the revisions and, hopefully, the advances in this edition, I have the pleasure of acknowledging, however briefly, the assistance of many people:

Andrew J. Berger generously provided the directions for dissecting the muscles of the Common Pigeon and guided the artist, Mrs. Barbara Downs, in making the drawings. The text carries his byline as does the literature on the muscular system which he updated from the previous edition. My debt to Dr. Berger is further deepened by his critical reading of the entire text on internal anatomy.

Peter Stettenheim, at the time a zoologist with the U.S. Department of Agriculture's Avian Anatomy Project, gave detailed attention to the section, "Feathers and Feather Tracts," clarifying or amplifying many aspects of feather growth, structure, coloration, and function, as well as directing the talented artist, Robert B. Ewing, who executed most of the accompanying illustrations. Dr. Stettenheim also perused the text on the pigeon skeleton and gave me the benefit of his judgment on several passages relating to morphology and function.

Richard L. Zusi provided under his byline an up-to-date listing of the more important literature on the skeletal system.

Paul A. Johnsgard shared with me his ideas of a sequential classification of mating displays and a suitable terminology. The result is a dual effort although I remain the target for any criticisms since the final decisions were mine.

Earl B. Baysinger took the time from his crowded schedule at the Bird Banding Laboratory, Patuxent, Maryland, to assemble much of the information on banding techniques, methods of capturing birds, and the procedures and regulations for obtaining permits to capture, band, mark, and collect wild birds.

Robert E. Beer at the University of Kansas contributed under his byline the review of the ectoparasites of birds, given in Appendix I.

John T. Emlen and Millicent S. Ficken reviewed the new section on behavior; Stephen T. Emlen and Bertram G. Murray, parts of the section on migration; Robert W. Storer and Pierce Brodkorb, the new section on origin, evolution, and decrease of birds; James C. Howell, the sections on territory and song; Kenneth C. Parkes, the part dealing with plumages; and Stephen I. Rothstein, the parts dealing with social parasitism. For their pertinent suggestions and criticisms I am grateful, although I alone take the responsibility for all statements now in the text.

From time to time during the tenure of the previous edition, numerous instructors and students pointed out certain weaknesses, errors, or inconsistencies or suggested improvements in content and presentation. Among these persons who were especially helpful and thus materially strengthened this edition are Albert J. Barden, Jr., Pershing B. Hofslund, Joseph C. Howell, Angela B. Kepler, Cameron B. Kepler, Heinz Meng, Miklos D. F. Udvardy, and George J. Wallace.

I acknowledge with appreciation the meticulous work of Eleanor R. Pettingill in preparing the index and of D. Jean Tate in checking references; the vast amount of time given by Walter J. Breckenridge to matters of format and illustration; and the assistance of Douglas A. Lancaster and James Tate, Jr., my colleagues at the Cornell Laboratory of Ornithology, in reading proof and checking on numerous details for accuracy.

Finally, a special word for Charles S. Hutchinson of the Burgess Publishing Company: from the day in 1939, when he agreed to publish the first edition, to this day, three editions and thirty years later, I have enjoyed every aspect of my association with him and his congenial, competent staff. No suggestion of mine has he ever ignored, no request of mine has he ever denied, on any matter that would improve this work. Indeed, his concern for the book in all its stages has been as personal as mine. What more can an author say in praise of his publisher?

Olin Sewall Pettingill, Jr.

Ithaca, New York
December 15, 1969

TABLE OF CONTENTS

BIRDS AND ORNITHOLOGY:
AN INTRODUCTION

Birds among all animals offer the most favorable combination of attributes for scientific study. They are numerous, abundantly diversified in form, and easily observed. They are highly organized and responsive with sensory capacities similar to man's and therefore understandable. Pleasing in colors and movements, they are also, with few exceptions, inoffensive in their habits and incapable of physically harming the investigator. Many adapt readily to experimentation. Little wonder that ornithology, the science of birds, boasts so many practitioners, and in turn contributes so significantly to modern concepts of evolution, speciation, behavior, and ecology.

Birds Defined

Birds are unique among all animals in being feathered. Like mammals, they too are warm-blooded, or homoiothermous, capable of regulating their body temperature. And like most of their vertebrate associates, excepting most mammals and a few others, they lay eggs.

Animals move from place to place by running, hopping, walking, crawling, swimming, gliding, and flying. Among birds, flight is the principal means of locomotion, even though some forms—for example, ostriches, kiwis, and penguins—in the course of evolution have lost their ability to fly. Therefore, one recognizes birds as birds because they are formed to fly.

The modern bird, like an airplane, is structurally and functionally efficient. A bird must be able to take flight, to stay aloft, and to reach its destination under the most adverse conditions.

Achievements for Flight

Several achievements have contributed to the bird's mastery of the air.

Lightness. Achieved by a covering of feathers—"the strongest materials for their size and weight known"—instead of a thick skin; by the loss of teeth and the heavy jaws to support them; by a reduction of the skeleton and by the hollowing, thinning, and flattening of the remaining bones; by a radical shortening of the intestine and the elimination of the urinary bladder; and by air spaces in the bones, body cavity, and elsewhere.

Streamlining. Also achieved by the feathers, overlapping and smoothing the angular, air-resistant surfaces and providing bays, wherein the feet may be withdrawn.

Centralization and Balance. Achieved by positioning all locomotor muscles toward the body's center of gravity—leaving the wings, like puppets, controllable by tendinous strings; and by positioning the gizzard, the avian substitute for teeth, and other heavy abdominal organs in the center of the body.

Maximum Power. Achieved by the combination of an exceptionally high, steady body temperature for aerial maneuvers in all extremes of climate and weather; by feathers, which aid in conserving the heat; by increased heart rate, more rapid circulation of the blood, and greater oxygen-carrying capacity of the blood stream; by a unique respiratory system, which permits a double tide of fresh air over the lung surfaces, synchronizes breathing movements with flight movements, cools the body internally, and eliminates excess fluids; and by a highly selective diet of energy-producing foods, which contain few indigestible substances to cause excess weight.

Visual Acuity and Rapid Control. Achieved by large eyes with a wide visual field and remarkable distance determination, and by a brain whose greatly enlarged visual and locomotor centers are capable of recording and transmitting nerve impulses with the speed of a seasoned pilot.

Range in Size

Birds range widely in size. The Ostrich *(Struthio camelus)*, standing between 8 and 9 feet tall and weighing nearly 350 pounds, is the largest. But it is, of course, flightless. Among the largest flying birds are the Wandering Albatross *(Diomedea exulans)*, with a wingspan of 11.5 feet, and the Andean and California Condors *(Vultur gryphus* and *Gymnogyps californianus)*, with wingspans of 10 feet and 9.5 feet, respectively. The Marabou Stork *(Leptoptilos crumeniferus)* may be the largest flying bird, if, as reported, its wingspan measures over 12 feet.

The smallest birds include numerous species of hummingbirds, the extreme being the Cuban Bee Hummingbird *(Calypte helenae)* that measures 2.25 inches from bill-tip to tail-tip and weighs less than 2 grams. Fourteen Bee Hummingbirds would weigh no more than an ounce.

Within a species there is often sexual difference in size, the males averaging slightly larger. In some species sexual dimorphism is very marked with the male about a third larger, as in the Wild Turkey *(Meleagris gallopavo)* and in the largest of all grouse, the Capercaillie *(Tetrao urogallus)*, or with the female a third larger as in the Sharp-shinned and Cooper's Hawks *(Accipiter striatus* and *A. cooperii)*.

There are limits to the size that flying birds may attain. They cannot be as small or as large as many other animals. Because they have a high rate of metabolism for support-ing a high body temperature, flight movements, and so on, birds need sufficient food to maintain this rate and at the same time conpensate for heat loss from body surfaces.

Theoretically, the smaller the bird, the greater is its relative body surface in relation to weight and the greater its heat loss. Consequently, the smaller the bird, the more it must eat in proportion to size. Again, theoretically, a bird smaller than kinglets and chickadees would have to eat all the time, night and day. Hummingbirds exist, small as they are, because they lower their body temperature—that is, become torpid—at night or at other times when they cannot eat. Thus they conserve energy.

The larger the bird, the faster it must fly to stay airborne. It needs bigger flight muscles for greater speed. This, in turn, means greater weight because flight muscles are heavy.

The larger birds have attained their size while retaining their ability to fly by develop-ing a dependence on air currents. Albatrosses and condors practically require winds and updrafts in order to fly at all.

Ornithology Defined

Ornithology, simply defined, is the science of birds. For a descriptive definition, there is none more suitable than the one written by Elliott Coues, the perceptive American ornithologist, over a half century ago:

Ornithology consists in the rational arrangement and exposition of all that is known of birds, and the logical inference of much that is not known. Ornithology treats of the physical structure, physiological functions, and mental attributes of birds; of their habits and manners; of their geographical distribution and geological succession; of their probable ancestry; of their every relation to one another and to all other animals, including man.

Ornithology Previewed

One must study ornithology in both laboratory and field because a knowledge of birds "in the hand" is incomplete without a knowledge of birds "in the wild," and *vice versa.*

Form, Structure, and Physiology

Basic to the study of ornithology is an introduction to the form, structure, and physiology of birds. This the student can best accomplish by making direct observations on the physical make-up of a "generalized" bird, such as the Common Pigeon *(Columba livia),* and by learning from a text the role of each organ system in the bird's way of life. Throughout the introduction he must center his attention on those features that will particularly enhance his appreciation of birds as biological entities. Frequently, he must compare certain features to their homologues in man, thereby making them more understandable to him.

The logical sequence in the introduction is, first, the identification of the different parts of the bird's topography, followed by a study of the bird's feather covering—how the feathers are structured and variously modified, how they develop, how they are colored, and how they are arranged on the body. A detailed knowledge of these exterior features is indispensable not only in describing birds and their actions, but in accounting for many of their adaptations.

With this knowledge, the student is then prepared to investigate the internal organ systems. As he proceeds, the text will point out lines of inquiry that he may pursue on his own. Avian anatomy and physiology, he will discover, offer many opportunities for research. Indeed, an increasing number of ornithologists specialize in one or both of these fields, dealing particularly with the adaptive and comparative aspects among different species of birds.

Species and Speciation

Although uniformly specialized for flight, birds have nonetheless radiated widely in form and action in order to live in particular environments.

Consider, for example, the adaptations for locomotion and feeding. Some species customarily fly swiftly; others fly slowly. Some hover; others soar. Some swim and dive; others wade. Some walk or hop; others climb. To get food, some species probe in the soil, others dabble in shallow water, scratch the ground, chisel holes in trees, make flying sorties, or hunt for prey in any number of different ways.

These adaptations and others, always in complex combination, account for the different shapes of wings, tails, bills, and feet and differences in body shape, plumages and coloration, breeding habits, seasonal movements, and general behavior. Or, to put it another way, thanks to adaptive radiation operating so vigorously in the descent of birds, there are some 8,600 different species today.

In studying birds one naturally thinks of them in terms of species. Therefore, the logical sequel to a knowledge of their form, structure, and physiology is an acquaintance with the many different species occurring in the student's immediate area. This requires understanding the concept of species and speciation and the methods of classifying, naming, and identifying.

Gaining a thorough acquaintance with the 150 to 300 species regularly occurring in the average study area of temperate North America demands a knowledge of the taxonomic characters and other means of recognizing species in both laboratory and field, together with an understanding of changes in plumage and plumage coloration among different species.

The identification of species is not an end in itself but a stepping stone to investigations of many aspects of bird life, or of biological problems in which birds play a role. Some students find speciation *per se* a challenging field since there is still much to be learned about the origin, status, and interrelationships of species.

Distribution

Although most modern species of birds can fly and thus can rove the earth, each species is confined to a particular geographical range, which may be from several hundred acres as on a sea island to one or more continents in size.

The ranges of species overlap so that in any one area there is an aggregation of species—an avifauna. Because the ranges of species are rarely or never identical, avifaunas vary markedly. Students over the years have given attention to the composition, comparison, and origin of avifaunas, yet there is much about them that remains to be investigated.

Geographical ranges are unstable due partly to the tendency among species to invade new areas. Cyclonic storms may help or hasten resettlement by moving individuals to a different place, where they survive and reproduce if the environment suits them. Man has a part in it, too, when, for example, he transports birds on his ships. House Sparrows *(Passer domesticus)* reached the Falkland Islands in the South Atlantic on ships that first stopped at Montevideo, Uruguay, where the birds, attracted to sheep-pens on deck, came aboard and remained until the ships reached the islands.

Any student, after having observed birds in a given area for a few years, is certain to note shifts in ranges and ponder the reasons. Modern ornithologists pay considerable attention to local distribution as the abundant literature on the subject clearly indicates.

Within its geographical range a species, if normally migratory, is seasonally distributed, appearing in one part of its range in one season, in another part in another season.

Within its geographical range a species is also ecologically distributed. It usually occupies a particular environment or habitat and shares this habitat with other organisms, plant and animal, all of which are adapted to the prevailing conditions of soil, air temperature, moisture, and light. All the organisms in a given habitat collectively comprise a biotic community, since they show relationships to one another.

When any two communities meet, more often than not, there is an area of mixture and overlap, or ecotone, in which the birds and other living forms, characteristic of these communities, are intermixed and in which are additional forms that, preferring this ecotone, seldom occur elsewhere.

A student soon becomes aware of the importance of habitat or community in accounting for the presence or absence of species. He learns to associate different species with particular environments—the Red-eyed Vireo *(Vireo olivaceus)* with the deciduous forest, the Horned Lark *(Eremophila alpestris)* with the prairie grassland, and the Verdin *(Auriparus flaviceps)* with the scrub desert. When he travels northward on the continent of North America or climbs a high mountain, he expects a sequence of species as he passes through one environment after another—the Olive-sided Flycatcher *(Nuttallornis borealis)* in the coniferous forest, the White-crowned Sparrow *(Zonotrichia leucophrys)* at the timberline ecotone, and ptarmigan *(Lagopus* spp.) on the tundra.

At the same time the student becomes conscious of several significant aspects of ecological distribution. Rarely does he find one species throughout a community, even though it may be characteristic of that environment. As a rule, it occupies merely a niche, and is adjusted to this position in structure, function, and behavior as no other species in the same community. The Red-eyed Vireo, for instance, occupies a treetop niche and is adjusted to this position in structure, function, and behavior as no other the forest floor. It would be unusual to see the Red-eyed Vireo on the ground or the Ovenbird in the treetops. While a species may appear to share its niche with other species, not one of these behaves exactly as another does or requires the same food, the same nesting site, and so on.

Species of birds occur in greater variety and density in ecotones than in the pure communities that border them. This phenomenon, called edge effect, is important to anyone wishing to see larger numbers of birds.

Edge effect results in a greater variety of vegetation—grasses, shrubs, and trees—providing a greater variety of food and cover for birds. For example, ecotones where field and forest merge have the plants characteristic of both field and forest and many additional shrubs. Thus they bring together birds of both field and forest and also attract species that require either shrublands or a combination of trees, shrubs, and grasses.

Some bird species are adapted so strongly to a special niche that they cannot live in a different situation. If an element in the niche on which they depend is destroyed or seriously altered, they are more likely to disappear than to make an adjustment. The Everglade Kite *(Rostrhamus sociabilis)* probably would disappear in Florida were disaster to befall the big freshwater snail, *Pomacea palludosa*, on which it feeds exclusively. It is likely that the Kirtland's Warbler *(Dendroica kirtlandii)* would disappear in northern Lower Michigan, where it breeds exclusively, if there were no more jack pines 6 to 18 feet high under which it almost invariably nests.

A good many bird species, on the other hand, are much more adaptable. Sometimes they are so widely tolerant of different situations that their precise niches are unrecognizable. The Blue Jay *(Cyanocitta cristata)*, Black-capped Chickadee *(Parus atricapillus)*, and Cedar Waxwing *(Bombycilla cedrorum)* are so adaptable that one may find them almost everywhere in wooded areas through their ranges.

The species that restrict themselves to narrowly prescribed niches generally have small populations within correspondingly small ranges. The species tolerant of environmental changes and variations are mainly the inhabitants of the ecotones; they have large populations and often range widely.

The underlying factors accounting for the ecological distribution of many species still remain to be determined. Here is a study with a degree of urgency. As man steadily destroys the natural environments, an understanding of a species' ecological requirements is the first step in preventing its decrease. The next step is to see that its requirements are maintained through intensive management and conservation practices.

Behavior

The behavior, or ethology, of birds has recently attracted scores of investigators. Birds are ideal animals for behavioral studies. Each species has an impressive repertoire of innate behaviors and, at the same time, its ability to learn compares favorably with that of most mammals. Thanks to a rich variety of bird species, each with a different mode of life, investigators have available for study a correspondingly rich variety of behaviors.

An understanding of the principles of bird behavior is essential for any beginning student, helping as it does to explain the basis of many avian activities. Even more important, an understanding of bird behavior illuminates many of the basic ethological principles applied to human life. Modern psychologists are now paying attention to such phenomena as individual distance and dominance relationships (first noted in birds!) that are so evident in urban societies. Continued, in-depth studies of avian behavior will, almost certainly, sharpen further man's perception of his own social problems.

The procedure in the study of behavior is to identify, describe, and name the behaviors of a species and then to determine what each behavior accomplishes, its significance to the species' survival, its causes, how it has evolved, and whether it is innate, learned, or both innate and learned. Many mating displays are actually derived from such maintenance activities as preening or scratching; or from displacement activities—for example, when a bird breaks off fighting and pecks at some object; redirected activities—when a bird redirects its attack to an object other than one which elicited the response; and intention movements—when a bird makes a move to fly but fails to do so, thereby performing an incomplete act.

Inherited behavior predetermines the extent to which learned behavior may develop. Learned behavior is actually adaptive behavior resulting from experience. A bird inherits the ability to fly, yet it must learn by experience to take off *into* the wind rather than *with* it, to choose the perch that will best accommodate its feet. This is called learning by trial and error. Other forms of learning are by habituation and by imprinting. A few birds show ability to learn by insight. The different methods of learning among birds demand much more research.

Investigators often give considerable attention to social behavior since most birds are by nature gregarious and have consequently developed many kinds of interactions related to attack, escape, defense, flocking, and reproduction. Although the literature on social behavior in birds is already enormous, the subject is still a fertile field for study.

Migration and Orientation

No aspect of bird life has so excited man's interest down through the centuries as the withdrawal of birds from an area in the colder seasons and the return to the same area when the seasons become warmer. In spite of a great store of knowledge on the initiation and procedure of migration among modern birds, the question of how and when migration originated still remains speculative—an ever-present challenge to one's thinking.

Experimental studies, started over 25 years ago, demonstrate that a specific day length in the spring stimulates the activity of a bird's endocrine glands and this stimulation brings the bird into a migratory state. Some external factor such as a sudden change in temperature then releases migratory behavior. In the fall, with a regression of endocrine activity, the bird reaches another migratory state ready for triggering by an outside cause.

The present wealth of information on the process of migration—starting and stopping times, rate, duration, distances covered, routes, and relation to weather—is due in a large measure to direct observations and record-keeping by hundreds of persons and to returns from many millions of banded birds.

Radar and radiotelemetry are useful tools in fathoming some of the "mysteries" of night migration and determining the speed, direction, and elevation of migratory flights.

Migrating birds have obvious navigational ability. Otherwise they could not return as they do to their nesting grounds after the winter spent hundreds, sometimes thousands, of miles away. Just how migrant birds orient themselves has been the object of numerous experiments. By using caged birds that display migratory activity by "fluttering" in the direction of migration in the wild, some investigators have demonstrated that birds migrating on clear days may be guided by the sun and on clear nights by star patterns. These and other experiments, although convincing, do not explain orientation by all birds under all circumstances. Undoubtedly different birds use different cues or different combinations of cues, depending on where and when they migrate and the prevailing weather conditions. The whole subject of orientation, complex and fascinating, beckons for much more research.

The Reproductive Cycle

The main stages of the reproductive cycle of most bird species are the establishment of territory, the coming together of the sexes, nest-building, egg-laying, incubation, hatching of the eggs, and the development and care of the young. Involved in the establishment of territory and the coming together of the sexes are two prominent activities —singing and mating displays.

In the past 50 years many investigators have studied the reproductive cycle of different species, resulting in the accumulation of a vast amount of data. Yet, surprisingly, detailed, comprehensive information is available on relatively few species. For only about 5 percent of North American species is the size of territory known; for about 10 percent, the average length of nestling life; for about 20 percent, the average incubation period; for about 30 percent, the full description of songs and mating displays.

Anyone beginning a study of birds should carefully observe the reproductive cycle of at least one species from territory establishment to fledging and dispersal of the young. The more detailed information he can obtain, so much the better. Hopefully, he will contribute to knowledge of the species, but whether he does or not, he is almost certain to profit by gaining an intimacy with the living wild bird, its behavior and problems of survival.

Longevity, Numbers, and Populations

How long do birds live? How many birds are there in given areas? What are the factors controlling the numbers of birds? These are questions that always fascinate anyone studying ornithology and the answers continue to be unsatisfactory in scope and often controversial.

The student should familiarize himself with the questions and gain some first-hand experience in estimating numbers of birds.

Direct counting of individuals of most species generally is futile because they are so numerous and widespread. The student's time is better used in measuring the populations of all species in a given area and understanding how their populations are controlled. This is a complex undertaking. It includes determining their reproductive rates;

the ratio of age groups and sexes; the annual fluctuations of their respective populations because of varying physical factors of the environment (air temperature, precipitation, and so on) and biological factors (predation, diseases, food supply); and ways in which their populations are controlled over long periods of time. Although populations normally fluctuate in numbers of individuals per year, they are remarkably stable over a period of, say, 50 years, if their habitat is unchanged. Annual fluctuations are scarcely more than wrinkles in the long history of a population.

The study of populations has endless opportunities for investigation. It is of vital—"vital" meaning life-or-death—importance at the present time as man hastens his encroachment upon and destroys the natural environment. Determining when certain populations are showing a sharp decline provides the basis for informing conservation agencies and urging remedial action.

Evolution

Where did birds come from? The story goes back to prehistoric time, in the Triassic Period some 200 million years ago, when birds arose from a somewhat specialized group of reptiles that had long hindlimbs. The avian line from this reptilian specialty may have begun as tree-climbing forms, which first jumped from branch to branch by using membranes stretched between the sections of their shorter and slightly flexed forelimbs.

As they gradually evolved the ability to fly farther, these arboreal forms acquired greater sailing surface through expansion and modification of the scales on the trailing edges of their forelimbs and along the outer edges of their long tails. At this point birds came into being, for of all the physical features of birds, none distinguishes them more sharply from all other creatures than these outgrowths of the skin.

The remarkable fossil *Archaeopteryx lithographica* possessed feathers and is thus recognized as the earliest known bird. This creature of the Jurassic Period, some 140 million years ago, may have been one of several kinds of similarly primitive birds already existing. Nobody knows. But in any case, one such primitive species, probably of either Eurasian or African origin, acquired the power of flight—that is, the ability to sustain itself in the air for indefinite periods by flapping its wings. And from this stock many species began to emerge as they spread out and filled more habitats and niches.

This evolutionary process, commonly called adaptive radiation, was slow at first but steadily quickened during the next 139 million years, through the Cretaceous and Tertiary Periods. Birds in time inhabited all the earth's great land masses and occupied most of the primitive environments.

But as the continents separated, merged, and separated, as mountain ranges rose and were worn away, as the climates shifted, and as plant forms evolved, flourished, and vanished, so did habitats for birds. The species, so precisely adapted to one habitat that they could live in no other, disappeared when the habitat disappeared. More species were always evolving, however, to fill new niches.

The primitive birds became extinct through the Cretaceous Period. The "new" birds began to look more and more like modern species, and many birds were recognizable by the end of the Tertiary as ostriches, pelicans, cranes, nuthatches, thrushes, and so on.

With the coming of the Pleistocene Epoch, or Ice Age, about a million years ago, the abundance of birds in number of species attained a peak that has never been exceeded. This period of prehistory could have been called the Age of Birds, had mammals not already taken the ascendancy in size and aggressiveness to dominate the earthly scene.

Toward the end of the Pleistocene and the start of the Recent Epoch, about 15 thousand years ago, bird species began disappearing more rapidly than they were evolving. The decrease of birds was underway. Man had not yet become a major destructive force in the avian environment. How, and how fast, that destructive force grows may determine how, and how fast, the presently extant 8,600 species of birds disappear.

The first bird species definitely known to have been eliminated by man was the Dodo *(Raphus cucullatus),* in 1681. Since that date, 78 species have become extinct over the world, nearly half of them destroyed by man. At this rate of disappearance, the future for bird life appears alarming.

And it is alarming! Every student of ornithology should keep this in mind as he investigates the attributes of birds, being constantly alert to discover ways and means that will insure their protection and continue their survival for centuries to come.

TOPOGRAPHY

The various parts of a bird's exterior are mapped out as the **topography.**

For convenience the description of the topographical parts are grouped below under seven titles: Head, Neck, Trunk, Bill, Wings, Tail, Legs and Feet. While studying each part, refer to specimens of the Common Pigeon *(Columba livia)* and the House Sparrow *(Passer domesticus)* and to the outline drawings (Plates I-VI). Follow the instructions for labeling the drawings. Write all labels outside the drawings and parallel to the top of the page. Use dotted (i.e., broken) leader lines.

Place a pigeon specimen on its back and observe its outline or **contour.** The body shape tapering at both ends is streamlined for cleaving the air in flight.

THE HEAD

The upper, or dorsal, part of the head is somewhat curved and composed of an anterior (forward) and a posterior (rear) part: the anterior part, the **forehead,** extends up and back from the bill to an imaginary line joining the anterior corners of the eyes; the remainder of the upper part of the head, the posterior, is the **crown.** (Some authorities call the sloping posterior portion of the crown the **occiput** or **hindhead.**) Below the lateral boundary of the forehead and crown is the **superciliary line,** distinctively colored in some birds but not in the Common Pigeon or male House Sparrow.

The side of the head is rather flat and divided into the **orbital** and **auricular regions.**

The orbital region includes the **eye, eyelids,** and **eye-ring.** The eye, as revealed through the circular eye-opening, consists only of the dark **pupil** and pigmented **iris.** (The eyeball, actually of great size, can be felt under the skin.) Note that the pigeon's iris is bright orange or yellow. What is the color of the sparrow's iris? The two eyelids are skin-folds, one above the eye and one below. In the pigeon they are unfeathered and red. How do they differ in the sparrow? In all birds, as in mammals and in many reptiles and amphibians, the lids close the eye; only in birds, however, do the lids close the eye at death. Observe that at closure, in the pigeon and sparrow, the lower lid comes up more than the upper lid comes down. This is the rule among most diurnal birds; in most nocturnal species (e.g., owls and goatsuckers) and in a few others, the upper lid is the more mobile, as in mammals and alligators. The anterior corner of the eye (toward the nostril), where the eyelids come together, is the **nasal canthus;** the posterior corner (near the temple in man) is the **temporal canthus.** Find the **nictitating membrane,** sometimes called the "third eyelid," a translucent, vertical fold under the lids on the side of the eye toward the bill. If the eye of a living bird is touched, the nictitating membrane—just before the lids close—slips obliquely across the exposed surface of the eye.

Ordinarily, the membranes of both eyes, and the lids of both, act together (consensually) even when only one eye is touched. Birds, like mammals, blink periodically. The pigeon and a few other species blink with both the nictitating membrane and the lids, but most species, including the House Sparrow, blink with the nictitating membrane alone, the lids—usually the lower—closing only in sleep or when the eye is menaced by foreign objects. In some birds (though not in the pigeon or sparrow) the feathers immediately around the eyelids are distinguished from the surrounding feathers by different color and are called collectively the eye-ring.

The auricular region is the area around the ear opening, concealed by a patch of feathers, the **auriculars.** (The temporal region, between the auriculars and the orbital region, is small. Generally, it is considered part of the auricular region and not used in describing birds. The area between the eyelid and the base of the upper part of the bill is the **lore** (between the eye-ring and the bill in those birds having an eye-ring).

The side of the head from the base of the lower part of the bill to the angle of the jaw (found by feeling for a bony prominence behind and below the ear) is the **malar region** (cheek). It is bounded above by the lore, orbital region, and auricular region and below by the edge of the lower jaw.

The under (ventral) part of the head is flat and divisible into an anterior part, the **chin,** a feathered area in the fork of the lower part of the bill; and a posterior part, the **gular region,** a continuation of the chin to an imaginary line drawn between the angles of the jaw.

THE NECK

The neck extends from the posterior margin of the crown to the trunk, and is divided into four regions: **nape, jugulum,** and **sides.** The upper, or dorsal, part is the nape. The lower, or ventral, part is the jugulum. (The term "throat," frequently used in descriptions of birds, includes the gular region of the head and the jugulum of the neck.) The side of the neck extends, between the nape and jugulum, from the posterior borders of the auricular and malar regions to the trunk.

THE TRUNK

The trunk is divided into two surfaces: the **upper parts** include all the trunk above an imaginary line drawn from the shoulder joint to the base of the outermost tail feathers; the **under parts** include all the trunk below this line. (Sometimes the terms "upper parts" and "under parts" are used to include the dorsal and ventral surfaces of the wing and tail, as well as those of the trunk.) The upper parts of the trunk are made up of the **back** and **rump.** The back is the anterior two-thirds of the area between the base of the neck and the base of the tail; the rump is the posterior one-third. The under parts are divided into **breast, abdomen, sides,** and **flanks.** The rounded portion of the under parts, beginning at the lower border of the jugulum, is the breast; the flatter portion ending in an imaginary line drawn across the vent is the abdomen. The breast and abdomen curve upward, forming the sides of the body. The parts lying between the posterior half of the abdomen and the rump are frequently termed the flanks. Although technically the sides of the body belong to both the upper parts and the lower parts, the imaginary line separating the two surfaces is so high on the trunk that the sides of the body are generally considered regions of the under parts only.

Label on Plate I: Eyelids; iris; pupil; nasal canthus; temporal canthus; nictitating membrane (approximate position).

Plate I

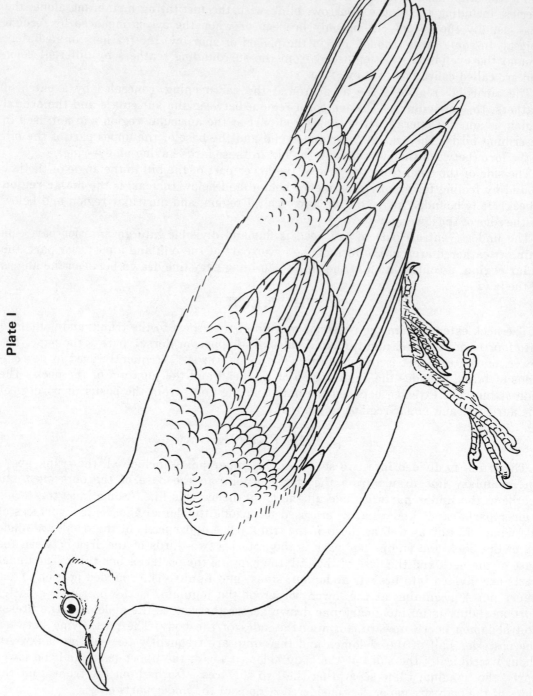

COMMON PIGEON

Plate II

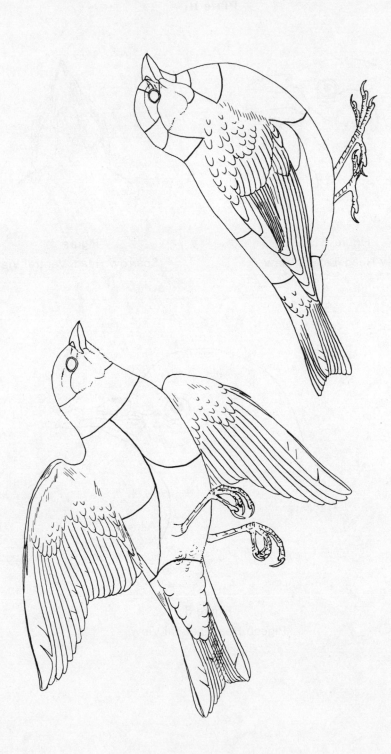

Figure 2

Figure 1

HOUSE SPARROW

Plate III

Figure 1
Sparrow Head, Lateral View

Figure 2
Sparrow Head, Ventral View

Figure 3
Pigeon Head, Lateral View

COMMON PIGEON AND HOUSE SPARROW

Label on Plate II, Figure 1: Head; neck; side of body; flanks; breast; abdomen.

Label on Plate II, Figure 2: Forehead; crown; occiput; superciliary line; auriculars; lore; malar region; chin: gular region; nape; jugulum; side of neck; back; rump; side of body; flanks; breast; abdomen.

THE BILL

The bill consists of an **upper** and a **lower mandible,** lying, as their names indicate, above and below the **mouth.** Each mandible is a bony modification of the skull covered with a durable horny sheath. Thus the bill is a more or less rigid structure; along most of its length it is rather hard.

Notice that in the pigeon and the sparrow the lower mandible is a little shorter and narrower than the upper and much shallower. The bill is also mapped in a number of parts.

Upper Mandible: The ridge of the upper mandible—the uppermost, central, longitudinal line—is the **culmen,** extending from the tip of the mandible back to the bases of the feathers. It is formed by fusion of the two rounded **sides of the upper mandible.** Seen in profile, the culmen is somewhat convex, particularly toward the tip of the bill. The cutting edges of the upper mandible are the **upper mandibular tomia** (singular, **tomium**). Toward the base of each side of the mandible is a **nostril.** In the pigeon, overarching the nostril posteriorly, is a soft, noticeably swollen structure characteristic of pigeons and called the **operculum.** The sparrow's nostril opens into a depression, the **nasal fossa,** common in the majority of small birds.

Lower Mandible: The cutting edges of the lower mandible are the **lower mandibular tomia.** They are overlapped slightly, when the bill is closed, by the upper mandibular tomia. Viewed from below, the bill has a prong-like projection extending posteriorly on each side of the jaw. This is the **mandibular ramus.** The lowermost ridge of the lower mandible is the **gonys,** formed by an anterior fusion of the rami. Like the culmen, its profile is somewhat convex. The **sides of the lower mandible** include not only the surfaces between the gonys and tomium but also the surfaces of the rami.

Several parts of the bill are evident when the two parts of the mandible are considered in relation to each other. The line along which the mandibles come together is the **commissure** or **gape.** (The term, "Commissure," is preferred to "gape" which often means the space between the opened mandibles.) The point on each side where the mandibles meet posteriorly is called the **commissural point** (angle of the mouth). The tomium of each mandible has two parts: the **tomium proper**—the hard cutting edge of the mandible; and the **rictus**—the softer, more fleshy, part of the tomium near the commissural point. The rictus is more prominent in the House Sparrow than in the pigeon.

Label on Plate III, Figure 1: Culmen, upper mandibular tomium, lower mandibular tomium, rictus, nasal fossa.

Label on Plate III, Figure 2: Mandibular ramus.

Label on Plate III, Figure 3: Gonys, operculum, commissural point, nostril, commissure, mandibular ramus.

THE WINGS

The wings are the appendages arising from the shoulder or pectoral girdle. Though homologous to the forelimbs of man and other vertebrates, they are specially adapted to flight, having a peculiar shape and a series of feathers arranged in a definite fashion.

Spread out the wings of the pigeon and note that the feathers belong to two main groups: the flight feathers, or **remiges** (singular **remex**), are the long stiff quills projecting posteriorly; the **coverts** are the smaller feathers overlying the bases of the remiges and covering the rest of the wing. Other groups of feathers are the **alular quills, scapulars, tertiaries,** and **axillars.**

To identify these groups of wing feathers, it is necessary to know from what parts of the wing they arise, and to understand the skeletal framework and external anatomy of the unfeathered wing.

Examine either an articulated human skeleton, or a detailed chart of one, and locate the following bones of the forelimb: humerus (upper arm bone), radius and ulna (forearm bones), 8 carpals (wrist bones), 5 metacarpals (hand bones), and 14 phalanges (digit or finger bones).

The bird's wing has the same skeletal plan and terminology as the forelimb (pectoral appendage) of man and other vertebrates, yet it shows certain striking differences. The skeleton of the bird's wing, like many other parts of its anatomy, is highly specialized for flight. Some of the bones in the human forelimb are lacking in the wing or fused with others; the movements of the various bones upon one another differ markedly. From the wrist outward the skeleton of the bird's wing is especially at variance with the skeleton of the human forelimb.

Examine a prepared articulated skeleton of a pigeon's wing. Count the number of bones. What is the difference between the number in the wing and the number in the human forelimb? Identify the following bones in the pigeon wing:

Humerus. A relatively short, thick bone which articulates, by means of a vertically elongated head, with the shoulder girdle. On the proximal ventral side is the opening of the **humeral pneumatic cavity** which receives one of the air sacs. The humerus widens out toward its distal end to form two large condyles which articulate with the radius and ulna.

Radius. A slender, rather straight bone which articulates with the external condyle of the humerus by a cup-like structure on its proximal end. Its outer posterior margin articulates with the ulna; its distal end fits into one of the carpals.

Ulna. A stouter bone than the radius and decidedly more curved. On the outer side is a row of small prominences, the points of attachment of the remiges. Proximally the ulna articulates with the internal condyle of the humerus and ends in an **olecranon process** to form the point of the elbow. Distally the ulna articulates with the two carpal bones.

Carpals. Two somewhat squarish bones. One, the **radiale,** is at the end of the radius; the other, the **ulnare,** is at the end of the ulna. They articulate proximally with the radius and ulna. The additional carpals, found in other vertebrates, are fused with the radiale and ulnare and with the adjoining metacarpals. Although present in the embryo of a bird, they are not distinguishable in the adult.

Metacarpals. The first and fifth metacarpals are wanting in the bird, but the second, third, and fourth persist, fusing with vestigial carpals to form the large composite bone called the **carpometacarpus** which articulates with the radiale and ulnare. The third, or median, metacarpal constitutes a large part of the carpometacarpus. Fused to its

Plate IV

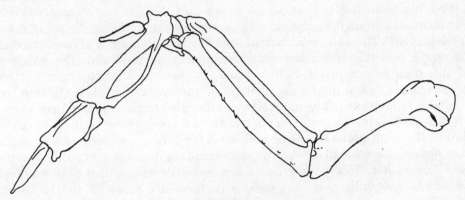

Figure 1
Skeleton of Wing

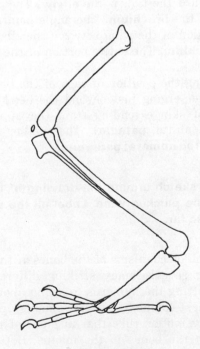

Figure 2
Skeleton of Leg and Foot

COMMON PIGEON

proximal end on the radial side is the small remnant of the second metacarpal. On the opposite side is the remnant of the fourth metacarpal, a slender bone nearly as long as the third metacarpal to which its ends are fused; the two bones are sometimes also joined laterally by a thin, bony membrane.

Phalanges. The pigeon has only four phalanges. They compose the skeletal structure of the three digits or fingers that persist in the bird. The second metacarpal bears the single phalanx of one digit (corresponding to the second, or index, finger in man); it is short and pointed. The third metacarpal has the two phalanges of another digit (the "third finger"), and this forms the main continuation of the hand. The proximal phalanx of this digit is very much flattened and its posterior margin sharply edged. The fourth metacarpal bears the single phalanx of the remaining digit ("fourth, or ring, finger") and is somewhat triangular. (Many authorities have considered the three digits of the bird's wing to be the first, second, and third fingers; these authorities refer to the first digit of the wing as the **thumb** or **pollex.**) A few birds, belonging mostly to the more primitive orders, have a nail or claw on the second digit of the wing and sometimes on the third (see Fisher, 1940). Wing claws are better developed in newly hatched birds than in adults—an indication that these structures are relics of the bird's reptilian ancestry. The young of the Hoatzin *(Opisthocomus hoazin),* a tree-inhabiting gallinaceous bird of South America, use the nails on their second and third digits for climbing.

Pluck one wing of the pigeon, leaving only the remiges. Note that the wing has two prominent angles giving it the shape of the letter Z written backwards. The angle nearest the trunk, pointing toward the tail, is the **elbow.** The portion of the wing between the trunk and the elbow is the **brachium.** The angle pointing forward is the **wrist,** or **bend of the wing.** The portion of the wing between the elbow and the bend of the wing is the **forearm,** or **antebrachium.** The entire portion of the wing beyond the bend is the **hand** or **manus.**

Locate on the plucked wing the position of each of the bones just described. Find the rudimentary second finger emerging just beyond the bend of the wing, on the anterolateral surface. The fold of skin extending from the upper arm to the entire antebrachium is the **patagium** (plural, **patagia**). The smaller fold of skin extending from the brachium to the trunk is the **humeral patagium.**

> **On Plate IV, Figure 1, sketch around the drawing of the wing bones of the pigeon an outline of the plucked wing. Label all the parts of the wing that have been mentioned so far.**

A comparative study of the mechanisms of the bones of the human arm and hand and those of the pigeon's wing reveals many striking differences. Whereas the human shoulder joint is free, permitting the humerus to swing about, the pigeon's is restricted and limits the humerus almost completely to movements up and down and to and from the body. There is no rotary motion (like that in man) of the radius around the ulna, and no movements between the bones in the manus. Both antebrachium and manus thus form a firm support for the flight feathers. The wrist joint in the bird does not permit the manus to swing about as in man; the manus can move only to and from the antebrachium and in the same plane. The only joint that allows the same (and no other) motion in man and pigeon is the elbow joint: it permits the antebrachium, or forearm, to move to and from the brachium, or upper arm, in the same plane as the brachium. The most plausible explanation for the sharp differences between the forelimb of man and that of the pigeon (and other birds) is that man's forelimb is "generalized" for a variety of functions, whereas the pigeon's is "specialized" for one function, namely flight.

Having studied the structure of the pigeon's wing, identify now the several sets of feathers, or topographical regions, of the wing. Turn to the unplucked wing of the pigeon and find:

Primaries. The remiges attached to the manus. They are counted and numbered from the inside out. How many are there? Which primary is the longest?

Secondaries. The remiges on the antebrachium and elbow; all are attached to the ulna. They are counted and numbered from the outside in. How many are there?

Tertiaries. Sometimes in descriptive ornithology, the feathers growing upon the adjoining portion of the brachium are called the tertiaries. They are not remiges. In certain species of birds the tertiaries are greatly modified and differ considerably from the secondaries.

Scapulars. A group of prominent feathers arising from the shoulder and adjoining portion of the upper surface of the brachium. They slightly overlap the tertiaries.

Alular Quills. Three feathers, stiffened like the remiges, springing from the second finger. They are known collectively as the **alula** (pronounced *al'-you-la*).

Wing Coverts. Feathers overlying the remiges on both the upper and under surfaces of the wing. They include all the feathers of the wing except the remiges and the alular quills. The upper wing coverts are as follows:

Greater Primary Coverts. The feathers overlying the bases of the primaries. There is one covert for each primary.

Median Primary Coverts. The shorter, less exposed feathers overlapping the greater primary coverts. There is one row of them. (Lesser primary coverts are wanting in the pigeon.)

Greater Secondary Coverts. The single row of feathers overlying the secondaries.

Median Secondary Coverts. A row of shorter, less exposed feathers overlapping the greater secondary coverts.

Lesser Secondary Coverts. Even shorter feathers lying in two or three rows directly over the median secondary coverts.

Alular Quill Coverts. Three small feathers, each one overlapping an alular quill at its base.

Marginal Coverts. The remaining coverts of the upper surface of the wing. They arise immediately anterior to the lesser secondary coverts and are indistinguishable from them. They are densely inserted on the patagium and along its extreme anterior border. They are also inserted along the outer surface of the manus and extend distally to the outermost median primary covert.

The under wing coverts are as follows:

Greater Primary Coverts. Overlying the primaries at their bases.

Greater Secondary Coverts. Overlying the secondaries at their bases.

Median, Lesser, and *Marginal Coverts.* Generally all of these coverts are referred to as the **lining of the wing.** They overlap each other in much the same fashion as the corresponding coverts on the upper surface of the wing. But the feathers of the three groups are much more alike and less distinctly arranged in rows. Therefore no attempt is made here to distinguish between them.

Axillars. These are under wing feathers lying close to the body in the axilla or "armpit." They are white, and both longer and stiffer than the coverts. In certain species of birds the axillars are even more peculiarly modified.

Plate V

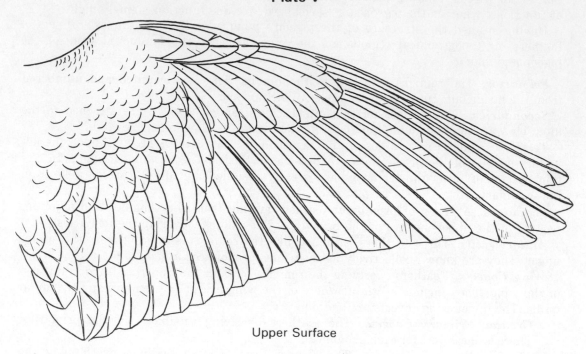

Upper Surface

Under Surface

COMMON PIGEON WING

Plate VI

Upper Surface

Under Surface

HOUSE SPARROW WING

Study the wing of the House Sparrow. Note that the tenth (outermost) primary, rudimentary and quite concealed, is a minute, narrowly pointed feather lying on the outside of the wing next to the outermost greater primary covert. How many secondaries are there?

Certain species of birds lack the tenth primary altogether, hence the outermost primary is the ninth. Certain species also have fewer inner secondaries. This explains why the primaries are counted from the inside out and the secondaries from the outside in.

On Plates V and VI, identify the several sets of wing feathers (except the scapulars and median primary coverts, which are not shown). Label completely, using dotted leader lines. Number the primaries and secondaries. Color with crayon the following regions of the upper wing: alular quills (alula) and alula quill coverts—*purple;* greater primary coverts—*green;* greater secondary coverts—*yellow;* median secondary coverts—*red;* lesser secondary coverts—*blue;* marginal coverts—brown.

On Plate I, color as above and label: primaries; secondaries; alula; greater primary coverts; greater secondary coverts; median secondary coverts; lesser secondary coverts; marginal coverts.

On Plate II, Figure 1, label: axillars; lining of the wing.

Consider now the wing as a flying mechanism.

Spread the unplucked pigeon wing to its full extent and notice that the anterior part containing the bones, muscles, and tendons is thicker than the posterior part, which bears the remiges. The basic structure is, therefore, roughly tapered or cambered from the leading edge to the trailing edge. Notice also that the remiges are supported by the wide membrane (mostly skin) from which they emerge and that, except for the tertiaries, they are further supported by being attached to the bones. On the spread wing the remiges point in different directions: the primaries outward, away from the body; the secondaries mostly backward, and the tertiaries toward the body. The scapulars, on the shoulder and brachium, are directed outward from their skin support and somewhat overlap the tertiaries. Each time the wing is spread, the remiges and scapulars, controlled by their supports, automatically take these directions. During flight the wing thus has a broad, continuous surface from its origin on the body to the tip.

Spread the unplucked wing and examine the upper surface. The marginal coverts arise straight up, then bend posteriorly to overlap one another. The posteriormost marginal coverts bend over the bases of the lesser coverts, and the lesser coverts bend over the median coverts, the median coverts over the greater coverts, and the greater coverts over the remiges in a similar manner. This arrangement, giving an even curve to the tapered structure of the wing, streamlines the wing. Since air has weight and exerts pressure, streamlining is an essential for flight. It allows air to flow smoothly over the wing surface, and at the same time prevents excessive pressure from building up in front, reduces pressure on the upper and lower surfaces, and lessens vacuum and turbulence behind.

The wing supports the bird in flight and moves it forward in the air. The proximal part of the wing, from body to wrist joint, supplies the principal support by giving **lift;**

Plate VII

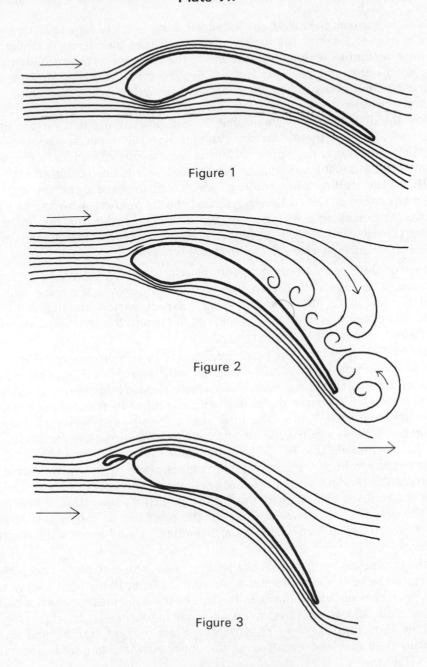

Figure 1

Figure 2

Figure 3

THE BIRD'S WING IN FLIGHT

Diagrams showing how a wing meets an oncoming air stream. Figure 1, during ordinary flapping flight. Figure 2, when the wing is greatly tilted. Figure 3, when the wing is greatly tilted and slotted.

(Adapted from Jones, "Elements of Practical Aero-dynamics," John Wiley and Sons, 1942.)

the distal part provides the forward motion by acting in a manner analagous to a propeller.

Note that the proximal half of the extended wing is tilted upward from trailing to leading edge so that more of the under surface will face the direction of flight. During ordinary flapping flight the under surface meets the pressure of the oncoming air stream, deflects it downward, and prevents it from flowing over the upper surface. (See Plate VII, Figure 1.) This gives lifting force because the pressure on the under surface is greater than the pressure on the upper surface, creating a "suction effect." The wing can increase the lifting force by increasing the tilt, but only up to a certain limit. If the tilt is too great, none of the air stream can slip over the upper surface (see Plate VII, Figure 2); the pressure on the upper surface is thus so diminished as to create a partial vacuum into which the surrounding air rushes and swirls, destroying the lift and causing a **stall.** At the trailing edge, eddies of air extend forward along the upper surface; at the wing tip, other eddies (collectively called the **tip vortex**), whirling at still greater speed, move inward along the upper surface. The explanation for this intrusion upon the upper surface is that the air, at higher pressure under the wing, tends to swirl up and over the upper surface where the air pressure is lower.

Turbulence of this sort not only drags on the wing, but, by extending up and over the upper surface, interferes with the smooth passage of air over the wing. The effects of turbulence are partially offset by the wing's **aspect ratio**—proportion of length to breadth. In other words, a wing is long and broad enough to allow for disturbances and yet has sufficient surface for lifting purposes.

The distal half of the wing, "the propeller," moves in semicircles. In ordinary flapping flight, the propeller moves in a half circle, forward and downward, then upward and backward, the tip describing a course like a figure eight. Hold the proximal half of the wing in one hand and move the distal half with the other in the manner described. The forward-downward thrust "pulls" the bird along; the upward-backward thrust, which is much quicker, presses against the air and "pushes" the bird along. In small, fast-moving birds, the downward-forward stroke is the main source of forward motion, while the upward-backward stroke is for recovery only and is made by partially folding the wing and separating the remiges so as to reduce pressure on the upper surface. The proximal half of the wing moves only slightly during the propeller's operation, since its primary function is to give lift and act as a shaft for the propeller. It thus has a steadying influence by preventing the bird's body from bounding up and down with the thrusts of the propeller.

Spread the unplucked wing so that the leading edge is on a straight line perpendicular to the body, and force the alula to stand out slightly from the wing. Observe that there are several apertures or "slots" between the tips of the outermost primaries and another between the alula and wing margin. Broad-winged, soaring birds such as eagles and vultures show slots that are very much larger. Slots prevent stalling and increase lift by making air flow fast and evenly over the upper surface, thus reducing turbulence. (See Plate VII, Figure 3.)

The wings of both the pigeon and the sparrow, relatively short and broad, have, consequently, a low aspect ratio. With so small a wing area in relation to body weight, the pigeon must flap its wings rapidly in order to gain sufficient lift. The Herring Gull *(Larus argentatus)* with long, narrow wings (high aspect ratio) need not move its wings so fast. In normal, unhurried flight the pigeon flaps its wings an average of 3.0 times per second, the Herring Gull only 2.3 times per second (Blake, 1947). The sparrow may attain a velocity of 39 miles per hour (Schnell, 1965); the pigeon, 47 mph (Meinertz-hagen, 1955); and the Herring Gull up to 49 mph when flying with a strong wind (Schnell, *op. cit.*).

THE TAIL

The bird's tail is actually a small bony and fleshy structure hidden by the feathers. In descriptive ornithology, however, the term "tail" has come to mean the feathers that arise from this structure.

Spread out the pigeon's tail and observe that it is fan-shaped, with a rounded posterior margin due to the graduated lengths of the feathers. Notice that the tail feathers are of two kinds, the **rectrices** (singular **rectrix**) and **tail coverts:**

Rectrices. The strong, conspicuous feathers whose outer ends form the posterior margin of the tail. They are the flight feathers of the tail and correspond to the remiges of the wing. They are paired, the number of rectrices growing on each side of the tail being equal.

How many rectrices are there in the pigeon? Note that the rectrices partially overlap each other. Which one is not overlapped? How many rectrices are there in the sparrow?

Tail Coverts. Similar in appearance to the coverts of the wing. They overlie and underlie the flight feathers in much the same way. In descriptive ornithology they are not divided into separate rows or groups. The under tail coverts, sometimes collectively known as the **crissum,** are separated from the feathers of the abdomen by an imaginary line drawn transversely through the vent. The **upper tail coverts** are less clearly marked off from the feathers of the rump. They may be considered to end at an imaginary line drawn transversely through a point on the upper surface directly over the vent.

> On Plate I and Plate II, Figure 2, label the upper tail coverts and the rectrices. On Plate II, Figure 1, label the crissum.

During flight a bird's tail, especially when spread, serves a variety of purposes. It supplements the lifting surface of the wings and forms a slot in conjunction with the trailing edge of the wings. It can serve as a rudder by steering the bird to left or right; as an elevator, by directing it up or down; or as a brake by retarding its forward speed.

THE LEGS AND FEET

The legs and feet of the bird are less specialized than the wings and show a greater diversity in structure corresponding with the varied habits of different species.

Examine either an articulated human skeleton, or a detailed chart of one, and locate the following bones of the hindlimb (pelvic appendage): femur (thigh bone), patella (knee-cap), tibia and fibula (shin bones), 7 tarsals (ankle bones), 5 metatarsals (foot bones), and 14 phalanges (toe bones).

The skeleton of a bird's pelvic appendage, like the skeleton of a bird's wing, is built upon the same plan as its homologue in man. But since it is more specialized than the human limb, the bones are less easily recognizable.

Examine now a prepared articulated skeleton of a pigeon's pelvic appendage. Count the separate bones. (The patella may be missing, having been lost when the skeleton was prepared.) What is the difference in the number of bones in the bird and in man? Identify the following bones in the pigeon's pelvic appendage:

Femur. A stout, cylindrical bone whose proximal part bends inward and has a prominent head that is received by the pelvic girdle. An irregular projection, the **trochanter,** extends beyond the shaft. The distal extremity of the femur has a pulley-shaped surface, which receives the patella, and two convex condyles, which articulate with the tibiotarsus and fibula.

Patella. A small bone found in front of the knee joint.

Tibiotarsus and *Fibula.* Two bones running parallel to each other. The tibiotarsus, a composite bone (see explanation to follow), is expanded at its proximal end, where it articulates with the inner condyle of the femur. It extends to the heel. The fibula articulates with the outer condyle of the femur but is a poorly developed bone extending only two-thirds of the way to the heel. Closely pressed to the outside of the tibiotarsus as a slender spicule, the fibula is partly fused with the shaft of the tibiotarsus.

Tarsals. The tarsals, or ankle bones, do not occur in the pigeon as separate elements. Some have fused with the lower end of the tibia and, because of this fusion, the tibiotarsus (tibia + tarsal elements) gets its name. Other tarsals have fused with the next bone of the foot to be considered.

Metatarsals. The second, third, and fourth metatarsals fuse to form one bone which is homologous to the human instep. To the proximal end of this composite bone are fused, as previously mentioned, certain tarsals. This bone is, therefore, properly called the **tarsometatarsus,** although it is more commonly referred to as the **metatarsus.** The proximal end of the metatarsus is very irregular and is provided with two concavities which articulate with the tibiotarsus. The three metatarsals making up this composite bone remain distinct at the distal end as three articular projections for the anterior toes. A rudimentary first metatarsal, sometimes known as the **accessory metatarsal,** is connected by a ligament with the inner and posterior aspect of the distal part of the metatarsus. The fifth metatarsal is absent in adult birds.

Phalanges. There are four toes or digits, three anterior and one posterior. Each is made up of a series of phalanges placed end to end; the proximal phalanges are nearest to the metatarsus, and each of the distal phalanges terminates in a strong, curved **claw.** The toe projecting posteriorly is the first toe, or **hallux** (toe No. 1). It is articulated with the accessory metatarsal and has two phalanges; it is homologous to the "big toe" in man. The three toes extending anteriorly are articulated with the three projections on the distal end of the metatarsus. The innermost of these three toes has three phalanges; the middle toe, four; and the outermost toe, five. They are, respectively, the second (Toe No. 2), third (Toe No. 3), and fourth (Toe No. 4) toes and are homologous to the second, third, and fourth toes of man. A fifth toe is not found in birds.

Turn now to the pigeon and pluck the feathers from the leg. Find the position of the bones studied; then become familiar with the following parts of the pelvic appendage:

Thigh. The proximal segment of the leg containing the femur. It was, prior to plucking, entirely hidden.

Crus. The distal segment of the leg containing the tibiotarsus. Sometimes called the *shank* and more popularly known as the "drumstick," it is entirely feathered.

Knee. The junction of the thigh and crus. It bends forward as in man.

Foot. The remaining portion of the pelvic appendage. It is divisible into two parts:

  Tarsus. The third segment of the pelvic appendage, between the crus and the bases of the toes. It is noticeably scaled except at the proximal end, where it is feathered. It contains the metatarsus and accessory metatarsal.

  Toes. Four in number.

Heel. The junction of the crus and foot and always bends backward. The bird is, therefore, **digitigrade,** walking on its toes with its heels in the air. Man is plantigrade, walking on the soles of his feet with his heels on the ground.

On Plate IV, Figure 2, sketch around the drawing of the bones of the leg and foot of the pigeon an outline of the plucked leg. Show and label all of the parts mentioned above, and number the toes.

Examine the unplucked leg of the pigeon. Locate the crus. In descriptive ornithology the crus, together with the feathers, is called the **tibia** and will be thus designated in later sections of this book.

On Plate I and Plate II, Figure 1, label the hallux, tarsus, and tibia. On Plate I, number the toes of one foot.

Flex the pigeon's pelvic appendage at the knee and heel, drawing the tarsus to the crus and the crus to the thigh; then extend the appendage and repeat the procedure. Note that when the appendage is flexed the toes assume a grasping position and that when the appendage is extended the toes straighten out. This action of the toes is brought about by tensions exerted upon them by muscles in the thigh and crus. When the appendage is flexed the tension brought to bear on the tendons causes the toes to bend. By means of this arrangement the toes automatically grasp and hold fast to a perch while the bird squats during rest or sleep.

REFERENCES

Beebe, C. W.
1906   The Bird: Its Form and Function. Henry Holt and Company, New York. (Dover reprint available. Chapters 10 through 15 deal with the bills, heads, necks, bodies, wings, feet, legs, and tails of birds.)

Bellairs, A. d'A., and C. R. Jenkin
1960   The Skeleton of Birds. In *Biology and Comparative Physiology of Birds.* Volume 1. Edited by A. J. Marshall. Academic Press, New York. (For the skeleton of the wing and hindlimb.)

Berger, A. J.
1961   Bird Study. John Wiley and Sons, New York. (Chapter 1 includes a good review of topography.)

Blake, C. H.
1947   Wing-flapping Rates of Birds. Auk, 64: 619-620.

Chapman, F. M.
1932   Handbook of Birds of Eastern North America. Second revised edition. D. Appleton and Company, New York. (Dover reprint available. Habits and structure; uses of bill, wings, and feet, pp. 111-116.)

Coues, E.
1903   Key to North American Birds. Fifth edition. Two volumes. Dana Estes and Company, Boston. (Volume 1, pp. 96-139, contains basic information on the topography of birds.)

Fisher, H. I.
1940   The Occurrence of Vestigial Claws on the Wings of Birds. Amer. Midland Nat., 23:234-243.

Hess, G.
1951   The Bird: Its Life and Structure. Greenberg, New York. (Chapter 3 deals in part with the external structure and the flight of birds.)

Jack. A.
1953   Feathered Wings: A Study of the Flight of Birds. Methuen and Company, London.

Meinertzhagen, R.
1955   The Speed and Altitude of Bird Flight (with Notes on Other Animals). Ibis, 97:81-117.

Rand, A. L.
 1954 On the Spurs on Birds' Wings. Wilson
       Bull., 66: 127-134. (The occurrence of
       wing spurs is noted for all species of
       screamers, Family Anhimidae, some
       plovers, two jacanas, and two ducks.
       They occur on different parts of the
       wing and involve the radius, the ra-
       diale, or the fused metacarpals, de-
       pending on the species. The structures
       are used in fighting. Wing spurs
       should not be confused with vestigial
       claws. See Fisher, 1940.)
Schnell, G. D.
 1965 Recording the Flight-speed of Birds
       by Doppler Radar. Living Bird, 4:79-
       87.
Storer, J. H.
 1948 The Flight of Birds Analyzed through

       Slow-motion Photography. Cranbrook
       Inst. Sci. Bull. No. 28, Bloomfield Hills,
       Michigan.
Thomson, A. L., Editor
 1964 A New Dictionary of Birds. McGraw-
       Hill Book Company, New York. (Var-
       ious external parts of birds are listed
       alphabetically and described.)
Walls, G. L.
 1942 The Vertebrate Eye and Its Adaptive
       Radiation. Cranbrook Inst. Sci. Bull.
       No. 19, Bloomfield Hills, Michigan.
Young, J. Z.
 1962 The Life of Vertebrates. Second edi-
       tion. Oxford University Press, New
       York. (Chapter 16 contains much use-
       ful information on the external fea-
       tures of the bird and on bird flight.)

A student should examine at least one authoritative work in which birds are described by topographical parts. He will then see for himself the application of a knowledge of topography. Among the several suitable works, the following series is available in most college and university libraries.

Ridgway, R.
 1901- The Birds of North and Middle Amer-
 19   ica: A Descriptive Catalogue of the
       Higher Groups, Genera, Species, and
       Subspecies of Birds Known to Occur
       in North America, from the Arctic
       Lands to the Isthmus of Panama, the
       West Indies and Other Islands of the
       Caribbean Sea, and the Galapagos
       Archipelago. Parts 1-8. Bull. U. S.
       Natl. Mus. No. 50.

Ridgway, R., and H. Friedmann
 1941- The Birds of North and Middle Amer-
 46   ica. Parts 9-10. Bull. U. S. Natl. Mus.
       No. 50.

Friedmann, H.
 1950 The Birds of North and Middle Amer-
       ica. Part 11. Bull. U.S. Natl. Mus. No.
       50.

FEATHERS AND FEATHER TRACTS

Feathers are peculiar to birds and constitute their principal covering. Like the sheath of the bill, the scales on the feet, and the claws on the toes, feathers are horny, keratinized outgrowths of the skin or integument. They develop from tiny pits or follicles in the skin, just as do the hairs of mammals.

The scales on the feet of birds are clearly of the reptilian type and feathers probably evolved from comparable scales, becoming lengthened and elaborated to occupy greater space. Except in initial development, scales and feathers bear little resemblance to each other, and there are no known structures in either reptiles or birds, living or extinct, that provide any evidence of linkage between scales and feathers.

The two most important functions of the bird's feathers are to provide insulation—thus reducing loss of body heat—and to make flight possible by giving a streamlined contour and increasing the surface of the wings and tail. Through their coloration, feathers also aid certain species in concealment, in sex and species recognition, and in numerous displays.

Structure of a Typical Feather

Remove from the unplucked wing of a Common Pigeon *(Columba livia)* one of the best-developed primaries. Look for the parts described below and locate them on Plate VIII. The terms "dorsal" and "ventral" refer to the upper and under surfaces of the feather itself without regard to the feather's position on the bird.

Shaft

The **shaft** is the axis or "tube" of the feather and has two parts:

Calamus. The proximal (lower) part of the shaft, without vanes. Remaining almost entirely in the skin follicle from which the whole feather developed, the calamus is a hollow and somewhat transparent barrel, circular in cross section, with the basal end tapering to a point. At this point is the **inferior umbilicus** where the nutrient pulp entered during the growth of the feather. It is now closed by a horny plate. The distal end of the calamus is marked by the **superior umbilicus**, a minute opening on the ventral side of the shaft—i.e., the side that faces toward the body of the bird when the wing is closed—between the points where the vanes begin. The superior umbilicus is the remnant of the open, upper end of the tube of epidermis in a growing feather.

With a sharp scalpel, scrape some of the covering or **cortex** from the calamus. Note within the hollow interior a series of downward projecting, cup-like structures seemingly fitted one into the other. Commonly called **internal pulp caps** they were formed from the cornification at regular points of the layer of epidermis enclosing the pulp, long since resorbed. Each cap consists of a dome and a side-wall that adheres to the calamus next to it. Note that the caps extend from the lower to the superior umbilicus and, though rather evenly spaced, become successively farther apart toward the superior umbilicus. Note also that a rod-like structure, the cornified remnant of the axial artery, passes through the centers of the caps.

Rachis. The distal part of the shaft supporting the vanes. It is a continuation of the feather tube above the dorsal side of the calamus, from the superior umbilicus to the tip of the feather. Roughly quadrangular in cross section, it has two layers: (1) an outer cortex, thin and transparent, with ridges projecting inward from the dorsal surface, and (2) an inner medulla comprised of pithy tissue, firm and opaque. Outside the rachis on the ventral side are a few **external pulp caps** that extend through the superior umbilicus. At one time these caps continued to the tip of the feather. Examine the ventral surface of the rachis and observe that it features a longitudinal groove with a ventral ridge on each side.

Note that the rachis is flexible from side to side but quite stiff dorso-ventrally. A cross section of the rachis will show septa in the medulla between the dorsal and ventral surfaces. These septa plus the longitudinal groove on the ventral surface assist in giving the primary feather the proper flexibility and stiffness to act as a resilient airfoil.

Vanes

The **vanes** are two in number. They are more or less flexible structures springing from the opposite sides of the rachis. The outer vane—i.e., the vane overlapping the next outer feather—is narrower. Each vane is made up of a series of more or less flattened, parallel plates set obliquely and closely on the rachis with their free ends sloping toward the distal end of the feather. The proximal part of each vane is rather downy.

Barbs

The plates, which can be plainly seen in the vanes, are termed the **barbs.** Each barb is roughly comma-shaped in cross section: the dorsal surface—the top of the comma—is more or less rounded; the two sides are compressed, with the side facing the tip of the feather more flattened or even concave than the opposite side, which is always convex; ventrally the two sides meet to form a ridge that tends to be pointed. The barbs of the outer vane are shorter and thicker than those of the inner and emerge from the rachis at a sharper angle.

Afterfeather

This structure, sometimes called the **hypoptile,** is not well developed in the pigeon. In fact, it is represented only by a few downy barbs emerging near the superior umbilicus.

Examine the large afterfeather on an abdominal feather of a grouse or pheasant. Note that it has a shaft, called the **aftershaft** or sometimes the **hyporachis,** from which barbs emerge just as they do from the shaft of the main feather. Afterfeathers enhance the insulative property of a bird's feathering and also pad or fill out the body contours. In emus and cassowaries the afterfeather so closely duplicates the main feather in size that the two look like a double feather.

> **Label on Plate VIII:** calamus; rachis; internal pulp caps; outer vane; inner vane.

Label on Plate IX: superior umbilicus; inferior umbilicus; external plup cap.

Continue the study of the same pigeon primary. Observe that the barbs adhere closely to each other, thus giving the feather resistance to the passage of air. Separate several of the barbs, then press them together again, noting that they adhere as before. This remarkable adaptation is due to the relationship of certain minute structures invisible to the naked eye.

From the same feather cut a section of the inner vane, about midway between the proximal and distal ends, and place it under a dissection microscope and consult Plate X, Figure 1.

Barbules

Under the microscope, note that each barb is composed of a main axis, the **ramus,** with closely parallel sets of branches, called **barbules,** on opposite sides. The barbules are of the pennaceous type (cf., plumulaceous type to be seen later) and cross obliquely the barbules of the neighboring barbs. The barbules on the proximal side of each barb—the **proximal barbules**—resemble long, slender, scroll-like plates with the upper edges folded into **flanges.** Toward its end in a fine point, the **pennulum,** each proximal barbule shows several teeth and spines, both very tiny outgrowths. The barbules on the distal side of the barb—the **distal barbules**—have several **hooklets,** about two-thirds of the distance from the barb, that project downward and reach across from one to four barbules of the next higher barb of the vane and grasp their flanges. Besides the hooklets, each distal barbule also shows toward its end several other tiny outgrowths, **ventral teeth** and **cilia,** before terminating in a slender point. The hooklets and all the other outgrowths—the teeth and spines of the proximal barbules and the teeth and cilia of the distal barbules—are collectively called **barbicels.** The spines, teeth, and cilia function in the interlocking mechanism of the barbules by maintaining proper spacing between the barbules and keeping the hooklets from sliding off the flanges.

Label on Plate X, Figure 1: proximal barbule, flange; pennulum; distal barbule, hooklet, ventral tooth, cilia. Below Figure 1 on Plate X, make a drawing (Figure 2) of two barbs and their interlocking barbules as seen under the microscope.

Pluck a typical feather from the body of the pigeon and study one of the downy barbs from its basal part. Notice that the ramus is more slender and flexible than that of the barb of the pigeon primary. The barbules are of the plumulaceous type, each with a relatively short, strap-like base and a long slender extension. The latter has swellings along its course, called **nodes,** and thus resembles a stalk of bamboo. The nodes are shaped in various ways that are characteristic of different orders of birds. (See the drawing on page 35 showing the downy barbule of the Common Pigeon.) Partly for this reason, downy barbules have been used more successfully than pennaceous barbules in identifying feather remains in archaeological finds and in the stomach contents of mammalian predators (see day, 1966).

Plate VIII

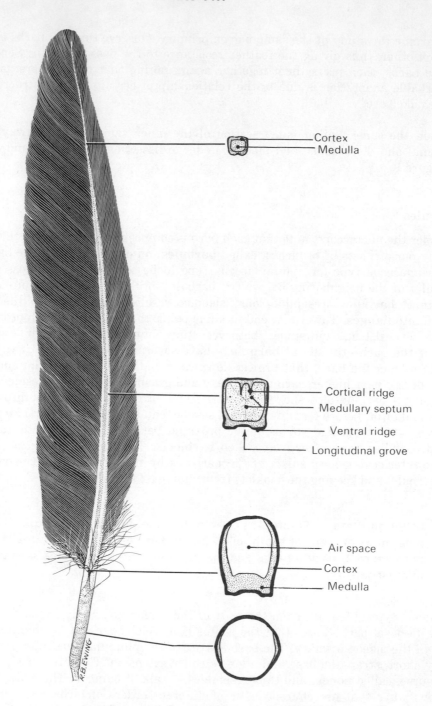

Cortex
Medulla

Cortical ridge
Medullary septum
Ventral ridge
Longitudinal grove

Air space
Cortex
Medulla

K.B.EWING

PRIMARY OF COMMON PIGEON

Whole feather with cross sections
of the shaft. Ventral view.

Plate IX

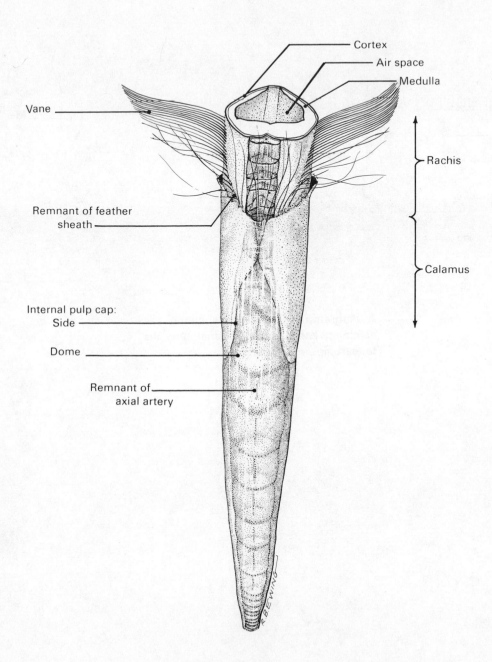

Cortex

Air space

Medulla

Vane

Rachis

Remnant of feather
sheath

Calamus

Internal pulp cap:
Side

Dome

Remnant of
axial artery

R.BEWING

PRIMARY OF COMMON PIGEON

Calamus and proximal end of
rachis. Oblique view.

Plate X

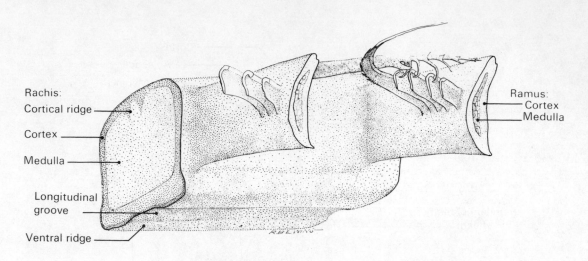

Rachis:
Cortical ridge
Cortex
Medulla
Longitudinal groove
Ventral ridge

Ramus:
Cortex
Medulla

Figure 1

A diagrammatic drawing showing two adjoining barbs, with proximal and distal barbules interlocking.

Figure 2

A drawing of two barbs and interlocking barbules as seen under the microscope.

INTERLOCKING MECHANISM OF BARBULES

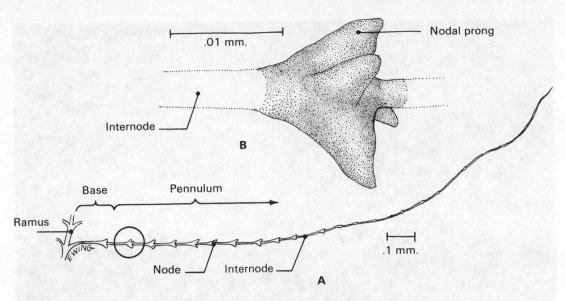

DOWNY BARBULE OF COMMON PIGEON

A.   Whole barbule.   B.   Magnified view of node encircled in A.

Feathers of Adult Birds

All feathers on adult birds are called **teleoptiles.** There are five kinds on the pigeon; a sixth kind occurs on many other groups of birds.

1. Contour Feathers

The primary of the pigeon is a good example of a contour feather. All other remiges, as well as rectrices and other exposed feathers, are contour feathers—i.e., they take part in forming the outline or contour of the bird.

Examine the contour feathers on different parts of the pigeon, observing especially the variation in size, shape, and texture. Compare with the drawing on page 36 of a typical contour feather of the body. In all flight feathers the vanes are of unequal width, the outer one being the narrower; from the innermost remiges and rectrices to the outermost, the outer vane becomes gradually narrower. The contour feathers making up the body plumage generally have vanes of equal width. Noteworthy is the fact that the coverts of the flight feathers show a complete transition in structure from that of a flight feather to that of a body feather. Thus a greater covert is more like a flight feather than a body feather, while a lesser covert is more like a body feather than a flight feather.

Among other species of birds are many peculiarly modified contour feathers:

Modified Remiges. The primaries of swans, geese, ducks, grouse, turkeys, a few species of pheasants and owls, and other birds have the basal one-third to two-thirds of the inner vane greatly stiffened by the expansion of the ventral ridge of each barb into a wide, thin, glistening flap, called the *tegmen,* which overlaps and presses against the barb distal to it; thus the ventral surfaces of the parts of the vanes involved have a glossy appearance. The tegmen is believed to act as a valve allowing air to pass through from above but not from below. The three outermost primaries of the American Woodcock *(Philohela minor)* are narrowed and stiff, for the production of sound during flight.

TYPICAL BODY CONTOUR FEATHER

The calamus of the innermost primary of the Standard-winged Nightjar *(Macrodipteryx longipennis)* is upturned. In the flightless cassowaries, the primaries are reduced to five or six black quills without vanes. The secondaries of waxwings have "waxy tips" (function unknown) formed by the fusion of the shaft with the outer vane.

Modified Rectrices. Woodpeckers and other birds which use their tails as supports while climbing have strengthened shafts, with the distal barbs greatly thickened and either without barbules or with a reduced number of barbules that fail to interlock. Some swifts (e.g., the Chimney Swift, *Chaetura pelagica*), whose tails function as supports in vertical roosting, have rectrices with shafts free of distal barbs and consequently projecting beyond the vanes as stiff spines. The central rectrices of the Pomarine Jaeger *(Stercorarius pomarinus)* are peculiarly twisted and those of most motmots (Momotidae) are exceedingly long with racket-shaped tips, caused by the loss of vanes along the subterminal section of each shaft. Male birds-of-paradise (Paradisaeidae) show remarkable modifications in all their rectrices but none more so than the male Superb Lyrebird *(Menura novaehollandiae)*: his outermost pair of rectrices, broad and S-shaped, produce the lyre-like form; the innermost pair, extremely narrow, cross each other soon after emerging from the skin and then curl forward near their tips; the remaining 14 pairs, delicate throughout, with hair-like barbs which are wide apart and without barbules, give a lacy effect to the tail when it is fully displayed.

Other Modified Contour Feathers. The auriculars in some birds—e.g., the Domestic Fowl *(Gallus gallus)*—are much longer around the anterior margin of the ear opening than those around the posterior margin and are supported on the tips of the posterior feathers. The anterior feathers with their widely separated barbs and short barbules adhering to the sides of the barbs form a screen over the ear opening that reduces interference with sound waves and at the same time prevents intrusion of foreign particles. Feathers of similar structure grow densely over the nostrils of such birds as crows, forming nasal tufts.

The feathers of many birds have become variously developed as ornamental plumes, and these may arise from almost any part of the bird, depending on the species or group of species. In egrets, the plumes are principally the back feathers which extend over the tail; in some species of birds-of-paradise, they are long flowing feathers that project from the sides and flanks. In both egrets and birds-of-paradise, the plumes are of soft texture, with the barbs filamentous and the barbules either absent or much reduced and not interlocking. Among the plumes of several species of birds-of-paradise are feathers whose shafts are devoid of barbs and greatly elongated or wire-like. Sometimes birds have the coverts of their flight feathers enormously developed. Thus in the male Quetzal *(Pharomachrus mocino)*, a trogon, the median coverts of the remiges are drooping plumes, while the upper tail coverts are long streamers, the central pair being four times the length of the tail; in the male peafowl (*Pavo* spp.) the lower rump feathers, as well as the upper tail coverts, are of striking form, owing to the uneven arrangement of the barbs and barbules, and are extended back over the rectrices as a long "train." Numerous species have plumes arising from the head as crests with varying conditions of shaft, barb, and barbule development. Both sexes of the peafowl, for instance, have crests of delicate feathers, each scantily barbed except at the tip, where there is a thick tuft of interlocked barbs.

Inspect a series of modified contour feathers of various birds and note the structural differences between these feathers and the primary of a pigeon. On page 38, titled "Modified Contour Feathers" make a series of sketches illustrating some of the extreme types.

MODIFIED CONTOUR FEATHERS

2. Semiplumes.

Hidden beneath the body feathers are small, white feathers called **semiplumes**. Search for them on the pigeon and pluck several. The semiplume has a downy texture—no interlocking barbs—and its rachis is longer than the longest barb. Structurally, the semiplume is intermediate between the contour feather and the next kind of feather to be considered; functionally, the semiplume assists in entrapping air for thermal insulation.

PIGEON SEMIPLUME

3. Adult Down Feathers

These feathers are not abundant on the adult pigeon but they may be found by careful searching under the contour feathers, particularly on the sides of the body. The down feather appears as a soft tuft. Pluck one, mount it on a slide, and examine it under a microscope. Characteristically soft throughout, it differs from the semiplume in having a rachis that is shorter than the longest barb. Like the semiplumes, down feathers assist in thermal insulation. They are especially abundant on waterfowl and many other aquatic birds.

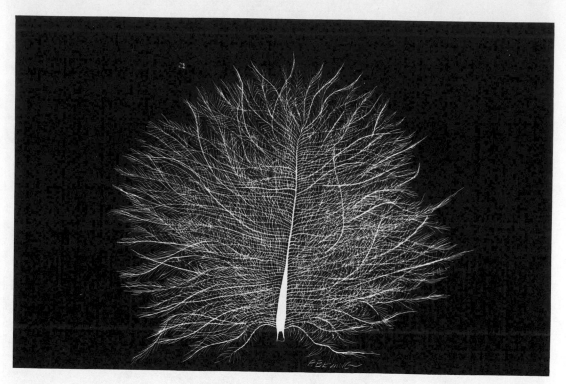

ADULT DOWN FEATHER

4. Filoplumes

These are very slender, hair-like feathers found all over the body of the pigeon, always accompanying other feathers including semiplumes and down feathers. Typically, there are one or two filoplumes with nearly every body feather, situated laterally and medially rather than dorsally at its base. There may also be several filoplumes around the rim of the follicle for each remex and rectrix. Pluck a filoplume—be careful that it does not break off at the superior umbilicus, mount it on a slide, and examine it under the microscope. Note that it has a distinct calamus and rachis, with a few barbs growing from near the tip of the rachis. A filoplume, broken at the superior umbilicus, might lead the student to believe that it consists only of a rachis.

Filoplumes are somewhat less slender, longer, and consequently easier to find on bigger birds, particularly at the bases of their remiges. In some passerine birds, filoplumes extend beyond the tips of the contour feathers on the back of the neck. Exposed filoplumes are unusually conspicuous in cormorants; white in males and brownish in fe-

FILOPLUME OF PIGEON
(Fourteen millimeters in actual length)

RICTAL BRISTLE OF ROBIN
(Turdus migratorius)
(Seven millimeters in actual length)

males, they appear on the back, rump, thighs, and less frequently on the under parts of the body.

The function of filoplumes is conjectural. Since they are situated at the bases of contour feathers which may receive vibrations from outside forces, possibly filoplumes serve a sensory function by monitoring the larger feathers (von Pfeffer, 1952) and stimulating appropriate action of the muscles controlling the larger feathers.

5. Bristles

Bristles do not occur in the pigeon. They fringe the rictal region as rictal bristles in such birds as the Whip-poor-will *(Caprimulgus vociferus)*, Brown Thrasher *(Toxostoma rufum)*, and New World flycatchers (Tyrannidae), and encircle the eyes as eyelashes in the Ostrich *(Struthio camelus)*, Marsh Hawk *(Circus cyaneus)*, and a few other kinds of birds. The typical bristle is characterized by a stiff, tapered shaft, occasionally with barbs along the proximal portion of the rachis; it has a brown to black coloration. In all probability, bristles have a tactile function, perhaps serving in the same manner as the whiskers of many mammals.

6. Powder-down Feathers

In many birds the plumage is dusted by powder which comes from the downy elements of contour feathers, semiplumes, and ordinary downy feathers. But the powder comes chiefly from specially modified down feathers, considered by some authorities as a separate kind, the powder-down feathers. They are notably well developed on such birds as herons and bitterns, being clustered in areas known as powder-down patches. Examine the breast of an American Bittern *(Botaurus lentiginosus)*, pushing aside the contour feathers. Here are two large, thick patches of yellowish feathers, the powder-downs. Their growth and the accompanying production of powder are said to be continuous.

Powder-down feathers are present in the pigeon, the Marsh Hawk, some parrots, and many other birds. However, they are scattered over the body, among the ordinary down feathers and are therefore difficult if not impossible to differentiate by superficial analysis. The powder is formed by the proliferation and keratinization of cells which accompany the barbs in the feather germ and is released while and after the feather emerges from its sheath. Because it has a waterproof quality, some authorities regard the powder as supplementing the function of the oil gland in providing a dressing for the feathers.

Feathers of Newly Hatched Birds

Certain kinds of young birds at hatching have natal down feathers called **neossoptiles.** They may be either thickly or sparsely distributed. Not long after hatching they are pushed out from their follicles by the ensheathed tips ("pinfeathers") of the next generation of feathers, the juvenal feathers. The neossoptiles remain attached to the tips of the juvenal pinfeathers for a short period only; by the time the pinfeathers have completely unfolded from their sheaths, the neossoptiles will have been dislodged.

Most neossoptiles differ from the adult down feathers already described. In gallinaceous birds the neossoptiles have a minute rachis or none, and no hyporachis. In ducks, on the other hand, the neossoptiles have a conspicuous rachis whereas adult downs have a very small one. Some natal downs are distinct feathers while others are modified tips of barbs of the next generation of feathers. The central barbs of unworn adult downs carry barbules to their tips but the barbs of natal downs are bare at the tips. The barbules themselves differ: those of natal downs do not show the nodes characteristic of adult downs.

Feather Development

The skin or integument of birds consists of two major parts: the dermis and the epidermis. The dermis, the inner part, is the nutrient tissue, carrying blood vessels. The epidermis, the outer part, consists of numerous cell layers: the innermost layers comprise the stratum germinativum, divisible (from the outside in) into the stratum transitivum, stratum intermedium, and stratum basale. These layers continually proliferate new cells outward. The outermost layers constitute the stratum corneum—horny tissue consisting of cells that have become flattened and forced outward by the new cells from the stratum germinativum. In the formation of a feather the dermis gives rise to the pulp whose blood vessels supply the nutrition for growth. The stratum corneum and stratum transitivum form the sheath of the feather and the outer layers of the calamus. The stratum intermedium is the source of nearly all parts of the feather including the inner layers of the calamus. The stratum basale forms the germinating ring at the base of the feather, the walls of the pulp caps, and a small amount of material that is discarded when the feather unfurls from its sheath.

Plate XI

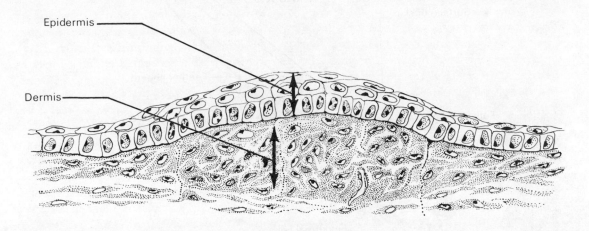

Epidermis

Dermis

Figure 1

Initial stage in embryonic formation of
a feather papilla.

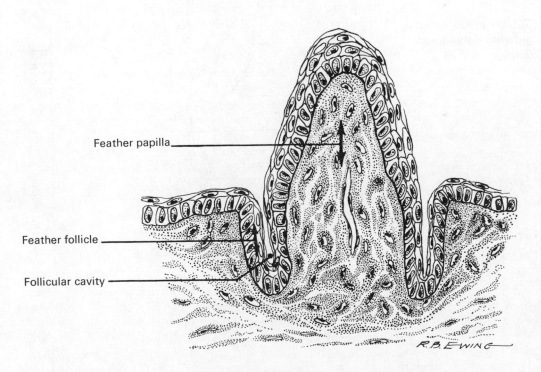

Feather papilla

Feather follicle

Follicular cavity

R.B. EWING

Figure 2

Late stage in embryonic formation of a
feather papilla.

FORMATION OF A FEATHER PAPILLA

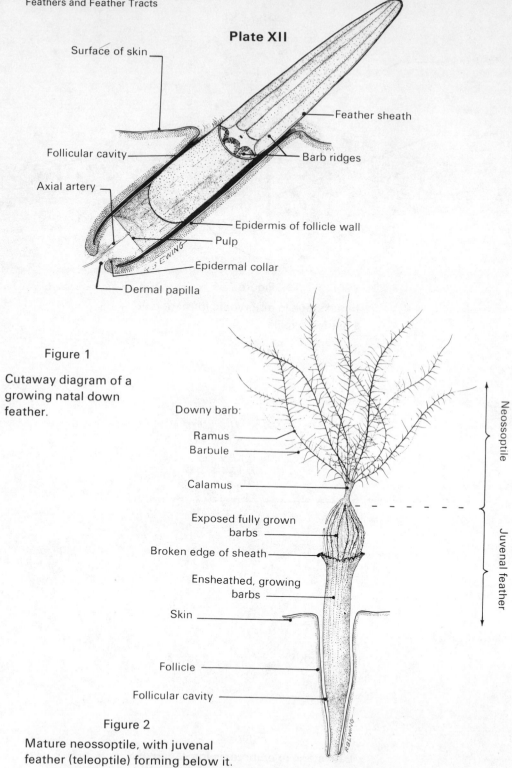

Plate XII

Surface of skin

Feather sheath

Follicular cavity

Barb ridges

Axial artery

Epidermis of follicle wall

Pulp

Epidermal collar

Dermal papilla

Figure 1

Cutaway diagram of a growing natal down feather.

Downy barb:

Ramus

Barbule

Neossoptile

Calamus

Exposed fully grown barbs

Broken edge of sheath

Juvenal feather

Ensheathed, growing barbs

Skin

Follicle

Follicular cavity

Figure 2

Mature neossoptile, with juvenal feather (teleoptile) forming below it.

GENERALIZED STRUCTURE OF GROWING AND MATURE NATAL DOWN FEATHERS (NEOSSOPTILES)

The growth of a feather or scale starts with a papilla which pushes up the overlying epidermis. (See Plate XI, Figure 1.) Thereafter the similarity between a feather and a scale ceases altogether. If the papilla is destined to form a feather, it becomes elongated and tubular; if a scale is to result, it soon becomes flattened and plate-like.

Development of a Neossoptile

The tubular papilla, which is the pulp of the developing feather, continues its outward growth and at the same time sinks into a pit, the future feather follicle. (See Plate XI, Figure 2.) Meanwhile the stratum germinativum of the overlying epidermis, through the outward proliferation of cells by the germinating ring around the base of the papilla, produces (1) a series of columns of barb ridges (the future barbs; also the rachis, if there is to be one, as in ducks) which run parallel from the ring to the tip of the papilla and are closely applied to the inner pulp, and (2) an outer, cone-shaped sheath. As a result of rapid growth, barb ridges each differentiate within the sheath into ramus and barbules and the distal end of the papilla soon projects from the follicle above the surface of the skin; the proximal end stays in the follicle as the calamus. (See Plate XII, Figure 1.) The production of keratin, a specific protein within the cells, gradually hardens the feather parts. Eventually the feather splits, beginning at the tip; the barb ridges separate as barbs and extend free, and the pulp disappears by resorption. By the time the growth and keratinization of the feather are completed, the entire sheath above the follicle has disintegrated into flakes and the barbs have spread out from their base on the calamus (see Plate XII, Figure 2); the pulp in the calamus has been resorbed and the inferior umbilicus no longer functions as an entrance for pulp and is closed by a pulp cap. The base of the calamus rests on the ensheathed tip of the next feather generation, the juvenal feather emerging from the same follicle.

Development of a Teleoptile

The early stages in the regeneration of a typical contour feather are like those of a neossoptile. From the bottom of the follicle the dermal papilla with its overlying epidermis proceeds to grow outward. Within the sheath the barb ridges first appear, as before, in parallel columns originating on the germinating ring; but after this a marked differentiation in development occurs. The ridges grow tangentially from the germinating ring. The distal end of each barb is the first part to be formed. Progressively more proximal portions are laid down in a half-spiral that curves dorsally around the pinfeather as it grows outward. The center of growth for each barb eventually reaches the mid-dorsal line where it meets the ridge of the future rachis. The bases of the barbs then fuse with the ridge of the rachis, "the rachidial ridge," at an oblique angle. (See Plate XIII, Figures 1 and 2.) Finally the sheath splits and disintegrates, allowing the feather (which has been rolled up in the sheath with its dorsal surface against the sheath and its ventral surface next to the pulp) to flatten out. The stratum intermedium soon stops proliferating barb ridges, leaving the base of the feather as the calamus; the pulp is resorbed and caps are formed from the innermost epidermal layers surrounding the pulp.

The parts of a barb—barbules, flanges, hooklets, etc.—develop by differentiation of the barbule cells and ramus cells while a feather is growing. They begin to fuse and to keratinize even before they have attained their final shape. The parts of a barb are fully formed by the time the feather sheath flakes away from it; they simply unfold into their final position.

If a feather is to have an afterfeather, this begins to form sometime after the start of the main feather from the germinating ring on either side of the mid-ventral line. New

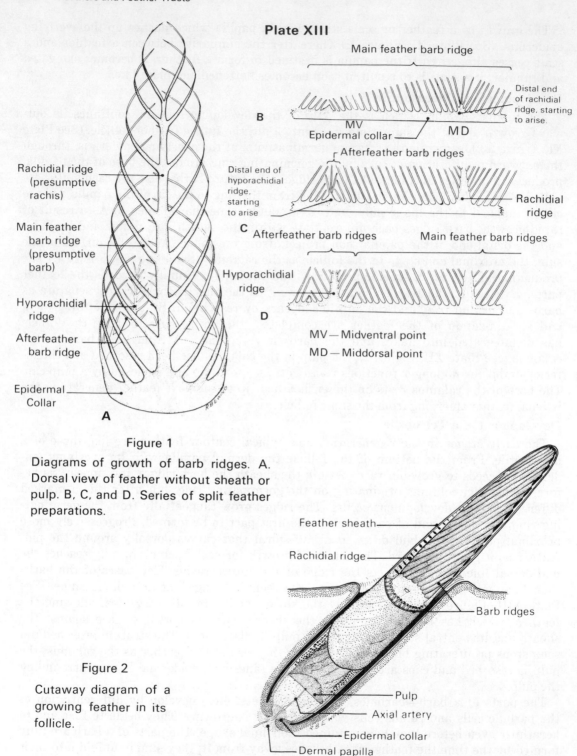

Plate XIII

Main feather barb ridge

B

Distal end of rachidial ridge, starting to arise.

MV
Epidermal collar ——— MD

Afterfeather barb ridges

Distal end of hyporachidial ridge, starting to arise

C

Rachidial ridge

Afterfeather barb ridges   Main feather barb ridges

Hyporachidial ridge

D

MV — Midventral point
MD — Middorsal point

Rachidial ridge (presumptive rachis)

Main feather barb ridge (presumptive barb)

Hyporachidial ridge

Afterfeather barb ridge

Epidermal Collar

A

Figure 1

Diagrams of growth of barb ridges. A. Dorsal view of feather without sheath or pulp. B, C, and D. Series of split feather preparations.

Feather sheath

Rachidial ridge

Barb ridges

Pulp

Axial artery

Epidermal collar

Dermal papilla

Figure 2

Cutaway diagram of a growing feather in its follicle.

GENERALIZED STRUCTURE OF GROWING CONTOUR FEATHERS (TELEOPTILES)

barbs continue to arise as before; some will make up the afterfeather while the others will contribute to the main feather. If the afterfeather is to have a hyporachis, it will arise and receive its barbs in the same way as the rachis of the main feather.

Examine a pinfeather 10 to 30 millimeters long from a molting bird. Carefully dissect it by cutting it free at the base of the follicle. Make an incision on one side along its full length. Unroll the feather tube, remove the core of pulp, and observe the barb ridges, the developing rachis, and the developing hyporachis if any.

The Coloration of Feathers

The colors of feathers are due to two primary factors: chemical substances and physical properties. Coloration resulting from chemical substances is commonly spoken of as **chemical coloration**, and coloration from physical properties as **structural coloration**.

Chemical coloration is caused by pigmentary compounds, or pigments, called **biochromes**, which absorb specific wave lengths within the visible spectrum and reflect the remaining light waves to the eye of the observer as color. The principal chemical colors of feathers are the following:

Red, Orange, and Yellow. Produced in most instances by carotenoids (formerly called lipochromes), which are fat-soluble pigments appearing in a diffused state rather than in discrete granules. Carotenoids occur in fat deposits, egg yolk, secretion of the oil gland, and bare skin, as well as in feathers. They are primarily synthesized by plants and appear in birds only after being modified from ingested food. Carotenoids within a feather occur primarily near the tip of the rachis and the rami of pennaceous barbs, and they may be removed by various organic solvents such as alcohol or ether.

Black, Gray, Brown, and Brownish-yellow (Tawny). Produced by melanins, which are relatively insoluble pigments appearing as granules. Unlike carotenoids, melanins are synthesized in special pigment cells (melanocytes) from amino acids. (See Plate XIV.) In feathers, melanins occur in the rachis and all parts of the barbs except the barbicels; elsewhere in the body they occur in the skin, the horny covering of the bill, scales of the feet, and certain internal organs. The differences in the melanistic colors depend in part on the amount of pigment deposition and the size of the pigment granules. For example, black is due to a large amount of pigment which absorbs all light waves, gray to a smaller amount which absorbs fewer light waves, etc. Melanins may mask carotenoids, making them invisible, or combine with them visually (not chemically) to create certain colors. Olive-green is caused by the combination of black melanin in the tips of the barbules and a yellow carotenoid in the bases of the barbules and in the rami. Quite apart from its color-producing function, melanin serves to increase a feather's resistance to wear. Thus a black or brown feather is less subject to abrasion than a white or brightly colored one; the black or brown part of a feather may remain intact after the white or brightly colored tip has worn away.

Green. Produced sometimes by iron-containing, non-granular pigments called porphyrins. Turacoverdin, one of the porphyrins, gives the green color in the feathers of turacos (Musophagidae). Other porphyrins produce colors such as red, brown, and buff in the feathers of birds in over a dozen orders.

Green in birds is often produced by pigments other than turacoverdin—for example, by the juxtaposition of melanin and yellow, as in parrots. Green is also produced by structural conditions (see below).

Plate XIV

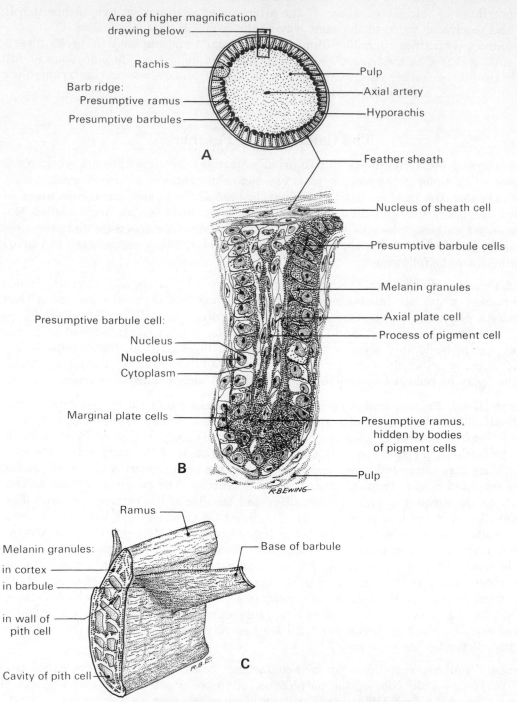

Area of higher magnification drawing below

Rachis

Barb ridge:
Presumptive ramus
Presumptive barbules

Pulp

Axial artery

Hyporachis

A

Feather sheath

Nucleus of sheath cell

Presumptive barbule cells

Melanin granules

Axial plate cell

Process of pigment cell

Presumptive barbule cell:
Nucleus
Nucleolus
Cytoplasm

Marginal plate cells

Presumptive ramus, hidden by bodies of pigment cells

B

Pulp

R.BEWING

Ramus

Base of barbule

Melanin granules:
in cortex
in barbule

in wall of pith cell

Cavity of pith cell

R.B.E.

C

MELANIN IN A CONTOUR FEATHER

A. Cross section of a growing feather showing the arrangement of melanin. B. Higher magnification of a part of the same feather showing the pigment cells in a barb ridge. C. Segment of a pennaceous barb showing the distribution of melanin granules.

Structural coloration is caused by the presence of minute structural elements, which produce colors by indirection. Biochromes, especially melanins, may combine with structural elements to modify or intensify structural colors. Some of the prominent structural colors of feathers are the following:

White. Feathers that appear white have little or no pigment. The outer covering or cortex of their rachises and rami is composed of numerous colorless cells, and their barbules are transparent and without important internal structure. Whiteness in feathers is due primarily to structural elements of the inner medulla or pith which reflect all light of all wave lengths within the visible spectrum.

Blue. Produced by the rami of the barbs. The combination of biochromes and structural elements causing blue in feathers of the Blue Jay *(Cyanocitta cristata)* is illustrated by the accompanying drawing of the ramus of a barb in cross section. Note that the ramus is covered by a transparent cortex (A). Under the cortex, in the medulla, are two types of so-called cloudy cells: Dorsally and laterally, a thin layer (B) of transparent, colorless, polyhedral cells, each containing a large, central, gas-filled vacuole; and elsewhere, large, closely packed cells (C), cuboidal and pervaded with many tiny vacuoles together with granules of melanin. Blue color results from a scattered reflection of blue light—the same phenomenon that causes the blue of the sky—by the vacuoles in combination with the melanin pigmentation which absorbs all light not reflected and thus permits only blue light to show. In the Blue Jay and most other blue-colored birds only the upper surfaces of the feathers are blue, owing to the dorsal and lateral position of the blue-producing elements, but in some species the under surfaces may also be blue because blue-producing elements are ventral in the barbs.

Non-iridescent Green. Feathers may appear green instead of blue, when the cortex above the cloudy cells is thicker and transparent yellow, due to diffuse carotenoids. The blue color from the blue-producing cells is changed to green by the yellow cortex. If the student will scrape the vane of a green feather with a sharp scalpel, thereby removing the yellow cortex from the barbs, the vane will appear blue.

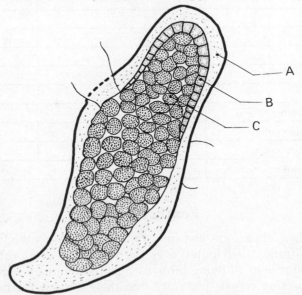

CROSS SECTION OF A BARB FROM A BLUE JAY FEATHER
(Modified from Gower, 1936)

Iridescent Colors. Produced by the barbules. A variety of iridescent or spectral hues may be produced by feathers having numerous overlapping barbules that are broadened and twisted and contain a great amount of dark, usually brown, melanin in the form of flat or rod-shaped granules. The colors result from the interference of light waves by the successive surfaces of the barbules while the melanin granules absorb the light waves not reflected. Spectral colors may also be produced by barbules—e.g., in the neck feathers of the Common Pigeon—having melanin in the form of spherical granules which are in close contact with the thin, overlying cortex. In this case, iridescence results when the granules reflect light waves which interfere with other light waves reflected by the cortex; the granules also absorb light waves not reflected. In effect, iridescence in feathers is the same interference color phenomenon that produces colors of soap bubbles and oil on water.

> Place under the microscope small feathers, or portions of feathers, mounted on slides, which show the colors discussed above. Include several feathers with iridescent colors. Study the colors, using first reflected light (i.e., light falling on the feathers from the side and reflected from them to the eye) and then transmitted light (i.e., light passing from the mirror and through the feathers to reach the eye). Record below the colors of these feathers under each set of conditions.

	REFLECTED LIGHT	TRANSMITTED LIGHT
Red feather		
Orange feather		
Yellow feather		
Black feather		
Gray feather		
Brown feather		
Tawny feather		
Green feather, from a turaco		
White feather		
Blue feather		
Green feather, not from a turaco		
Feather with iridescent color		
Feather with iridescent color		
Feather with iridescent color		

Feather Tracts

Penguins, ostriches, rheas, cassowaries, emus, and a few other flightless birds have feathers rather uniformly distributed over the body. The same condition would superficially appear to be true of all flying birds such as the Common Pigeon and House Sparrow, but close inspection reveals instantly that their contour feathers arise from tracts or **pterylae** (singular **pteryla**) on the skin and that there are certain areas or **apteria** (singular **apterium**) where contour feathers do not occur. This does not mean that apteria are necessarily bare, as they may have down feathers and semiplumes. The apteria are normally concealed because they are overlapped by the adjoining pterylae. The arrangement and distribution of feathers in tracts is technically spoken of as the **pterylosis**.

Begin a study of the pterylosis of the House Sparrow. In this bird the apteria are proportionately larger than in the pigeon; consequently, the pterylae are somewhat narrower and more distinct. If possible, use a nestling sparrow with feathers only partially developed; the tracts show clearly without need of special preparation. If a nestling sparrow is not available, use an adult bird. With scissors, closely clip the entire plumage with the exception of one wing. Note that the contour feathers do not simply "fill" the different tracts but that they are evenly spaced in regular rows. Determine the following pterylae and their associated regions and apteria:

Capital Tract. The pteryla extending over the top of the head from the base of the upper mandible posteriorly to the point where the head and neck join. It is bounded on each side by an imaginary line passing from the mandibular ramus to the angle of the jaw.

Spinal Tract. The pteryla extending posteriorly from the capital tract to the upper tail coverts. Along the neck it is bordered on each side by a **cervical apterium;** along the trunk it is bordered on each side by a large **lateral apterium**. The spinal tract is divisible into four regions identified mainly by their shape and locations. Thus the narrow **cervical region** extends from the head to the trunk; the narrow **interscapular region** extends posteriorly between the shoulder blades (scapulae); the saddle-shaped **dorsal region** extends from the shoulder blades to a point approximately halfway to the tail; the broad **pelvic region**, lying between the hips and extending from the dorsal region to the tail coverts, completes the spinal tract.

Humeral Tract. A narrow pteryla found on each wing running obliquely backward on the brachium from the anterior part of the shoulder, where it barely merges with a feather tract below. The feathers arising from this pteryla are the scapulars.

Femoral Tract. Another narrow pteryla; it extends along the outer surface of each thigh from a point near the knee to the vent. When the leg is drawn up, this tract is almost parallel to the spinal tract.

Crural Tract. The remaining feathers of the leg. It is separated from the femoral tract by a narrow apterium.

Ventral Tract. The pteryla beginning at the junction of the mandibular rami and extending posteriorly to the circlet of feathers around the vent. It encloses a longitudinal **mid-ventral apterium**. It is bounded by the capital tract and the cervical and lateral apteria. In some birds the ventral tract is not continuous as in the sparrow; instead it may be more or less divisible into such tracts as the cervical, pectoral, sternal, and abdominal.

Caudal Tract. Includes the rectrices and the upper and under tail coverts (see under "Topography" in this book, page 25); also, the circlet of feathers around the vent. The oil gland, which is located dorsally at the base of the tail, is not feathered in the sparrow. When feathers do occur on it, as in certain kinds of birds, the feathers are considered part of the caudal tract.

Alar Tract. Includes the remiges, all their coverts, and the other feathers arising on the wing except the feathers of the humeral tract which is separated by a narrow, unnamed apterium. See under "Topography" in this book, page 19, for details on the arrangement of the feathers on the wing.

Turn to the two drawings of the plucked sparrow (Plate XV). By stippling, indicate the feather tracts, using approximately as many dots as there are

Plate XV

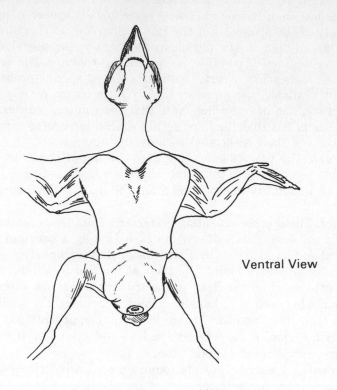

Ventral View

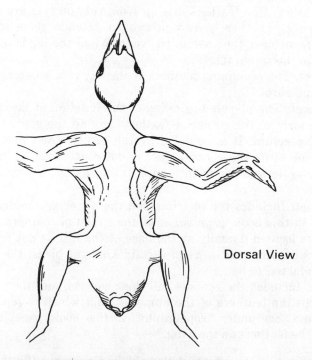

Dorsal View

PLUCKED HOUSE SPARROW

feathers. Take care, particularly in the case of the ventral tract, to show how the feathers are arranged in rows. Then label all pterylae, regions of the spinal pteryla, and apteria.

The pterylosis of the House Sparrow must not be considered the standard or typical pattern for all birds as there is no typical pattern. If possible, study and compare the pterylosis of several widely different species of birds, adult as well as natal forms. Use the papers by Humphrey and Clark (1961) and Wetherbee (1957) as models for detailed study.

Among different species, and between groups of species, a considerable variation exists with respect to the size and shape of pterylae, presence of certain apteria, the density of feathers on certain pterylae, and the number of remiges and rectrices and their coverts. Within a species, however, or a group of species, the pterylae and apteria show generally constant characteristics. For example, many families of birds have wings that are diastataxic (from diastataxy = "arrangement with separation") because the fifth secondary is absent even though its major covert remains. The House Sparrow and other passerine birds, as well as still other groups of species, have the full complement of secondaries and are therefore eutaxic.

Pterylography. The study and description of pterylae and their intervening apteria is useful in taxonomic investigations. By comparing the pterylosis of species apparently similar, it is often possible to determine whether or not a close relationship actually exists. The works by Ames, Heimerdinger, and Warter (1968), Berger (1960), and Compton (1938) illustrate how pterylography can be applied to taxonomy. Data on pterylosis can also be applied to studies of molts and age determination, enabling the investigator to identify and compare feathers from birds of different ages and species. See, for example, the papers by Aldrich (1956), Foster (1967), Mewaldt (1958), and Naik and Andrews (1966).

The Numbers of Contour Feathers on a Bird

Birds of the same species in the same area and season generally show only slight individual variation in number of contour feathers. The main exceptions are species in which the sexes differ greatly in size and those in which the males have a more elaborate feather arrangement; in these species a correlation between the number of feathers and sex may be expected. Sharp seasonal differences in the number of feathers on birds of a species occur when the individuals inhabit a region with marked seasonal changes. As a rule, the number of feathers is much greater in winter than in summer. For example, the contour feathers of three House Sparrows collected in Michigan during January and February numbered 3,546, 3,615, and 3,557 respectively; on two taken in July in the same area the totals were 3,138 and 3,197. The authority for these figures (Staebler, 1941) calculated that there was a loss from winter to summer of 11.5 percent of the feathers. By contrast, in Pretoria, South Africa, where the year-round temperature is milder than in Michigan, the counts of feathers on 11 specimens of the Laughing Dove *(Streptopelia senegalensis)* taken in different seasons showed no apparent variation (Markus, 1965).

Among species a considerable variation occurs in the number of contour feathers, a fact borne out in counts on birds from Michigan by Ammann (1937), from Florida by Brodkorb (1949), and from the vicinity of Washington, D. C., by Wetmore (1936). The lowest number found was 940, on a Ruby-throated Hummingbird *(Archilochus colubris)* collected in June (Wetmore), while the highest was 25,216, on a Whistling Swan *(Cygnus columbianus)* taken in November (Ammann). On 74 species of passerine birds (fly-catchers, jays, chickadees, vireos, blackbirds, etc.) the numbers ranged from 1,119 (a Ruby-crowned Kinglet, *Regulus calendula*, taken in October—Wetmore) to 4,607 (an Eastern Meadowlark, *Sturnella magna*, taken in February—Brodkorb). Some of the counts on non-passerine birds were as follows:

15,016	Pied-billed Grebe *(Podilymbus podiceps)*	December	Brodkorb
3,867	Least Bittern *(Ixobrychus exilis)*	May	Brodkorb
14,914	Pintail *(Anas acuta)*	January	Brodkorb
7,224	Clapper Rail *(Rallus longirostris)*	April	Brodkorb
13,913	American Coot *(Fulica americana)*	November	Brodkorb
4,480	Least Sandpiper *(Erolia minutilla)*	April	Brodkorb
2,635	Mourning Dove *(Zenaidura macroura)*	June	Wetmore
9,206	Barred Owl *(Strix varia)*	June	Brodkorb
3,332	Common Nighthawk *(Chordeiles minor)*	April	Brodkorb
3,665	Red-bellied Woodpecker *(Centurus carolinus)*	April	Brodkorb

It is not surprising that the diminutive hummingbird has relatively few feathers and that a big bird, such as a swan, has so many, or that small birds generally have fewer feathers than large birds. But body size is not the sole criterion in the number of feathers. Ammann found that nearly 80 percent of the feathers on the Whistling Swan were on the head and neck, indicating that a species' physical peculiarity (e.g., an exceptionally long neck) may account for a large number of feathers. The number of feathers may be attributed also to the structure of feathers and the particular uses which feathers serve. As an illustration, the Pintail has short feathers with tightly interlocking barbs to resist water and the Barred Owl has long, loose-textured feathers to enable silent flight. Though the two birds have nearly the same body size, the duck has more than double the number of feathers.

Two investigators, Hutt and Ball (1938), demonstrated that a small bird has more feathers per unit of body surface than a large bird, and that among land species the number of feathers per unit of body surface actually increases with decreasing body weight. The basis for this phenomenon is that the amount of heat lost by a bird, or any warm-blooded animal, is directly proportional to the surface area of its body. As a consequence, a small bird has greater difficulty in maintaining a body temperature above that of the environment than a large one because the surface area per unit of weight is much greater. A small bird must, therefore, have relatively more feathers for insulation. The weight of the plumage in relation to body weight shows a similar trend. Turček (1966) found that the relative weight of the plumage is lighter in heavier birds in spite of an increase in the absolute number of feathers.

Anything approaching a thorough investigation of the number of feathers among all species and on individuals within a species has not yet been undertaken. Do the Hutt-Ball findings apply to aquatic birds? Does the presence of afterfeathers in any way affect the number of feathers per unit of surface area or weight? Does the number of down feathers (teleoptiles) increase or decrease in proportion to the number of contour feathers? Does the number of filoplumes increase or decrease in the same way? These are only a few of the questions to be answered by additional feather counts.

Feather Parasites

All birds are hosts to parasites on their feathers. While this is a sweeping statement, it is undoubtedly a true one as investigations have borne out.

There are two principal groups of feather parasites: the feather lice, order Mallophaga, of the arthropod class Insecta, and the feather mites of the arthropod class Arachnida. The feather lice are the more prominent in numbers of species. Many are parasitic on particular species, or closely allied species, of birds. Considering the extent of their distribution and their host specificity, feather lice are believed to have become parasitic on birds at a very early stage in their evolution and therefore are useful to ornithologists in determining the taxonomic relationships of birds. Feather mites are much less well known since their study is in its infancy.

See Appendix I for a brief review of these two groups of parasites on birds.

REFERENCES

Aldrich, E. C.
1956   Pterylography and Molt of the Allen Hummingbird. Condor, 58: 121-133.

Allen, G. M.
1925   Birds and Their Attributes. Marshall Jones and Company, Boston. (Dover reprint available. Chapter 2, "Feathers.")

Ames, P. L., M. A. Heimerdinger, and S. L. Warter
1968   The Anatomy and Systematic Position of the Antpipits *Conopophaga* and *Corythopis*. Peabody Mus. Postilla No. 114. (Includes the use of pterylosis.)

Ammann, G. A.
1937   Number of Contour Feathers of Cygnus and Xanthocephalus. Auk, 54: 201-202.

Auber, L.
1957a The Distribution of Structural Colours and Unusual Pigments in the Class Aves. Ibis, 99: 463-476.
1957b The Structures Producing "Non-iridescent" Blue Colour in Bird Feathers. Proc. Zool. Soc. London, 129: 455-486.

Beebe, C. W.
1906   The Bird: Its Form and Function. Henry Holt and Company, New York. (Dover reprint available. A lucid discussion of structure, development, and arrangement of feathers in Chapter 2.)

Bell, E., and Y. T. Thathachari
1963   Development of Feather Keratin during Embryogenesis. Jour. Cell. Biol., 16: 215-223.

Berger, A. J.
1953   The Pterylosis of *Coua caerula*. Wilson Bull., 65: 12-17.
1960   Some Anatomical Characters of the Cuculidae and the Musophagidae. Wilson Bull., 72: 60-104. (Includes pterylosis.)

Berger, A. J., and W. A. Lunk
1954   The Pterylosis of the Nestling *Coua ruficeps*. Wilson Bull., 66: 119-126.

Boulton, R.
1927   Ptilosis of the House Wren *(Troglodytes aedon aedon)*. Auk, 44: 387-414. (A detailed classification and description of feather tracts.)

Brodkorb, P.
1949   The Number of Feathers in Some Birds. Quart. Jour. Florida Acad. Sci., 12: 1-5.
1955   Number of Feathers and Weights of Various Systems in a Bald Eagle. Wilson Bull., 67: 142. (The total count of contour feathers was 7,182.)

Brush, A. H.
1967   Pigmentation in the Scarlet Tanager, *Piranga olivacea*. Condor, 69: 549-559.

Burt, W. H.
1929   Pterylography of Certain North American Woodpeckers. Univ. California Publ. in Zool., 30: 427-442. (A good example of a comparative study of feather arrangement. The paper is based on 23 species and subspecies.)

Chandler, A. C.
1914   Modifications and Adaptations to Function in the Feathers of *Circus hudsonius.* Univ. California Publ. in Zool., 11: 329-376. (A study of the range of variation and degrees of development displayed in the plumage of a single bird.)
1916   A Study of the Structure of Feathers, with Reference to Their Taxonomic Significance. Univ. California Publ. in Zool., 13: 243-446. (One of the most important studies ever made on the subject.)

Compton, L. V.
1938   The Pterylosis of the Falconiformes with Special Attention to the Taxonomic Position of the Osprey. Univ. California Publ. in Zool., 42: 173-212. (Another good example of a comparative study of feather arrangement.)

Coues, E.
1903   Key to North American Birds. Fifth edition. Two volumes. Dana Estes and Company, Boston. (In Volume 1, pp. 81-90: neossoptiles and teleoptiles; development, structure, types, and arrangement of feathers.)

Day, M. G.
1966   Identification of Hair and Feather Remains in the Gut and Faeces of Stoats and Weasels. Jour. Zool., 148: 201-217.

Desselberger, H.
1930   Ueber das Lipochrom der Vogelfeder. Jour. f. Ornith., 78: 328-376.

Dorst, J.
1951   Recherches sur la Structure des Plumes des Trochilidés. Mus. Natl. d'Hist. Nat. Mém. (Ser. A Zool.), 1:125-260. (An important paper on structure of feathers in many birds, besides hummingbirds.)

Durrer, H.
1965   Bau und Bildung der Augfeder des Pfaus *(Pavo cristatus* L.). Rev. Suisse Zool., 72: 263-411.

Fisher, H. I.
1943   The Pterylosis of the King Vulture. Condor, 45: 69-73.

Foster, M. S.
1967   Pterylography and Age Determination in the Orange-crowned Warbler. Condor, 69: 1-12.

Fox, H. M., and G. Vevers
1960   The Nature of Animal Colours. Macmillan Company, New York.

Frank, F.
1939   Die Färbung der Vogelfeder durch Pigment und Struktur. Jour. f. Ornith., 87: 426-523.

Gladstone, J. S.
1918   A Note on the Structure of the Feather. Ibis (Ser. 10), 6: 243-247. (Concerns the tegmen in certain feathers.)

Gower, C.
1936   The Cause of Blue Color as Found in the Bluebird *(Sialia sialis)* and the Blue Jay *(Cyanocitta cristata).* Auk, 53: 178-185.

Greenewalt, C. H.
1960   Hummingbirds. American Museum of Natural History, New York. (Contains a fine, uniquely illustrated account of the technicalities of iridescence in this group of birds.)

Greenewalt, C. H., W. Brandt, and D. D. Friel
1960   Iridescent Colors of Hummingbird Feathers. Jour. Optical Soc. Amer., 50(10): 1005-1013.

Hamilton, H. L.
1940   A Study of the Physiological Properties of Melanophores with Special Reference to Their Role in Feather Coloration. Anat. Rec., 78: 525-547.
1952   Lillie's Development of the Chick. Third edition. Henry Holt and Company, New York.

Humphrey, P. S., and G. A. Clark, Jr.
1961   Pterylosis of the Mallard Duck. Condor, 63: 365-385.

Hutt, F. B., and L. Ball
1938   Number of Feathers and Body Size in Passerine Birds. Auk, 55: 651-657.

Jones, L.
1907   The Development of Nestling Feathers. Lab. Bull. No. 13, Oberlin College.

Kent, F. W.
1955   Feathers in Detail. Atlantic Nat., 10: 186-196. (Excellent photomicrographs of an eagle primary and other kinds of feathers.)

Lillie, F. R.
1942   On the Development of Feathers. Biol. Rev., 17: 247-266.

Lubnow, E.
1963   Melanine bei Vögeln und Säugetieren. Jour. F. Ornith., 104: 69-81.

Madsen, H.
1941   Hvad gør Fuglenes Fjer-Dragt vandskyende? Dansk Ornith. Foren. Tidsskr., 35: 49-59. (A good discussion on the question of waterproofing in feathers.)

Markus, M. B.
1965   The Number of Feathers on Birds. Ibis, 107: 394.

Mason, C. W.
1923 Structural Colors in Feathers. I and II. Jour. Phys. Chem., 27: 201-251; 401-447.

Mayaud, N.
1950 Teguments et Phanères. In *Traité de Zoologie*. Edited by P.-P. Grasse. Volume 15. Masson et Cie, Paris. (Highly useful information on the integument, on the structure, development, varieties, and coloration of feathers, and on pterylosis; with excellent illustrations.)

Mewaldt, L. R.
1958 Pterylography and Natural and Experimentally Induced Molt in Clark's Nutcracker. Condor, 60: 165-187.

Miller, A. H.
1928 The Molts of the Loggerhead Shrike *Lanius ludovicianus* Linnaeus. Univ. California Publ. in Zool., 30: 393-417. (Contains a detailed description of pterylae arrangement.)

Miller, W. DeW.
1915 Notes on Ptilosis, with Special Reference to the Feathering of the Wing. Bull. Amer. Mus. Nat. Hist., 34: 129-140.
1924a Further Notes on Ptilosis. Bull. Amer. Mus. Nat. Hist., 50: 305-331.
1924b Variations in the Structure of the Aftershaft and Their Taxonomic Value. Amer. Mus. Novitates, No. 140: 1-7.

Morlion, M.
1954 Pterylography of the Wing of the Ploceidae. Gerfaut, 54: 111-158. (A helpful paper for studying the pterylosis of the House Sparrow.)

Naik, R. M., and M. I. Andrews
1966 Pterylosis, Age Determination and Moult in the Jungle Babbler. Pavo, 4: 22-47.

Nitzsch, C. L., and C. C. H. Burmeister
1840 System der Pterylographie. English translation by W. S. Dallas; edited by P. L. Sclater and published in 1867 by the Ray Society, London. (The classic work on the subject of feather arrangement.)

Pitelka, F. A.
1945 Pterylography, Molt, and Age Determination of American Jays of the Genus Aphelocoma. Condor, 47: 229-260.

Rawles, M. E.
1960 The Integumentary System. In *Biology and Comparative Physiology of Birds*. Volume 1. Edited by A. J. Marshall. Academic Press, New York.

Richardson, F.
1942 Adaptive Modifications for Tree-trunk Foraging in Birds. Univ. California Publ. in Zool., 46: 317-368. (Contains a discussion of the stiffening of tail feathers for climbing.)

Rutschke, E.
1966 Die Submikroskopische Struktur Schillernder Federn von Entenvögeln. Ztschr. f. Zellforsch., 73: 432-443. (Consult for iridescent colors.)

Schmidt, W. J.
1961 Über Luftführende Federstrahlen beim Blutfasan *(Ithaginis sinensis)* nebst Bemerkungen über Luftgehalt von Federn Überhaupt. Jour. f. Ornith., 102: 34-40.

Schmidt, W. J., and H. Ruska
1962 Über das Schillernde Federmelanin bei Heliangelus und Lophophorus. Ztschr. f. Zellforsch., 57: 1-36. (Modifications for iridescent coloration.)

Schüz, E.
1927 Beitrag zur Kenntnis der Puderbildung bei den Vögeln. Jour. f. Ornith., 75: 86-224. (A comprehensive treatise on powder-down feathers.)

Sick, H.
1937 Morphologisch-funktionelle Untersuchungen über die Feinstruktur der Vogelfeder. Jour. f. Ornith., 85: 206-372.

Staebler, A. E.
1941 Number of Contour Feathers in the English Sparrow. Wilson Bull., 53: 126-127.

Steiner, H.
1956 Die Taxonomische und Phylogenetische Bedeutung der Diastataxie des Vogelflügels. Jour. f. Ornith., 97: 1-20. (A good review of the significance of diastataxy/eutaxy.)

Stettenheim, P., and others
1963 The Arrangement and Action of the Feather Muscles in Chickens. Proc. XIIIth Internatl. Ornith. Congr., pp. 918-924.

Strong, R. M.
1902a The Metallic Colors of Feathers from the Neck of the Domestic Pigeon. Biol. Bull. Woods Hole, 3: 85-87.
1902b The Development of Color in the Definitive Feather. Bull. Mus. Comp. Zool., 40: 147-185.
1903 The Metallic Colors of Feathers from the Sides of the Neck of the Domestic Pigeon. Mark Anniversary Volume, Article 13, pp. 263-277.

Test, F. H.
1942 The Nature of the Red, Yellow, and Orange Pigments in Woodpeckers of the Genus Colaptes. Univ. California Publ. in Zool., 46: 371-390.

Turcek, F. J.
1966 On Plumage Quantity in Birds. Ekolog. Polska (Ser. A), 14:617-633.

Voitkevich, A. A.
1966 The Feathers and Plumage of Birds. Sidgwick and Jackson, London. (Originally published in Russian. Concerned mainly with the physiology of growth and development of feathers and plumage.)

Völker, O.
1961a Bemerkungen über ein Grünes, nicht Carotinoides Pigment in Kleingefieder von *Rollulus roulroul* (Galli). Jour. f. Ornith., 102: 270-272.
1961b Die Chemische Charakterisierung roter Lipochrom im Gefieder der Vogel. Jour. f. Ornith., 102: 430-438.

von Pfeffer, K.
1952 Untersuchungen zur Morphologie und Entwicklung der Fadenfedern. Zool. Jahrb., abt. f. Anat. u. Ontog. d. Tiere, 72: 67-100. (Original research on filoplumes.)

Watson, G. E.
1963 The Mechanism of Feather Replacement during Natural Molt. Auk, 80: 486-495.

Watterson, R. L.
1942 The Morphogenesis of Down Feathers with Special Reference to the Developmental History of Melanophores. Physiol. Zool., 15: 234-259.

Wetherbee, D. K.
1957 Natal Plumages and Downy Pteryloses of Passerine Birds of North America. Bull. Amer. Mus. Nat. Hist., 113: 341-436.

Wetmore, A.
1920 The Function of Powder Downs in Herons. Condor, 22: 168-170.
1936 The Number of Contour Feathers in Passeriform and Related Birds. Auk, 53: 159-169. (The results of painstaking work in counting feathers in many species.)

With, T. K.
1967 Frei Porphyrine in Federn. Jour. f. Ornith., 108: 480-483.

Wood, H. B.
1950 Growth Bars in Feathers. Auk, 67: 486-491.

Ziswiler, V.
1962 Die Afterfeder der Vögel: Untersuchungen zur Morphogenese und Phylogenese des Sogenannten Afterschaftes. Zool. Jahrb., Abt. f. Anat., 80: 245-308. (Morphology and function of the afterfeather.)

ANATOMY AND PHYSIOLOGY

A study of the organ systems of the bird, except the integumentary system already treated at length, is undertaken in the ensuing pages. Part I deals with the anatomy of the Common Pigeon *(Columba livia)* as revealed by observation and dissection. In the directions for study are a few comments relating to the function of certain organs or their parts. Part II takes up some of the more important aspects of avian anatomy and physiology with a view to stimulating the student's interest in further reading and investigation.

Part I
THE ANATOMY OF THE PIGEON
SKELETAL SYSTEM

The skeletal system of a bird has two notable characteristics: (1) a strong tendency toward fusion of adjacent bones; (2) a lightness resulting from the pneumaticity of many of the bones. As in other vertebrates, the skeleton of a bird is divisible into the axial skeleton (bones lying along the central axis of the body—skull, vertebral column, ribs, and sternum) and the appendicular skeleton (bones of the pectoral and pelvic girdles and their limbs). For convenience, the skeleton will be considered here as made up of three parts: (1) bones of the limbs; (2) bones of the trunk; (3) bones of the head.

The Bones of the Limbs

Since the bones of the limbs were studied in connection with the external anatomy of the wings, legs, and feet, they will not be considered again.

The Bones of the Trunk

The trunk comprises the vertebral column, ribs, sternum, and the pectoral and pelvic girdles. Use two prepared skeletons of the pigeon, one in which the bones are completely articulated and one in which the bones of the trunk and limbs are disarticulated and may be handled at will. Use a human skeleton, or a detailed chart of one, for comparative purposes.

Vertebral Column

The vertebral column provides a base for the bones of the trunk and limbs and is the main support of the head. It is made up of a chain of bony elements called **vertebrae.**

Vertebrae conform to one general plan in all animals bearing them. Study a human cervical vertebra, preferably the fifth. Viewed from the front, there is a body or **centrum** (plural, **centra**) surmounted on the dorsal side by a **neural arch** which surrounds a **neural canal** for the passage and protection of the spinal cord. The neural arch is in reality a composite structure made up of two plate-like masses meeting at a median line above the canal to form a **neural spine.** Vertebrae commonly have seven processes. The neural spine, already mentioned, is one. The **transverse processes** are two. They are located on either side of the neural arch and project laterally; at the base of each transverse process is the **vertebrarterial canal.** The remaining four processes bear surfaces for articulation with adjoining vertebrae: two, the **prezygapophyses,** extend anteriorly from either side of the arch and face upward; two, the **postzygapophyses,** extend posteriorly from either side of the arch and face downward. The prezygapo- physes of any given vertebra rest on the corresponding postzygapophyses of the vertebra next in front, thus allowing a certain amount of movement between them.

Label the drawings of a cervical vertebra of man (Plate XVI, Figure 1).

The vertebrae have different characteristics in the different regions of the body in which they occur; they are, therefore, classified into: **cervical** (neck), **thoracic** (chest), **lumbar** (loin), **sacral** (pelvis), and **caudal** (tail) **vertebrae.** In man and other mammals the vertebrae in these groups are relatively distinct, but in birds they have undergone considerable modification, certain groups having been reduced, crowded, and even fused together. All the cervical vertebrae remain completely distinct and freely movable. They warrant special study here, for they show many significant specializations.

Cervical Vertebrae

The cervical vertebrae lie between the skull and the first vertebra that is connected with the sternum by a pair of complete ribs. In the pigeon, there are fourteen such vertebrae, but this number is not constant among birds; it ranges from thirteen to twenty-five. Man and most mammals have but seven cervical vertebrae.

Study and compare the sixth cervical vertebra of the pigeon with the fifth cervical vertebra of man. Observe particularly the following parts:

Centrum. In the bird the anterior end of the centrum is saddle-shaped; that is, it is convex above (dorso-ventrally) and concave from side to side. The posterior end shows the reverse condition. A centrum of this sort is said to be **heterocoelous.** The human centrum is acoelous, for its anterior and posterior ends are flattish. From each side of the pigeon centrum, near the base of the transverse process, there projects posteriorly, and somewhat medially, a process which, with its fellow of the other side, forms on the forward end of the centrum a canal for the carotid artery. Does it occur in man?

Transverse Process. The transverse process in both bird and man is pierced by an opening, called the **vertebrarterial canal,** through which the vertebral artery and vein pass.

Neural Spine. In man the neural spine is bifurcated. Does it differ in the bird?

Zygapophyses. Both the pre- and postzygapophyses of the bird are large in proportion to the size of the vertebra, being generally broader. Thus the zygapophyses overlap one another more extensively than in man.

The other cervical vertebrae have many features in common with the ones just studied. The first and second vertebrae in both bird and man are more specialized than

Plate XVI

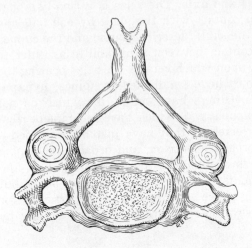

Anterior View   Side View

Figure 1

Cervical Vertebra of Man

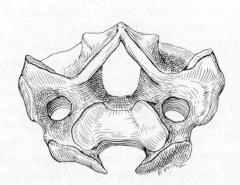

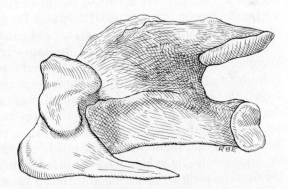

Anterior View   Side View

Figure 2

Cervical Vertebra of Common Pigeon

VERTEBRAE

the others and are called respectively the **atlas** and **axis.** The atlas is a ring-like bone without a centrum. In the bird the atlas is in contact with the skull by one ball and socket joint formed by the occipital condyle of the skull (described later) and the cupped articular surface of the atlas. How does the contact between skull and atlas differ in man? The axis possesses the **odontoid process,** upon which rotates the atlas bearing the skull. Some of the cervical vertebrae of the bird have small median spines, **hypapophyses,** extending ventrally from their centra. Others have small ribs which do not reach the sternum. Which vertebrae have hypapophyses? Which have incomplete ribs?

Many of the features of the avian cervical vertebrae that have just been studied— e.g., the large number of vertebrae, the heterocoelous centrum, and broad, overlapping zygapophyses—are specializations for greater flexibility. The neck in the pigeon and most birds has three functionally distinct segments: (1) anterior, for bending below but not above a straight line, (2) middle, for bending above and below, and (3) posterior, for bending slightly below but mostly above. These properties are determined mainly by the angles of the pre- and postzygapophyses.

Label the drawings of the cervical vertebra of the pigeon (Plate XVI, Figure 2). Also label cervical vertebrae, including atlas and axis, in the drawing of the pigeon skeleton (Plate XVII).

Thoracic, Lumbar, Sacral, and Caudal Vertebrae

In the pigeon, there are five thoracic vertebrae. They succeed the cervical vertebrae posteriorly and are fused together (1) by centra and zygapophyses and (2) by several processes: the neural spines which form a dorsal ridge; the transverse processes which, though lacking as in mammals the vertebrarterial canals, form thin lateral plates, the hypapophyses of the first three vertebrae which form a ventral ridge. The thoracic vertebrae each possess a pair of complete ribs passing to the sternum. The side of each vertebra thus has two articular surfaces for a rib, one on the transverse process and one on the centrum. The lumbar, sacral, and caudal vertebrae are three, four, and twelve in number respectively and succeed the thoracic vertebrae in that order. All of them, with the exception of the last six caudal vertebrae, and the most posterior thoracic vertebra form a fused series, the **synsacrum.** No attempt will be made either to distinguish between them or to note their individual modifications. It will suffice to point out the long dorsal ridge, formed by the fusion of the neural spines, and the crossbars seen from below, which represent transverse processes emerging from their closely fused vertebrae and joining the inner walls of the pelvic girdle. The last six caudal vertebrae are more or less freely movable, with unfused neural spines and transverse processes. The terminal vertebra represents the fusion of several caudal vertebrae. In the shape of a ploughshare, it is appropriately called the **pygostyle.** It serves as a base for the rectrices.

How many thoracic vertebrae does man have? Lumbar? Sacral? Caudal (coccygeal)?

Ribs

A typical rib, such as that found in man, is a flattened arch of bone attached to the transverse process and centrum of a vertebra by two heads and to the sternum by one cartilaginous extremity. In man how many ribs are there? How many are unattached to the sternum?

In the pigeon there are seven ribs on each side. The first two are articulated with the cervical vertebrae and do not reach the sternum. The next four are articulated with the thoracic vertebrae and sternum. The last is articulated with a thoracic vertebra but has its ventral extremity attached to the ventral end of the rib in front, instead of to the sternum. All but the first two ribs are jointed and are made up of two pieces: the dorsal or **vertebral rib** and the ventral or **sternal rib.** The sternal rib corresponds to the cartilaginous portion of the human rib. From the posterior margins of all but the first and last vertebral ribs project **uncinate processes**—crossbones which weld the thorax into a firm unit and provide surface for the attachment of costal muscles which, in the bird, play a prominent role in respiratory movements.

In the articulated skeleton of the pigeon, observe the relative lengths and positions of the ribs. The third, fourth, fifth, and sixth ribs tend to increase in length successively, and their vertebral and sternal parts tend, at the same time, to meet at decreasingly acute angles. This arrangement permits the sternum to move downward and forward, then upward and backward, in breathing.

Sternum

The sternum or breastbone is one of the most highly specialized parts of the avian skeleton. Two kinds of sterna exist in birds: **ratite** and **carinate.** The ratite sternum has the ventral surface flattened, like the bottom of a raft, and occurs in flightless birds such as the ostrich. The carinate sternum has the ventral surface keeled, like the bottom of a sail boat. Flying birds have this kind; it permits more surface for the origin of the all-important muscles that operate the wings. The pigeon has the carinate sternum. This will vary in size and shape among different birds in accordance with the type of flight and the arrangement of muscles needed for it.

Viewed from above, the sternum of the pigeon is somewhat oblong, with the longer lateral borders more or less parallel but converging posteriorly. Several noteworthy structures are evident. On the anterior border are two deep, smooth-faced grooves which receive the expanded portions of the coracoid bones of the shoulder girdle and thus provide a base for the bones of this region. These articular surfaces are properly called the **coracoidal facets.** Between them a small forked process, the **rostrum,** projects anteriorly. On the anterior lateral borders are small pits, each one of which accommodates the sternal end of a true rib. They are the **costal facets.** Just behind the costal facets appears a conspicuous backward-pointing process, the **external-lateral xiphoid process.** This is followed by a deep **notch.** The posterior border of the sternum is somewhat rounded and is formed by the fusion of the posterior ends of two more backward-pointing processes, namely, the **internal-lateral xiphoid process** and the **single median xiphoid process.** This fusion forms a bony bar, leaving an opening or **fenestra** just anterior to it.

The lateral and posterior borders of the sternum vary considerably in different birds. In some cases all three xiphoid processes are unfused at their ends, leaving two notches between them. A sternum is then **double-notched.** In other cases all three processes may be fused at their tips, making the sternum **bi-fenestrate;** or completely fused along their entire lengths, making the sternum **entire.** The pigeon, of course, has a **single-notched uni-fenestrate** sternum.

Viewed from the side, the sternum shows an enormous keel, the **carina,** extending ventrally along the median line. The carina drops down abruptly in front, with an anteriorly concave vertical border, to a prominent **apex.** It then curves gradually upward to the level of the posterior end of the sternum.

Pectoral Girdle

The pectoral or shoulder girdle resembles a tripod supporting the pectoral appendage or wing. One "leg" is the **coracoid,** which finds its base on the anterior end of the sternum; the second "leg" is the **scapula,** which rests upon the ribs; and the third "leg" is the **clavicle,** which is supported by and fused to its fellow on the opposite side. The three legs do not share equally in direct support. Only the coracoid and scapula form the cup-like articular surface, the **glenoid cavity,** for the head of the humerus. The clavicle is actually attached to a special process of the coracoid (see below) just anterior to the glenoid cavity.

Coracoid. The strongest bone of the girdle. It is a short, thick cylinder with one end, the **foot,** expanded to fit the coracoidal facet in the sternum and with the other end, the **head,** likewise expanded to form three important areas: a portion of the glenoid cavity on its lateral surface; a roughened surface for fusion with the scapula on the inner dorsal aspect; and a terminal **clavicular process** on the inner ventral aspect on which is an articular surface to receive the clavicle. The clavicle and scapula come together here with the coracoid in such a fashion as to form the **foramen triosseum** through which passes the tendon of an important flight muscle, the supracoracoideus, to its insertion on the humerus.

Scapula. A flattened sabre-like bone whose anterior end or head expands to meet the coracoid in the manner mentioned above. The outer surface of the head comprises the remaining portion of the glenoid cavity. The inner surface extends forward slightly to form the **acromion process,** which is united by a ligament to the clavicle. The blade portion of the scapula extends posteriorly, more or less parallel to the vertebral column, to rest upon the ribs.

Clavicle. A thin, rod-like bone whose upper or dorsal end expands considerably to form the **epicleidium.** This is articulated with the clavicular process of the coracoid and is further held in place by ligaments from the acromion process of the scapula. The lower or ventral end of the clavicle is fused with its fellow of the opposite side at an acute angle, thus forming a laterally compressed process, the **hypocleidium.** The clavicles, considered together, constitute the **furcula,** the popularly known "wishbone." It acts as a spring-like connection between the two shoulder girdles and, consequently, lends them necessary support. A ligament connects the hypocleidium with the apex of the sternum, thus lending still additional support. In some birds, such as pelicans, the hypocleidium is fused with the sternum.

Examine the shoulder girdle of man. Observe the shape of the scapula or shoulder blade. Identify its acromion process. The coracoid does not exist in man as an individual bone but becomes the coracoidal process of the scapula. Note the clavicle or collarbone and its manner of articulation with both sternum and scapula.

Pelvic Girdle

The pelvic girdle gives support to the pelvic appendages, or legs. While the pectoral girdle is quite different in general appearance from its homologue in man, it rather closely resembles man's, being made up of three elements, the **ilium, ischium,** and **pubis.** Although these bones originated separately, they are fused and meet in a deep concavity, the **acetabulum.** This structure is a socket for the head of the femur. It is not completely ossified within, thus leaving a small foramen.

Ilium. The largest and longest bone of the girdle. It is joined firmly to the transverse processes of all the vertebrae which make up the synsacrum. For descriptive purposes

Plate XVII

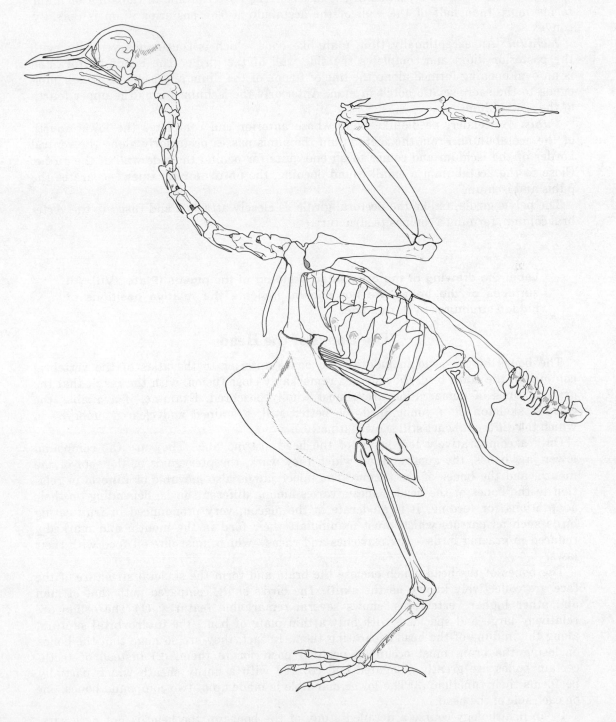

SKELETON OF COMMON PIGEON

the ilium may be considered as divisible into two halves, the **anterior ilium** and **posterior ilium.** The anterior ilium lies in advance of the acetabulum and is dorsally concave. Beneath it passes a portion of the last rib. The posterior end of the anterior ilium forms more than half of the wall of the acetabulum. The posterior ilium is dorsally convex.

Ischium. An exceptionally thin, plate-like bone which is continuous ventrally with the posterior ilium and completes the side wall of the girdle. The **ischiadic foramen** is an oval opening formed along the line of fusion of the ilium and ischium. It is homologous to the sacro-sciatic notch in man. Anteriorly the ischium forms the upper fourth of the acetabulum.

Pubis. A slender, needle-like bone whose anterior end completes the lower fourth of the acetabulum. From the acetabulum the pubis passes posteriorly along the ventral border of the ischium and comes to an end quite far behind the side wall of the girdle. Close to the acetabulum a small round opening, the **obturator foramen,** separates the pubis and ischium.

The pelvic girdle, unlike the pectoral girdle, is closely attached and fused to the vertebral column, forming a composite structure.

Label the drawing of the articulated skeleton of the pigeon (Plate XVII). All surfaces of the bones are not shown. Indicate the relative positions of hidden structures by dotted lines.

The Bones of the Head

The bones of the head include all the bones anterior to the atlas of the vertebral column. In the adult bird, they show a remarkably close fusion, with the result that the original lines of demarcation are almost wholly obscured. Examine, if available, the cleaned skeleton of a young bird—or better still, a stained and cleared embryo—in which the different bones still show distinctly.

Only a comparatively few bones of the head are movable. They are the compound lower jaw bones, the quadrates on which they work, the pterygoids of the roof of the mouth, and the bones of the tongue. The upper jaw is also movable or **kinetic** in relation to the bones of the head. Kinesis varies among different birds, depending on their adaptations for feeding. It is moderate in the pigeon, very pronounced in fruit-eating birds such as parrots which must manipulate their food in the mouth, and markedly reduced in grazing birds—e.g., ostriches and rheas—which must bite off food with their jaws.

The bones of the head which enclose the brain and form the skeletal structure of the face are collectively known as the **skull.** The bird's skull, compared with that of man and other higher vertebrates, shows several remarkable features: (1) The orbits are relatively large and spacious, with only a thin plate of bone, the **interorbital septum,** along the midline of the head separating them. In fact, they are so huge that the bones enclosing the brain must occupy a position posterior to them. (2) Instead of teeth, certain bones are greatly elongated and covered with a horny sheath which partially performs their function. (3) The lower mandible is made up of two compound bones, one on each side of the head.

In an ornithology course a detailed study of the bones of the head is not necessary. Many of the bones are quite small and inconspicuous and have no place in descriptive works. A student should be able to recognize, however, the bones that are prominent and of paramount importance in the classification of birds.

For convenience, consider the bones of the head in three groups: (1) bones of the cranium; (2) bones of the face; (3) bones of the tongue.

Bones of the Cranium

The bones of the cranium enclose the brain and form the so-called "brain box." None is freely movable. Compare the shape of cranium in the pigeon with that in man.

Occipital. The bone forming the base of the cranium. A large opening, the **foramen magnum,** permits the passage of the spinal cord (actually the medulla) to the brain. The occipital is composed of four bones with indistinct boundaries. They are: the **supraoccipital,** which forms the upper boundary of the foramen; the **exoccipitals,** which form the lateral boundaries of the foramen and extend forward on each side to the ear opening; and the **basioccipital,** which forms the lower boundary of the foramen and bears the ball-like **occipital condyle** for articulation with the atlas.

Parietals. A pair of broad, squarish bones, fused along their medial borders. They continue up over the back of the skull from the occipital. They are bounded anteriorly by the frontals and laterally by the squamosals. The parietals, together with the occipital, roof over the posterior portion of the brain.

Frontals. A pair of bones indistinctly fused along their medial borders. They continue forward from the parietals, forming (1) the roof of the anterior portion of the brain and (2) the roof and superior margins of the orbits.

Squamosals. A pair of bones, one on each side of the head. They form the posterior margins of the orbits and join the lateral borders of the frontals and parietals to complete the roof of the cranium.

The lower border of each squamosal joins the occipital, while its lowermost angle provides an eave for the ear opening and possesses an articular depression for the quadrate bone.

Periotic Capsules. Paired skeletal structures containing the organs of hearing. Each capsule has three bones: the **prootic, epiotic,** and **opisthotic.** They are homologous to the petrosal bone of man.

The periotic capsules are within the cranium and are indistinguishably fused with the exoccipitals and basisphenoid bone.

Sphenoid. A large bone lying in the center of the cranium and providing a base for many bones of the skull. It is commonly considered as made up of four parts: the **basisphenoid,** the **alisphenoids (laterosphenoids),** the **presphenoids,** and the **orbitosphenoids.**

The basisphenoid forms the greater part of the cranial floor. Viewed from below, it assumes the shape of a rough triangle with its base implanted on the basioccipital and more or less covered over by a bone of membranous origin. the **basitemporal.** Its apex is noticeably prolonged into a sharp-pointed structure, the **basisphenoidal rostrum,** which is wedged between the bones of the palate. It thus forms the central axis of the base of the skull. Laterally, the basisphenoid joins the periotic capsule and bounds the ear opening in front; it also joins the squamosals. Anteriorly, the rostrum articulates by special **basipterygoid processes** with the pterygoid bones and still farther forward it articulates with the palatine bones. Dorsally the rostrum joins the alisphenoids, presphenoids, and ethmoid (see below) to form the lower portion of the interorbital septum.

The alisphenoids are wing-like structures which extend dorsally from either side of the basisphenoid to form the lower posterior walls of the orbits and to enclose that part of the brain which is not already encased by the bones mentioned above. The alisphenoids pass upward to meet the frontals and squamosals, and far enough forward to form the inner walls of the orbits up to and including the hind margin of the **optic foramen** —the large oval opening for the passage of the optic nerve.

The presphenoids and orbitosphenoids are continuations of the alisphenoids and rostrum forward and upward beyond the optic foramen. Save at their posterior ends they are fused medially, their right and left halves being closed together by the greatly enlarged eyeballs. The central portion of the interorbital septum is a thin plate made up of these two bones. In two places (postero-dorsally and postero-ventrally) it is incomplete, leaving openings between the orbits. It is bounded dorsally by the frontals, anteriorly by the ethmoid, and ventrally by the rostrum.

Ethmoid. This bone, a perpendicular plate sometimes known as the **mesethmoid,** is a continuation of the interorbital septum forward from the sphenoidal elements and thus completes the septum. Ventrally, the ethmoid unites with the rostrum. Dorsally and somewhat anteriorly, the ethmoid sends out a pair of lateral plates which form an **orbitonasal septum** separating the orbits from the nasal cavities. The ethmoid consequently forms a portion of the anterior orbital wall and is bounded antero-dorsally by the nasal and frontal bones and laterally by the lacrimal bones. A part of the mesethmoid passes farther forward through the orbitonasal septum to separate the nasal cavities into right and left halves.

Bones of the Face

The facial bones either compose, or are directly associated with, the skeleton of the upper and lower mandible.

Prevomer. Not found in the pigeon. When present in birds it is on the midline of the palate immediately in front of, or sometimes attached to, the basisphenoidal rostrum.

Quadrates. Right and left bones connecting the lower mandible with the cranium. As the name implies, each quadrate is somewhat quadrangular but with a constriction in the middle making two expanded ends. The end toward the cranium freely articulates with (1) a depression in the lowermost angle of the squamosal and (2) the basisphenoid just anterior and slightly dorsal to the ear opening. The end of the quadrate pointing in the opposite direction freely articulates with (1) the adjoining side of the lower mandible, (2) the zygomatic bar, and (3) the pterygoid. A process extending from the quadrate toward the orbit is for the attachment of muscles.

The quadrate bone does not exist as such in man, but becomes a part of the temporal bone and forms the bony tube of the ear opening.

Quadratojugals and *Jugals.* The major part of the slender **zygomatic bar** on either side of the head, below the orbit, is composed of these two bones. The posterior portion is the quadratojugal which articulates with the quadrate. It joins the jugal in front by an oblique suture. The jugal is a very small scale-like bone which is in turn obliquely sutured in front to the posterior process of the maxillary bone. Frequently a portion of the quadratojugal bone extends far enough forward to be sutured also to the maxillary process.

Maxillae. Right and left bones, each one consisting of three processes: (1) a **posterior process** which unites obliquely with the jugal and quadratojugal and thus completes the zygomatic bar; (2) an **anterior** or **dentary process** which is sutured laterally to the premaxillary and nasal bones; (3) a broad **ventral process,** sometimes called the **maxillopalatine process,** which descends downward and inward. The ventral process ends blindly and does not meet its fellow of the opposite side. A cleft is therefore formed in this region of the palate.

In man, the maxilla joins with its fellow to form a prominent bone for holding the upper jaw. No cleft normally exists between them.

Pterygoids. A pair of short, rather thick, rod-like bones articulating with the quadrates and the palatines. Their long axes are obliquely disposed between these bones. The middle posterior surfaces of the pterygoids articulate with the basipterygoid processes of the basisphenoidal rostrum.

Palatines. A pair of bones forming the greater portion of the palate of birds. In the pigeon they are long and slender bones and lie along the midline of the mouth parallel to each other. Anteriorly they unite with the palatal processes of the premaxillary bones; posteriorly they become considerably flattened horizontally and rest on the basisphenoidal rostrum, articulating at their extreme ends with the pterygoids.

Premaxillae. These two bones together form the tip of the upper mandible. Each premaxilla composes one-half of it and has three backward-projecting processes which form the main bulk of the mandible. One, the **frontal** or **nasal process,** passes to the frontal bone and fuses with its fellow to form the culmen of the beak. The second, the **dentary process,** extends horizontally to join the anterior process of the maxilla and forms half the tomium of the mandible. The third, the **palatal process,** contributes to the formation of the palate by extending along the roof of the mouth to join with the anterior end of the palatine bone of the same side.

Nasals. Paired bones extending forward from the frontal bones and fusing along their medial borders. They rest on the ethmoid bone below. Each nasal bone forms the posterior boundary of a nostril by dividing into two processes: one, the **superior process,** passes along the medial side of the nostril to meet laterally the frontal process of the premaxilla; the other, the **inferior process,** descends to fuse with the anterior process of the maxilla.

Lacrimals. Paired bones (sometimes called the **prefrontals**) situated in the anterior portions of the orbits. Each is attached above to the frontal and nasal bones and descends, somewhat flattened, toward the dorsal surface of the zygomatic bar, bounding the orbit anteriorly.

Lower Jaw Bones. The lower mandible of the bird is made up of two jaw bones which fuse anteriorly to form a V-shaped structure. Each lower jaw bone is actually a composite affair, being made up of five bones immovably fused together—the **dentary, splenial, angular, surangular,** and **articular.** No attempt will be made here to distinguish between them. The posteriormost of the five bones, the articular, has a double-cupped superior surface for articulation with the quadrate.

Bones of the Tongue

The bones of the tongue, sometimes called the **hyoid apparatus,** are divisible into two groups: median and paired. Of the median group there are three bones. The anteriormost, the **glossohyal,** serves as the skeleton for the main bulk of the tongue. A small piece of cartilage projects from its forward end. Loosely articulated to the posterior end of the glossohyal is the **basihyal.** The third median bone, the **basibranchial,** is closely fused to the posterior end of the basihyal; it is usually recognized as the tapering portion of the two united bones. Coming off near the postero-lateral portions of the basihyal are the "horns" of the hyoid apparatus. The first bones are the slender, paired **ceratobranchials.** To their posterior ends are articulated the equally slender **epibranchials.** (One pair of bones not clearly seen are the **ceratohyals,** which emerge from the anterior end of the basihyal. For present purposes they are not important.)

Plate XVIII

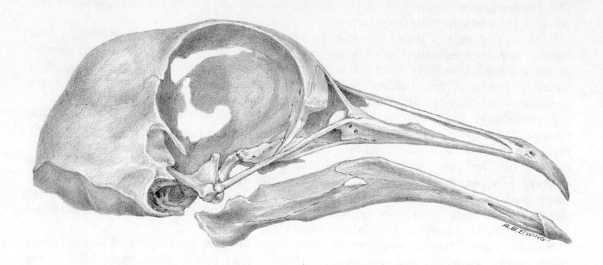

Figure 1
Viewed from Side

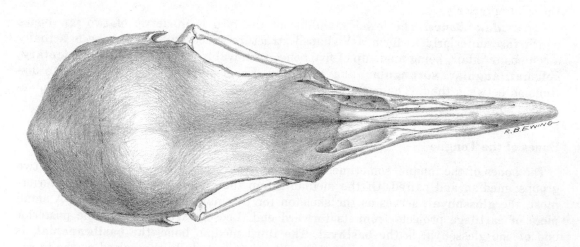

Figure 2
Viewed from Above

SKULL OF COMMON PIGEON

Plate XIX

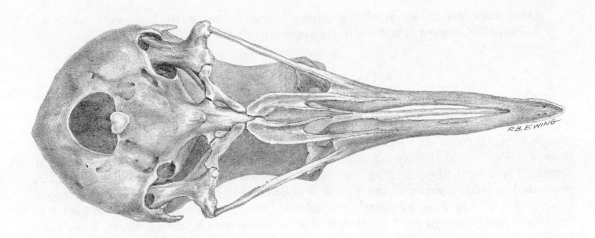

Figure 1

Ventral View of Skull with Lower Jaw Removed

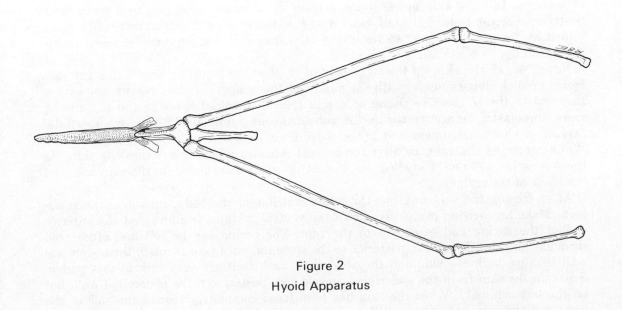

Figure 2

Hyoid Apparatus

SKULL AND HYOID APPARATUS OF COMMON PIGEON

Examine the hyoid apparatus of a woodpecker, which shows extraordinary development of the branchials. These greatly elongated bones and the muscles attached to them are an important part of the mechanism that permits the woodpecker to extend and retract the tongue when getting insects from holes in trees.

Label fully the drawings of the pigeon's skull (Plate XVIII, Figures 1 and 2; Plate XIX, Figure 1) and the hyoid apparatus (Plate XIX, Figure 2).

MUSCULAR SYSTEM
by Andrew J. Berger

The muscular system has 175 different muscles, most of which are paired—i.e., each muscle is represented on both the right and left sides of the body. A thorough study of these muscles would require far more time than the average course in ornithology will permit. The aim in the following text is to introduce the student to some of the more important muscles used in systematics and to point out special features of avian musculature. The student desiring descriptive information on all the muscles of the bird may refer to *Avian Myology* by J. C. George and A. J. Berger (Academic Press, New York, 1966).

Obtain a specimen of a pigeon and prepare to dissect the muscles.

The first step is to remove the skin. As in most small birds, it is relatively very thin. Therefore, be careful when taking off the skin not to cut the underlying muscles.

Place the specimen on its ventral side and make a two-inch incision in the dorsal midline of the neck. Then separate the skin from the underlying muscles by inserting the blunt handle of a scalpel (not the blade!) between the skin and the superficial layer of muscles. Lift the skin upward—away from the muscles—and continue the incision posteriorly to the base of the tail. Stay in the middorsal line. This can be readily determined by feeling the neural spines of the vertebrae with one's fingers. Proceed slowly and carefully.

Now, "work" the skin on the right side of the body outward by separating the skin from the underlying muscles with the handle of the scalpel. In most regions the skin is fastened to the **connective tissue** or **fascia** (plural, **fasciae**) covering the muscles by loose fibroelastic connective tissue, the **subcutaneous connective tissue** or **superficial fascia.** A blunt instrument will break these fibers but will not cut into the muscles. While removing the skin, be alert for **dermal muscles** which insert into the skin. As these muscles will not be studied, cut them from their attachment to the skin by using the blade of the scalpel.

After freeing the skin to about the midlateral line of the body, turn over the specimen. Make an incision in the skin down the ventral midline, beginning at the anterior end of the carina and extending to the vent. The carina can be felt and often seen through the skin. Be careful, posterior to the sternum, not to cut through the abdominal wall because both the skin and the abdominal wall itself are very thin in this region. Separate the skin from the underlying muscle and, posteriorly, the abdominal wall, but on the left side only. When the skin has been freed completely around the side of the body, cut the skin and remove it. Take special care when removing the skin posterior to the humerus, from elbow to shoulder. Look for the tiny, shiny tendon of the **expansor secundariorum** (a muscle to be dissected later) between the two layers of skin. Try not

Plate XX

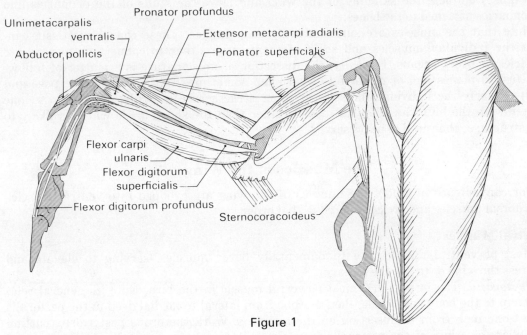

Ulnimetacarpalis ventralis

Abductor pollicis

Pronator profundus

Extensor metacarpi radialis

Pronator superficialis

Flexor carpi ulnaris

Flexor digitorum superficialis

Flexor digitorum profundus

Sternocoracoideus

Figure 1

A ventral view of the sternum and right wing. The bird's right pectoralis muscle has been removed to reveal the underlying supracoracoideus and coracobrachialis posterior muscles. The humerocarpal band has been removed to show the relationships of the flexor carpi ulnaris and flexor digitorum superficialis muscles.

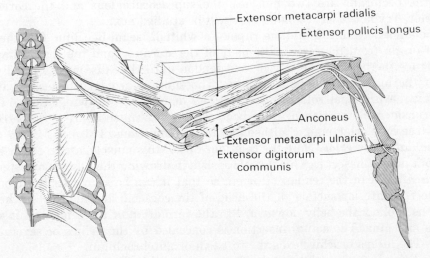

Extensor metacarpi radialis

Extensor pollicis longus

Anconeus

Extensor metacarpi ulnaris

Extensor digitorum communis

Figure 2

A dorsal view of the muscles of the right shoulder and wing as seen after the skin and fascia have been removed. A flap has been reflected in the posterior part of the rhomboideus superficialis muscle to show the deeper rhomboideus profundus muscle. The posterior aponeurotic extension of the insertion of the tensor patagii brevis muscle on the belly of the extensor metacarpi radialis muscle has been removed to show better the relationships of the proximal forearm muscles.

MUSCLES OF THE WING, COMMON PIGEON

to cut it when removing the skin. Examine the feather follicles on the deep surface of the skin. Complete the skinning of the wing and leave the stubs of the primaries and secondaries attached to the bones.

Note that the muscles are surrounded by deep fasciae. These connective tissues invest the individual muscles and some of them serve either to connect or to separate muscles or other organs. Basically, the dissection of musculature is a matter of following fascial planes and of removing fascia—and sometimes fat—in order to expose and define clearly the individual muscles or other structures. As a rule, dissection is done better with the blunt handle of a scalpel rather than the sharp blade which is likely to cut structures that need to be preserved.

The Muscles of the Wing

For convenience in study, the muscles of the wing are grouped into ventral muscles and dorsal muscles and are considered in that order.

Ventral Muscles

Most of ventral muscles are fundamentally flexor muscles, serving to elevate and depress the wing or to flex the forearm.

Pectoralis. The largest and most powerful muscle in the bird. Note its general relationship to the body as a whole. Pass a probe from lateral to medial deep to the pectoralis about one inch from its insertion on the humerus. Work the probe posteriorly, cutting the fibers of origin of the pectoralis from the carina and the lateral border of the sternum. Scrape the muscle attachments from the clavicle and from the membrane running from the clavicle to the coracoid. After all the attachments have been cut, reflect the pectoralis outward toward its insertion but do not cut the fibers of the insertion. Lying deep under the pectoralis are two muscles, the supracoracoideus and the coracobrachialis posterior. Try not to cut them as they will be studied next.

Supracoracoideus. Note the midline raphe—a whitish, seam-like line. Cut the muscle's fibers of origin from the carina, the body of the sternum, and the coracoclavicular membrane. Trace the belly of the muscle and tendon until they disappear dorsally. The actual site of the insertion will be seen later. The supracoracoideus together with the pectoralis play the principal role in elevating and depressing the wing. Both muscles constitute a considerable portion of total body weight—as much as 34 to 36 percent in some species (see F. A. Hartman, Smithsonian Misc. Collections, 143: 1-91, 1961).

Coracobrachialis Posterior. Note the relationships of this muscle lateral to the supracoracoideus but do not dissect it. This muscle assists in drawing the humerus posteriorly.

Biceps Brachii. Clean the tendon of origin so that it can be seen clearly. It has one attachment to the anterior surface of the head of the coracoid and another to the head of the humerus. Follow the belly downward to the formation of the tendon of insertion on the radius and ulna. The actual insertion is concealed by the bellies of several forearm muscles. The biceps brachii flexes the forearm or antebrachium.

The Biceps Slip. Note the relationship of the belly of this fleshy muscle to the biceps brachii, from which it arises, and to the tendon of insertion of a muscle, the **tensor patagii longus.** The presence or absence of the biceps slip in different groups of birds is commonly used as a diagnostic character.

Expansor Secundariorum. Look for the belly of this muscle posterior to the elbow. It is a peculiar muscle composed entirely of smooth rather than striated muscle fibers. The insertion is on the calami of several of the secondary feathers. From the proximal end of the belly, trace the fine scapular tendon upward to the armpit or axilla. If necessary, cut the tendon of the biceps and reflect the belly outward. The scapular tendon is attached to the dorsomedial edge of the scapula; it is reinforced by a second tendon which arises from the fascial envelope surrounding the distal portion of the belly of the dorsalis scapulae, to be studied later. The expansor secundariorum also has a short tendinous origin from the distal end of the humerus. The expansor secundariorum serves to draw the proximal secondaries medially and downward.

Triceps Brachii. Note the general relationships of this large extensor muscle along the posterior surface of the humerus. The muscle consists of two distinct parts: the **scapulotriceps** and the **humerotriceps.** The scapulotriceps arises from the lateral surface of the scapula just posterior to the glenoid cavity; it inserts on the dorsal surface of the base of the ulna (it will be dissected later). The humerotriceps arises from most of the posterior surface of the humerus. The only indication of two heads for this muscle is found in the pneumatic fossa of the humerus. The humerotriceps muscle inserts by tendinous fibers on the proximal end of the ulna. The triceps brachii extends the forearm.

Other Ventral Muscles. Observe the following five muscles on the ventral surface of the antebrachium but do not dissect them except to trace the tendon of the biceps brachii to its insertion: the **extensor metacarpi radialis,** the **pronator superficialis,** the **pronator profundus,** the **flexor digitorum superficialis,** and the **flexor carpi ulnaris.** What do the names tell you about these muscles and their actions? Be alert for the **humerocarpal band** covering (and concealing) parts of the bellies of the flexor digitorum superficialis and the flexor carpi ulnaris.

Label the drawing of ventral muscles (Plate XX, Figure 1).

Dorsal Muscles

Most of the dorsal muscles are extensor muscles serving to extend and elevate the wing.

Latissimus Dorsi. The most superficial muscle in the back. There is a single belly, the **pars anterior,** in the pigeon. The latissimus arises from the neural spines of the last cervical and the first two thoracic vertebrae. There are four parts in some birds: besides the pars anterior, the **pars posterior, pars metapatagialis,** and **pars dorsocutaneous.** At this time, trace the belly laterally only until it disappears deep to the muscles of the arm. The latissimus serves to draw the humerus posteriorly and medially.

Tensor Patagii Longus et Brevis (Propatagialis Longus et Brevis). This complex in the pigeon consists of a single hypertrophied belly; it arises from the apex or epicleidium of the clavicle and from the acromion process of the scapula. (There are two separate bellies in most birds.) The anterior, thinner part of the belly is the tensor patagii longus; the posterior, thicker part of the belly represents the tensor patagii brevis. Clean and study the entire muscle, including the insertion of the brevis tendon on the belly of the extensor metacarpi radialis (see above). The brevis part of the tensor patagii assists in flexing and elevating the wing. Note the relationship of the longus tendon to the biceps slip and trace the tendon to its insertion at the wrist. The longus part of the tensor patagii together with the biceps slip assist in tensing the patagium (see page 18). Cut the origin of the belly of the tensor patagii from the epicleidium of the clavicle and the acromion process of the scapula; then turn the belly outward. Do not cut the tendons of insertion.

Deltoideus Major. This has a small **anterior head** and a large **posterior head.** Note the extent of each. Cut each head at its origin from the anterior end of the scapula and remove the anterior head. Identify now the tendon of insertion of the supracoracoideus. The deltoideus major elevates the brachium and draws it posteriorly.

Scapulotriceps. This is a large head of the triceps brachii. Note its extent and relationships, and then cut the muscle at its origin from the inferolateral surface of the scapula and from the inferior margin of the posterior lip of the glenoid cavity. Turn the belly down toward the elbow. Now trace the latissimus dorsi pars anterior to its insertion on the humerus.

Rhomboideus Superficialis and *Rhomboideus Profundus.* Identify these two muscles running between the neural spines and the medial border of the scapula. These muscles serve to stabilize the scapula or to draw it toward the vertebral column. Do not remove them.

Dorsalis Scapulae (Scapulohumeralis Posterior). Note this large muscle, arising from the lateral surface of the scapula. The belly passes forward, ends on a tendon, and inserts on the proximal end of the humerus. This muscle serves to draw the brachium medially and rotates it so that the leading edge of the wing is turned downward. Do not dissect it.

Other Dorsal Muscles. Examine, but do not dissect, the following extensor muscles on the dorsal surface of the forearm: **extensor metacarpi radialis, extensor digitorum communis, extensor metacarpi ulnaris,** and **anconeus.**

Label the drawing of dorsal muscles (Plate XX, Figure 2).

The Muscles of the Pelvic Appendage

Beginning in the dorsal midline of the synsacrum, remove the skin from the lateral surface of the left thigh and leg; then remove the skin from the medial surface. Remember that the skin is very thin in these areas. For convenience, the muscles are considered in two groups, the superficial muscles and the formula muscles.

Superficial Muscles

The superficial muscles must be removed in order to expose the formula muscles.

Sartorius (Extensor Iliotibialis Anterior). The most anterior muscle on the anterolateral surface of the thigh. Trace the belly from its origin on the anterior end of the ilium to its insertion on the patellar ligament at the proximal end of the tibiotarsus. The sartorius serves to extend both the thigh and the leg. Leave it in place but separate it from the muscle lying immediately posterior to it, the iliotibialis.

Iliotibialis (Extensor Iliotibialis Lateralis). This, the most superficial muscle on the lateral surface of the thigh, is a very thin layer of muscle and aponeurosis (dense connective tissue). Considerable care is required in removing it without damaging the underlying muscles. Cut the aponeurosis of origin from the anterior and posterior iliac crests of the ilium and from the intervening median dorsal ridge of the synsacrum. Pull the aponeurosis and belly outward and downward toward the knee. Observe the three parts of the iliotibialis complex: anterior and posterior fleshy bellies and an aponeurotic central sheet in approximately the distal three-fourths of the muscle. The aponeurotic portion probably will be fused with the underlying muscles and will have to be shaved from them. The muscle ends in a tough aponeurosis, which forms the anterior layer of

Plate XXI

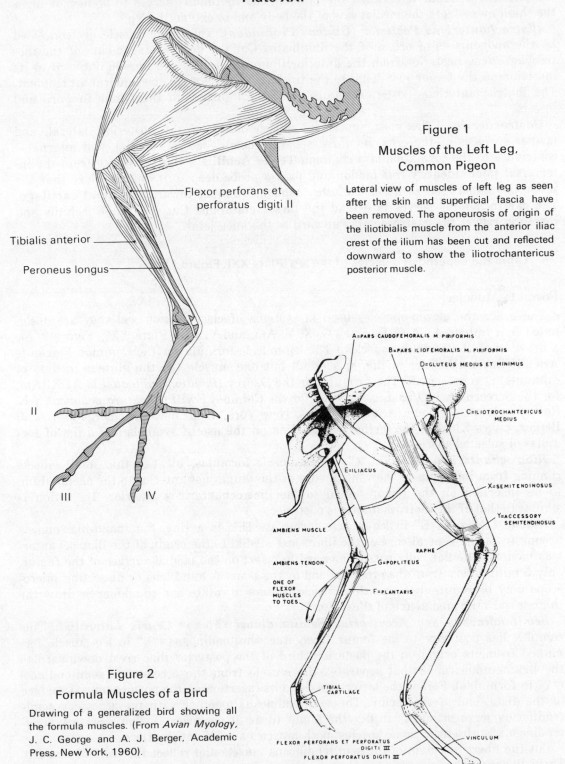

Figure 1
Muscles of the Left Leg, Common Pigeon

Lateral view of muscles of left leg as seen after the skin and superficial fascia have been removed. The aponeurosis of origin of the iliotibialis muscle from the anterior iliac crest of the ilium has been cut and reflected downward to show the iliotrochantericus posterior muscle.

Flexor perforans et perforatus digiti II

Tibialis anterior

Peroneus longus

II

I

III

IV

A=PARS CAUDOFEMORALIS M PIRIFORMIS

B=PARS ILIOFEMORALIS M PIRIFORMIS

D=GLUTEUS MEDIUS ET MINIMUS

C=ILIOTROCHANTERICUS MEDIUS

E=ILIACUS

X=SEMITENDINOSUS

Y=ACCESSORY SEMITENDINOSUS

AMBIENS MUSCLE

RAPHE

AMBIENS TENDON

G=POPLITEUS

ONE OF FLEXOR MUSCLES TO TOES

F=PLANTARIS

TIBIAL CARTILAGE

FLEXOR PERFORANS ET PERFORATUS DIGITI III

FLEXOR PERFORATUS DIGITI III

VINCULUM

Figure 2
Formula Muscles of a Bird

Drawing of a generalized bird showing all the formula muscles. (From *Avian Myology*, J. C. George and A. J. Berger, Academic Press, New York, 1960).

the **patellar tendon.** It encloses the patella. The iliotibialis serves to abduct or draw the thigh away from the medial axis of the body and to extend the leg.

Iliotrochantericus Posterior (Gluteus Profundus). This large muscle lies concealed by the aponeurosis of origin of the iliotibialis. Cut the fleshy attachments of the iliotrochantericus posterior from the anterior ilium and reflect the muscle outward to its insertion on the femur just distal to the trochanter. Do not cut the femoral attachment. The iliotrochantericus posterior rotates the lateral surface of the femur forward and inward.

Gastrocnemius. The most superficial muscle mass on the posterior, lateral, and medial surfaces of the crus. Its three separate heads—lateral, medial, and internal—contribute to the formation of a common **Tendo Achillis** a short distance from the intertarsal joint. Identify this tendon and pass a probe deep to it, making sure that the probe lies superficial to all of the other tendons running toward the **tibial cartilage.** The gastrocnemius serves to extend the tarsometatarsus. Cut the Tendo Achillis and carefully reflect the gastrocnemius upward to the knee joint.

Label the superficial muscles of the leg (Plate XXI, Figure 1).

Formula Muscles

These are the eleven muscles used in systems of classification and they are designated by symbols A, B, C, D, E, F, G, X, Y, Am, and V. (See Plate XXI, Figure 2, for a drawing of the formula muscles.) The leg-muscle formula for the Common Pigeon is ABCEFGXYAmV since in this species all but one muscle (D, the gluteus medius et minimus) is represented. The formula for the Osprey *(Pandion haliaetus)* is ADEBAm; for the Screech Owl *(Otus asio),* ADEG; for the Chimney Swift *(Chaetura pelagica),* AE. Refer to *Avian Myology* (Academic Press, New York, 1966) by J. C. George and A. J. Berger, pages 233-236, for further explanation on the use of symbols and a list of formulas of selected species.

Iliotrochantericus Medius. "C" in leg-muscle formulas. Identify this small muscle running from its origin on the ventral edge of the ilium, just anterior to the acetabulum, to its insertion on the femur dorsal to the iliotrochantericus anterior. Its action is similar to that of the iliotrochantericus posterior.

Iliacus (Psoas). "E" in leg-muscle formulas. This is a tiny, flat, band-like muscle arising from the ventral edge of the ilium just medial to the origin of the iliotrochantericus medius. The flat belly passes outward to insert on the medial surface of the femur, only 5 millimeters from the proximal end of the bone. A hand lens or dissecting microscope may be required to find this muscle. It acts to rotate and to adduct or draw the thigh toward the medial axis of the body.

Semitendinosus and *Accessorius Semitendinosi (Flexor Cruris Lateralis).* This complex lies posterior to the femur. Trace the semitendinosus ("X" in leg-muscle formulas) from its origin on the posterior third of the posterior iliac crest downward to the ligamentous raphe that separates the muscle from the accessorius semitendinosi ("Y" in formulas). Follow the latter muscle to its insertion on the postero-lateral surface of the distal end of the femur. The semitendinosus together with the accessory semitendinosus serve primarily to flex the femur (draw it posteriorly). The accessory semitendinosus is absent in some species; both muscles are absent in others.

Cut the fibers of origin of the semitendinosus muscle and reflect the belly downward. This will expose the following muscle.

Plate XXII

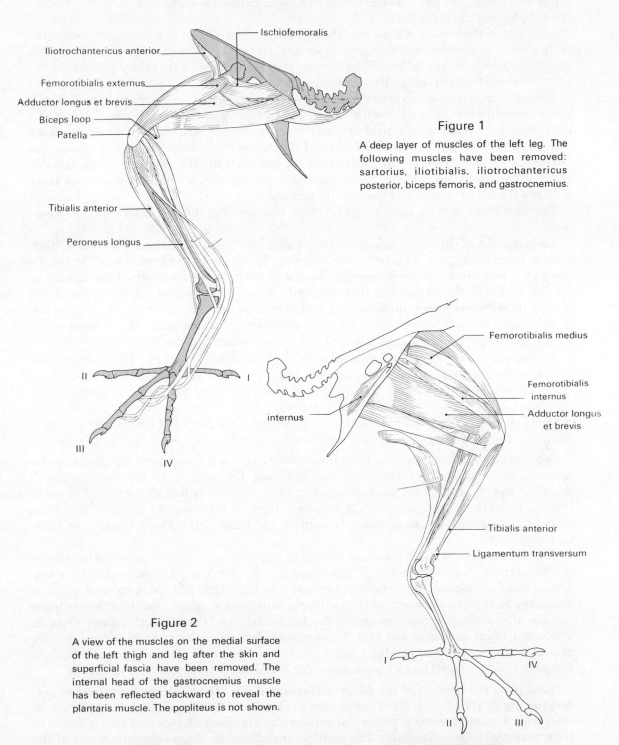

Ischiofemoralis

Iliotrochantericus anterior

Femorotibialis externus

Adductor longus et brevis

Biceps loop

Patella

Tibialis anterior

Peroneus longus

II

I

III

IV

Figure 1

A deep layer of muscles of the left leg. The following muscles have been removed: sartorius, iliotibialis, iliotrochantericus posterior, biceps femoris, and gastrocnemius.

Femorotibialis medius

Femorotibialis internus

Adductor longus et brevis

internus

Tibialis anterior

Ligamentum transversum

I

IV

II

III

Figure 2

A view of the muscles on the medial surface of the left thigh and leg after the skin and superficial fascia have been removed. The internal head of the gastrocnemius muscle has been reflected backward to reveal the plantaris muscle. The popliteus is not shown.

MUSCLES OF THE LEG AND THIGH, COMMON PIGEON

Piriformis (Caudofemoralis and Iliofemoralis). This muscle in the pigeon has two separate bellies: the **pars iliofemoralis** and **pars caudofemoralis** ("B" and "A" respectively in leg-muscle formulas).

The pars iliofemoralis arises from the ventral surface of the posterior iliac crest. The belly passes downward and forward, superficial to the pars caudofemoralis, and inserts on the postero-lateral surface of the femur. Cut the muscle at its origin and reflect the belly forward, thus exposing all of the next part.

The pars caudofemoralis arises from the ventral surface of the pygostyle and from a dense aponeurosis associated with the under tail coverts. The strap-like belly passes anteriorly between the semitendinosus (already reflected) and the **semimembranosus** (a muscle consisting of a flattened band of fleshy fibers running from its origin on the postero-inferior surface of the ischium to its insertion on the tibiotarsus; do not dissect it). The fleshy belly of the pars caudofemoralis ends on a small tendon which fuses with the tendon of insertion of the pars iliofemoralis.

The piriformis assists in flexing the thigh and moving the tail laterally and downward.

Ambiens. "Am" in leg-muscle formulas. This is the most medial muscle of the thigh. Follow the muscle from its aponeurotic origin on the pubis to the knee. Trace the tendon through a compartment in the patellar ligament to the lateral surface of the knee. The tendon of the ambiens muscle then descends deep to the tendon of insertion of the **biceps femoris (extensor iliofibularis)** and serves as part of the origin of three of the flexor muscles to Toes No. 2, 3, and 4, thus reinforcing their actions. The pattern varies in other bird species. This muscle is found in some reptiles but not in mammals.

Plantaris. "F" in leg-muscle formulas. It may be necessary to cut the internal head of the gastrocnemius in order to study the plantaris muscle. The plantaris arises from the postero-medial surface of the tibiotarsus, beginning just below the proximal articular surface. The belly extends about halfway down the crus, tapers to a flattened tendon, and inserts on the proximal end of the tibial cartilage. The plantaris serves to draw the tibial cartilage proximally and to aid in extension of the tarsometatarsus.

Popliteus. "G" in leg-muscle formulas. This is the deepest muscle on the posterior surface, proximal end, of the crus. Carefully cut the origins of the flexor muscles posterior to the head of the tibiotarsus and reflect the bellies distad. Identify the popliteus as a small, rectangular-shaped muscle extending between the head of the fibula and the tibiotarsus. Its weak action is to draw the head of the fibula toward the tibiotarsus.

The Vinculum. "V" in leg-muscle formulas. As they pass down the posterior side of the tibiotarsus, the two tendons of the flexor muscles for Toe 3 are connected by a tendinous band or vinculum. In order to expose the vinculum and the two tendons, it is necessary to remove the fibers of the gastrocnemius tendon which insert on the posterior surface of the tibiotarsus. Now grasp the Tendo Achillis (already cut) a short distance above the tibial cartilage and pull it downward sharply in order to break the attachments to the bone. Then carefully separate the tendons on the posterior surface of the tibiotarsus and locate the two tendons to Toe 3 and the interconnecting vinculum.

Now trace the tendons of the **flexor perforatus digiti III** and **flexor perfornas et perforatus digiti III** to their insertion on the phalanges of Toe 3 and study the pattern of insertion. A similar general pattern of relationships to both fingers and toes is found in most mammals, including man. The multiple insertions on the several phalanges of the fingers in man, plus his opposable thumb, account for man's great manual dexterity.

Label the drawings of the deep muscles of the leg and medial surface muscles of the thigh (Plate XXII, Figures 1 and 2).

RESPIRATORY AND DIGESTIVE SYSTEMS

These two organ systems may be conveniently considered together, since they are closely associated with each other. Both systems in the bird show marked modifications and reductions of organs or parts of organs.

Obtain a specimen of a pigeon and prepare to dissect it internally.

Lay the left side of the mouth open by cutting the angle of the jaw and continuing the incision down the side of the neck. Study the following regions.

Mouth. The mouth is a cavity, sometimes called the **buccal cavity,** between the upper and lower mandibles. Its roof or **palate** is hard and horny anteriorly and somewhat softer posteriorly, resembling the soft palate of many mammals. A median slit, the **choana,** separates the palate into two longitudinal **palatal folds** with small backward-projecting horny papillae. A cleft palate of this sort is characteristic of birds. Separate the palatal folds and observe the **nasal cavity.** Antero-dorsally it is divided by the **nasal septum** into right and left nasal cavities which communicate directly with the corresponding right and left **anterior nares,** or **nostrils,** to be seen externally at the base of the upper mandible.

The floor of the mouth is occupied by the **tongue,** whose shape conforms generally to that of the lower mandible. It is attached by only a small part of its under surface. Its covering is thick and horny. Anteriorly it is sharp-pointed but posteriorly it is forked, and bears, like the roof of the mouth, backward-projecting horny papillae. **Taste buds,** said to number between 25 and 60, occur at the base of the tongue, and a few more lie in the softer, posterior part of the palate. Small **salivary glands,** four pairs, empty into the floor of the mouth.

Pharynx. The pharynx is a continuation of the mouth posteriorly, beginning at the posterior end of the palatal slit. In contrast to the mouth, its walls are more or less muscular. In the middle of the dorsal wall, and almost continuous with the palatal slit, is a smaller slit-like opening common to the paired Eustachian tubes, which connect the pharynx with the middle ears. In mammals these tubes enter the pharynx separately, but in birds they enter together. In the middle of the floor of the pharynx are two **laryngeal folds** which bound a relatively narrow, slit-like opening, the **glottis.** Its margins bear horny papillae, but there is nothing to represent the "trapdoor," or epiglottis, which in mammals protects the opening. The posterior dorsal and ventral walls of the pharynx have pairs of membranous folds with horny papillae, the **dorsal** and **ventral pharyngeal folds.**

Hyoid Apparatus. The hyoid apparatus has already been studied in detail. However, determine its location and function by carefully dissecting away (1) the covering of the tongue, to find the median bones, and (2) the skin of the malar and auricular regions, to find the paired bones which pass toward the ears. Pull the horns of the hyoid apparatus to see how they work.

Continue the dissection down the side of the neck by pushing aside the skin along the incision. Identify the soft, thin-walled food tube, or **esophagus,** which connects the pharynx to the stomach, and the windpipe, or **trachea,** with the stiffened rings in its wall, which connects the pharynx with the bronchi of the lungs. Study these structures further.

Trachea. The rings in the walls of the trachea are bony on their ventral sides but cartilaginous on their dorsal sides. This is in marked contrast to the rings of the human trachea, which are cartilaginous ventrally and membranous dorsally.

The anterior end of the trachea is expanded to form the **larynx.** It is not a sound-producing organ as in mammals. The larynx supports the laryngeal folds in the floor of the pharynx and the glottis opens into it. Dissect out the larynx by cutting around it and freeing it from the pharynx. Then cut it open on one side and identify the cartilages within. Conspicuous ventrally is the large triangular cartilage, the **cricoid,** with lateral processes which bend around dorsally, coming to narrowed ends in back of the larynx. Between these two dorsal ends of the cricoid is a median piece of cartilage, the **procricoid.** At the base of the procricoid are attached a pair of slender, curved, somewhat bony cartilages, the **arytenoids,** which extend anteriorly along the upper parts of the larynx and form the skeletal structure of the margins of the glottis. Anteriorly, the arytenoid cartilages are attached to the upper, inner surface of the cricoid cartilage. There are no true vocal cords in birds.

Esophagus. This is a destensible tube lying dorsal to the trachea and following a relatively straight course to the stomach. Trace it posteriorly. Just before it enters the thoracic cavity it becomes dilated into a bi-lobed sac, the **crop.** Here food is detained before it is passed to the stomach. All pigeons and gallinaceous birds possess a crop.

Air Sacs and *Body Cavity.* The lungs of birds feature outpocketings filled with air which extend between various organs and penetrate certain bones; they have few blood vessels and no respiratory surfaces. The sacs are noticeably thin-walled and resemble soap bubbles. In the pigeon one single and four paired air sacs are recognized. To be studied successfully the respiratory system should be artificially inflated by cutting the trachea, inserting a tight-fitting glass tube, blowing through it, and then immediately tying off the trachea to prevent the air from escaping. In this way all the parts of the respiratory system, including the air sacs, will be distended and made more prominent. Another way to study the air sacs is to inject them with Woods Alloy and obtain casts (for the method, see paper by P. W. Gilbert, *Auk,* 56: 57-63, 1939). Latex is also a good injection medium (see D. H. Tompsett, *Ibis,* 99: 614-620, 1957).

Inflate the respiratory system as directed and identify the **interclavicular** and **cervical air sacs** (see Plate XXIII) in the vicinity of the crop. The interclavicular sac is directly dorsal to the two clavicles of the furcula, touches the dorsal side of the crop, and surrounds the posterior end of the trachea and bronchi. It is the only single sac of the respiratory system and has on each side a diverticulum which sends branches to the shoulder region and into the sternum, clavicle, coracoid, and humerus. The cervical sacs are dorsal and paired; they are antero-dorsal prolongations of the interclavicular sac supplying the cervical and thoracic vertebrae.

With the scissors, extend the incision in the side of the neck ventrally to the vent, keeping it just to the left of the median line. In the breast region cut through the large flight muscles and sternum, keeping close to the keel. In the abdominal region cut through the thinner muscular layers. Be careful not to cut too deeply, thus injuring the organs below. Spread apart the edges of the incision and examine the body cavity.

As in the higher vertebrates the body cavity is readily divisible into the **thoracic** and **abdominal cavities.** The thoracic cavity is located dorsal to the sternum and contains three smaller divisions. Medially there is the **pericardial cavity**—a space between the prominent **heart** and the thin sac surrounding the heart. This pericardial sac or **pericardium** is in contact ventrally with the inner surface of the sternum and dorsally and laterally with the inner surfaces of the body cavity. Only the posterior part of the pericardium is free. Laterally and somewhat anteriorly are two **pleural cavities** containing the lungs. These fill in the remainder of the thoracic cavity. The abdominal cavity is posterior to the sternum. In it is the large chocolate-colored **liver,** the tightly coiled **intestine,** and, to the left, the enormous **stomach.** Observe that there is no diaphragm separating the thoracic from the abdominal cavity. Instead there is a membranous, double-walled partition extending obliquely backward between the two from the points where the pericardium meets the body walls laterally. This is called the **oblique septum.** The part of the bird corresponding to the diaphragm of the mammal is a thin sheet of muscle arising from the inner surfaces of the ribs and bodies of the vertebrae and closely attached to the ventral surfaces of the pleural cavities.

Lying on each side of the pericardial cavity is a small **anterior thoracic air sac** (see Plate XXIII), while between each lung and the liver inside the double-walled oblique septum is a **posterior thoracic air sac.** On each side of the abdominal cavity is a large **abdominal air sac** which passes between the various organs of the cavity.

Prepare to study in more detail the organs of the thoracic and abdominal cavities. The air sacs may now be punctured.

Return to the trachea and follow it backward to the point where it bifurcates to form the **right** and **left bronchus.** Each bronchus passes directly to the lung on the corresponding side. The **syrinx,** or voice organ, is at the point where the trachea divides and was enveloped by the interclavicular air sac. The syrinx itself appears as a laterally compressed area between the last tracheal rings and the first bronchial rings. See the drawings of the syrinx below.

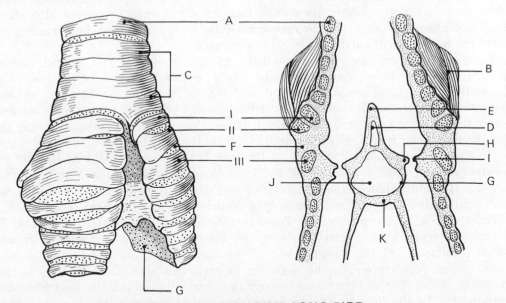

SYRINX OF A PASSERINE SONG BIRD

At left, the syrinx viewed externally; at right, the syrinx with ventral side removed. A, tracheal ring; B, syringeal muscles; C, tympanum; D, pessulus; E, semilunar membrane; F, external tympaniform membrane; G, internal tympaniform membrane; H, internal labium; I, external labium; J, projection of anterior thoracic air sac; K, bronchiodesmus. I, II, III, bronchial rings. (Redrawn from *Der Gersang der Vögel* by V. Hacker, 1900, as reproduced in *Bird Song: Acoustics and Physiology* by C. H. Greenewalt, 1968, Smithsonian Institution Press, Washington, D. C.)

Syrinx. Study the external portion of the syrinx, noting the two **sternotracheal muscles,** which are inserted on the trachea and pass posteriorly to their origin on the sternum. Note also the two smaller **bronchiotracheal muscles,** which originate on the syrinx at the bronchi and pass anteriorly to their origin on the trachea. Cut open the ventral side of the syrinx and compare it with the drawings on the preceding page. The last tracheal rings support the **tympanum,** and the space within is called the **tympanic cavity.**

Between the two bronchial openings a small septum extends forward into the trachea. The free margin of the septum is a crescent-shaped membrane and known as the **semilunar membrane.** It is supported by a small ridge, the **pessulus.** On each side of the septum is a thin, elastic membrane extending downward and forming the inner wall, which is without cartilaginous rings. It is really a continuation posteriorly of the semilunar membrane into the bronchus and is appropriately called the **internal tympaniform membrane.** Note that it bears a membranous lip, the **internal labium,** projecting into the lumen. Attached to the walls of the tympanum, directly opposite each side of the semilunar membrane, is the **external tympaniform membrane** with a prominent fold bearing a lip, the **external labium,** directly opposite the internal labium. Both labia collectively form a slit-like opening of the bronchus. Externally, connecting the medial walls of the two bronchi well below their fusion, is a broad sheet of tissue, the **bronchiodesmus.** The space between the bronchiodesmus and the fusion of the bronchi is occupied by a projection of the anterior thoracic air sac.

Bronchi. The bronchi have their outer walls strengthened by half-rings of cartilage; their inner walls are membranous only. Each bronchus enters the ventral surface of a lung and passes through it as the **mesobronchus.** As the mesobronchus proceeds posteriorly the half-rings of cartilage gradually disappear.

Lungs. The lungs are covered ventrally by the linings of the pleural cavities, or **pleura,** and the rudimentary diaphragm. Note that each lung is bright red, owing to its containing a large amount of blood, and that it is somewhat flattened against the dorsal wall of the cavity where it is not invested by pleura. Closer examination will show that the lung fits into the spaces between the ribs and vertebrae so that the impressions of these bones are visible on its surface. How far back does the lung extend?

Within each lung, leading off from the mesobronchus (see Plate XXIII), are several **ventrobronchi,** from which extend the interclavicular, cervical, and anterior thoracic air sacs (for convenience called the anterior sacs), and two rows of several **dorsobronchi.** The ventro- and dorsobronchi branch into innumerable **parabronchi** (not visible to the unaided eye) of uniform diameter. These minute tubes connect with one another freely, forming a network of air capillaries, in the meshes of which is a similar network of blood capillaries. At the posterior end of the lung the mesobronchus divides into two tubes going, respectively, to the posterior thoracic and abdominal air sacs (called the posterior sacs). The anterior sacs (except the cervical sac) and the posterior sacs are reconnected to the lung by the **recurrent bronchi,** which join the parabronchi inside the lung.

Gaseous exchange or respiration occurs in the parabronchi. By forming a network of air capillaries, the parabronchi permit a continuous circuit of air through the lung. This is not the case in the lung of a mammal, where small branches, or bronchioles, arising from the bronchi end blindly in alveoli, making such a circuit impossible.

Much of the air, on entering the lung from the bronchus, passes through the mesobronchus (a) to the posterior sacs, (b) back through recurrent bronchi to the parabronchi, (c) then to the anterior sacs by way of the ventrobronchi, and (d) finally out through

Plate XXIII

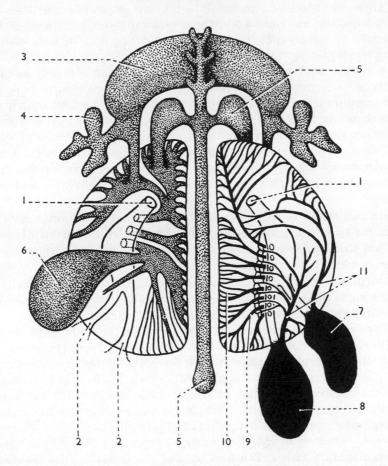

DIAGRAM OF THE LUNGS AND AIR SACS OF THE BIRD

The ventral view is at the left, the dorsal view at the right. The parts of the system concerned with inspiration are in black, the parts with expiration are stippled.

1. Mesobronchi
2. Opening of the mesobronchi into the air sacs
3. Interclavicular air sac
4. Diverticulum of the interclavicular sac to the sternum, coracoid, clavicle, and humerus
5. Cervical sac
6. Anterior thoracic air sac
7. Posterior thoracic air sac
8. Abdominal air sac
9. Dorsobronchi
10. Parabronchi
11. Recurrent bronchi

(From Portmann, 1950, in *Traité de Zoologie,* edited by Grassé, Volume 15, Figure 200, Masson et Cie; after Brandes and Hirsch.)

recurrent bronchi and parabronchi to the bronchus. Some of the air, however, on entering the lung from the bronchus, passes into the mesobronchus and then out through the ventrobronchi to the anterior sacs without making the circuit through the posterior sacs. The direction of the air movement through the lung and air sacs is controlled to a large extent by a complicated valvular system.

The relation of air movement to inspiration and expiration is briefly as follows: When the body cavity is enlarged at inspiration, the posterior sacs receive pure air from the bronchus and mesobronchus, and the anterior sacs receive partly vitiated air from the parabronchi and partly pure air from the bronchus, mesobronchus, and ventrobronchi. As the body cavity decreases in volume at expiration, all the air sacs expel air through the recurrent bronchi and parabronchi to the bronchus. During both inspiration and expiration a gaseous exchange takes place in the parabronchi, but the exchange is much less during inspiration because much less air passes through the parabronchi at this time.

Pay particular attention to the following organs in the abdominal cavity. Note that the abdominal cavity is lined, as in other vertebrates, by a thin membrane, the **peritoneum,** and that this membrane is deflected at certain points to cover the organs and to form **mesenteries** which hold the organs in place.

Liver. This organ monopolizes the anterior end of the abdominal cavity. Two lobes are present: the right one extends far back into the abdominal cavity. The forward end of the liver is somewhat ventral to the heart and partly conceals it.

Stomach. The stomach is obvious on the left side of the abdominal cavity posterior to and partially covered over by the left lobe of the liver. Two portions of the stomach are recognized: the **proventriculus,** which is the soft, glandular anterior portion continuous with the esophagus, and the **gizzard,** which is the hard, muscular portion. The gizzard is the conspicuous portion of the stomach; the proventriculus appears to be little more than an enlargement of the esophagus before entering the gizzard. A small constriction marks the union of proventriculus and gizzard.

Small Intestine. The small intestine emerges from the gizzard near the inner side where the proventriculus enters. The first part of the intestine, the **duodenum,** makes a long, U-shaped loop posteriorly and is easily distinguished for this reason. A thin, lobulated **pancreas** occupies the main area within the loop. Three **pancreatic ducts** pass from the right side of the pancreas into the right side of the duodenal loop. Into the duodenal loop pass also two **bile ducts** from the deep depressions in the dorsal surface of the right lobe of the liver. One enters the left side of the loop just beyond the gizzard; the other enters the opposite side. No gall bladder is present in the pigeon. The remaining parts of the small intestine, the **jejunum** and **ileum,** cannot be distinguished. They are greatly coiled and are suspended from the dorsal wall of the cavity by a mesentery.

Large Intestine. The large intestine is relatively reduced in the bird and does not differ markedly from the small intestine. It is merely a continuation of the small intestine, without enlargement, from the middle of the abdominal cavity straight back to a point just ventral to the vertebral column. No attempt is made to distinguish between the colon and rectum.

Caeca. Where the small and large intestines merge, appear two lateral pouches, or diverticula, called **caeca** (singular **caecum**). These structures show great variation in birds, ranging from bud-like objects to ones of great length.

Cloaca. This is a tubular cavity common to the digestive and urogenital systems and opening exteriorly through the **vent.** It receives the large intestine on its median ventral surface. (See urogenital system.)

Separate the stomach from the adjacent mesenteries and lift it forward. Running along its dorsal surface to the intestine will be found the **spleen,** a rather round, reddish organ. Remove the stomach from the body cavity by severing the esophagus and small intestine. Note that the gizzard is flattened and rounded like a bi-convex lens but with one curved surface greater than the others. The glistening effect of the gizzard is due to the many tendons of the outer muscular layer. Cut the stomach in two, anteroposteriorly. The proventriculus, being highly glandular, presents a spongy appearance within. The gizzard, on the other hand, shows a thick horny lining, raised in hard ridges. It contains many small pebbles which the bird has swallowed. Notice that the walls of the gizzard are not uniformly muscular. Thus the center of each right and left half contains no muscle, while the anterior and posterior ends of each half contain powerful masses of muscle. These are the **lateral muscles.** Observe that their inner horny walls almost meet each other, leaving little space in the gizzard. When food is taken into the gizzard, it is immediately pressed between the walls. By alternate movements of the lateral muscles the food is rubbed against the hard walls and pebbles, and thus ground into fine particles.

Remove the intestine. How many times longer than the body of the bird is it? In relation to the body length the intestine is proportionately shorter in the bird than in man.

Label the drawing of the digestive system (Plate XXIV, page 89).

CIRCULATORY SYSTEM

Several major features of the circulatory system should be observed.

The **heart** of the bird is proportionately large and conical, with its apex, in the pigeon and most other species, pointing posteriorly. The chambers of the heart, like those of the mammal, are four in number and entirely separate from one another. Two of the chambers, the **atria** (singular **atrium**), occupy the anterior end or base of the heart; both are thin-walled. The other two chambers, the **ventricles,** comprise the remaining part of the heart and have thick, muscular walls. The greater bulk of the heart is, therefore, ventricular.

The vessels leaving and entering the heart are essentially the same in their form, distribution, and function as the mammalian vessels.

The **aorta,** on emerging from the left ventricle, turns to the right, instead of to the left as in mammals. Arising from the aorta, where it curves posteriorly to form the **aortic arch,** are two **brachiocephalic arteries,** large vessels carrying oxygenated ("pure") blood to the wings and to the anterior thoracic, neck, and head regions. Follow either one laterally and a little anteriorly to the point where it divides into a **common carotid artery** to the head and a **subclavian artery** to the adjacent wing. In the angle formed by this division and lying against the common carotid artery is the **thyroid gland,** an oval body with a reddish color. Just lateral to the thyroid gland and separate from it are two very small **parathyroid glands.** The thyroid and two parathyroid glands are paired endocrine structures occurring in the same position in the pigeon's right and left sides.

Near the point where the aorta leaves the heart, find the **pulmonary artery** which proceeds from the right ventricle in a left direction, dorsal to the aorta. It takes the venous ("impure") blood to the lungs.

Now follow the aorta as the **dorsal aorta.** This leads posteriorly from the aortic arch, dorsal to the heart and between the lungs, through the oblique septum, and into the abdominal cavity. The dorsal aorta carries oxygenated blood to the tail and posterior appendages, giving off during its course numerous branches which distribute similarly "pure" blood to the organs and walls of the posterior thoracic and abdominal regions.

Venous blood is returned to the heart from the anterior region of the body by two **precavae** and from the posterior region by the **postcava.** Lift up the ventricular part of the heart and identify these big veins as they approach the right atrium: the precavae from a right and left direction, respectively, and the postcava from the liver through which it passes. The postcava enters the right atrium between the two precavae.

The pulmonary veins bear oxygenated blood from the lungs to the left atrium. They are small and difficult to find.

Remove the heart from the body cavity by severing the connecting vessels. Make a cut across the ventricles, midway between the atria and apex. Note the crescent-shape of the right ventricle as it tends to overlap the larger, more rounded left ventricle; and note also that only the left ventricle extends to and includes the apex of the heart. The wall of the right ventricle is thinner than that of the left. Lay open the right ventricle further and observe the crescent-shape of the opening into the right atrium, also the muscular fold or flap which acts as a valve. In the mammalian heart the opening into the right atrium is controlled by the more elaborate tricuspid valve. Open the other ventricle and observe two valves, both thin and membranous, which may close the more circular opening into the corresponding atrium. These two valves, present in the mammalian heart as well as the avian, comprise the **bicuspid valve.**

UROGENITAL SYSTEM

The urinary and reproductive systems may be conveniently studied together. Dissect first the system in the pigeon being used, then obtain a specimen of the opposite sex. In each case, remove all the digestive organs, except the cloaca, from the body cavity.

Female Urogenital System

Two kidneys are present in the bird. They are just posterior to the lungs and securely placed in the deep depression formed mainly by the synsacrum. In the pigeon each kidney is trilobed. From the medial border of each kidney, between the anterior and and median lobes, emerges a narrow **ureter** which goes directly to the cloaca. This bears the urine to the cloaca for temporary storage. There is no bladder in the bird. Lying immediately anterior to each kidney is a small orange-yellow body, roughly oval in shape. This is the **adrenal gland,** a part of the endocrine system.

Ventral to the left kidney at its antero-medial end and very near the adrenal gland is a single **ovary,** whitish in color. It shows many rounded follicles of different sizes containing ova. A small mesentery holds the ovary in place. Slightly posterior and lateral to the ovary a convoluted **oviduct,** supported by another mesentery, leads to the cloaca; it is not attached to the ovary but begins beside it with a funnel-shaped dilation or **ostium** (also called the **infundibulum**). An ovum, on being released by a follicle, is engulfed by this wide opening and is pushed by peristaltic movements through the long winding course of the oviduct to the cloaca.

Most adult female birds possess but one ovary and one fully developed oviduct, and these are always on the left side. In female embryos a right ovary and oviduct are present, but in most birds they disappear or nearly so, before hatching. A remnant of the

Plate XXIV

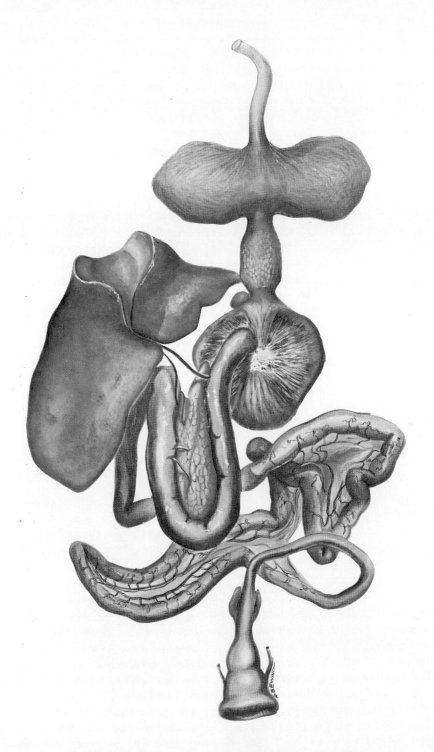

DIGESTIVE SYSTEM OF COMMON PIGEON

right oviduct often stays attached to the right side of the cloaca. Such a vestige is usually present in the pigeon.

When the female is in laying condition both ovary and oviduct are more obvious. Many follicles in the ovary are very big and contain correspondingly big ova holding great quantities of yolk material. The oviduct, which now occupies a considerable space ventral to the left kidney, has its walls considerably lengthened and thickened by enlarged muscles and glands. Besides the ostium, the oviduct has four parts; they are (from ostium to cloaca) the **magnum, isthmus, uterus,** and **vagina.** Though they are distinctive in their histological composition and function, they do not differ markedly in their gross anatomy, except in relative size. As the name indicates, the magnum has the greatest size—always much longer than any other part of the oviduct. It is in the magnum that a passing ovum, having been picked up by the ostium, replete with yolk material, receives by glandular secretion its principal coatings of albumen ("white of egg"). The isthmus, the next part, can often be distinguished by being slenderer than the magnum and having fewer folds in its internal wall. In the isthmus both the outer and inner shell membranes are added, covering the albumen. In the uterus, which is conspicuously dilated and somewhat bulbous in shape, substances are secreted to form the calcareous shell and its pigments and to contribute further to the albumen through the porous shell membranes. The terminal part of the oviduct, the vagina, is noticeably constricted and has no known function other than to direct the completely formed egg to the cloaca.

Make an incision along the right side of the cloaca, opening it fully. Note that it is divisible into three parts. The first, the **coprodeum,** is continuous with the large intestine. It is the largest part and is situated ventrally. The second, the **urodeum,** is the middle part of the cloaca and is situated above the coprodeum, being separated from it by a membranous fold. Through its lateral walls the right and left ureters enter, while above the left ureter the oviduct enters. The third, the **proctodeum,** is more posterior and somewhat dorsal to the second; it is smaller and opens directly to the vent. (In the dorsal wall of the proctodeum of the young pigeon an opening leads to a blind, unpaired sac, the **bursa of Fabricius.** Though it seems to be active in the early life of the bird, its function is obscure. (See Appendix A, page 443, for further information on the bursa.)

Male Urogenital System

The urinary system of the male is similar to that of the female. The preceding description of the system as it occurs in the female should, therefore, be read and applied to the male.

The **testes** are paired, ellipsoid organs, whitish in color, located on the kidneys in positions similar to that of the ovary in the female. They show asymmetry, however, one, generally the left, being larger than the other. During the breeding season the testes increase greatly in circumference, more than doubling their size. The medial border of each testis is rather concave, with a minute projection, barely visible, called the **epididymis.** From this springs the **vas deferens** (plural, **vasa deferentia**), a small convoluted duct which passes directly to the cloaca lateral to the ureters. It gradually widens into a **seminal vesicle** as it approaches the cloaca. In life the vas deferens bears the sex cells or spermatozoa from the testis, and the seminal vesicle stores them.

The cloaca is practically the same as in the female save that it is smaller, while the lips of the vent are thicker and tend to protrude in a more conspicuous manner. No penis occurs in the pigeon. (See Appendix A, page 442, for a description in several groups of birds.) The spermatozoa are passed to the female when the lips of the cloacae of the two sexes meet during copulation. The urodeum receives the two vasa deferentia, instead of an oviduct; these ducts enter the chamber laterally, just above the ureters.

Label the drawings of the male and female urogenital systems (Plate XXV, page 92) which show the genital systems in full breeding condition.

If a specimen of the Common Pigeon or any other bird is to be dissected for the sole purpose of examining the **gonads** (ovary or testes), a longitudinal incision should be made on the left side of the abdominal cavity, halfway between the mid-ventral and mid-dorsal lines and through the posteriormost ribs. Then the viscera ventral to the left kidney should be lifted up and pressed aside. This will bring into view either the ovary or the larger testis. Should the specimen be immature and not otherwise in breeding condition, the gonad will be very small. Consequently, care must be taken not to confuse it with the adrenal gland, which may be more prominent than the gonad and may, in some species, be shaped like an ovary. Avoid confusion by noting color. Gonads are always whitish; adrenal glands are more highly colored—usually orange-yellow, though occasionally light yellow and in a few instances pink or red.

NERVOUS SYSTEM

Remove the remaining organs from the body cavity and note the ventral branches of the spinal nerves passing over the dorsal wall. Trace them dorsally and note the ganglia of the sympathetic nervous system where they converge to go between the vertebrae to the spinal cord.

Sense Organs

A knowledge of three organs of special sense is necessary in understanding the behavior of a bird.

Organs of Smell. Cut into one of the nasal cavities by making an incision in the palate slightly to the side of the midline. Find three **conchae** or **turbinals** extending into the nasal cavity from its lateral wall. (See Plate XXVI, Figure 1.) They are in line from front to back and are consequently referred to as the anterior, median, and posterior. The last is the most superior in position and is also the smallest, being quite rounded. It supports the **olfactory membrane,** which is connected to the brain by the olfactory nerve. In man there are three conchae, which occupy a similar position in the nasal cavity; the olfactory membrane is supported by the nasal septum for a short distance as well as by the posterior concha.

Organs of Sight. The eyes of birds are highly specialized and deserve careful attention. Cut away the tissues surrounding one of the eyes and chip away the roof of the orbit. Note the peculiar turnip-shaped eyeball: the whitish, tapering sclerotic part and the transparent, more curved **cornea.** Move it, and notice that there are six eye muscles attached to it. Find the optic nerve entering on the inner side. Ventral to one of the eye muscles on the anterior part of the eyeball is a whitish mass, the **Harderian gland.** Dissect away from the eyeball the tissues below and find a small **lacrimal gland.**

Plate XXV

Oviduct

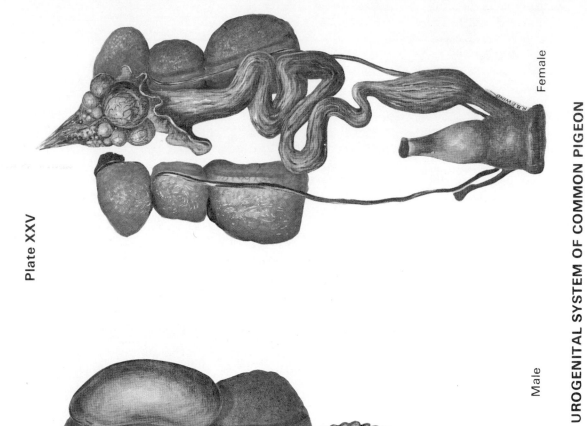

Female

R.B. EWING

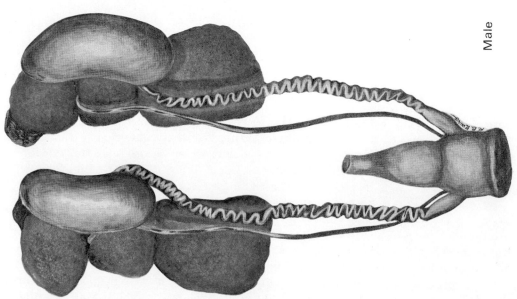

Male

R.B. EWING

UROGENITAL SYSTEM OF COMMON PIGEON

Plate XXVI

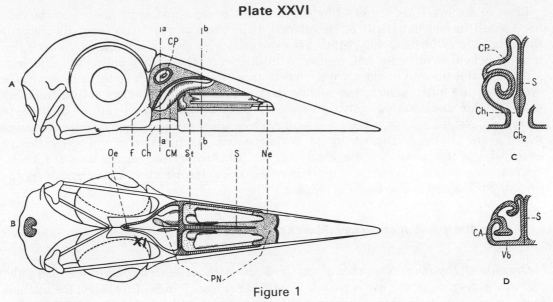

Figure 1

Nasal Cavity of the Bird

A, lateral view; B, ventral view; C, cross section of the cavity (a-a of A); D, cross section of the cavity (b-b of A); CA, anterior concha; Ch, choana; Ch 1, primary choana; Ch 2, secondary choana; CM, median concha; CP, posterior concha; f, connection between the nasal cavity and the infraocular air space; Ne, nostril (anterior naris); Oe, opening of the Eustachian tube; PN, floor of the nasal cavity; S, nasal septum; St, transverse fold in the floor of the nasal cavity; Vb, floor of anterior nasal cavity.

(From Portmann, 1950, in *Traité de Zoologie*, edited by Grassé, Volume 15, Figure 148, Masson et Cie; after Technau.)

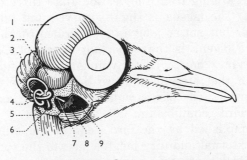

Figure 2

Ear of Common Pigeon

Lateral view of the head of a Common Pigeon showing the position and principal parts of the ear. 1, cerebral hemisphere; 2, optic lobe; 3, cerebellum; 4, semicircular canals; 5, neck; 6, lagena; 7, columella in middle ear; 8, tympanum; 9, external auditory meatus.

(From Portmann, 1950, in *Traité de Zoologie*, edited by Grassé, Volume 15, Figure 153, Masson et Cie; after Krause.)

Remove the eye from the head by severing the muscles and optic nerve and (holding the eyeball in one hand) cut off the dorsal wall. Observe the large **lens.** Its shape is characteristic of birds, being rather flat externally and convex internally. As in mammals, it is held in place by, and focused with the aid of, the **ciliary body.** The chambers of the eye and the two humors are similar to those in mammals. The three layers in the eye's wall, the outer **sclera,** the middle dark **choroid,** and the inner **retina,** together with the thin **conjunctiva** which passes over the cornea, are also similar to those in mammals. Note, however, a brown vascular fringe which projects from the lower medial wall toward the lens. This structure is the **pecten.** Its function is unknown. Scrape away a bit of the tissue of the eyeball near the point where the cornea joins the sclera. Here a bony ring encircles the eye. It is called the **sclerotic ring** and strengthens the eyeball. Such a ring is also found in reptiles.

Label the drawing of the eye (Plate XXVII, Figure 2).

Organs of Hearing. The ears of the bird are without external appendages. In the pigeon the opening of the ear is rounded and covered by a fringe of feathers, the **auriculars;** it leads directly into a passage, the **external auditory meatus,** situated below and behind the eye. (See Plate XXVI, Figure 2.) Cut into the meatus and locate the transparent **tympanum** or **eardrum.** This membrane marks the inner boundary of the **external ear.** Cut through the tympanum and thus expose the **middle ear.** Here observe a rod-like bone, the **columella,** one end of which was attached to the membrane before it was cut through. The other end meets the **stapes,** a tiny bone which is in contact with an oval opening, the **fenestra vestibuli,** in the inner wall. These two bones have a function corresponding to that of the three bones of the middle ear of man. From the ventral floor of the middle ear the Eustachian tube leads to the pharynx. The **internal ear** is embedded in the periotic capsule medial to the fenestra vestibuli. While it is too small and complicated to be dissected, it will nevertheless be briefly described.

The inner ear has three **semicircular canals** which emerge from a central chamber, the **vestibule.** Also arising from the vestibule are three chambers, the **utriculus, sacculus,** and **lagena.** The lagena is slightly curved and relatively longer than the same structure in reptiles but not so long as the spiraled cochlea of mammals, to which it is homologous. In the lagena is the important **organ of Corti** with its basement membrane of ciliated cells ("hair cells"). Lymphatic fluids fill the semicircular canals and all the chambers of the internal ear, also the spaces between their walls and the surrounding bony capsule. The semicircular canals have as their principal function the maintaining of the bird's equilibrium.

Hearing is accomplished by the bird's ear in the following manner: Sound waves picked up by the external auditory meatus pass to the tympanum. The resulting vibrations of this membrane are transported across the middle ear by the columella and stapes to the fenestra vestibuli. Vibrations in the fluids of the internal ear are then set up which reach the hair cells of the organ of Corti and activate them. The vibrations are converted to nerve impulses which are carried to the sensory endings of the auditory nerve and thence to the brain.

Brain

Remove the bones of the roof of the skull, exposing the brain dorsally and laterally. Note the membranous coverings or **meninges.** Several features are worthy of attention.

Plate XXVII

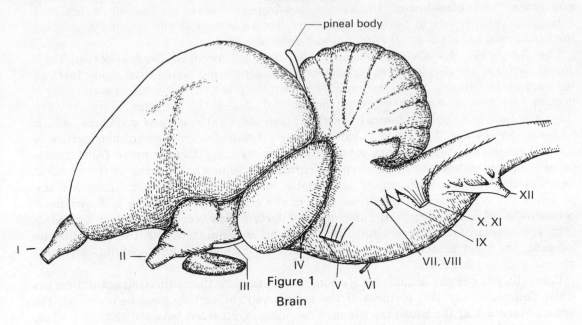

Figure 1
Brain

The twelve cranial nerves are identified by Roman numerals.

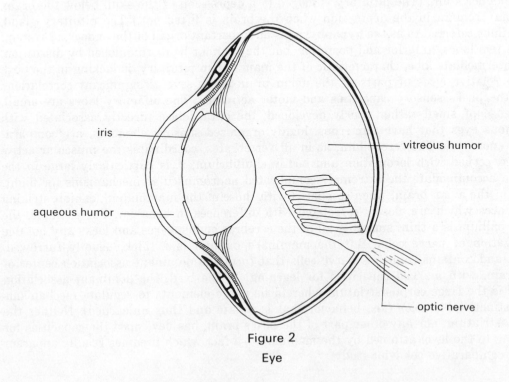

Figure 2
Eye

BRAIN AND EYE OF COMMON PIGEON

The brain is divided, as in other vertebrates, into three main parts: the **fore-brain, mid-brain,** and **hind-brain.** Unlike the quadrupeds, however, the bird's brain is noticeably curved, due to flexures. The chief flexure occurs in the mid-brain, causing the hind-brain to be almost at right angles to the fore-brain.

The fore-brain has the conspicuous right and left **cerebral hemispheres.** Their dorsal surfaces are decidedly convex and smooth, there being none of the many furrows that are so prominent in the mammalian brain. They are, nonetheless, separated by a deep dorsal fissure. Each hemisphere has a roof, called the **pallium,** which is very thin; a hollow interior (the **lateral ventricle**); and a floor (the **corpus striatum**), which is unusually thick and tends to bulge upward. In front of each cerebral hemisphere is a small projection, the **olfactory lobe,** which receives the olfactory nerve from the olfactory membrane of the nasal cavity. Ventral and somewhat lateral to the cerebral hemispheres are the exceptionally large **optic lobes** of the mid-brain. There are the principal centers for visual reception. Back of the cerebral hemispheres is the rounded **cerebellum** with its very noticeable middle portion, the **vermis,** marked by many transverse fissures. The cerebellum lies upon and is connected with the **medulla oblongata,** the most posterior part of the brain. Both cerebellum and medulla oblongata are parts of the hind-brain.

Carefully lift out the brain from its bony base, severing the connecting cranial nerves while doing so. Only two features of the brain's ventral surface need be observed: the **optic chiasma** and the **pituitary gland.** The chiasma, situated between the optic lobes, consists of the crossed tracts of optic nerves coming from the eyes. The pituitary gland is just posterior to the chiasma, where it is connected to the **hypothalamus** of the brain by a slender stalk. (The pituitary is lodged in a depression of the skull below the brain, and may remain in the depression when the brain is lifted out.) The pituitary gland, sometimes referred to as the **hypophysis,** is an important organ of the endocrine system. It has two lobes, anterior and posterior, but they cannot be distinguished by dissection. The intermediate lobe, characteristic of the mammalian pituitary, is lacking in the bird.

The relative sizes of parts of the avian brain show several significant correlations with the bird's sensory capacities and motor activities. The olfactory lobes are small, the sense of smell rather poorly developed; the optic lobes, directly associated with enormous eyes that have numerous, highly organized sensory elements, are comparatively enormous. The cerebellum, as in all vertebrates, coordinates the muscular activities concerned with locomotion and bodily equilibrium; it is particularly large in the bird to accommodate the extremely complicated neuromuscular mechanisms for flight. Parts of the avian brain, when compared with those of the mammalian, exhibit striking differences which are closely correlated with differences in behavior. For example, the bird's pallium is a thin, smooth wall of the cerebral hemispheres and lacks any notable aggregation of nerve cells. But the mammal's pallium is a thick, usually furrowed, cortex and contains countless nerve cells that form the dominant association center of the brain, with marked capacities for learning. In the bird, the dominant association center is the large corpus striatum wherein lie nerve elements to regulate mechanisms for instinctive behavior (i.e., behavior that is innate and thus unlearned). Neither the corpus striatum, nor any other part of the bird's brain, has developed the capacities for learning to the level attained by the mammal's—a fact which becomes readily apparent in any comparative behavior study.

Label the drawing of the brain (Plate XXVII, Figure 1).

Part II
SOME ANATOMICAL AND PHYSIOLOGICAL CONSIDERATIONS

LITERATURE

There is no one treatise on avian anatomy, either gross or microscopic, to which the student may readily refer for desired details on all the organ systems. The most extensive works on anatomy relate to domestic birds and are intended for use in poultry husbandry, aviculture, and veterinary medicine. The only books on physiology *per se* (Horton-Smith and Amoroso, 1966; Sturkie, 1965) are largely concerned with the Domestic Fowl *(Gallus gallus)*.

Listed below are general works that should be of value to a student in one way or another. Works that have to do with specific aspects of avian anatomy and physiology are given later under appropriate headings.

Beddard, F. E.
  1898   The Structure and Classification of Birds. Longmans, Green, and Company, London. (Though somewhat out of date, this is still a good book on anatomy.)
Berger, A. J.
  1961   Bird Study. John Wiley and Sons, New York. (The chapter, "Structure and Function," provides the most concise review of the subject available; recommended for introductory reading.)
Bradley, O. C., and T. Grahame
  1960   The Structure of the Fowl. Fourth edition. Oliver and Boyd, Edinburgh.
Coues, E.
  1903   Key to North American Birds. Fifth edition. Two volumes. Dana Estes and Company, Boston. (Volume 1, pp. 139-233, has one of the best general summaries on the organ systems in the English language.)
Darling, L. and L.
  1962   Bird. Houghton Mifflin Company, Boston. (Designed for the lay reader; separate chapters for each organ system with simplified drawings for quick understanding.)
Fisher, H. I.
  1955   Avian Anatomy, 1925-1950, and Some Suggested Problems. In *Recent Studies in Avian Biology*. Edited by A. Wolfson. University of Illinois Press, Urbana. (Highly recommended to anyone interested in anatomy as a field of investigation.)
Fürbringer, M.
  1888   Untersuchungen zur Morphologie und Systematik der Vögel, zugleich ein Beitrag zur Anatomie der Stütz- und Bewegungsorgane. Two volumes. T. J. van Holkema, Amsterdam. (Particularly good for the anatomy of the appendages.)
Gadow, H., and E. Selenka
  1891   Vögel. In *Bronn's Klassen und Ordnungen der Thier-Reichs*. Sechster Band, vierte Abtheilung, I. Anatomischer Theil. Leipzig. (A good source of useful information.)
Grasse, P. -P., Editor
  1950   Traite de Zoologie: Anatomie, Systématique, Biologie. Volume 15. Oiseaux. Masson et Cie, Paris. (By far the best source of information on anatomy since Stresemann's work.)
Hamilton, H. L.
  1952   Lillie's Development of the Chick. Third edition. Henry Holt and Company, New York. (Especially helpful in understanding how the skeleton and body cavities develop.)
Harvey, E. B., H. E. Kaiser, and L. E. Rosenberg
  1968   An Atlas of the Domestic Turkey *(Meleagris gallopavo):* Myology and Osteology. U. S. Atomic Energy Commission; Division of Biology and Medicine. (A well-illustrated and extensively labeled work, usable as a guide to the dissection of many species. Includes a muscle synonymy.)

Horton-Smith, C., and E. C. Amoroso, Editors
1966 Physiology of the Domestic Fowl. Oliver and Boyd, Edinburgh. (Contains 34 papers, grouped in five sections: "Reproductive Physiology and Endocrinology," "Metabolism and Nutrition," "Calcium Matabolism and Eggshell Formation," "Environmental Physiology," "Pharmacology and General Physiology.")

Humphrey, P. S., and G. A. Clark, Jr.
1964 The Anatomy of Waterfowl. In *The Waterfowl of the World*, Volume 4, by J. Delacour. Country Life, London. (A concise, comprehensive review, with numerous drawings.)

Hyman, L. H.
1942 Comparative Vertebrate Anatomy. Second edition. University of Chicago Press, Chicago. (A laboratory manual for dissection, but contains much textual material. Directions for the dissection of the bird refer mainly to the pigeon.)

Lucas, A. M., and P. Stettenheim
1965 Avian Anatomy. In *Diseases of Poultry*. Fifth edition. Edited by H. E. Biester and L. H. Schwarte. Iowa State University Press, Ames. (A review of the subject; highly recommended.)

Marshall, A. J., Editor
1960- Biology and Comparative Physiology of Birds. Two volumes. Academic Press, New York.
61 (Organ systems reviewed by different authors in separate chapters.)

Newton, A.
1896 A Dictionary of Birds. Adam and Charles Black, London. (Still an indispensable work, briefly covering many parts of organ systems.)

Romanoff, A. L.
1960 The Avian Embryo. Macmillan Company, New York.

Romer, A. S.
1955 The Vertebrate Body. Second edition. W. B. Saunders Company, Philadelphia. (Good reading for the student desiring to know how the bird relates anatomically to the other vertebrates.)

Stresemann, E.
1927- Aves. In *Kükenthal u. Krumbach, Handbuch der Zoologie*. Volume 7, part 2. De Gruyter,
34 Berlin. (The coverage on anatomy is the best ever published in any language. Unfortunately long out of print and difficult to obtain.)

Strong, R. M.
1939- A Bibliography of Birds. Field Mus. Nat. Hist. Zool. Ser., 25. Parts 1-4. (One of the best
59 sources for anatomical and physiological literature up to 1926.)

Sturkie, P. D.
1965 Avian Physiology. Second edition. Cornell University Press, Ithaca, New York.

Thomson, A. L., Editor
1964 A New Dictionary of Birds. McGraw-Hill Book Company, New York. (A good reference for general information.)

Wallace, G. J.
1963 An Introduction to Ornithology. Second edition. Macmillan Company, New York. (The chapter, "Internal Features and Their Function," provides excellent introductory reading.)

Welty, J. C.
1962 The Life of Birds. W. B. Saunders Company, Philadelphia. (A fine survey in five chapters.)

SKELETON

by Richard L. Zusi

The skeleton in birds differs between taxonomic groups chiefly in the number of vertebrae; in the loss of certain bones or processes of the skull, pectoral appendage, or foot; in the configuration of the palatal complex, sternum, humerus, and pelvis; and in the degree of ossification of nasal cartilages and of tendons. Such differences may reflect different evolutionary origins; they may, along with other differences in the shape and proportions of skeletal elements within orders or families of birds, also reflect adaptations to different ways of life. The skeleton is thus a rich source both for systematic studies and for the study of functional anatomy and structural adaptation.

Although the literature on descriptive and functional osteology of birds is voluminous, there is no one source for a comprehensive survey of the avian skeletal system. The accounts of the skeleton in Newton (1896), Coues (1903), and Marshall (1960)—see the preceding bibliography of general works on anatomy—nonetheless provide an adequate background for understanding the more specialized papers listed below. Most of the papers include bibliographies on the subjects concerned.

Adams, C. T.
  1955 Comparative Osteology of the Night Herons. Condor, 57: 55-60. (A taxonomic evaluation of the osteological characters of the Black-crowned Night Heron, *Nycticorax nycticorax*, and Yellow-crowned Night Heron, *Nyctanassa violacea*.)
Ashley, J. F.
  1941 A Study of the Structure of the Humerus in the Corvidae. Condor, 43: 184-195. (A comparative study of humeral configuration in crows, jays, and magpies. Shows osteological differences at the generic level. Includes a drawing of the brachial plexus and a description of the muscles at the head of the humerus in the Common Crow, *Corvus brachyrhynchos*.)
Barnikol, A.
  1952 Korrelationen in der Ausgestaltung der Schädelform bei Vögeln. Morph. Jahrb. 92: 373-414. (A study of the different forms of the skull in birds in relation to the position and size of the eye, brain, and jaw muscles.)
Beecher, W. J.
  1962 The Bio-mechanics of the Bird Skull. Bull. Chicago Acad. Sci., 11 (2): 10-33. (A good introduction to the subject of avian kinesis.)
Berger, A. J.
  1952 The Comparative Functional Morphology of the Pelvic Appendage in Three Genera of Cuculidae. Amer. Midland Nat., 47: 513-605. (Includes a discussion of variation and functional aspects of the postcranial skeleton and shows positions of muscle attachments on bones of the hindlimb.)
  1955 Suggestions Regarding Alcoholic Specimens and Skeletons of Birds. Auk, 72: 300-303. (Discusses aspects of the skeleton that need special care in preparation; gives many important points for the production of a scientifically useful bird skeleton.)
Boas, J. E. V.
  1933 Kreuzbein, Becken und *Plexus lumbosacralis* der Vögel. Kongl. Danske Videnskab. Selsk., Naturvidensk. og Mathem. Afd., Ser. 9.5 (1): 5-74. (A detailed account of the pelvic girdle and its associated nerves.)
Bock, W. J.
  1960 Secondary Articulation of the Avian Mandible. Auk, 77: 19-55. (An account of a poorly understood feature of the skull.)
  1962 The Pneumatic Fossa of the Humerus in the Passeres. Auk, 79: 425-443.
  1963 The Cranial Evidence for Ratite Affinities. Proc. XIIIth Internatl. Ornith. Congr., pp. 39-54. (Presents an argument opposed to that of McDowell, 1948).
  1966 An Approach to the Functional Analysis of Bill Shape. Auk, 83: 10-51.
Bowman, R. I.
  1961 Morphological Differentiation and Adaptation in the Galápagos Finches. Univ. California Publ. in Zool., 58: 1-302. (A detailed study of feeding adaptations.)
Curtis, E. L., and R. C. Miller
  1938 The Sclerotic Ring in North American Birds. Auk, 55: 225-243.
Dilger, W. C.
  1956 Adaptive Modifications and Ecological Isolating Mechanisms in the Thrush Genera *Catharus* and *Hylocichla*. Wilson Bull., 68: 171-199. (The adaptive significance of small differences in limb proportions.)
Dullemeijer, P.
  1951- The Correlation between Muscle System and Skull Structure in *Phalacrocorax carbo*
  52   *sinensis* (Shaw and Nodder). Proc. Koninklijke Nederlandse Akad-van Wetenschappen,

Series C,I,54(3): 247-259. II,54(4): 400-404. III,54(5): 533-536. IV,55(1): 95-102. (One of numerous papers on the skull, jaw muscles, and functional complexes of the head, published by Dullemeijer and Van der Klaauw and their students.)

Eaton, S. W., and others
  1963   Some Osteological Adaptations in Parulidae. Proc. XIIIth Internatl. Ornith. Congr., pp. 71-83. (Adaptations of limbs and eyes of some wood warblers.)

Engels, W. L.
  1938   Variation in Bone Length and Limb Proportions in the Coot *(Fulica americana)*. Jour. Morph., 62: 599-607. (A statistical analysis of variation in skeletal elements and their proportions.)
  1940   Structural Adaptations in Thrashers (Mimidae:Genus *Toxostoma*) with Comments on Interspecific Relationships. Univ. California Publ. in Zool., 42: 341-400. (Includes an important discussion of pitfalls in the interpretation of differences in proportions of skeletal elements.)

Fisher, H. I.
  1944   The Skulls of the Cathartid Vultures. Condor, 46: 272-296. (A study of skull variation with a view to determining relationships within a family at the generic level.)

Heimerdinger, M. A., and P. L. Ames
  1967   Variation in the Sternal Notches of Suboscine Passeriform Birds. Peabody Mus. Postilla. No. 105: 1-44.

Hofer, H.
  1945   Untersuchungen über den Bau des Vogelschädels, besonders über den der Spechte und Steisshühner. Zool. Jahr. (Anat.), 69: 1-158.
  1955   Neuere Untersuchungen zur Kopfmorphologie der Vögel. Proc. XIth Internatl. Ornith. Congr., pp. 104-137. (These and other papers by Hofer present descriptive and functional aspects of the avian skull in considerable detail.)

Howard, H.
  1929   The Avifauna of Emeryville Shellmound. Univ. California Publ. in Zool., 32: 301-394. (Contains an excellent series of labeled drawings of the skeletons of the Golden Eagle, *Aquila chrysaëtos*, and Snow Goose, *Chen hyperborea*. A standard reference for the names of bones and parts of bones.)

Huggins, R. A., and others
  1942   Ossification in the Nestling House Wren. Auk, 59: 532-543.

Huxley, T. H.
  1867   On the Classification of Birds; and on the Taxonomic Value of the Modifications of Certain of the Cranial Bones Observable in That Class. Proc. Zool. Soc. London, 1867: 415-472. (The classical work on palatal structure.)

Jollie, M. T.
  1957   The Head Skeleton of the Chicken and Remarks on the Anatomy of This Region in Other Birds. Jour. Morph., 100: 389-436. (Describes and illustrates all bones of the skull as they appear in the young bird.)

Larson, L. M.
  1930   Osteology of the California Road-runner, Recent and Pleistocene. Univ. California Publ. in Zool., 32: 409-428.

Lindsay, B.
  1885   On the Avian Sternum. Proc. Zool. Soc. London, 1885: 684-716. (A study of embryology and homology.)

Linsdale, J. M.
  1928   Variations in the Fox Sparrow *(Passerella iliaca)* with Reference to Natural History and Osteology. Univ. California Publ. in Zool., 30: 251-392. (Includes a rare example of the study of skeletal variation at the subspecies level.)

Lowe, P. R.
  1926   More Notes on the Quadrate as a Factor in Avian Classification. Ibis, 1926: 152-188.
  1931   An Anatomical Review of the "Waders" *(Telmatomorphae)*, with Special Reference to the Families, Sub-families, and Genera within the Sub-orders *Limicolae, Grui-limicolae,* and *Lari-Limicolae.* Ibis, 1931: 712-771. (A fine study of relationships based in part on a comparative osteological survey.)

McDowell, S.
  1948   The Bony Palate of Birds. Part I, The Palaeognathae. Auk, 65: 520-549. (Argues that the

so-called "palaeognathous" palate of the ostrich-like birds represents four morphological types. See also Bock, 1963.)

Nero, R. W.
1951   Pattern and Rate of Cranial "Ossification" in the House Sparrow. Wilson Bull., 63: 84-88. (A useful paper for determining age of bird specimens by skull condition. Other important papers on the subject are cited.)

Owre, O. T.
1967   Adaptations for Locomotion and Feeding in the Anhinga and the Double-crested Cormorant. Amer. Ornith. Union, Ornith. Monogr. No. 6. (Analyzes the proportions of the skeleton and muscles in relation to the skeleton, thus showing adaptations for two different kinds of feeding under water.)

Parker, W. K.
1879   On the Structure and Development of the Bird's Skull. Part 2. Trans. Linnaean Soc. London, 1876: 99-154. (One of the many fine papers by this author, covering a wide range of families and stressing the development of cranial ossification.)

Pycraft, W. P.
1907   Contributions to the Osteology of Birds. Part IX. *Tyranni; Hirundines; Muscicapa; Lanii;* and *Gymnorhines.* Proc. Zool. Soc. London, 1907: 352-379. (A good example of the author's many notable contributions on the osteology of birds.)

Rooth, J.
1953   On the Correlation between Jaw Muscles and the Structure of the Skull in *Columba palumbus palumbus* L. K. Nederl. Akad. Wetenschap. Proc. Ser. C., 56 (2): 251-264. (An example of the many good papers on the skeleton and musculature of the head and neck in various birds, published by students of Van der Klaauw, University of Leiden.)

Schoonees, J.
1963   Some Aspects of the Cranial Morphology of *Colius indicus.* Ann. Univ. Stellenbosch, Vol. 38, Ser. A, No. 7: 215-246. (Particularly useful in presenting serial cross sections through the complicated nasal region.)

Shufeldt, R. W.
1909   Osteology of Birds. New York State Mus. Bull. No. 130. (Includes a list of the author's many papers up to this date. Shufeldt's papers dealt with a vast array of birds in a manner that was for the most part purely descriptive.)

Simonetta, A. M.
1960   On the Mechanical Implications of the Avian Skull and Their Bearing on the Evolution and Classification of Birds. Quart. Rev. Biol., 35(3): 206-220.

Simpson, G. G.
1946   Fossil Penguins. Bull. Amer. Mus. Nat. Hist., 87: 1-99. (The relationships of fossil forms are very convincingly worked out through the osteology of Recent species.)

Sushkin, P. P.
1905   Zur Morphologie des Vogelskelets. Vergleichende Osteologie der Normalen Tagraubvögel (Accipitres) und die Fragen der Classification. Teil 1, Grundeinteilung der Accipitres. Teil 2, Falken und Ihre Nächsten Verwandten. Nouv. Mem. Soc. Nat. Moscou, 16: 1-247. (An important, detailed study of the skeleton and its application to the classification of the diurnal birds of prey, the Accipitridae and Falconidae.)

Sy, M.
1936   Funktionell-anatomische Untersuchungen am Vogelflügel. Jour. f. Ornith., 84: 199-296. (The essential paper for understanding the mechanics of a bird's wing.)

Tiemeier, O. W.
1941   Repaired Bone Injuries in Birds. Auk, 58: 350-359. (Over four percent of some 6,000 skeletons of wild birds examined showed mending as a result of injuries. Careful analysis of an unusual survey.)
1950   The Os Opticus of Birds. Jour. Morph., 86: 25-46.

Tordoff, H. B.
1954   A Systematic Study of the Avian Family Fringillidae. Based on the Structure of the Skull. Univ. Michigan Mus. Zool. Misc. Publ. No. 81: 1-41. (An important approach to an understanding of the relationships of the Fringillidae and closely allied families.)

Verheyen, R.
1958   Analyse du Potentiel Morphologique et Projet d'une Nouvelle Classification des Charadriiformes. Bull. Inst. Royal Sci. Nat. Belgique, Vol. 34, No. 18: 1-35. (One of many similar

papers by this author using skeletal features and other characters for the classification of birds.)

Woolfenden, G. E.
   1961 Postcranial Osteology of the Waterfowl. Bull. Florida State Mus., 6 (1): 1-129.
Zusi, R. L.
   1962 Structural Adaptations of the Head and Neck in the Black Skimmer *Rynchops nigra* Linnaeus. Publ. Nuttall Ornith. Club, No. 3, Cambridge, Massachusetts.
   1967 The Role of the Depressor Mandibulae Muscle in Kinesis of the Avian Skull. Proc. U. S. Natl. Mus., 123 (3607): 1-28. (Discusses functional properties of the jaw articulation.)
Zusi, R. L., and R. W. Storer
   1969 Osteology and Myology of the Head and Neck of the Pied-billed Grebes *(Podilymbus)*. Univ. Michigan Mus. Zool. Misc. Publ. No. 139: 1-49. (Includes drawings of skull and neck vertebrae with processes labeled and muscle attachments indicated.)

MUSCULATURE
by Andrew J. Berger

In spite of a number of significant studies on the anatomy and function of avian muscles, there is still a great deal to be learned. Many more comparative studies are necessary if one is to know how much variation occurs in the muscle pattern and in the relative development of muscles within a family. Such data are prerequisite to a better understanding of the significance of differences in the muscles among the various families of birds. Much additional information is needed on both the wing musculature and on Garrod's leg-muscle formulas as expanded by Hudson (1937) and Berger (1959). Several of the papers listed below will clearly demonstrate how research on muscle homologies, adaptations, and functions serves to further a better understanding of phylogenetic relationships, as well as of locomotion and other body activities.

Banzhaf, W.
   1929 Die Vorderextremität von *Öpisthocomus cristatus* (Vieillot). Zeits. Morph. Ökol. Tiere, 16: 113-233.
Beecher, W. J.
   1950 Convergent Evolution in the American Orioles. Wilson Bull., 62: 51-86.
   1951a Adaptations for Food-getting in the American Blackbirds. Auk, 68: 411-440.
   1951b Convergence in the Coerebidae. Wilson Bull., 63: 274-287.
   1953 A Phylogeny of the Oscines. Auk, 70: 270-333. (The above four studies use dietary adaptations in jaw musculature as a means of determining evolutionary lines and relationships in closely allied species.)
Berger, A. J.
   1952 The Comparative Functional Morphology of the Pelvic Appendage in Three Genera of Cuculidae. Amer. Midland Nat., 47: 513-605.
   1953 On the Locomotor Anatomy of the Blue Coua, *Coua caerulea*. Auk, 70: 49-83.
   1954 The Myology of the Pectoral Appendage of Three Genera of American Cuckoos. Univ. Michigan Mus. Zool. Misc. Publ. No. 85: 1-35.
   1955 Suggestions Regarding Alcoholic Specimens and Skeletons of Birds. Auk, 72: 300-303.
   1956a The Expansor Secundariorum Muscle, with Special Reference to Passerine Birds. Jour. Morph., 99: 137-168.
   1956b Anatomical Variation and Avian Anatomy. Condor, 58: 433-441.
   1957 On the Anatomy and Relationships of *Fregilupus varius*, an Extinct Starling from the Mascarene Islands. Bull. Amer. Mus. Nat. Hist., 113: 225-272.
   1959 Leg-muscle Formulae and Systematics. Wilson Bull., 71: 93-94.
   1960a Some Anatomical Characters of the Cuculidae and the Musophagidae. Wilson Bull., 72: 60-104.
   1960b The Musculature. In *Biology and Comparative Physiology of Birds*. Volume 1. Edited by A. J. Marshall. Academic Press, New York.

1968 Appendicular Myology of Kirtland's Warbler. Auk, 85: 594-616.
1969 Appendicular Myology of Passerine Birds. Wilson Bull., 81: 220-223.

Boas, J. E. V.
1929 Biologisch-anatomische Studien über den Hals der Vögel. Danske Videnskabs Selskabets Skrifter, Naturvidensk. og Methem. Afd., 9, Raekke, I, 3: 105-222. (Covers both the myology and osteology of many different birds in a comprehensive manner.)

Bock, W. J., and W. DeW. Miller
1959 The Scansorial Foot of the Woodpeckers, with Comments on the Evolution of Perching and Climbing Feet in Birds. Amer. Mus. Novitates No. 1931.

Buri, R. O.
1900 Zur Anatomie des Flügels von Micropus melba und einigen anderen Coracornithes, zugleich Beitrag zur Kenntnis der systematischen Stellung der Cypselidae. Jena Zeit. Naturwiss., 23: 361-610. (An excellent study of the shoulder muscles of a swift, with comparative analysis concerning systematic relationships.)

Burt, W. H.
1930 Adaptive Modifications in the Woodpeckers. Univ. California Publ. in Zool., 32: 455-524. (Contains considerable information on myology, also excellent illustrated dissection of the limb and tongue muscles of the Pileated Woodpecker, *Dryocopus pileatus*.)

Fiedler, W.
1951 Beiträge zur Morphologie der Kiefermuskulatur des Oscines. Zool. Jahrb., Abt. Anat. Ontog. Tiere, 71: 235-288. (A classic study of the jaw muscles in the oscines.)

Fisher, H. I.
1946 Adaptations and Comparative Anatomy of the Locomotor Apparatus of New World Vultures. Amer. Midland Nat., 35: 545-727. (An excellent example of comparative work on all the genera of one family, in this case the Cathartidae.)
1957 The Function of M. depressor caudae and M. caudofemoralis in Pigeons. Auk, 74: 479-486.
1966 Hatching and the Hatching Muscle in Some North American Ducks. Trans. Illinois State Acad. Sci., 59: 305-325.

Fisher, H. I., and D. C. Goodman
1955 The Myology of the Whooping Crane, *Grus americana*. Illinois Biol. Monogr., 24, No. 2. (An excellent study of all the muscles of this species.)

Forbes, W. A.
1885 The Collected Scientific Papers of the Late William Alexander Forbes. Edited by F. E. Beddard. R. H. Porter, London. (Contains all the works of this British anatomist.)

Fürbringer, M.
1885 Ueber Deutung und Nomenklatur der Muskulatur des Vogelflügels. Morph. Jahrb., 11: 122-125. (Homologies of the wing musculature.)
1888 Untersuchungen zur Morphologie und Systematik der Vögel, zugleich ein Beitrag zur Anatomie der Stütz- und Bewegungsorgane. Two volumes. Van Holkema, Amsterdam. (The first thorough presentation of the anatomy of birds as a basis for classification.)
1902 Zur vergleichenden Anatomie des Brustschulterapparates und der Schultermuskeln. Jena Zeit. Naturwiss., 36: 289-736. (A classic study of the muscles of the shoulder and arm in birds.)

Gadow, H.
1882 Observations in Comparative Myology. Jour. Anat. and Physiol., 1882: 493-514. (One section deals with homologies of the hind limb musculature.)

Garrod, A. H.
1873- On Certain Muscles of the Thigh of Birds, and Their Value in Classification. I and II.
74 Proc. Zool. Soc. London, 1873: 626-664; 1874: 111-123. (The classic work on thigh muscle formulas.)
1881 The Collected Scientific Papers of the late Alfred Henry Garrod. Edited by W. A. Forbes. R. H. Porter, London. (A valuable collection of the works of this pioneering British anatomist who first proposed the use of leg-muscle formulas.)

Gaunt, A. S.
1969 Myology of the Leg in Swallows. Auk, 86: 41-53.

George, J. C., and A. J. Berger
1966 Avian Myology. Academic Press, New York. (Summarizes available information on the histochemistry and gross anatomy of the avian muscular system, and points up where additional studies are needed.)

Goodman, D. C., and H. I. Fisher
  1962 Functional Anatomy of the Feeding Apparatus in Waterfowl. Aves: Anatidae. Southern
        Illinois University Press, Carbondale. (Deals particularly with jaw musculature.)
Hartman, F. A.
  1961 Locomotor Mechanisms of Birds. Smithsonian Misc. Coll., 143: 1-91.
Holmes, E. B.
  1963 Variation in the Muscles and Nerves of the Leg in Two Genera of Grouse (Tympanuchus
        and Pedioecetes). Univ. Kansas Publ. Mus. Nat. Hist., 12: 363-474.
Howell, A. B.
  1937 Morphogenesis of the Shoulder Architecture: Aves. Auk, 54: 364-375. (A valuable work
        which homologizes avian shoulder muscles with those of other vertebrates on the basis
        of innervation; based on the Domestic Fowl, *Gallus*.)
  1938 Muscles of the Avian Hip and Thigh. Auk, 55: 71-81. (Like the above work, based on
        Gallus.).
Hudson, G. E.
  1937 Studies on the Muscles of the Pelvic Appendage in Birds. Amer. Midland Nat., 18: 1-108.
        (An important work, based on dissections of muscles in 16 bird orders.)
  1948 Studies on the Muscles of the Pelvic Appendage in Birds II: The Heterogeneous Order
        Falconiformes. Amer. Midland Nat., 39: 102-127. (Contains comparisons of the Secretary-
        bird, *Sagittarius*, and several types of falconiform birds from North America.)
Hudson, G.   and P. J. Lanzillotti
  1955 Gross Anatomy of the Wing Muscles in the Family Corvidae. Amer. Midland Nat., 53:
        1-44.
  1964 Muscles of the Pectoral Limb in Galliform Birds. Amer. Midland Nat., 71: 1-113.
Hudson, G. E., P. J. Lanzillotti, and G. D. Edwards
  1959 Muscles of the Pelvic Limb in Galliform Birds. Amer. Midland Nat., 61: 1-67.
Hudson, G. E., R. A. Parker, J. Vanden Berge, and P. J. Lanzillotti
  1966 A Numerical Analysis of the Modifications of the Appendicular Muscles in Various
        Genera of Gallinaceous Birds. Amer. Midland Nat., 76: 1-73.
Hudson, G. E., S. Y. C. Wang, and E. E. Provost
  1965 Ontogeny of the Supernumerary Sesamoids in the Leg Muscles of the Ring-necked
        Pheasant. Auk, 82: 427-437.
Kuroda, N.
  1962 On the Cervical Muscles of Birds. Misc. Rept. Yamashina's Inst. for Ornith. and Zool., 3:
        189-211.
Lakjer, T.
  1926 Studien über die Trigeminus-versorgte Kaumuskulatur der Sauropsiden. C. A. Reitzel,
        Copenhagen. (An excellent treatise on the homologies of the jaw musculature in birds
        and reptiles.)
Merz, R. L.
  1963 Jaw Musculature of the Mourning and White-winged Doves. Univ. Kansas Publ. Mus.
        Nat. Hist., 12: 521-551.
Mitchell, P. C.
  1894 On the Perforated Flexor Muscles in Some Birds. Proc. Zool. Soc. London, 1894: 495-498.
        (About the relationship between flexores perforatus and ambiens.)
  1901 On the Anatomy of Gruiform Birds; with Special Reference to the Correlation of Mod-
        ifications. Proc. Zool. Soc. London, 1901 (2): 629-655. (Largely myological.)
  1913 The Peroneal Muscles in Birds. Proc. Zool. Soc. London, 1913: 1039-1072. (A good treatise
        on the peroneus longus and p. brevis in different groups of birds.)
Mudge, G. P.
  1903 On the Myology of the Tongue of Parrots, with a Classification of the Order, Based upon
        the Structure of the Tongue. Trans. Zool. Soc. London, 16: 211-278.
Müller, J.
  1878 On Certain Variations in the Vocal Organs of the Passeres That Have hitherto Escaped
        Notice. Oxford University Press, London.
Owre, O. T.
        Adaptations for Locomotion and Feeding in the Anhinga and the Double-crested Cormor-
        ant. Amer. Ornith. Union, Ornith. Monogr. No. 6. (Deals extensively with myological
        adaptations for feeding under water.)

Palmgren, P.
1949 Zur Biologischen Anatomie der Halsmuskulatur der Singvögel. In *Ornithologie als Biologische Wissenschaft . . . Erwin Stresemann.* Carl Winter, Heidelberg.

Richardson, F.
1942 Adaptive Modifications for Tree-trunk Foraging in Birds. Univ. California Publ. in Zool., 46: 317-368.

Romer, A. S.
1927 The Development of the Thigh Musculature of the Chick. Jour. Morph., 43: 347-385. (Contains much valuable material on homologies and innervation.)

Scharnke, H.
1931 Beiträge zur Morphologie und Entwicklungsgeschichte der Zunge der Trochilidae, Meliphagidae und Picidae. Jour. f. Ornith., 79: 425-491. (An important study of the structure and development of the tongue in hummingbirds, honey-eaters, and woodpeckers.)

Shufeldt, R. W.
1890 The Myology of the Raven *(Corvus corax sinuatus).* Macmillan and Company, London. (The first book to deal with all the muscles of a given species. Because of errors and omissions, it is unreliable for the beginning student; see Hudson and Lanzillotti, 1955).

Stallcup, W. B.
1954 Myology and Serology of the Avian Family Fringillidae: A Taxonomic Study. Univ. Kansas Publ. Mus. Nat. Hist., 8: 157-211. (Deals in detail with the myology of the pelvic appendage.)

Steinbacher, J.
1957 Über den Zungenapparat einiger Neotropischer Spechte. Senckenbergiana Biol., 38: 259-270. (A study of the tongue in neotropical woodpeckers.)

Sullivan, G. E.
1962 Anatomy and Embryology of the Wing Musculature of the Domestic Fowl *(Gallus).* Australian Jour. Zool., 10: 458-518.
1967 Abnormalities of the Muscular Anatomy in the Shoulder Region of Paralysed Chick Embryos. Australian Jour. Zool., 15: 911-940.

Swinebroad, J.
1954 A Comparative Study of the Wing Myology of Certain Passerines. Amer. Midland Nat., 51: 488-514.

Weymouth, R. D., R. C. Lasiewski, and A. J. Berger
1964 The Tongue Apparatus in Hummingbirds. Acta Anat., 58: 252-270.

Wilcox, H. H.
1952 The Pelvic Musculature of the Loon, *Gavia immer.* Amer. Midland Nat., 48: 513-573.

Zusi, R. L.
1959 The Function of the Depressor Mandibulae Muscle in Certain Passerine Birds. Auk, 76: 537-539.
1962 Structural Adaptations of the Head and Neck in the Black Skimmer, *Rynchops nigra* Linnaeus. Publ. Nuttall Ornith. Club, No. 3, Cambridge, Massachusetts.

Zusi, R. L., and R. W. Storer
1969 Osteology and Myology of the Head and Neck of the Pied-billed Grebes *(Podilymbus).* Univ. Michigan Mus. Zool. Misc. Publ. No. 139: 1-49.

THE OIL GLAND

In birds, as in reptiles, skin glands of any kind are rare; sweat and sebaceous glands, so common in mammals, are completely lacking. The only integumentary gland of any prominence or known importance in birds is the **oil gland,** sometimes referred to as the **uropygial** or **preen gland,** situated dorsally at the base of the tail and concealed by contour feathers. Most of the following discussion is based on a summary of the literature by Elder (1954).

The oil gland secretes a substance containing much fatty acid plus some fat and wax. This the bird smears on its bill and head plumage; then rubs it off on the various feathers over the body and on the wings. Secretion in the gland is probably stimulated, through a reflex mechanism, by the act of preening.

Generally the gland is a relatively large structure, terminating in a nipple-like opening. Depending on the species of bird, there may be one to eight outlets (Grassé, 1950) and there may be a tiny cluster of slender feathers around the nipple which lengthen the nipple into a brush for dispensing the secretion. The gland reaches its largest size in aquatic birds. Among parrots and doves, it may be absent in some species and in others show various stages in development, from the rudimentary to the fully functional. It is entirely wanting in ostriches, rheas, emus, and cassowaries and in certain species of the orders Galliformes, Gruiformes, Caprimulgiformes, and Apodiformes.

Elder (1954) conducted experiments on the function of the oil gland in ducks and arrived at a number of conclusions. Several are given below more or less verbatim. The secretion maintains the water-repellent quality of feathers either directly or by preserving their physical structure. Without this secretion the feathers lose much of their efficiency in their normal functions as a flight mechanism and as a heat-insulating medium. It seems unlikely that a bird rendered glandless could survive in the wild. Degenerative plumage changes following removal of the glands are more pronounced in waterfowl than in chickens (Domestic Fowl) and more pronounced in chickens than in pigeons. This seems reasonable in view of relative gland size and probable need for "waterproofing." The secretion is used to anoint the bill and maintain its surface structure and gloss; without the secretion the bill becomes dry and shows some sloughing.

Elder, W. H.
  1954   The Oil Gland of Birds. Wilson Bull., 66: 6-31. (Includes an extensive bibliography on the subject.)
Grassé, P. -P.
  1950   La Glande Uropygienne. In *Traité de Zoologie*. Volume 15. Masson et Cie, Paris.

DIGESTIVE SYSTEM

Mouth Cavity. Birds may have as many as seven different pairs of **salivary glands** (see Farner, 1960). In general, such glands are well developed in birds which eat seeds and other vegetable food that require moistening, and are reduced in aquatic birds whose food is pre-moistened; they are sometimes completely absent in the Pelecaniformes (Portmann, 1950). Besides lubricating the mouth with mucous and perhaps secreting a digestive enzyme (ptyalin), they may occasionally have special functions. In woodpeckers a pair of large salivary glands, whose ducts enter the floor of the mouth, coats the tongue with a sticky fluid to assist that organ in retrieving insects. Gray Jays *(Perisoreus canadensis)* have similarly large salivary glands in the same position to coat the tongue (Bock, 1961). The copious mucous in this case is presumably used by the bird in making a bolus of food so sticky that it will adhere to any surface where the bird may choose to store it (Dow, 1965). Some species of swifts have salivary glands that become especially enlarged in the breeding season, producing mucous for nest construction (see page 337). The Black Swift *(Cypseloides niger)*, however, shows no such modifications since it nests in situations where saliva is unnecessary (Johnston, 1961).

Most birds have no soft palate, but pigeons at least show a softer condition of the posterior palate, enabling them to swallow successive boluses of water, as hoofed mammals do, without lifting the head.

The Rosy Finch *(Leucosticte tephrocotis)* has an extension of the mouth cavity in the form of two well-developed sacs or pouches, each with a separate opening on either side of the tongue and glottis (Miller, 1941). During the nesting season they use the pouches for carrying large quantities of food to the young, which are usually in high barren places, such as lofty cliffs, far from the food supply. The Pine Grosbeak *(Pinicola eneucleator)* has pouches apparently identical with those of the Rosy Finch (French, 1954). Such structures are unusual; most birds bear food to their young either in the mouth (by distending the floor), in the esophagus, or in the crop.

Modifications of the Esophagus. The esophagus is a distensible tube, lined with mucous epithelium. Though unmodified along its course in most birds, it nevertheless serves in all birds as a temporary reservoir for food. Geese may fill it so greatly from pharynx to stomach as to make the neck bulge perceptibly. Occasionally the esophagus will have a simple dilation, as in cormorants, or be considerably dilated ventrally, as in diurnal birds of prey. The Common Redpoll *(Acanthis flammea)* has a single enlargement which extends "laterally around the right of the vertebral column and forms a lesser enlargement on the left side of the column" (Fisher and Dater, 1961). Similarly, enlargements occur in other seed-eating birds. Only in a few groups of birds—e.g., gallinaceous birds, pigeons, and parakeets—does the esophagus show a true crop, an outpocketing of the ventral wall with a more or less constricted connection. The crop of pigeons is characteristically bi-lobed.

Any dilation of the esophagus, or any crop, is primarily for food storage. While ptyalin from the mouth may act upon food in the crop slightly and mucous and the muscular action of the crop's wall may soften the food partially, the crop actually plays a very minor part in digestion.

In pigeons during the breeding season the crop lining of both sexes produces, by the proliferation and sloughing of its epithelial cells, a white substance ("pigeon's milk") rich in fats and proteins. After being mixed with food received by the crop, the substance is regurgitated into the mouths of the young birds.

The Greater Prairie Chicken *(Tympanuchus cupido)* and possibly a few other species have the anterior end of the esophagus modified as a vocal sac or resonating chamber. In the Heath Hen, the recently extinct eastern subspecies of the Greater Prairie Chicken, Gross (1928) found that air is forced into the sac directly from the trachea by way of the pharynx. As the sac is inflated, sound waves from the syrinx strike the tense walls, which act as resonators.

Stomach. The proventriculus provides the gastric juice (mainly mucous, pepsin, and hydrochloric acid); in this respect it corresponds closely with the fundus region of the mammalian stomach. In point of development, however, the proventriculus is believed to be a differentiation peculiar to birds, whereas the other division of the stomach, the gizzard, is a reptilian legacy. The waxy, musky oil which petrels eject when disturbed is a secretion from glands of the proventriculus (Matthews, 1949).

The gizzard tends to be very strongly muscular in birds that eat seeds and herbage, much less so in birds that subsist chiefly on fruits and animal flesh. This difference in muscularity depends on the amount of mechanical action required to macerate the particular type of food. Mucous glands are present in the gizzard walls. In the highly muscular gizzard of seed-eaters these glands secrete a horny lining with ridges to facilitate the process of maceration. In the less muscular gizzard of fruit-eaters the glands secrete only mucous and the gizzard itself has the appearance inside and out of being simply a larger adjunct to the proventriculus.

In addition to preparing food for digestion, the gizzard serves as the principal barrier to indigestible materials such as feathers, fur, bones, animal shells, and so on. These are usually ejected via the esophagus and mouth; certain groups of birds, notably hawks and owls, eliminate them in spindle-shaped wads called pellets. Pebbles, normally present in most gizzards as an aid to maceration, may eventually be discharged through the esophagus and mouth or, if worn down sufficiently, voided by way of the intestinal tract.

Liver. The two-lobed liver, the largest visceral organ of the body, is larger in birds than in mammals of equivalent body size. Its main contribution to the digestive process is the secretion of bile into the small intestine by way of two bile ducts. In most birds a reservoir for bile, the gall bladder, is present along the course of the duct from the larger (right) lobe. The gall bladder is particularly large in penguins and plantain-eaters (Musophagidae). Only in pigeons and relatively few other groups of birds—ostriches, rheas, parakeets, and hummingbirds—is it absent.

Pancreas. This organ is relatively large in birds; as a rule, it is larger per unit of body weight in small birds than in large (Sturkie, 1965). Apparently the avian pancreas once consisted of three lobes, one dorsal and two ventral; though they are now united to form one glandular mass, their ducts still persist separately. Through these ducts important digestive ferments or enzymes (e.g., amylase, trypsin, and lipase) are secreted by the pancreas into the small intestine. In what was probably the dorsal lobe are the islets of Langerhans which secrete the hormone, insulin.

Small Intestine. This is the principal organ for the digestion and absorption of food. Here food is received from the stomach, where it has been mixed with, and perhaps slightly changed by, gastric juices and has been extensively macerated. Once in the intestine it undergoes complete chemical alteration by the action of enzymes accumulated from the mouth, proventriculus, and pancreas, by bile from the liver, and by hydrochloric acid from the proventriculus. Possibly the intestine itself contributes enzymes, but they have not yet been satisfactorily determined (Sturkie, 1965). Subjected to the peristaltic movements of the intestinal wall, the intestinal contents are mixed thoroughly and moved along. Gradually the end-products of this digestive process are absorbed by the intestinal mucosa and reach the blood stream. The process of absorption is practically completed by the time the remainder of the intestinal contents reaches the large intestine.

The intestine is generally longer in seed- and herbage-eaters than in flesh-eaters, for the food received is more bulky in proportion to its nutritive content; the intestine thus needs more space to accommodate it and more inner surface for absorbing the end-products of digestion. In some groups of birds—ostriches, tinamous, waterfowl (swans, geese, and ducks), and gallinaceous birds, all primarily vegetarians—the intestinal caeca are greatly elongated. Not only do they serve to increase absorptive surface, but they assist digestion by lodging high concentrations of bacteria which reduce cellulose.

Rate of Digestion. The physiology of avian digestion, which includes both physical processes (taking and swallowing of food, storing food in crop and esophagus, and macerating food in the gizzard) and chemical processes (alteration of food by the action of juices from the mouth, proventriculus, liver, pancreas, and intestine, and by the action of bacteria), has yet to be thoroughly studied. Basically it is the same as in mammals, though the rate of digestion is probably more rapid. In the Domestic Fowl *(Gallus)* grain has been found to pass through the digestive tract in 2.5 to 12 hours, depending on the type and amount of food and the physiological state of the individual (Sturkie, 1965). Ordinarily a bird digests animal food more rapidly. For example, a magpie *(Pica)* may digest a mouse in three hours (Hewitt, 1948).

Beams, H. W., and R. K. Meyer
 1931   The Formation of Pigeon "Milk." Physiol. Zool., 4: 486-500.

Bhaduri, J. L., and B. Biswas
  1947   Caeca of Some Indian Birds. Jour. Bombay Nat. Hist. Soc., 46: 645-649. (Information based on 52 species belonging to 23 families.)
Bock, W. J.
  1961   Salivary Glands in the Gray Jays *(Perisoreus).* Auk, 78: 355-365.
Dow, D. D.
  1965   The Role of Saliva in Food Storage by the Gray Jay. Auk, 82: 139-154.
Farner, D. S.
  1960   Digestion and the Digestive System. In *Biology and Comparative Physiology of Birds.* Volume 1. Edited by A. J. Marshall. Academic Press, New York.
Fisher, H. I., and E. E. Dater
  1961   Esophageal Diverticula in the Redpoll, *Acanthis flammea.* Auk, 78: 528-531.
French, N. R.
  1954   Notes on Breeding Activities and on Gular Sacs in the Pine Grosbeak. Condor, 56: 83-85.
Gier, L. J., and O. Grounds
  1944   Histological Study of the Digestive System of the English Sparrow. Auk, 61: 241-243.
Gross, A. O.
  1928   The Heath Hen. Mem. Boston Soc. Nat. Hist., 6: 487-588.
Hewitt, E. A.
  1948   Digestion. In *Diseases of Poultry.* Edited by H. E. Biester and L. H. Schwarte. Second edition. Iowa State University, Ames.
Hickey, J. J., and H. Elias
  1954   The Structure of the Liver of Birds. Auk, 71: 458-462.
Johnston, D. W.
  1958   Sex and Age Characters and Salivary Glands of the Chimney Swift. Condor, 60: 73-84. (Tentatively suggested that "the *modus operandi* for the enlargement and possible secretion of the glands is mediated via hormones, namely, the combined actions of testosterone, thyroxine, and hormone(s) from the pituitary gland.")
  1961   Salivary Glands in the Black Swift. Condor, 63: 338. (The salivary glands are smaller than in some other swifts.)
Leopold, A. S.
  1953   Intestinal Morphology of Gallinaceous Birds in Relation to Food Habits. Jour. Wildlife Management, 17: 197-203.
Lucas, F. A.
  1897   The Tongues of Birds. Rept. U. S. Natl. Mus. for 1895, pp. 1003-1020.
Matthews, L. H.
  1949   The Origin of Stomach Oil in the Petrels, with Comparative Observations on the Avian Proventriculus. Ibis, 91: 373-392.
Miller, A. H.
  1941   The Buccal Food-carrying Pouches of the Rosy Finch. Condor, 43: 72-73.
Portmann, A.
  1950   Le Tube Digestif. In *Traité de Zoologie.* Edited by P. -P. Grassé. Volume 15. Masson et Cie, Paris.
Stevenson, J.
  1933   Experiments on the Digestion of Food by Birds. Wilson Bull., 45: 155-167.
Sturkie, P. D.
  1965   Avian Physiology. Second edition. Cornell University Press. Ithaca, New York. (Chapters 10, 11, and 12.)

RESPIRATORY SYSTEM

The Syrinx and Its Function. The syrinx or voice organ is situated at a point where the trachea divides into the two bronchi and involves the last three treacheal rings (see the drawings, page 83). Between groups, even species, of birds, the syrinx varies in form and the number of muscles acting upon it. The "higher" families of passerine birds, which include most of the notable singers, may have as many as seven to nine pairs of syringeal muscles, while nearly all the other birds have fewer, often only two pairs. In general, birds possessing more muscles are able to produce a wider variety of notes than those having fewer muscles (see Miskimen, 1951).

Within a species there may be sexual differences in the syrinx. In females that do not sing the syrinx is less well developed (Portmann, 1950). When neither sex sings, but gives different calls, the syrinx may show differences in structure (Gullion, 1950).

Sound is produced, according to well-documented investigations by Greenewalt (1968), in each bronchus, thereby enabling the bird to give two notes or phrases simultaneously. Herewith is a partial extract of some of his findings.

The sound arises in each bronchus when the internal tympaniform membrane—on being forced into the passageway of the bronchus by the pressure of air within the surrounding interclavicular sac and subjected to opposing tension by the syringeal muscles —constricts the passage such that air from the lungs causes the membrane to vibrate as it flows past. This produces the sound waves or song. An increase in the tension of the membrane causes simultaneously an increase in the frequency of the sound waves (i.e., the pitch) and the diameter of the constricted passage. As the diameter of the passage enlarges from a very small value, the amplitude (the volume or range) of the sound waves increases; thus an increase in frequency comes in association with amplitude. As the diameter of the passage enlarges further with continued tension on the membrane, a point is reached when the flow of air no longer causes the membrane to vibrate and the amplitude begins decreasing. When the passage is fully opened, air passes through without producing sound. During its production by the tympaniform membranes the sound may be modified by the external labium lying opposite.

Whistled ("sinusoidal") song, such as given by a thrush, is produced when the tympaniform membranes vibrate rather slowly and evenly; song with harmonics or overtones such as given by a nuthatch is when the membranes vibrate rather rapidly in ripples. Some birds may give both sounds. For instance, a chickadee's clear *fee-bee* is whistled song, its familiar *dee-dee-dee* is a series of harmonics.

Song, whether whistled or with harmonics, comes from two separate and independent sources, one in each bronchus. One source may operate without the other—i.e., tension may be applied to the membrane in one bronchus but not to the membrane in the corresponding bronchus. Or both sources may operate at the same time, tension being applied to the membranes in both bronchi. Using oscillograms of different songs, Greenewalt convincingly demonstrated that a bird's song or call note—indeed its whole vocal repertoire—is dual in origin. The sound that one hears comprises the overlapping of two different sounds.

Modifications of the Trachea. Among different groups of birds are cases where the trachea is elaborated. A particularly unusual case occurs in two diverse groups of birds, the cranes and swans, represented by the Whooping Crane *(Grus americana)* and the Trumpeter Swan *(Cygnus buccinator)*. Through parallel evolution, each has developed an enormously long trachea that is coiled and fitted into the bony mass of the sternum. A somewhat similar case appears in the bird-of-paradise, *Phonygammus keraudrenii,* in which an exceedingly elongated trachea is coiled in the breast region between the skin and the big flight muscles. The males of many ducks have the trachea dilated to form a bulbous area, either along its course or where it bifurcates to form the bronchi.

Most ornithologists have long accepted the assumption that the trachea in all birds modifies the sound produced in the syrinx and that its elaborations, such as those just described, further modify the sound, either by increasing the volume of the sound by resonance or by altering the quality of the tone. After studying the syringeotracheal system of sound production in geese which shows no special elaborations, Sutherland and McChesney (1965) likened the effect of the trachea to that of an open pipe as in a trumpet. Other ornithologists, from their investigations of systems without elaborations, have likened its effect to that of a closed or reed pipe. Greenewalt (1968), however, has presented strong evidence that the trachea plays no role in sound modulation and rejects the function attributed to any of its elaborations. Why then the elaborations? Their function or the cause of their development calls for renewed investigation.

The Function of the Air Sacs. Air sacs play a multifarious role in the lives of birds. Among the many functions credited to them, the following seem the most acceptable:

1. Air sacs moisten air, a function that the bird's greatly reduced nasal cavities cannot effectively perform.

2. Air sacs provide complete ventilation, an advantage to the bird's flying mechanisms. With each inspiration and each expiration, air sweeps over the lung's respiratory surfaces, providing a "double tide" of air (see pages 84-85 for the anatomical arrangement in the lung permitting this), whereas in the mammal there is always incomplete ventilation, there being ever present in the lung a considerable residuum of vitiated air.

3. Air sacs play a role in the regulation of body temperature by permitting loss of heat through radiation and vaporization. Thus the bird, with its feather-covered, dry skin—without the sweat glands of the mammal—is able to cool itself by internal radiation and "prespiration."

4. Air sacs serve in an excretory capacity by being the principal outlets for all waste body fluids that are eliminated by vaporization.

5. The interclavicular air sac functions in sound production by effecting air pressure on the syrinx.

6. The cervical air sacs in certain species are elaborated for display purposes (e. g., in a frigatebird, *Fregata*) or sound production (in the Brown Jay, *Psilorhinus morio;* see Sutton and Gilbert, 1942).

7. In pelicans (e. g., Brown Pelican, *Pelecanus occidentalis),* and possibly in a few other water-plunging species, the interclavicular sac is connected to specially developed subcutaneous air sacs which act as a cushion to reduce the force of impact with the water when the bird is diving from a height.

The concept that air sacs lessen specific gravity of a bird by harboring a great amount of warm air and making it more buoyant seems to have no basis in fact. There is actually not enough warm air in a bird's air-sac-system to cause any material effect.

Mechanics of Respiration. Certain muscles that act in breathing also move the wings. A considerable controversy exists as to whether the movements are synchronized, performing the two functions simultaneously. According to Zimmer's (1935) experiments on a crow, and Tomlinson and McKinnon's (1957) experiments on the Common Pigeon, the bird breathes a complete cycle with every wing-beat. But Lord, Bellrose, and Cochran (1962) found that Mallards *(Anas platyrhynchos)* breathe every other wing-beat and may make three wing-beats per breath. Tomlinson (1963) has since shown that, among birds heavier and larger than the Common Pigeon, breathing is actually less ordered, with the number of beats per breath tending to increase. Thus current evidence points away from complete synchronization. This would probably be deemed correct by Salt and Zeuthen (1960) who feel that a bird must breathe independently of its flight movements.

Akester, A. R.
    1960   The Comparative Anatomy of the Respiratory Pathways in the Domestic Fowl *(Gallus domesticus),* Pigeon *(Columba livia),* and Domestic Duck *(Anas platyrhyncha).* Jour. Anat. London, 94:487-505.

Beard, E. B.
  1951   The Trachea of the Hooded Merganser. Wilson Bull., 63: 296-301.
Beddard, F. E.
  1898   The Structure and Classification of Birds. Longmans, Green, and Company, London. (Contains a good review of tracheal structure in various birds.)
Blake, C. H.
  1958   Respiration Rates. Bird-Banding, 29: 38-40.
Chamberlain, D. R., and others
  1968   Syringeal Anatomy in the Common Crow. Auk, 85: 244-252.
Cover, M. S.
  1953   Gross and Microscopic Anatomy of the Respiratory System of the Turkey. III. The Air Sacs. Amer. Jour. Vet. Res., 14: 239-245.
Delphia, J. M.
  1961   Early Development of the Secondary Bronchi in the House Sparrow *Passer domesticus* (Linnaeus). Amer. Midland Nat., 65: 44-59.
Gier, H. T.
  1952   The Air Sacs of the Loon. Auk, 69: 40-49.
Gilbert, P. W.
  1939   The Avian Lung and Air-sac System. Auk, 56: 57-63.
Greenewalt, C. H.
  1968   Bird Song: Acoustics and Physiology. Smithsonian Institution Press, Washington, D. C.
  1969   How Birds Sing. Scientific Amer., 221 (5): 126-139. (Largely a summary of the book cited above.)
Gullion, G. W.
  1950   Voice Differences between Sexes in the American Coot. Condor, 52: 272-273.
Hazelhoff, E. H.
  1951   Structure and Function of the Lung of Birds. Poultry Sci., 30: 3-10.
Humphrey, P. S.
  1958   The Trachea of the Hawaiian Goose. Condor, 60: 303-307.
King, A. S.
  1956   The Structure and Function of the Respiratory Pathways of *Gallus domesticus*. Vet. Rec., 68: 544-547.
Lord, R. D., Jr., F. C. Bellrose, and W. W. Cochran
  1962   Radiotelemetry of the Respiration of a Flying Duck. Science, 137 (3523): 39-40.
McLeod, W. M., and R. P. Wagers
  1939   The Respiratory System of the Chicken. Jour. Amer. Vet. Med. Assoc., 95: 59-70.
Miller, A. H.
  1963   The Vocal Apparatus of Two South American Owls. Condor, 65: 440-441. (This is the final paper in a series by the author on the syringes of owls. His earlier papers are cited.)
Miller, W. DeW.
  1926   Structural Variations in the Scotors. Amer. Mus. Novitates No. 243: 1-5. (Deals in part with the trachea and its use in systematics.)
Miskimen, M.
  1951   Sound Production in Passerine Birds. Auk, 68: 493-504.
  1963   The Syrinx in Certain Tyrant Flycatchers. Auk, 80: 156-165.
Müller, B.
  1908   The Air-sacs of the Pigeon. Smithsonian Misc. Coll., 50: 365-414.
Myers, J. A.
  1917   Studies on the Syrinx of *Gallus domesticus*. Jour. Morph., 29: 165-215.
Portmann, A.
  1950   Les Organes Respiratoires. In *Traité de Zoologie*. Edited by P.-P. Grassé, Volume 15. Masson et Cie, Paris.
Richardson, F.
  1939   Functional Aspects of the Pneumatic System of the California Brown Pelican. Condor, 41: 13-17.
Rüppell, W.
  1933   Physiologie und Akustik der Vögelstimme. Jour. f. Ornith., 81: 433-542.

Salt, G. W., and E. Zeuthen
  1960   The Respiratory System. In *Biology and Comparative Physiology of Birds*. Volume 1. Edited by A. J. Marshall. Academic Press, New York.
Sturkie, P. D.
  1965   Avian Physiology. Second edition. Cornell University Press, Ithaca, New York.
Sutherland, C. A., and D. S. McChesney
  1965   Sound Production in Two Species of Geese. Living Bird, 4: 99-106.
Sutton, G. M., and P. W. Gilbert
  1942   The Brown Jay's Furcular Pouch. Condor, 44: 160-165.
Tomlinson, J. T.
  1963   Breathing of Birds in Flight. Condor, 65: 514-516.
Tomlinson, J. T., and R. S. McKinnon
  1957   Pigeon Wing-beats Synchronized with Breathing. Condor, 59: 401.
Tompsett, D. H.
  1957   Casts of the Pulmonary System of Birds. Ibis, 99: 614-620.
Wetherbee, D. K.
  1951   Air-sacs in the English Sparrow. Auk, 68: 242-244.
Zeuthen, E.
  1942   The Ventilation of the Respiratory Tract in Birds. Det Kgl. Danske Vid. Selsk., Biol. Medd., 17 (1): 1-50. (An extensive account of several physiologic aspects of respiration, mainly in the Domestic Fowl. Written in English.)
Zimmer, K.
  1935   Beiträge zur Mechanik der Atmung bei den Vögeln in Stand und Flug. Zool. Stuttgart, 33: 1-69.

CIRCULATORY SYSTEM

Blood Cells and Their Numbers. Birds share with mammals the distinction of having the "richest" blood—i. e., blood with the highest number of erythrocytes, or red blood cells, per unit of measure. According to Stresemann (1927-34), the figures among different species of birds range from 1,500,000 to 5,500,000 in a cubic millimeter, in mammals from 2,000,000 to 18,000,000. In man the count is 4,500,000 for adult females, 5,000,000 for adult males. The number of erythrocytes tends to be higher in small birds than in large. Thus, Nice, Nice, and Kraft (1935) found that the number in 15 species of passerine birds averaged from 4,200,000 to 6,055,000; the lowest figure, 3,930,000, was for a Tufted Titmouse *(Parus bicolor)*, the highest, 7,645,000, for a Slate-colored Junco *(Junco hyemalis)*. Portmann (1950) gives the number of erythrocytes found in a hummingbird *(Chrysolampis elatis)* as 6,590,000. The figures published by these authorities actually extend Stresemann's range, previously mentioned. In contrast to the figures for small birds are the following for large birds, given by Ponder (1924): 1,620,000, Ostrich *(Struthio camelus)*; 2,189,000, White Stork *(Ciconia ciconia)*; 2,547,000, Peregrine Falcon *(Falco peregrinus)*. As a rule, when the erythrocytes are higher in number, they are smaller and have more hemoglobin (the oxygen-carrying pigment). Such a condition, characteristic of smaller birds, indicates greater efficiency with respect to the oxygen-carrying capacity of the blood.

Within a bird species the number of erythrocytes is usually higher in the male. Domm, Taber, and Davis (1943) found the normal male Domestic Fowl to have 3,600,000 and the female 2,700,000—a difference of about a million. They concluded that gonadal hormones were responsible for this difference. Their experimental evidence showed that the higher erythrocyte number in the male is due to the presence of androgen and the lower number in the female is caused either by a lack in androgen or by the presence of estrogen. Within a species there may also be a seasonal variation in the number of erythrocytes. Counts reported by Riddle and Braucher (1934) in the Common Pigeon and by

Young (1937) in the Canary *(Serinus)* show that the higher figures appear in the cooler months of the year.

The Heart Rate. The rate of the heart beat, long considered an indicator of the physiological activity in animals, is higher in most birds than in mammals of equivalent size; moreover, it is higher in small birds than in larger ones (Sturkie, 1965). The rate in birds varies widely, being affected by such features as muscular movements, mental activity, feeding, and air temperature. Captive birds show a lower rate than wild birds (Odum, 1941), and probably the same condition occurs in domesticated birds. Woodbury and Hamilton (1937) found the average rate (i. e., heart beats per minute) of the adult Common Pigeon at rest to be 221. Much higher rates in smaller birds were found by Odum, who captured wild specimens and made measurements while the birds were kept undisturbed and in darkness. Some of his figures on adult male specimens were: 614, Ruby-throated Hummingbird *(Archilochus colubris)*; 522, Black-capped Chickadee *(Parus atricapillus)*; 455, House Wren *(Troglodytes aedon)*; 480, Yellow Warbler *(Dendroica petechia)*; 305, House Sparrow; 391, Cardinal *(Richmondena cardinalis)*. For comparison, about 72 times per minute is the rate of heart beat of the normal human adult at rest. While a bird is incubating, the rate of heart beat is apparently higher than when it is resting on a perch. In the House Wren, for example, Odum determined the rate to be from 550 to 650 on the nest during the day and as high as 701 on the nest late at night. This difference between day and night rates is correlated with changes in air temperature, the higher rate occurring late at night as the air temperature dips to its lowest point.

Heart Size. Birds as a group exhibit relatively larger hearts than mammals (Clark, 1927), while among different bird species the size of the heart varies inversely with the size of the body. Thus small birds have proportionately larger hearts. Hummingbirds, the smallest and probably the most active of all birds, have by far the largest hearts (Hartman, 1954). Species of birds which fly less frequently than others have correspondingly smaller hearts (Johnston, 1963). Heart size may therefore be just as important an indicator of physiological activity in birds as heart rate (see Brush, 1966).

Heart size may vary within bird genera and species. Thus species and subspecies in northern regions appear to have proportionately larger hearts than their counterparts in more southerly latitudes (Johnston, 1963; Hartman, 1955); and there is evidence that populations of a species living at high altitudes show an increase in heart size relative to body size (Rensch, 1931; Norris and Williamson, 1955).

Bartsch, P., and others
   1937   Size of Red Blood Corpuscles and Their Nucleus in Fifty North American Birds. Auk, 54: 516-519.
Brush, A. H.
   1966   Avian Heart Size and Cardiovascular Performance. Auk, 83: 266-273.
Clark, A. J.
   1927   Comparative Physiology of the Heart. Macmillan Company, New York.
Davies, F.
   1930   The Conducting System of the Bird's Heart. Jour. Anat. London, 64: 129-146.
Domm, L. V., E. Taber, and D. E. Davis
   1943   Comparison of Erythrocyte Numbers in Normal and Hormone-treated Brown Leghorn Fowl. Proc. Soc. Exp. Biol. and Med., 52: 49-50.
Fisher, H. I.
   1955   Major Arteries near the Heart in the Whooping Crane. Condor, 57: 286-289.
Garrod, A. H.
   1873   On the Carotid Arteries of Birds. Proc. Zool. Soc. London, 1873: 457-472. (A classic work on the subject.)

Glenny, F. H.
1953 A Systematic Study of the Main Arteries in the Region of the Heart. Aves XIX. Apodi-formes, Part 1. Ohio Jour. Sci., 53: 367-369. (One of a long series of papers by the author on his studies, chiefly of the carotid arteries, in different groups of birds. The information is mainly descriptive. For a listing of Glenny's papers, see his *Antarctica as a Center of Origin of Birds*, Ohio. Jour. Sci., 54: 307-314, 1954.)
1955 Modifications of Pattern in the Aortic Arch System of Birds and Their Phylogenetic Significance. Proc. U. S. Natl. Mus., 104 (3346): 525-621.

Groebbels, F.
1932 Der Vögel. Band 1: Atmungswelt und Nahrungswelt. Verlag von Gebrüder Borntraeger, Berlin.

Hartman, F. A.
1954 Cardiac and Pectoral Muscles of Trochilids. Auk, 71: 467-469.
1955 Heart Weight in Birds. Condor, 57: 221-238.

Hartman, F. A., and M. A. Lessler
1963 Erythrocyte Measurements in Birds. Auk, 80: 467-473. (Measurements in 124 species among 46 families in Panama and the United States.)

Jenkinson, M. A.
1964 Thoracic and Coracoid Arteries in Two Families of Birds, Columbidae and Hirundinidae. Univ. Kansas Publ. Mus. Nat. Hist., 12: 553-573.

Johnston, D. W.
1963 Heart Weights of Some Alaskan Birds. Wilson Bull., 75: 435-446. (Weights from 563 individuals representing 77 species.)

Kern, A.
1926 Das Vogelherz. Untersuchungen an *Gallus domesticus* Briss. Morph. Jahrb., 56: 264-315.

Lucas, A. M., and C. Jamroz
1961 Atlas of Avian Hematology. U. S. Dept. Agric., Agric. Monogr. 25. (Concerns mainly the Domestic Fowl.

Nice, L. B., M. M. Nice, and R. M. Kraft
1935 Erythrocytes and Hemoglobin in the Blood of Some American Birds. Wilson Bull., 47: 120-124.

Norris, R. A.
1963 A Preliminary Study of Avian Blood Groups with Special Reference to the Passeriformes. Bull. Tall Timbers Res. Sta. No. 4.

Norris, R. A., and F. S. L. Williamson
1955 Variation in Relative Heart Size of Certain Passerines with Increase in Altitude. Wilson Bull., 67: 78-83.

Odum, E. P.
1941 Variations in the Heart Rate of Birds: A Study in Physiological Ecology. Ecol. Monogr., 3: 299-326.

Olson, C., Jr.
1948 Avian Hematology. In *Diseases of Poultry*. Second edition. Edited by H. E. Biester and L. W. Schwarte. Iowa State College Press, Ames.

Ponder, E.
1924 Erythrocytes and the Action of Simple Haemolysis. Oliver and Boyd, Edinburgh. (Cited by Nice, Nice, and Kraft, 1935.)

Portmann, A.
1950 Les Organes de la Circulation Sanguine. In *Traité de Zoologie*. Edited by P.-P. Grassé. Volume 15. Masson et Cie, Paris.

Powell, J. R., and J. D. Burke
1966 Avian Blood Oxygen Capacity. Amer. Midland Nat., 75: 425-431. (In Common Pigeon and 25 species of wild birds. Smaller birds maintain a higher metabolic rate by virtue of their higher blood oxygen capacity per gram of body weight.)

Rensch, B.
1931 Der Einfluss des Tropenklimas auf den Vögel. Proc. VIIth Internatl. Ornith. Congr., pp. 197-205.

Riddle, O., and P. F. Braucher
1934 Hemoglobin and Erythrocyte Differences According to Sex and Season in Doves and Pigeons. Amer. Jour. Physiol., 108: 554-566.

Simons, J. R.
  1960   The Blood-vascular System. In *Biology and Comparative Physiology of Birds*. Volume 1. Edited by A. J. Marshall. Academic Press, New York.
Spanner, R.
  1925   Der Pfortaderkreislauf in der Vogelniere. Jahrb. Morph. Mikrosk. Anat., 54: 560-632.
Stresemann, E.
  1927- Aves. In *Kükenthal u. Krumbach, Handbuch der Zoologie*. Volume 7, Part 2. De Gruyter,
  34   Berlin.
Sturkie, P. D.
  1965   Avian Physiology. Second edition. Cornell University Press, Ithaca, New York. (Chapters 1-5.)
Williamson, F. S. L., and R. A. Norris
  1958   Data on Relative Heart Size of the Warbling Vireo and Other Passerines from High Altitudes. Wilson Bull., 70: 90-91. (Data on 73 specimens representing 24 species.)
Woodbury, R. A., and W. F. Hamilton
  1937   Blood Pressure Studies in Small Animals. Amer. Jour. Physiol., 119: 663. (Cited by Sturkie, 1965.)
Young, M. D.
  1937   Erythrocyte Counts and Hemoglobin Concentration in Normal Female Canaries. Jour. Parasitology, 23: 424-426.

BODY TEMPERATURE

The body temperature of passerine birds varies normally between 102.0 degrees F and 112.3 degrees F under natural conditions (Kendeigh, 1934). The Common Pigeon has a temperature roughly between 105.0 degrees F and 108.4 degrees F (Groebbels, 1932). In large birds, especially those of lower orders, the limits of variation are correspondingly lower, probably between 100.0 degrees F and 106.0 degrees F. According to data gathered by Portmann (1950) on body temperatures of birds, a kiwi *(Apteryx)* has the lowest temperature, namely 100.0 degrees F. Other large birds listed as having low temperatures are the Ostrich *(Struthio camelus)*, 104.0 degrees F; a penguin *(Aptenodytes)*, 100.7 degrees F; an albatross *(Diomedea)*, 105.3 degrees F; and the Mute Swan *(Cygnus olor)*, 105.8 degress F. There is no bird whose temperature is regularly as low as 98.6 degrees F, the normal temperature for an adult human being.

In their study of the body temperature in passerine birds, Baldwin and Kendeigh (1932) concluded that the normal temperature is highly variable. Such factors as emotional excitement, muscular activity, high air temperature, and the digestion of food cause a bird's temperature to rise, while low air temperature and lack of food bring about a decrease in temperature. The same two investigators found a daily rhythm in body temperature. Briefly, the temperature rises gradually during the morning from the beginning of the day's activities until the middle of the day; it decreases during the late afternoon. When the bird settles on the nest for the night, the temperature first falls rapidly, then decreases gradually until about midnight. Thereafter the body temperature fluctuates more or less until the short period, just before the bird leaves the nest for the first time in the morning, when there is a rapid rise in body temperature.

Adult birds, being warm-blooded (homoiothermal), are able to remain active under varying environmental conditions. By temperature-control mechanisms centered in the brain, they can maintain their normal body temperature even when the surrounding air temperature undergoes marked fluctuations. If the weather becomes extremely cold, they can conserve their body heat by lifting up or "fluffing" the feathers, thereby widening the insulating layer of air between feathers and skin; they can increase their body heat by stimulating muscle activity through shivering. On the other hand, if it gets

;mely warm, they can press the feathers against the skin to eliminate the insulat-
layer of air and thus facilitate reduction of their body heat by radiation; they can
.at—i.e., increase respiratory movements and consequently increase the amount of
internal radiation of heat and vaporization ("perspiration") through the lungs and air
sacs. Some birds such as pelicans, boobies, and nighthawks lower body heat by "gular
flutter"—fluttering their gular area—instead of panting.

In the course of their early development, all birds pass through a cold-blooded (poikilo-
thermal) stage, in which the temperature-control mechanisms are lacking. In precocial
birds the stage is completed in the egg; by hatching time, temperature control is already
partially established. A day-old chick of the Domestic Fowl has an average body temperature of 103.4 degrees F, which is about 3 degrees F below that of the adult (W. F.
Lamoreux and F. B. Hutt, *vide* Sturkie, 1965); at 20 days of age its temperature averages
the same as the adult's, and temperature control is fully established. Altricial young at
hatching time are still distinctly cold-blooded. In the House Wren *(Troglodytes aedon)*
temperature control is gradually established after hatching; when nine days old the
young bird obtains full temperature control (Baldwin and Kendeigh, 1928); thereafter
it is able to maintain an average temperature of 104.7 degrees F (Baldwin and Kendeigh,
1932), the same as that of the adult. When young birds, whether precocial or altricial,
have acquired temperature control they are able to tolerate moderate air temperatures,
but for a few days they may still require brooding or other form of protection if air temperatures are extreme.

Torpidity in Birds. A number of different kinds of birds—e. g., goatsuckers, swifts,
and hummingbirds—enter a torpid or hibernating state during which the body temperature is lowered to a level near that of the air, breathing is sometimes indiscernible,
and reactions to handling are slow, if they occur at all. Their general physiological condition often closely resembles hibernation in certain mammals, especially bats. Jaeger
(1949) made observations on a Poor-will *(Phalaenoptilus nuttallii)* in profound torpidity
for a known winter period of 85 days. Its body temperature, taken on five different days,
ranged from 64.4 degrees F to 67.6 degrees F, while the surrounding air temperature on
the same days ranged from 63.5 degrees F to 75.3 degrees F. During this winter period
insects—the bird's usual food—were scarce, thus suggesting the possibility that torpidity, at least in the case of this species, is an adaptation for tiding the bird over the
period of the year when food is unobtainable.

Baldwin, S. P., and S. C. Kendeigh
   1928   Development of Temperature Control in Nestling House Wrens. Amer. Nat., 62: 249-278.
   1932   Physiology of the Temperature of Birds. Sci. Publ. Cleveland. Mus. Nat. Hist., 3: i-x; 1-196.
Barth, E. K.
   1949   Kroppstemperatur hos Fugler og Pattedyr. Fauna och Flora, 1949: 163-177. (See review
          in Bird-Banding, 21: 65, 1950.)
Bartholomew, G. A., and T. J. Cade
   1957   The Body Temperature of the American Kestrel, *Falco sparverius.* Wilson Bull., 69: 149-
          154. (Evaporative heat loss possible from the cere, and perhaps the cornea.)
Bartholomew, G. A., and W. R. Dawson
   1952   Body Temperatures in Nestling Western Gulls. Condor, 54: 58-60.
   1958   Body Temperatures in California and Gambel's Quail. Auk, 75: 150-156. (A conspicuous
          diurnal cycle of body temperature correlates with activity in both juveniles and adults at
          moderate environmental temperatures.)
Bartholomew, G. A., T. R. Howell, and T. J. Cade
   1957   Torpidity in the White-throated Swift, Anna Hummingbird, and Poor-will. Condor, 59:
          145-155.

Bartholomew, G. A., J. W. Hudson, and T. R. Howell
  1962 Body Temperature, Oxygen Consumption, Evaporative Water Loss, and Heart Rate in the Poor-will. Condor, 64: 117-125.
Brauner, J.
  1952 Reactions of Poor-wills to Light and Temperature. Condor, 54: 152-159.
Brenner, F. J.
  1965 Metabolism and Survival Time of Grouped Starlings at Various Temperatures. Wilson Bull., 77: 388-395. (Starling enters no torpid state. Habit of roosting together may have evolved to reduce heat loss in cold weather.)
Dawson, W. R., and H. B. Tordoff
  1959 Relation of Oxygen Consumption to Temperature in the Evening Grosbeak. Condor, 61: 388-396. (A study to determine how small birds cope with extreme cold in nature.)
  1964 Relation of Oxygen Consumption to Temperature in the Red and White-winged Crossbills. Auk, 81: 26-35. (A continuation of authors' 1959 study, this time using crossbills.)
Eklund, C. R.
  1942 Body Temperatures of Antarctic Birds. Auk, 59: 544-548.
Farner, D. S., and D. L. Serventy
  1959 Body Temperature and the Ontogeny of Thermoregulation in the Slender-billed Shearwater. Condor, 61: 426-433. (Adults in burrows have lower temperature than those on surface due "to quiet burrow life.")
French, N. R., and R. W. Hodges
  1959 Torpidity in Cave-roosting Hummingbirds. Condor, 61: 223.
Groebbels, F.
  1932 Der Vögel. Band 1: Atmungswelt und Nahrungswelt. Verlag von Gebrüder Borntraeger, Berlin.
Heath, J. E.
  1962 Temperature Fluctuation in the Turkey Vulture. Condor, 64: 234-235. (An example of lowered body temperature normally occurring in a large bird.)
Howell, T. R.
  1959 A Field Study of Temperature Regulation in Young Least Terns and Common Nighthawks. Wilson Bull., 71: 19-32.
  1961 An Early Reference to Torpidity in a Tropical Swift. Condor, 63: 505.
Howell, T. R., and G. A. Bartholomew
  1959 Further Experiments on Torpidity in the Poor-will. Condor, 61: 180-185.
  1961a Temperature Regulation in Nesting Bonin Island Petrels, Wedge-tailed Shearwaters, and Christmas Island Shearwaters. Auk, 78: 343-354.
  1961b Temperature Regulation in Laysan and Black-footed Albatrosses. Condor, 63: 185-197. (Young birds effect heat loss by exposing vascularized foot webbing to the air.)
  1962a Temperature Regulation in the Red-tailed Tropic Bird and the Red-footed Booby. Condor, 64: 6-18. (The importance of panting in the tropicbird and gular flutter in the booby is emphasized.)
  1962b Temperature Regulation in the Sooty Tern *Sterna fuscata*. Ibis, 104: 98-105. (Chicks able to lose heat by vigorous panting and seeking shade.)
Huxley, J. S., C. S. Webb, and A. T. Best
  1939 Temporary Poikilothermy in Birds. Nature, 143: 683-684.
Jaeger, E. C.
  1949 Further Observations on the Hibernation of the Poor-will. Condor, 51: 105-109.
Kendeigh, S. C.
  1934 The Role of the Environment in the Life of Birds. Ecol. Monogr., 4: 299-417.
  1945 Resistance to Hunger in Birds. Jour. Wildlife Management, 9: 217-226.
  1961 Energy of Birds Conserved by Roosting in Cavities. Wilson Bull., 73: 140-147.
Lasiewski, R. C.
  1963 Oxygen Consumption of Torpid, Resting, Active, and Flying Hummingbirds. Physiol. Zool., 36: 122-140.
McAtee, W. L.
  1947 Torpidity in Birds. Amer. Midland Nat., 38: 191-206. (Swallows included in list of birds showing torpidity.)
Marshall, J. T., Jr.
  1955 Hibernation in Captive Goatsuckers. Condor, 57: 129-134.

Morrison, P.
    1962   Modification of Body Temperature by Activity in Brazilian Hummingbirds. Condor, 64: 315-323. (Tropical hummingbirds appear to be more sensitive to the lowering of the body temperature than do temperate species.)
Odum, E. P.
    1942   Muscle Tremors and the Development of Temperature Regulation in Birds. Amer. Jour. Physiol., 136: 618-622.
Pearson, O. P.
    1950   The Metabolism of Hummingbirds. Condor, 52: 145-152.
    1954   The Daily Energy Requirements of a Wild Anna Hummingbird. Condor, 56: 317-322.
    1960   Torpidity in Birds. Bull. Mus. Comp. Zool., 124: 93-103.
Portmann, A.
    1950   La Température du Corps et l'Homéothermie. In *Traité de Zoologie*. Edited by P.-P. Grassé. Volume 15. Masson et Cie, Paris.
Robard, S.
    1950   Weight and Body Temperature. Science, 111: 465-466.
Sladen, W. J. L., J. C. Boyd, and J. M. Pedersen
    1966   Biotelemetry Studies on Penguin Body Temperatures. Antarctic Jour. U. S., 1: 142-143.
Stoner, D.
    1937   Records of Bird Temperatures. New York State Mus. Circ. 19.
Sturkie, P. D.
    1965   Avian Physiology. Second edition. Cornell University Press, Ithaca, New York. (Chapter 8.)
Udvardy, M. D. F.
    1951   Heat Resistance in Birds. Proc. Xth Internatl. Ornith. Congr., pp. 595-599.
    1963   Data on the Body Temperature of Tropical Sea and Water Birds. Auk, 80: 191-194.
Warren, J. W.
    1960   Temperature Fluctuation in the Smooth-billed Ani. Condor, 62: 293-294.

UROGENITAL SYSTEM

Kidneys and Urine Elimination

The avian kidneys are relatively larger than those of reptiles and mammals, ranging from one to two percent of the body weight, and are more voluminous in aquatic birds (Benoit, 1950). Structurally they are more reptilian, since they show no demarcation between cortex and medulla as in mammals. The glomeruli, located in the cortical area, range in number from 30,000 in small passerine birds to 200,000 in the Domestic Fowl; in general, they are small and more numerous than those in mammals. The functional unit of the kidneys, the uriniferous tubule or nephron, closely resembles the mammal's.

Besides regulating the salts and liquids of the body, the kidneys eliminate the end products of protein metabolism. The urine which the nephron produces by filtration consists chiefly of water but also contains excess salts in solution and nitrogenous wastes from protein metabolism. In bird urine, as in reptile urine, the nitrogenous wastes are in the form of uric acid, a practically insoluble substance of high nitrogen concentration, whereas in mammal urine the wastes comprise urea, which has half the concentration.

Once the urine is in the cloaca some of its water content is absorbed by the cloaca. The residue is distinctly whitish, semi-solid, and usually encapsulated with a mucous material. Eventually it is voided, along with the fecal material, but nevertheless remains recognizable by its color.

About 98 percent of the water filtered by the kidney is reabsorbed either in the kidney or the cloaca (Sturkie, 1965). In this way the bird conserves water for further use. Thus the bird's urine when voided lacks the watery consistency of the mammal's. Actually the bird's body loses more water through vaporization, probably through the lungs and air sacs, than through the urine.

Reproductive System

Many studies have been made on the reproductive system of domestic birds, comparatively few on wild birds. Most of the data in the ensuing paragraphs, unless otherwise indicated, are drawn from studies on the Domestic Fowl as summarized by Sturkie (1965).

Ovary and Ova. The avian ovary consists of two parts: outer cortex and inner medulla. The cortex contains the ova, enormous in number and only a few of which ever reach maturity. In one ovary as many as 1,906 ova have been counted with the unaided eye and an additional 12,000 under the microscope. Each ovum develops within a follicle of its own. At the time of ovulation (i.e., the release of the ovum from the follicle) a follicle may have a diameter of 40 mm. Following ovulation the ovum begins its passage through the oviduct, taking about 4½ hours to reach the uterus and remaining in the uterus 18 to 20 hours. During the last 5 hours of its stay in the uterus it receives its coloration (pigment). From the uterus the egg is forced by contraction through the vagina, cloaca, and vent; it is not held for any length of time in either the vagina or cloaca.

Testes and Spermatozoa. Birds' testes have a regular shape, which is usually oval or ellipsoid. Within each testis are the seminiferous tubules, lined with epithelium in which the spermatozoa are produced. In birds the testis is not divided into lobes as it is in mammals; instead, the interior of the testis is one chamber in which the tubules form a complicated network.

In the majority of birds the seminal vesicles, which store spermatozoa, are in the body cavity, but in some passerine birds during the breeding season they take up a position in a cloacal protuberance (see Appendix A) outside the body cavity. Here the body temperature is lower, as in a mammalian scrotum. Possibly the spermatozoa ultimately mature in the seminal vesicles of the protuberance, in which case the lower temperature might assist normal spermatogenesis (Wolfson, 1954).

The spermatozoa in different birds vary considerably in form; unlike those of mammals, most have elongated heads and extremely long, whip-like tails with undulating membranes. There are generally said to be two types of spermatozoa in birds: sauropsidian and passeriform (Benoit, 1950). The first, common to both reptiles and all birds except passerine species, has a more or less curved head; the second, found in passerine species only, has a spiraled head.

The number of spermatozoa in bird semen from a single ejaculation during copulation has been estimated to range from 1,700,000,000 to 3,500,000,000. In man the number from one ejaculation is usually estimated at 300,000,000. By their own motility, the spermatozoa pass from the cloaca up the oviduct to the magnum or isthmus where, presumably, the ovum is fertilized. On the average, the time between copulation and fertilization is about 72 hours, but it may be as short as 19.5 hours.

Asymmetry in the Reproductive System

Asymmetry in the reproductive system of the bird has long been an intriguing subject for comment and investigation, and its probable cause has aroused much speculation. Though the male often shows asymmetries with respect to the size and location of the gonads, it is the female which has the more marked variations, ranging from almost perfect bilateral symmetry in a few birds (e. g., hawks) to a unilateral condition in most birds in which the right ovary is missing and the right oviduct reduced to a mere vestige. Witschi (1935a), who studied the embryonic development of gonadic asymmetry in the Domestic Fowl, House Sparrow, and Red-winged Blackbird *(Agelaius phoeniceus)*, found the germ cells evenly distributed on both the right and left sides up to the end of the third day of incubation, but by the end of the fourth day asymmetry was established, the germ cells from the right side having migrated to the left. He concluded that asymmetry in the bird is due to a genetically inherited deficiency or inability on the part of the ovarian cortex to attract and keep the germ cells. The same author has put forward the provocative idea that "asymmetry in the female oviducts might have evolved as an adaptation to the aviatic life habits of the birds, while the more or less complete reduction of the right ovary followed later, in order that eggs may more safely reach the one remaining fallopian tube (oviduct)."

Benoit, J.
    1950   Organes Uro-génitaux. In *Traité de Zoologie*. Edited by P.-P. Grassé. Volume 15. Masson et Cie, Paris.
Blanchard, B. D.
    1941   The White-crowned Sparrows *(Zonotrichia leucophrys)* of the Pacific Seaboard: Environment and Annual Cycle. Univ. California Publ. in Zool., 46: 1-178. (Has a good account of the testes in various stages of development.)
Dominic, C. J.
    1960   On the Secretory Activity of the Funnel of the Avian Oviduct. Current Science, 29: 274-275.
Fitzpatrick, F. L.
    1930   Bilateral Ovaries in Cooper's Hawk, with Notes on Kidney Structure. Anat. Rec., 46: 381-383.
Huber, G. C.
    1916   A Note on the Morphology of the Seminiferous Tubules of Birds. Anat. Rec., 11: 177-180.
Retzius, G.
    1909   Die Spermien der Vögel. Biol. Untersuch., 14: 89-122.
Romanoff, A. L., and A. J. Romanoff
    1949   The Avian Egg. John Wiley and Sons, New York. (Chapter 4 has an extensive description of the female reproductive system of the Domestic Fowl.)
Sperber, I.
    1960   Excretion. In *Biology and Comparative Physiology of Birds*. Volume 1. Edited by A. J. Marshall. Academic Press, New York.
Stanley, A. J.
    1937   Sexual Dimorphism in North American Hawks. I. Sex Organs. Jour. Morph., 61: 321-339.
Sturkie, P. D.
    1965   Avian Physiology. Second edition. Cornell University Press, Ithaca, New York. (Chapters 13, 15, and 16.)
Watson, A.
    1919   A Study of the Seasonal Changes in Avian Testes. Jour. Physiol., 53: 86-91.
Witschi, E.
    1935a   Origin of Asymmetry in the Reproductive System of Birds. Amer. Jour. Anat., 56: 119-141.
    1935b   Seasonal Sex Characters in Birds and Their Hormonal Control. Wilson Bull., 47: 177-188. (Has illustrations of the House Sparrow's urogenital system in active and quiescent stages.)
Wolfson, A.
    1954   Sperm Storage at Lower-than-body Temperature outside the Body Cavity in Some Passerine Birds. Science, 120: 68-71.
    1960   The Ejaculate and the Nature of Coition in Some Passerine Birds. Ibis, 102: 124-125.

SALT GLANDS

Many aquatic birds have a large pair of multilobular glands on the skull above the orbits; each lies in a crescentric depression and sends a duct into the nasal cavity. Called the supraorbital or nasal glands, their function remained undetermined until 1957-1958 when Knut Schmidt-Nielsen and his co-workers made a surprising discovery. In the Double-crested Cormorant *(Phalacrocorax auritus)*, these glands excrete salt (sodium chloride) in concentrated solution (Schmidt-Nielsen, Jorgensen, and Osaki, 1958); in the Humboldt Penguin *(Spheniscus humboldti)*, they could perform the same function (Schmidt-Nielsen and Sladen, 1958). Since these initial observations it has become apparent that the glands—now more properly called salt glands—have a true excretory function in all marine birds as well as in other species which normally acquire large amounts of salt in their food or drinking water. Excess salt, picked up by the blood stream and transferred to the salt glands, is conveyed to the nasal cavity and out through the nostrils or (particularly in the case of pelecaniform birds with occluded nostrils) into the mouth cavity through the choana and thence to the tip of the bill. In 10 to 15 minutes after a bird feeds on a salty substance, a clear fluid appears at the nostrils or bill tip in large droplets, and the bird dispatches them by vigorously shaking its head.

Salt glands are subject to variation within a species depending on the amount of salt which particular individuals or populations normally acquire. Domestic ducks *(Anas platyrhynchos)* drinking only salt water developed larger glands than control birds which were allowed to drink only fresh water (Schmidt-Nielsen and Kim, 1964). The size of the gland therefore appears to be determined by individual adaptation rather than genetically.

Cooch, F. G.
  1964   A Preliminary Study of the Survival Value of a Functional Salt Gland in Prairie Anatidae. Auk, 81: 380-393. (Ducks and other water birds inhabiting the Great Plains have a salt gland that definitely has survival value for birds living in an alkaline environment.)
Schmidt-Nielsen, K., and R. Fange
  1958   The Function of the Salt Gland in the Brown Pelican. Auk, 75: 282-289.
Schmidt-Nielsen, K., C. B. Jorgensen, and H. Osaki
  1958   Extrarenal Salt Excretion in Birds. Amer. Jour. Physiol., 193: 101-107.
Schmidt-Nielsen, K., and Y. T. Kim
  1964   The Effect of Salt Intake on the Size and Function of the Salt Gland of Ducks. Auk, 81: 160-172.
Schmidt-Nielsen, ., and W. J. L. Sladen
  1958   Nasal Salt Secretion in the Humboldt Penguin. Nature, 181: 1217-1218.

ENDOCRINE ORGANS

The endocrine organs are small, ductless glands which pass their secretions (called hormones) into the blood or lymph. Once in the circulatory system, the hormones are taken to various parts of the body where they effect changes in the cells or tissues. The organs form no real "system" in the anatomical sense, as they are widely scattered over the body and sometimes far removed from structures on which they nonetheless exercise profound influence. Briefly outlined below are the principal endocrine organs of the bird, their better known hormones, and some of the more important functions attributed to them.

Pituitary Gland. Anterior Lobe, Gonadotrophic Hormones: Stimulate growth of ovarian follicles, growth of seminiferous tubules, production of spermatozoa, and production of gonadal hormones. Adrenocorticotrophic Hormone: Stimulates development of adrenal gland and controls function of adrenal cortex. Thyrotrophic Hormone: Stimulates secretion of the thyroid gland. Lactogenic Hormone or Prolactin: Stimulates production of "milk" in the pigeon's crop, possibly induces broodiness, and inhibits gonadal activity. Posterior Lobe. Vasopressin: Increases blood pressure. Oxytocin: Causes premature expulsion of eggs from oviduct. Pitressin: Retards flow of urine.

Adrenal Gland. Adrenalin (from the medullary portion): Stimulates body activity (e.g., increases blood pressure and heart rate) and inhibits digestion. Presumably the cortical portion of the adrenal gland provides hormones which influence carbohydrate and salt metabolism and activities of the kidney, liver, intestine, and other organs.

Thyroid Gland. Thyroxin: Stimulates body growth (by increasing metabolic rate); induces molting; increases growth rate of feathers; influences size and shape of feathers; affects coloration (small doses increase melanins, while heavy doses may inhibit coloration or cause decoloration).

Parathyroid Gland. Parathormone: Regulates calcium and phosphorus metabolism; may, in the laying female, synergize with estrogen (see below) in the mobilization of calcium for egg shells.

Pancreas. Insulin: Regulates carbohydrate metabolism.

Gonads. Androgen (from testis): Stimulates development of accessory reproductive organs, secondary sex characters (e.g., bright coloration, tarsal spurs, plumage for display purposes), and courtship behavior patterns; influences number of erythrocytes and amount of hemoglobin in blood; stimulates general body growth. Estrogen (from ovary): Stimulates development of accessory reproductive organs and incubation patch, secondary sex characters, and breeding behavior patterns; increases blood calcium, phosphorus, and lipids; depresses secretion of prolactin from pituitary gland.

For about forty-five years a significant line of research has been conducted on the role of light in stimulating the activity of avian gonads. William Rowan (1925) was primarily responsible for initiation of the work; he found that the inactive gonads of male Slate-colored Juncos *(Junco hyemalis)* could be stimulated in winter to produce spermatozoa by exposure of the birds to an additional several hours per day of (artificial) light. His work was later confirmed by others, notably by Bissonnette (1937) in experiments on the Starling *(Sturnus vulgaris)*. From these investigations has come the generalization that increased day length induces sexual activity prior to the nesting season. The gonadotrophic hormones from the pituitary gland are now believed to be the factors which stimulate gonadal activity, but just how light influences the pituitary to secrete the hormones has not been determined. Recently attention has been given to factors which bring about, after the nesting season, a so-called "refractory" period during which light cannot induce gonadal activity. In some way, apparently, gonadotrophic hormones from the pituitary are blocked. Bailey (1950) demonstrated that prolactin is capable of inhibiting light-induced gonadal activity and is probably one of the hormones responsible for the refractory period. Estrogen is probably another. The stimulus that causes the secretion of prolactin is yet to be found. Students interested in the work that has been done on photoperiodism in birds will find the reviews by Burger (1949), Wolfson (1952a), Höhn (1961), and Farner (1967) particularly instructive.

Bailey, R. E.
   1950   Inhibition with Prolactin of Light-induced Gonad Increase in White-crowned Sparrows. Condor, 52: 247-251.
   1952   The Incubation Patch of Passerine Birds. Condor, 54: 121-136.
Bartholomew, G. A., Jr.
   1949   The Effect of Light Intensity and Day Length on Reproduction in the English Sparrow. Bull. Mus. Comp. Zool., 101: 433-477.

Benoit, J.
  1950   Les Glandes Endocrines. In *Traité de Zoologie*. Edited by P.-P. Grassé. Volume 15. Masson et Cie, Paris.
Bissonnette, T. H.
  1937   Photoperiodicity in Birds. Wilson Bull., 49: 241-270.
Boss, W. R.
  1943   Hormonal Determination of Adult Characters and Sex Behavior in Herring Gulls *(Larus argentatus)*. Jour. Exp. Zool., 94: 181-203.
Burger, J. W.
  1949   A Review of Experimental Investigations on Seasonal Reproduction in Birds. Wilson Bull., 61: 211-230.
Davis, J., and B. S. Davis
  1954   The Annual Gonad and Thyroid Cycles of the English Sparrow in Southern California. Condor, 56: 328-345.
Dawson, W. R., and J. M. Allen
  1960   Thyroid Activity in Nestling Vesper Sparrows. Condor, 62: 403-405.
Domm, L. V.
  1955   Recent Advances in Knowledge Concerning the Role of Hormones in the Sex Differentiation of Birds. In *Recent Studies in Avian Biology*. Edited by A. Wolfson. University of Illinois Press, Urbana.
Farner, D. S.
  1967   The Control of Avian Reproductive Cycles. Proc. XIVth Internatl. Ornith. Congr., pp. 107-133. (Includes hypothalamus, anterior pituitary, and gonads.)
Farner, D. S., and L. R. Mewaldt
  1955   The Natural Termination of the Refractory Period in the White-crowned Sparrow. Condor, 57: 112-116.
Flickinger, D. D.
  1959   Adrenal Responses of California Quail Subjected to Various Physiologic Stimuli. Proc. Soc. Exp. Biol. and Med., 100: 23-25.
Forsyth, D.
  1908   The Comparative Anatomy, Gross and Minute, of the Thyroid and Parathyroid Glands in Mammals and Birds. Jour. Anat. London, 42: 141-169, 302-319. (Information on many kinds of birds.)
Greeley, F., and R. K. Meyer
  1953   Seasonal Variation in Testis-stimulating Activity of Male Pheasant Pituitary Glands. Auk, 70: 350-358.
Hartman, F. A., and R. H. Albertin
  1951   A Preliminary Study of the Avian Adrenal. Auk, 68: 202-209.
Hartman, F. A., and K. A. Brownell
  1961   Adrenal and Thyroid Weights in Birds. Auk, 78: 397-422.
Höhn, E. O.
  1950   Physiology of the Thyroid Gland in Birds: A Review. Ibis, 92: 464-473.
  1961   Endocrine Glands, Thymus and Pineal Body. In *Biology and Comparative Physiology of Birds*. Volume 2. Edited by A. J. Marshall. Academic Press, New York.
King, J. R., L. R. Mewaldt, and D. S. Farner
  1960   The Duration of Postnuptial Metabolic Refractoriness in the White-crowned Sparrow. Auk, 77: 89-92.
Kirkpatrick, C. M., H. E. Parker, and J. C. Rogler
  1962   Some Comparisons of Thyroid Glands of Bobwhites and Japanese Quail. Jour. Wildlife Management, 26: 172-177.
Knouff, R. A., and F. A. Hartman
  1951   A Microscopic Study of the Adrenal of the Brown Pelican. Anat. Rec., 109: 161-178.
Lofts, B.
  1962   Photoperiod and the Refractory Period of Reproduction in an Equatorial Bird, *Quelea quelea*. Ibis, 104: 407-414.
Marshall, A. J.
  1958   The Role of the Internal Rhythm of Reproduction in the Timing of Avian Breeding Seasons, Including Migration. Proc. XIIth Internatl. Ornith. Congr., pp. 475-482.

Miller, A. H.
    1948   The Refractory Period in Light-induced Reproductive Development of Golden-crowned Sparrows. Jour. Exp. Zool., 109: 1-11.
    1954   The Occurrence and Maintenance of the Refractory Period in Crowned Sparrows. Condor, 56: 13-20.
    1959   Response to Experimental Light Increments by Andean Sparrows from an Equatorial Area. Condor, 61: 344-347.

Miller, R. A., and O. Riddle
    1942   The Cytology of the Adrenal Cortex of Normal Pigeons and in Experimentally Induced Atrophy and Hypertrophy. Amer. Jour. Anat., 71: 311-341.

Payne, F.
    1942   The Cytology of the Anterior Pituitary of the Fowl. Biol. Bull., 82: 79-111.

Rahn, H., and B. T. Painter
    1941   A Comparative Histology of the Bird Pituitary. Anat. Rec., 79: 297-311.

Riddle, O., R. W. Bates, and E. L. Lahr
    1935   Prolactin Induces Broodiness in Fowl. Amer. Jour. Physiol., 111: 352-360.

Riddle, O., and P. Frey
    1925   The Growth and Age Involution of the Thymus in Male and Female Pigeons. Amer. Jour. Physiol., 71: 413-429.

Rowan, W.
    1925   Relation of Light to Bird Migration and Developmental Changes. Nature, 115: 494-495.
    1929   Experiments in Bird Migration. I. Manipulation of the Reproductive Cycle: Seasonal Histological Changes in the Gonads. Proc. Boston Soc. Nat. Hist., 39: 151-208.
    1938   Light and Seasonal Reproduction in Animals. Biol. Rev., 13: 374-402.

Sauer, F. C., and H. B. Latimer
    1931   Sex Differences in the Proportion of the Cortex and Medulla in the Chicken Suprarenal. Anat. Rec., 50: 289-295.

Schildmächer, H.
    1937   Histologische Untersuchungen an Vogelhypophysen. I. Die Zelltypen der Amsel, *Turdus merula* L. Jour. f. Ornith., 85: 586-592. (See Auk, 55: 307, 1938, for a review.)

Schooley, J. P., and O. Riddle
    1938   The Morphological Basis of Pituitary Function in Pigeons. Amer. Jour. Anat., 62: 313-349.

Sturkie, P. D.
    1965   Avian Physiology. Cornell University Press, Ithaca, New York. (Chapters 17, 18, 19, 20 and 21.)

Taber, E., D. E. Davis, and L. V. Domm
    1943   Effect of Sex Hormones on the Erythrocyte Number in the Blood of the Domestic Fowl. Amer. Jour. Physiol., 138: 479-487.

Wallin, H. E., and S. C. Kendeigh
    1966   Seasonal and Taxonomic Differences in the Size and Activity of the Thyroid Glands in Birds. Ohio Jour. Sci., 66: 369-379.

Wilson, A. C., and D. S. Farner
    1960   The Annual Cycle of Thyroid Activity in White-crowned Sparrows of Eastern Washington. Condor, 62: 414-425.

Wingstrand, K. G.
    1951   The Structure and Development of the Avian Pituitary from a Comparative and Functional Viewpoint. C. W. K. Gleerup, Lund.

Witschi, E.
    1935   Seasonal Sex Characters in Birds and Their Hormonal Control. Wilson Bull., 47: 177-188.

Wolfson, A.
    1945   The Role of the Pituitary, Fat Deposition, and Body Weight in Bird Migration. Condor, 47: 95-127.
    1952a   Day Length, Migration, and Breeding Cycles in Birds. Sci. Monthly, 74: 191-200.
    1952b   The Occurrence and Regulation of the Refractory Period in the Gonadal and Fat Cycles of the Junco. Jour. Exp. Zool., 121: 311-326.
    1954   Production of Repeated Gonadal, Fat, and Molt Cycles within One Year in the Junco and White-crowned Sparrow by Manipulation of Day Length. Jour. Exp. Zool., 125: 353-376.
    1958   Role of Light in the Photoperiodic Responses of Migratory Birds. Science, 129 (3360): 1425-1426.

SPECIAL SENSES AND THEIR ORGANS

Taste

Taste buds in birds, owing to the extensive horny covering of the tongue and palate, are restricted to the soft tissue at the base and sides of the tongue and to the softer region of the palate. Probably the number of taste buds seldom exceeds 100 in any species and averages about 40 in most species. Compared to man, whose taste buds approximate 9,000, birds are ill equipped to detect food by taste.

Birds vary widely in their response to different substances. Nectar-feeding species such as hummingbirds and many fruit-eating species show a strong preference for sugary substances, while insect- and seed-eating species are indifferent to anything particularly sweet. Birds which drink or obtain their food in salt or brackish water show considerable tolerance for salty food, and certain finches actually seek salt as part of their diet. By contrast, many species reject food or water with a strong salty taste. In much the same degree, some birds tolerate food with a sour or bitter taste while others avoid it.

The broad generalization can be made that all birds are able to detect certain substances by taste, but in selecting their food they depend more on its visual properties.

Smell

Practically all birds have olfactory organs that appear to be functional in detecting odors. Yet the evidence from observations on birds generally is contradictory. Some species are credited with a keen sense of smell, others with none whatever.

Kiwis (*Apteryx* spp.) no doubt have a sense of smell far superior to most birds. With nostrils (anterior nares) at the tip of the bill, large nasal cavities, and greatly elaborated posterior conchae (Portmann, 1950), they have an olfactory apparatus remarkably like a mammal's. Supposedly, this condition enables kiwis, nocturnal birds with notoriously poor vision (see below), to orient themselves by odors.

Vultures (Cathartidae) are almost certainly assisted at times in finding carrion by a sense of smell. Chapman (1929, 1938), testing the ability of the Turkey Vulture *(Cathartes aura)* to locate animal matter by smell, hid dead mammals completely from view at Barro Colorado Island (Panama) where this species occurs. As soon as the carcasses produced odors through advanced decay, numerous vultures arrived at the spot of concealment. When Chapman similarly hid decaying fish with strong odors, no birds showed up. Chapman concluded that the Turkey Vulture is dependent for its existence on a discriminating use of its senses of smell and sight. More recently Owre and Northington (1961) proved from tests that Turkey Vultures have at least some olfactory ability. There is also ample evidence, according to Bang (1960) and Stager (1964), that Turkey Vultures have a well-developed sense of smell.

Other birds have been tested for their sense of smell. Strong (1911) reported the Common Pigeon able to recognize the oil of bergamot by smell. Calvin, Williams, and Westmoreland (1957) found that pigeons could not be trained to odor alone as a cue in conditioned experiments, but Michelson (1959) reported that pigeons could be stimulated by an odor. Frings and Boyd (1952) noted the ability of the Bobwhite *(Colinus virginianus)* in captivity to distinguish between two feeders by odor and consequently develop a preference for one. In such birds as cormorants, whose nostrils are normally occluded on approaching adulthood, one might expect the loss of an olfactory sense, but observations indicate that the nasal cavities of these birds may receive odors by way of the mouth and choana (Portmann, 1950).

Vision

Vision reaches such a high state of development in birds that it is worth more than passing attention. Most of the data and interpretations in the paragraphs that follow are drawn from Walls' *The Vertebrate Eye*, a thorough treatise, to which a student should refer for further details and additional information.

Eye Glands. Lacrimal and Harderian glands vary in their size and amount of secretion in accordance with the habits of the bird. Aquatic birds have very small lacrimals, the reason being perhaps that their secretions for moistening the surface of the eyes are not greatly needed in a watery environment. On the other hand, Harderian glands are apt to be very large in marine birds (e. g., cormorants), for their thick oily secretions serve to protect the exposed surface of the eyes from the harmful effects of salt water.

Size and Shape of Birds' Eyes. The eyes of birds are notoriously large; they are the biggest structures of the head, and they often weigh more than the brain. Large hawks and owls have eyes whose size in diameter rivals a full-grown man's; ostriches, with an eye diameter of about two inches, have the largest eyes among birds and, reputedly, among all land vertebrates.

Avian eyes may be classified on the basis of shape as flat, globose, and tubular. Flat eyes, characteristic of pigeons and the majority of birds, including pigeons, have relatively slight convexities on their corneal surfaces, while the distance through the eye from cornea to retina is shorter than from dorsal to ventral walls—in other words, flat eyes are broader than deep. Globose and tubular eyes, found in birds of prey, have much greater corneal convexities; in globose eyes, either distance through them is about the same whereas in tubular eyes the distance from cornea to retina tends to be longer and the sides of the eyes are noticeably concave. Both globose and tubular eyes, with greater distances between the lens and retina, broaden and sharpen the image thrown on the retina. This affords better vision at longer distances—an adaptation with obvious advantages for a bird such as an eagle which often seeks food from great heights.

Accommodation. In the eye of any land vertebrate light waves from a given object are bent or refracted as they pass through the cornea and lens, and to a slight extent as they pass through the two humors. In order for the light waves to strike the retina in proper focus and thus form a satisfactory image of the object, the eye must make certain adjustments. This is called **accommodation**. In most birds, as well as in reptiles and mammals, accommodation is effected by changing the curvature of the lens—flatter for distant objects, more rounded for near objects; in a few birds accommodation is assisted by changing the curvature of the cornea. The amount of range of accommodation of which an eye is capable is expressed in diopters, one diopter being equal to the reciprocal of the focal length in meters. A lens of one diopter will focus on an object a meter away, a lens of two diopters one-half meter away, a lens of four diopters a quarter of a meter away, and so on. Land birds vary considerably in their range of accommodation. Generally they have an accommodation of only a few diopters: from 8 to 12 in the Common Pigeon and Domestic Fowl; from 2 to 4 in owls and most other nocturnal birds. (Adult man has an accommodation of about 10.) On the other hand, aquatic diving birds (e. g., cormorants) have 40 to 50 diopters, a remarkably extensive range. Vision in the majority of birds is probably emmetropic—i. e., when the eye is resting it is focused on distant objects. A few birds are myopic or nearsighted under certain conditions. Penguins, for example, cannot see far when out of water though their eyes are very sensitive to light, enabling them to detect movements quite readily. The kiwis, nocturnal birds, are perhaps the most myopic of all avian species, especially during the day when they can scarcely see anything unless it is very near them.

The structure chiefly responsible for accommodation in birds is the ciliary body whose muscles bring pressure to bear on the lens via the suspensory ligament which holds the lens in place. The ciliary body is very much like that of reptiles but has a few minor modifications. Its muscles are strongly striated, are capable of very quick action; moreover, they are completely separated into two parts. Brücke's muscle, the posterior part, acts on the lens—usually soft in birds—increasing its curvature, particularly that of its outer surface, by pulling it forward, for near vision. Crampton's muscle, the anterior part, aids further in accommodation by changing the shape of the cornea. This double action is most fully developed in birds such as hawks whose eye must adjust rapidly from very far to very near vision during pursuit of prey. In aquatic diving birds, whose corneas are thick and inflexible, Crampton's muscle is all but absent, whereas Brücke's muscle is notably large, able to effect, in conjunction with a strong iris muscle, an extreme lens curvature for very near vision under water.

The Birds' Visual Field. The eyes of birds are more or less fixed in their sockets and so can be moved only to a very limited degree. Change in direction of vision is accomplished mainly by turning the head and neck. An owl can rotate its head through 270 degrees or more. The bird's visual field—i. e., the area which the eyes cover when the head is stationary—varies among different species, depending on the shape of the eyes and on their position in the head. Flat eyes allow for the coverage of wide areas; globose and tubular eyes have more restricted views. When the eyes are in the sides of the head so as to be directed laterally and give what is called monocular vision, the visual field of the two eyes is very wide, but when the eyes are farther forward in the head so as to be aimed anteriorly and give binocular vision—both eyes taking in part of the same area—the visual field is consequently more limited.

Monocularity and Binocularity. Probably very few birds have strictly monocular vision. Though the majority of birds have eyes in the sides of the head and quite commonly cock their heads to examine an object on the ground, looking with one eye while vision in their other eye is suppressed, they can nevertheless view binocularly by peering straight ahead at an object with both eyes. The Common Pigeon, a good example of a bird with eyes in the sides of the head, enjoys a total visual field of 340 to 342 degrees and a binocular field—the field on which both eyes may be aimed—of 24 degrees. Those birds with eyes situated more frontally have a greater binocular field, which aids in the pursuit of prey by allowing better distance-determination. The insect-catching goatsuckers and swallows have a wider binocular field than birds which normally pick food from the ground. Hawks and owls have a still wider field—35 to 50 degrees in hawks, 60 to 70 degrees in owls. Usually binocularity is more or less limited to objects in front of the head, but there are some species of birds whose head structure and eye positions give them binocularity for objects in other directions. An American Woodcock *(Philohela minor),* with eyes far back and far up in the head, no doubt can watch the sky or overhanging tree canopy for enemies while probing the ground for food with its long bill. The eyes of bitterns, which are long-billed birds, are directed somewhat "under their chins," so that they can see food on the ground when the bill is pointed ahead, and look ahead when the bill is pointed upward.

Binocular vision, in addition to providing better distance-determination, allows a keener perception of depth and solidity. In other words, binocular vision enhances stereopsis—gives a stronger third-dimensional effect. Many birds with limited binocularity and wide monocularity have developed actions which serve as substitutes for stereopsis. For example, the Common Pigeon in walking, or the American Coot *(Fulica americana)* in swimming, thrusts the head forward, then jerks it backward, with brief pauses between each motion; meanwhile the body moves forward steadily, and thus, presumably, the eyes do not go backward through space. With each pause, the bird gets a good lateral view; with each head-motion, vision is temporarily suppressed. The result is a quick succession of views at rapidly changing angles and distances that gives the bird a better estimation of relief or spatial relationships in its immediate environment. The bobbing of the foreparts, including the head, of the Killdeer *(Charadrius vociferus)*, and similar motions in other shore birds may produce the same effect. Another substitute for stereopsis is rapid peering (Grinnell, 1921). Thus, when birds are about to pick up motionless food, such as seeds, they frequently tilt or cock their heads at different angles to get a better notion of the shape of the food—perhaps necessary for identification—before seizing it.

Retinal Cells, Their Function and Distribution. The retina, which is the bird's all-important cellular layer for receiving visual stimuli and transferring them to the central nervous system, contains the receptor cells called rods and cones. The rods are effective in dull light and receive blurred images; cones function in bright light, give sharp visual details, and play a major role in color vision. In birds, both rods and cones have oil-droplets and the rods have the pigment rhodopsin.

Diurnal birds have an abundance of cones and relatively few rods, whereas nocturnal birds have many more rods. The cones, in both diurnal and nocturnal birds, are sparingly distributed over most of the retina except in certain spots, called foveas, where the cones are highly concentrated to form points of sharpest vision. Most mammals have but one, the fovea centralis, which is centrally located in the retina; most birds have a central fovea and some have another, the temporal fovea, in the posterior part or temporal quadrant of the retina. The central fovea serves in monocular vision of birds by registering the lateral field, while the temporal assists in binocular vision by covering the field that lies ahead. This "visual trident" (three simultaneous views, two lateral and one forward) is well developed in hawks and other fast-flying, predaceous birds and allows wide views of the field as a whole and concentrated, distance-discriminating views of the field into which the birds are flying. Paradoxically, owls which would seemingly need the trident have only the temporal fovea for forward, binocular vision. The lack of the central fovea is, however, satisfactorily compensated for by the ability of the birds to sweep an extraordinarily wide area with their eyes by turning their heads.

Visual Acuity. Bird eyes have a remarkable visual acuity or resolving power—the capacity to form distinguishable, unblurred images of objects as they become smaller and come closer together. The large size of the eyes and the big images which they cast partly account for this characteristic; other factors are the high concentration of visual cells and the high ratio of nerve fibers to the cells. In the human fovea there are only 200,000 visual cells per square millimeter, but in the House Sparrow the number is presumably 400,000. The hawks have the highest known concentrations. In the Buzzard *(Buteo buteo)*, whose eye is about the size of a man's, but whose visual acuity seems to be many times greater than man's, the fovea has about a million per square millimeter.

Nocturnal Vision. The abundance of rods in eyes of nocturnal birds is not unexpected; however, there may also be considerable numbers of cones, even in the eyes of owls. The importance of rods in nocturnal vision lies in their particular sensitivity to dim light through the presence in their outer parts of a purple-red pigment called rhodopsin ("visual purple"). Rhodopsin, containing a protein plus a derivative of vitamin A, breaks down (bleaches out) in the presence of bright light into its two components, reducing or destroying the responsiveness of the retina to light waves; but in the presence of dim light or darkness the components recombine by synthesis, increasing retinal sensitivity.

If a bird moves suddenly from a bright light to a dim light, there is a brief period of blindness while the rhodopsin resynthesizes. It hardly need be said that the greater the amount of rhodopsin in the rods, the greater their sensitiveness to light.

Nocturnality in birds is possible not only because of the presence of rhodopsin in the retina but also by the dilation of the pupils, giving the retina access to as much light as possible.

Color Vision. Careful work by Watson (1915) and later by Lashley (1916) revealed that color vision in the Domestic Fowl is, as in man, trichromatic— i. e., the retina is sensitive to mixtures of light waves or wave lengths from the middle to both sides of the spectrum. The longest and slowest light waves give the sensation of red, the shorter and faster waves give the sensations of orange, yellow, green, blue, and violet. Probably most diurnal birds have trichromatic vision, but subject to adaptive modifications in different species with respect to the distribution of oil-droplets in the retina. Oil-droplets (they are not present in man and the other higher mammals) are of two types, the colorless and the colored. The colored are red, orange, and yellow; all act as color filters by cutting out violet and blue light, thereby reducing chromatic aberration and glare, but the red droplets are more effective filters than orange, and the orange more than yellow. About 60 percent of the droplets in a kingfisher are red, a presumed adaptation for reducing glare on the water when looking for food below the surface. Red droplets are known to make red more distinct. On the basis of this fact, it is tempting to assume that an abundance of red droplets occurs in the retina of those species of hummingbirds which are particularly attracted to red flowers. Passerine birds, with an average of 20 percent red droplets, and hawks, with about 10 percent, probably have a color perception that is more like man's. Strongly nocturnal birds such as owls are believed to have achromatic vision in which the retina possesses merely colorless or faintly pigmented droplets and is sensitive only to black, gray, and white.

The Pecten. This conspicuous structure of the avian eye, projecting into the lower half of the posterior chamber from the head of the optic nerve, has long aroused much speculation as to its function. It is pigmented and highly vascular (consists mostly of minute blood vessels); it is undoubtedly a derivative of the retina. In reptiles it is a simple cone. Some thirty or so functions have been attributed to the pecten by various authorities. Though Walls (1942) has discussed those which seem the most plausible, he has found none entirely acceptable and concludes that its true function is yet unknown.

Hearing

The avian ear, like the mammalian ear, is sensitive to a wide range of sounds, or sound waves—i.e., vibratory disturbances in the air. The number of vibrations—very often called cycles—per second is referred to as the frequency and establishes the pitch. A sound of high frequency, or a great many vibrations per second, produces a high pitch, just as a sound of low frequency causes a low pitch.

The hearing of birds, as shown by data gathered by Schwartzkopff (1955), ranges from 40 cycles per second in the Budgerigar *(Melopsittacus undulatus)* to 29,000 c.p.s. in the Chaffinch *(Fringilla coelebs).* The upper limit, however, seldom exceeds 20,000 c.p.s. Brand and Kellogg (1939a) found that the Starling *(Sturnus vulgaris)* has a hearing range of 700 to 15,000 c.p.s.; the House Sparrow 675 to 11,500; and the Common Pigeon 200 to 7,500. All the birds became gradually less sensitive to sounds as the extremes in frequencies were approached. Comparing the hearing of these birds to that of man, who has a hearing range of about 16 to 17,000 c.p.s. (roughly nine octaves), man can hear about four octaves lower than the Common Pigeon and five lower than the Starling and House Sparrow; and he can hear as high as the Starling and House Sparrow and about an octave higher than the Common Pigeon. Owls show an exceptionally wide hearing range. The Great Horned Owl *(Bubo virginianus)* can respond to sounds as low as 60 c.p.s. (Edwards, 1943), while the Barn Owl *(Tyto alba)*, which catches mice in total darkness by detecting their highly pitched squeaks, is sensitive to sounds above 8,500 c.p.s. (Payne, 1962).

For a critical review of the findings on the hearing of birds by different authors, see Schwartzkopff, 1968.

Bang, B. G.
  1960 Anatomical Evidence for Olfactory Function in Some Species of Birds. Nature, 188: 547-549. (In the Turkey Vulture, *Cathartes aura;* Oilbird, *Steatornis caripensis;* Laysan and Black-footed Albatrosses, *Diomedea immutabilis* and *D. nigripes.)*
Bang, B. G., and S. Cobb
  1968 The Size of the Olfactory Bulb in 108 Species of Birds. Auk, 85: 55-61.
Bartholomew, G. A., and T. J. Cade
  1958 Effects of Sodium Chloride on the Water Consumption of House Finches. Physiol. Zool., 31: 304-310. (Salinity definitely affects fluid consumption.)
Bartholomew, G. A., and R. E. MacMillen
  1961 Water Economy of the California Quail and Its Use of Sea Water. Auk, 78: 505-514. (Change in salinity shows no significant change in fluid consumption. When sufficiently dehydrated, the species will drink even 70 percent sea water.)
Beecher, W. J.
  1952 The Role of Vision in the Alighting of Birds. Science, 115: 607-608. (According to the author, birds tend to alight by visual clues without regard to either wind velocity or wind direction.)
  1969 Possible Motion Detection in the Vertebrate Middle Ear. Bull. Chicago Acad. Sci., 11: 155-210.
Brand, A. R., and P. P. Kellogg
  1939a Auditory Responses of Starlings, English Sparrows, and Domestic Pigeons. Wilson Bull., 51: 38-41.
  1939b The Range of Hearing of Canaries. Science, 90: 354.
Cade, T. J., and G. A. Bartholomew
  1959 Salt Water and Salt Utilization by Savannah Sparrows. Physiol. Zool., 32: 230-238. (Salinity definitely affects fluid consumption.)
Calvin, A. D., C. M. Williams, and N. Westmoreland
  1957 Olfactory Sensitivity in the Domestic Pigeon. Amer. Jour. Physiol., 188: 255-256. (The species cannot learn to detect odor alone.)
Chapman, F. M.
  1929 My Tropical Air Castle: Nature Studies in Panama. D. Appleton and Company, New York.
  1938 Life in an Air Castle: Nature Studies in the Tropics. D. Appleton-Century Company, New York.
Donner, K. O.
  1951 The Visual Acuity of Some Passerine Birds. Acta Zool. Fennica, 66: 1-40.
Duncan, C. J.
  1960 Preference Tests and the Sense of Taste in the Feral Pigeon (*Columba livia* var. Gmelin). Animal Behaviour, 8: 54-60.
  1962 Salt Preference of Birds and Mammals. Physiol. Zool., 35: 120-132.
Edwards, E. P.
  1943 Hearing Ranges of Four Species of Birds. Auk, 60: 239-241.
Frings, H., and W. A. Boyd
  1952 Evidence for Olfactory Discrimination by the Bobwhite Quail. Amer. Midland Nat., 48: 181-184.
Frings, H., and B. Slocum
  1958 Hearing Ranges for Several Species of Birds. Auk, 75: 99-100.

Grinnell, J.
  1921   The Principle of Rapid Peering in Birds. Univ. California Chronicle, 23: 392-396.
Gurney, J. H.
  1922   On the Sense of Smell Possessed by Birds. Ibis, 1922: 225-253.
Hamrum, C. L.
  1953   Experiments on the Senses of Taste and Smell in the Bob-white Quail *(Colinus virginianus virginianus)*. Amer. Midland Nat., 49: 872-877.
Hocking, B., and B. L. Mitchell
  1961   Owl Vision. Ibis, 103a: 284-288. (Specializations for vision in weak light.)
Hussey, R. F.
  1917   A Study of the Reactions of Certain Birds to Sound Stimuli. Jour. Animal Behavior, 7: 207-219.
Lashley, K. S.
  1916   Color Vision in Chickens: The Spectrum of the Domestic Fowl. Jour. Animal Psychol., 6: 1-26.
Michelsen, W. J.
  1959   Procedure for Studying Olfactory Discrimination in Pigeons. Science, 130 (3376): 630-631. (Pigeons can discriminate by an olfactory sense.)
Moore, C. A., and R. Elliot
  1946   Numerical and Regional Distribution of Taste Buds on the Tongue of the Bird. Jour. Comp. Neurol., 84: 119-131.
Owre, O. T., and P. O. Northington
  1961   Indication of the Sense of Smell in the Turkey Vulture, *Cathartes aura* (Linnaeus), from Feeding Tests. Amer. Midland Nat., 66: 200-205.
Payne, R. S.
  1962   How the Barn Owl Locates Prey by Hearing. Living Bird, 1: 151-159. (The owl orients the head in such a way as to hear all frequencies, audible to it in a complex sound, at maximum intensity in both ears. When it has achieved such an orientation, it will automatically be facing the source of the sound with a theoretical accuracy of less than one degree.)
Portmann, A.
  1950   Les Organes des Sens. In *Traité de Zoologie*. Edited by P.-P. Grassé. Volume 15. Masson et Cie, Paris.
Pumphrey, R. J.
  1948   The Sense Organs of Birds. Ibis, 90: 171-199.
  1961   Sensory Organs: Vision and Hearing. In *Biology and Comparative Physiology of Birds*. Volume 2. Edited by A. J. Marshall. Academic Press, New York.
Rochon-Duvigneaud, A.
  1950   Les Yeux et La Vision. In *Traité de Zoologie*. Edited by P.-P. Grassé. Volume 15. Masson et Cie, Paris.
Schwartzkopff, J.
  1955   On the Hearing of Birds. Auk, 72: 340-347.
  1963   Morphological and Physiological Properties of the Auditory System in Birds. Proc. XIIIth Internatl. Ornith. Congr., pp. 1059-1068.
  1968   Structure and Function of the Ear and of the Auditory Brain Areas in Birds. In *Hearing Mechanisms in Vertebrates*. Edited by A. V. S. de Reuck and J. Knight. Little, Brown and Company, Boston.
Stager, K. E.
  1964   The Role of Olfaction in Food Location by the Turkey Vulture *(Cathartes aura)*. Los Angeles County Mus., Contrib. in Sci., No. 81. (From the evidence it is concluded that, among the cathartine vultures, the Turkey Vulture "possesses and utilizes a well-developed olfactory food-locating mechanism.")
Strong, R. M.
  1911   On the Olfactory Organs and the Sense of Smell in Birds. Jour. Morph., 22: 619-658.
Sturkie, P. D.
  1965   Avian Physiology. Second edition. Cornell University Press. Ithaca, New York. (Chapter 14.)

Walls, G. L.
  1942   The Vertebrate Eye and Its Adaptive Radiation. Cranbrook Inst. Sci. Bull. No. 19, Bloom-
         field Hills, Michigan.
Watson, J. B.
  1915   Studies on the Spectral Sensitivity of Birds. Papers Dept. Marine Biol., Carnegie Inst.
         Washington, 7: 87-104.
Wood, C. A.
  1917   The Fundus Oculi of Birds, Especially as Viewed by the Ophthalmoscope. Lakeside Press,
         Chicago. (A classic work on the subject.)

THE BRAIN

Relatively few studies have been made on the avian brain. The following works will introduce the student to available literature.

Kappers, C.U.A., G. C. Huber, and E. C. Crosby
  1936   The Comparative Anatomy of the Nervous System of Vertebrates, Including Man. Mac-
         millan Company, New York. (Contains a wealth of information on the avian brain, as
         well as on other parts of the nervous system.)
Portmann, A.
  1946- Études sur la Cérébralisation chez les Oiseaux. I-III. Alauda, 14: 2-20; 15: 2-15, 161-171.
    47   (Research on the relative degree of development of the brain in birds; a rather unusual
         approach to the study of avian evolution.)
Portmann, A., and W. Stingelin
  1961   The Central Nervous System. In *Biology and Comparative Physiology of Birds*. Volume
         2. Edited by A. J. Marshall. Academic Press, New York.
Sutter, E.
  1951   Growth and Differentiation of the Brain in Nidifugous and Nidicolous Birds. Proc. Xth
         Internatl. Ornith. Congr., pp. 636-644. (Concerned chiefly with postembryonic
         development.)

SYSTEMATICS

Systematics, or **taxonomy**, is essentially the classification of organisms into categories or **taxa** (singular, **taxon**) and includes such procedures as identifying and naming. In ornithology, systematics has reached a relatively advanced status; birds are now probably better known taxonomically than any other group of animals. There are recognized today some 8,600 species in the world, and it is estimated that less than one percent of the total number of species still remain unknown.

THE SPECIES

The **species** (plural, also **species**) is one of the several taxa into which all organisms are classified. In ornithology—and probably all other branches of biology—it is by far the most important taxon, because student and practicing scientist alike work at one time or another with species. No person can hope to be an ornithologist without knowing different species of birds, though he may succeed without a knowledge of other categories, such as genera and subspecies. For the beginning student in ornithology an understanding of the nature of species is a prerequisite.

A species may be defined as a population, or populations, of mutually fertile individuals, reproductively isolated from individuals of other populations and possessing in common certain characters which distinguish them from any other similar population, or populations. If cross-breeding of two species occurs, the offspring are often sterile.

Species, like other categories into which organisms are classified, are distinguished by combinations of inherent peculiarities called **taxonomic characters**. In avian systematics some of the commonly used taxonomic characters for species are morphological, such as those having to do with minor details of size, shape, and color; other characters are ecological requirements (e.g., type of niche), reproductive traits (kind of song, type of nest), and general behavior patterns. The student, in learning to identify different species of birds, will be inclined at first to pay special attention to morphological characters, particularly size, form, and color. Gradually he will discover, however, that morphological characters alone are not always helpful, for he will meet certain species in the field that he can recognize better by the way they sing or the way they behave than by their appearance. The student must, therefore, be prepared to identify species by characters other than morphological.

SUBSPECIES

Any species shows variation among the individuals comprising it. When a species is spread over a wide area, the variations seldom, if ever, occur uniformly, but tend to be grouped in local populations. A widespread species, then, is actually comprised of numerous local populations, each one with a variable combination of characters by which it differs from all other populations in the species. A local population with a combination of characters making it sufficiently distinct from other populations is called a **subspecies** (sometimes called a **race**). The species comprising these subspecific divisions is **polytypic**. On the other hand, local populations, not sufficiently distinct, have no categorical designation and are considered **monotypic**. Some examples of monotypic species are the American Woodcock *(Philohela minor)*, Ruby-throated Hummingbird *(Archilochus colubris)*, Piñon Jay *(Gymnorhinus cyanocephalus)*, and Bobolink *(Dolichonyx oryzivorus)*.

A subspecies may be defined as a geographically limited population whose members possess in common certain taxonomic characters which distinguish them from all other populations in the species. All the subspecies of a species are mutually fertile; hence interbreeding occurs where two or more subspecies meet. Any one subspecies usually consists of a group of local populations differing slightly from one another unless it is confined to a small, isolated area, such as an island, in which case the characters may be rather uniform among all the individuals.

Subspecies normally replace one another geographically; their ranges adjoin and frequently produce a zone where individuals show marked intergradation through interbreeding. If the subspecies replace one another over a wide area, a progressive change in certain taxonomic characters may be evident, often correlated with a progressive change in one or more environmental factors such as climate. This kind of character gradient is referred to as a **cline**.

Probably clines, occurring as they do with a gradual change in environment, represent adaptive change. Sometimes the succession of changes producing a cline conforms so closely in different species that systematists have established several so-called rules, examples of which follow:

Bergmann's Rule. Body size tends to be larger in cooler climates, smaller in warmer climates.

Gloger's Rule. Coloration tends to be darker in humid climates, lighter in arid climates. An increase in melanins produces the darker coloration.

Allen's Rule. Bills, tails, and other extensions of the body tend to be longer in warmer climates, shorter in cooler climates.

The differences between subspecies are often average differences, and only rarely can a single individual be identified to subspecies. To an experienced systematist the morphological differences between one subspecies population and another are soon apparent, but to the unpracticed eye of a student or to an ornithologist not engaged in systematics, most subspecies "look alike." This is to be expected. Unfortunately, because the subspecies is an established category in systematic ornithology, many students and ornithologists feel that, in order to be accurate, they must try to recognize subspecies, rather than just species, in their laboratory and field work. If they fail, as is often the case, they "bluff" their identifications by using the available information on the ranges which the subspecies are supposed to occupy. Such a procedure is wholly unscientific and is in no sense a contribution to knowledge. In the foregoing paragraphs the statement was made that the species is the most important category. It should be emphasized here that the subspecies is a category of special interest to systematists and those scientists engaged in the study of speciation (see below). The beginning student and most ornithologists should not expect to identify subspecies and should never pretend to except by proper scientific analysis.

SPECIATION

As indicated above, a widespread species is usually divided into many local populations, and each one has its own variable combination of characters. Ordinarily there is little outbreeding in any one population. With the passage of time a population may become split by some geographical factor into two wholly or partially isolated populations, and each of these populations may build up distinct characters. Eventually the distinctions may become sharp enough to designate each population as a subspecies. If the isolating factor continues to operate, further distinctions may accumulate and establish a physiological barrier that eliminates any further possibility of interbreeding. The two populations are then reproductively isolated from one another and each may be termed a species.

If the two species continue to be separate geographically, even though their ranges may be contiguous, they are **allopatric**. If, however, the two species eventually come together so that their breeding ranges merge and even overlap entirely, they are **sympatric**. But they will not ordinarily interbreed since they are now so distinctive in one or more ways—morphologically, physiologically, behaviorally (including voice), or ecologically (including choice of habitat)—as to be reproductively isolated.

In southeastern Canada and northeastern United States, sympatry is illustrated among three species of small flycatchers of the genus *Empidonax*—the Least Flycatcher *(E. minimus)*, Traill's Flycatcher *(E. traillii)*, and Yellow-bellied Flycatcher *(E. flaviventris)*. All three appear very much alike, yet they are prevented from interbreeding by virtue of their different vocalizations as well as by their habitats—the Least preferring park-like woodlands; the Traill's, willow and alder swales; and the Yellow-bellied, deeply shaded boreal forests of conifers. The Eastern and Western Meadowlarks *(Sturnella magna* and *S. neglecta)* exhibit sympatry in the grasslands of midwestern North America. Here their ranges meet and overlap broadly, yet the two species do not normally interbreed since not only their songs and calls differ markedly, but also their habitats—the Eastern Meadowlark generally occupying the more moist lowlands and the Western the drier uplands.

Occasionally, when two closely related species become sympatric with their ranges partly overlapping, they are apt to differ more sharply from each other where their ranges overlap than where their ranges are still separate. Called **character displacement**, this phenomenon in which two species may so rapidly diverge possibly reduces the chances of the two species interbreeding and/or minimizes competition when the two species are together. Parkes (1965) explained character displacement at some length and demonstrated it in two species of cuckoos inhabiting the Philippine archipelago.

Species formation, or **speciation**, is the result of splitting of an evolutionary (phyletic) line over a prolonged period of time. According to this concept, local populations are incipient subspecies, and isolated (not clinal) subspecies are incipient species. For information on some of the geographical and ecological factors affecting isolation, see the section in this book titled "Distribution," pages 200 and 208.

One should bear in mind that speciation is a continuous evolutionary process going on today as in the past—and much more rapidly than once supposed. A good example is the rate at which adaptive differentiation has occurred in the House Sparrow *(Passer domesticus)* since its introduction to North America from England and Germany in 1851. The species breeds now across the continent in extremes of environment to which the original introduced stock was never accustomed; and, like many American polytypic species, it has adapted itself clinally in size, color, and length of body extensions in accordance, respectively, with Bergmann's, Gloger's, and Allen's rules (Johnston and

Selander, 1964; Packard, 1967). These adaptations have evolved in no more and probably fewer than 90 generations or—assuming one generation a year—90 years (Packard, *op. cit.*).

In order to determine the taxonomic rank of closely allied populations of birds and their mode of evolution, the systematist must analyze the morphological characters of many hundreds of representative specimens and make detailed observations in the field on ecological preferences, reproductive traits, and so on. Any study of speciation involves many complex problems.

One of the problems is that of deciding whether certain populations deserve subspecific rank. Criteria vary. For example, if a systematist finds that 75 percent of the specimens examined in one population, or an assemblage of populations, of a species are distinctly separate from all specimens of other populations of the species, he may consider it a subspecies. This is the application of the **75 percent rule**. See Amadon (1949) for a discussion of the rule and an example of its application.

Another problem is that of deciding whether certain populations are subspecies of a polytypic species and therefore **conspecific** or are distinct allopatric species. The answer lies largely in what the systematist can learn about their origin and their degree of isolation. If the systematist finds evidence that certain allopatric species were once subspecies of a single species but have since achieved sufficient differentiation and isolation to be accorded the status of species, he may designate them **allospecies** and group them under the heading **superspecies**. Consult Amadon (1966) for details. The superspecies is not an established category in the system of classification. It is merely a convenient term to express the concept that a group of species are very closely related, yet not closely enough to be given subspecific rank. Some examples of superspecies are the Great-tailed and Boat-tailed Grackles *(Cassidix major* and *C. mexicanus)*, the Golden-fronted and Red-bellied Woodpeckers *(Centurus aurifrons* and *C. carolinus)*, the Barred and Spotted Owls *(Strix varia* and *S. occidentalis)*, and the American Red-tailed Hawk and Common Buzzard of Eurasia *(Buteo jamaicensis* and *B. buteo)*. Perhaps as many as a third of the bird species in the world may be components of superspecies.

Still another example of speciation problems is that of evaluating the occurrence of **hybrids** and **hybridization**. Hybrids result from the crossing of two taxonomically unlike individuals. **Interspecific hybrids** are produced by the crossing of individuals of two different species. Such hybrids are usually designated by writing the names of the two parent species one after the other with a cross between. For example: *Passerina cyanea* × *Passerina amoena*. Hybridization commonly means, in systematics, the crossing between individuals of two different species.

Various isolating mechanisms ordinarily prevent extensive hybridization between species though sometimes two species may hybridize to a limited extent. In Saskatchewan, Pettingill (Sibley and Pettingill, 1955) collected a hybrid between the Chestnut-collared and McCown's Longspurs *(Calcarius ornatus* and *Rynchophanes mccownii)*, two sympatric species normally separated by their habitat, the Chestnut-collared preferring the long-grass prairie, the McCown's the short-grass. Hybridization of this sort between sympatric species is unusual. Much more common is hybridization between allopatric species in areas where their ranges adjoin. A well-known case is that of two essentially allopatric species, the Blue-winged and Golden-winged Warblers *(Vermivora pinus* and *V. chrysoptera)*. The Blue-winged Warbler has a more southern breeding range, and the Golden-winged a more northern, but their ranges overlap extensively (see Parkes, 1951)—hence they are not typical allopatric species. In the zone of overlap they produce hybrids of two distinct types, the so-called Brewster's Warbler and the

Lawrence's Warbler. The hybrids, however, are relatively few, considering the size of the parent populations, and there is no combining of characters of the two species to form a true hybrid population.

In the western Great Plains of North America the ranges of several eastern species— e. g., the Baltimore Oriole *(Icterus galbula)* and Indigo Bunting *(Passerina cyanea)*— meet the ranges of their western allopatric counterparts—the Bullock's Oriole *(I. bullockii)* and Lazuli Bunting *(P. amoena)*. Here is a zone where the interbreeding of these allopatric species forms a large hybrid population. Some modern systematists believe that the presence of such a large population indicates that the parent populations are subspecies of a polytypic species rather than separate species. Thus, in their view, the Baltimore and Bullock's Orioles should be considered one species rather than two, and so should the Indigo and Lazuli Buntings. The presence of an occasional hybrid, as in the case of the Chestnut-collared Longspur × McCown's Longspur, or a few hybrids as in the case of the Blue-winged Warbler × Golden-winged Warbler, is not sufficient evidence to prove that the parent stocks are subspecies of a single species.

For papers dealing with studies of hybridization, the following are particularly recommended: Huntington (1952), Short (1963, 1965), Sibley (1950, 1954, 1957, 1961), Sibley and Short (1959), Sibley and West (1958, 1959), West (1962).

SOME EXAMPLES OF BIRD SPECIATION

Speciation, as already explained, is dependent on the isolation of populations long enough for them to build up distinctive characters that will insure a reproductive barrier. Oceanic archipelagos, remote from continents with many small islands separated by channels, often provide adequate isolation for terrestrial animals and consequently make ideal laboratories for the study of speciation. This is the case with the Galapagos and Hawaiian Islands.

Rising from the Pacific Ocean some 500 miles west of Ecuador, the Galapagos Islands are of volcanic origin with no history of a connection to South America. How long they remained unexploited by plants and animals, nobody knows. In due course—and quite by chance—terrestrial life arrived and, stranded, proceeded to adapt to the peculiarities of an unused environment. The few bird species existing on the Galapagos, when discovered by man, included Darwin's finches (an endemic subfamily Geospizinae of the family Fringillidae)—so called because they were largely responsible for stirring Darwin's thoughts that led to his theory of evolution.

Darwin's finches, numbering 14 species (Lack, 1947), probably stemmed from a finch-like form that arrived from South America and established a population on one of the islands. From this population, stragglers reached other islands where they in turn established populations. In time, as these populations adapted themselves to the new and unoccupied environments, they evolved distinctions as subspecies and eventually species. This diverging or branching out from one species into several to suit different ecological niches is a good example of **adaptive radiation**, a common evolutionary process. But the story does not end here. Having evolved on certain islands, these new finch species reached other islands where they entered into sympatry with other new finch species and proceeded to compete with them for habitat, food, and so on. This sort of rivalry between sympatric species intensified or reinforced their distinctions. Today, each of the principal Galapagos Islands has at least three finch species and some have as many as ten. In summary, the formation of Darwin's finches is the result, first, of initial isolation of populations on separate islands and, second, of competition between species when sympatry occurred.

Measuring four to eight inches in length, Darwin's finches are generally drab in appearance; both sexes are colored alike in browns and grays and occasionally black. Their primary distinctions are in the bill which varies for different kinds of foraging (Lack, 1945, 1947; Bowman, 1961). One type of bill persists finch-like for seed-eating; other types range from long and down-curved for obtaining nectar from cactus flowers and parrot-like for fruit-eating to small and somewhat chickadee-like or even warbler-like for insect-eating. One species, the Woodpecker Finch *(Cactospiza pallida)*, has a bill suited to excavating in wood and to manipulating a cactus spine as a probe for extracting insects from holes and crevices (see the section in this book titled "Behavior," page 248).

Like the Galapagos, the Hawaiian Islands in the north-central Pacific are of volcanic origin and even more remote from continental life. Here a comparable development by adaptive radiation took place among the honeycreepers, an endemic family (Drepanididae). See Amadon (1950) and Baldwin (1953).

The colonizing form was perhaps a nectar-feeding bird such as a honeycreeper (Coerebidae) or tanager (Thraupidae) from the New World tropics. From this evolved as many as 23 species, of which nine are now extinct. Although similar in size to Darwin's finches, adapative radiation among them proceeded further, producing not only more species but greater distinctions in color and more pronounced extremes in the shape of the bills. Some species are predominantly green or yellow; others are bright red or black; one species is mainly gray, with red streaks or spots, and features a crest. The sexes may or may not be alike. In some species the bill is long, thin, and remarkably down-curved for extracting nectar and taking insects from flowers; in other species it is grosbeak-like for seed- and fruit-crushing, or woodpecker-like for obtaining insects from bark and wood.

The necessary isolation for speciation on the continents is not as obvious as in oceanic archipelagos. Nevertheless, widely separated physiographic features such as high mountains or mountain ranges or widely spaced areas with sharp climatic differences can be veritable islands, providing adequate isolation.

A good example of speciation in a wide-ranging continental group of birds is among the juncos (genus *Junco*) of North America. After studying many hundreds of representative specimens and observing many forms—species and subspecies—in their natural environment, Miller (1941) found in Mexico and Central America that some forms resulted from populations restricted to different mountains where there was no opportunity for interbreeding. Farther north, where the range of juncos is less interrupted by physiographic features, he noted that different climatic conditions, correlated with particular geographical areas, served as isolating factors. If different forms of juncos came in contact, as was sometimes the case, each tended to remain attached to its own habitat. When interbreeding occurred, hybridization rather than intergradation resulted, provided the distinctive characters had become stabilized. The hybrids were different from their parents and formed a distinctive and self-perpetuating population.

Speciation in the large family of American wood warblers (Parulidae) offers a fruitful field for speculation. Probably the family originated in the North American tropics during the Miocene (see the Geologic Time Scale, page 414) and by the early Pliocene early forms were well established in the temperate ancestral deciduous forest of eastern North America. Mengel (1964) has postulated that speciation leading to the vast array of present-day parulids began with the advent of the Pleistocene or Ice Age. His hypothesis follows briefly: Parulid forms in the temperate ancestral deciduous forest in eastern North America developed adaptations to the northern coniferous (boreal) forest when it was forced deep into the southeast by the first glacial advance. Upon glacial recession a transcontinental coniferous forest formed in the wake of the retreating ice and was soon occupied by the newly formed parulids. When the next glacial advance separated the continent-spanning coniferous forest into eastern and western parts, the parulids separated into corresponding eastern and western populations, each developing its own distinctions. In the west, the process of further separation, isolation, and adaptive radiation continued among the parulids through the warm interglacial period for at this time the birds were forced into the "islands" of coniferous forest still remaining high on mountain slopes. Repetition of the process in subsequent glacial cycles—there were four altogether in the Pleistocene—very likely completed the formation of the present-day western parulids. Most of the present-day eastern species seem to have descended directly from the original parulid stock in the eastern deciduous forest.

THE HIGHER CATEGORIES

All species are classified into higher taxa on the basis of presumed "blood" relationships. This is a "natural" system because it relies on **phylogeny**—the evolution of related groups of organisms. For the beginning student, as well as the biologist, it brings order into an otherwise confusing array of species; at the same time, it provides a means of showing species relationships.

Frequently, some relationships are obscured by **divergence**, others by **convergence**. Two closely related groups of species may in outward appearances seem distantly related owing to their having diverged greatly in their structural and behavioral adaptations; on the other hand, two distantly related groups may have developed very similar adaptations, thereby converging in their outward appearances. Swifts, for example, seem more closely related to swallows than to hummingbirds. Yet such is not the case. Swifts and hummingbirds presumably diverged from common ancestry. Swifts and swallows, which have very different ancestry, have converged in form and habits essential for the aerial pursuit of insects.

No single study can provide a sufficient basis for establishing relationships. Various studies are necessary to meet and solve the problems posed by divergence and convergence.

Of paramount importance are studies of anatomy, particularly the more stable features of the skeleton and other organ systems; basic reproductive traits and behavior patterns; and geographical ranges and habitat preferences. See Brodkorb (1968) for many of the taxonomic characters denoting the higher categories of birds.

Useful in determining taxonomic relationships of birds is a study of their external parasites, most notably lice of the insect order Mallophaga (see Appendix I). As a rule, mallophagans are host-specific, each species being restricted and adapted to a single bird species or group of closely related bird species. With little opportunity to parasitize other bird species, since different species of birds rarely contact one another enough to allow any significant interchange of their ectoparasites, mallophagans have evolved along with their hosts. Thus the mallophagans are themselves as closely related as their hosts and consequently provide clues to the phylogeny of their hosts.

More recently a study of bird proteins, which are genetically controlled and therefore specific to a given group of birds, has proven highly useful in determining taxonomic relationships. By a fractionating process, known as electrophoresis carried out in a special laboratory apparatus, one may analyze and compare, for example, the egg-white proteins of different birds by observing their electrophoretic "profiles." Frequently the pictures corroborate morphological evidence of phylogenetic relationships and in some instances show relationships not previously detected. A great advantage of this means of taxonomic study is that it yields quantities of data with great precision. See Sibley (1960, 1962, and 1967) and Peakall (1962) for further information.

The three higher taxa most commonly used in the classification of birds warrant discussion here; they are, in their line of ascending rank or hierarchy, the genus, the family, and the order.

The **genus** (plural **genera**) embraces one or more species exhibiting a combination of taxonomic characters shared with no other taxon of the same rank. Some of the characters are morphological (e. g., the general color and details of shape and structure), some are ecological (the type of breeding habitat such as a forest, grassland, or shore), and some are behavioral (the type of displays and nesting habits). The identification of a polytypic genus (i. e., a genus containing two or more species) is a matter of determining the peculiarities common to different species, whereas the identification of a species is a problem of denoting distinctiveness between it and related species. In a few cases of monotypic genera, the identification of the single species of each genus in turn identifies the genus. The genus is the only higher category whose name is part of a species' technical name (see below).

The **family** embraces one genus, or two or more genera, exhibiting a combination of obvious morphological peculiarities such as the shape of the bill, presence of notches on the bill, the nature of the tarsal covering, and the number of primaries. See Storer (1960) for a review of world families that include some distinguishing external characteristics. Relatively few families contain only one genus. Usually a family has a large number of genera which show a variety of habitat preferences. The fact that peculiarities of a family are obvious is an indication that the families have diverged widely in the evolution from a common type. All family names end in **-idae,** making the categorical designation easy to recognize.

The **order** embraces one or more families exhibiting peculiarities of the skeleton and other parts of the anatomy that have been least modified by adaptive change and are consequently more basic, or stable. Whereas a family is often confined in its distribution to a continent or neighboring continents, an order is often world-wide in its distribution. All names of bird orders end in **-iformes.**

When any category contains a large number of taxonomic groups from the category below, systematists have sometimes found it desirable to express the relationships of these groups in a more precise manner by using intermediate categories whose names bear super- or sub- as a prefix. If, for example, a particular family includes a large number of genera which seems to show a natural division into two or more groups, then the groups are considered subfamilies. Among the various intermediate categories, two have standard endings for their names, *viz.*, **-oidea** for superfamily and **-inae** for subfamily. Below are given in hierarchic order all the categories used in systematic ornithology. To their right is the complete classification of the common subspecies of Mallard *(Anas platyrhynchos platyrhynchos),* the eastern subspecies of Belted Kingfisher *(Megaceryle alcyon alcyon),* the western subspecies of Robin *(Turdus migratorius propinquus),* and the Wood Thrush *(Hylocichla mustelina),* a monotypic species. Note that certain intermediate categories are not always used.

CLASS

Subclass	Neornithes	Neornithes	Neornithes	Neornithes
Superorder	Neognathae	Neognathae	Neognathae	Neognathae
Order	Anseriformes	Coraciiformes	Passeriformes	Passeriformes
Suborder	Anseres	Alcedines	Passeres	Passeres
Superfamily		Alcedinoidea		
Family	Anatidae	Alcedinidae	Turdidae	Turdidae
Subfamily	Anatinae	Cerylinae		
Genus	*Anas*	*Megaceryle*	*Turdus*	*Hylocichla*
Subgenus		Streptoceryle		
Species	*platyrhynchos*	*alcyon*	*migratorius*	*mustelina*
Subspecies	*platyrhynchos*	*alcyon*	*propinquus*	

A classification of the living and fossil birds of the world is given in the following pages. Categories in which fossil birds are placed are shown only to orders; categories in which living birds are placed are shown down to families except in the case of the family Anatidae, in which categories are shown down to subfamilies. For the sake of simplification, most all intermediate categories are omitted, as are all families not found in North America north of Mexico.

The student should become familiar with this classification, for it is widely used. Birds, whenever listed in modern guides and check-lists, are grouped under these categories and, in American works, presented in sequence from the oldest—or most like the ancestral form—to the most advanced—or least like the ancestral form. The student should, therefore, know the names of the orders and families, and determine their meaning and pronunciation by consulting *Webster's Third New International Dictionary of the English Language Unabridged* (1966). He should also learn the sequence of orders and the sequence of families in orders so that he may find a category quickly without scanning an entire list or referring to an index.

Class Aves, Birds.
    Subclass Archaeornithes, Ancestral Birds.
        Order Archaeopterygiformes, *Archaeopteryx* (fossil).
    Subclass Neornithes, True Birds.
        Superorder Odontognathae, New World Toothed Birds.
            Order Hesperornithiformes, *Hesperornis* and other fossils.
            Order Ichthyornithiformes, *Ichthyornis* and other fossils.
        Superorder Impennes, Penguins.
            Order Sphenisciformes, Penguins.
        Superorder Neognathae, Typical Birds.
            Order Struthioniformes, Ostriches.
            Order Rheiformes, Rheas.
            Order Casuariiformes, Cassowaries and Emus.
            Order Aepyornithiformes, Elephant birds (extinct).
            Order Dinornithiformes, Moas (extinct).
            Order Apterygiformes, Kiwis.
            Order Tinamiformes, Tinamous.
            Order Gaviiformes, Loons.
                Family Gaviidae, Loons.
            Order Podicipediformes, Grebes.
                Family Podicipedidae, Grebes.
            Order Procellariiformes, Tube-nosed Swimmers.
                Family Diomedeidae, Albatrosses.
                Family Procellariidae, Shearwaters and Fulmars.
                Family Hydrobatidae, Storm Petrels.

Order Pelecaniformes, Totipalmate Swimmers.
        Family Phaëthontidae, Tropicbirds.
        Family Pelecanidae, Pelicans.
        Family Sulidae, Boobies and Gannets.
        Family Phalacrocoracidae, Cormorants.
        Family Anhingidae, Anhingas
        Family Fregatidae, Frigatebirds.
Order Ciconiiformes, Deep-water Waders.
        Family Ardeidae, Herons and Bitterns.
        Family Ciconiidae, Storks and Wood Ibises.
        Family Threskiornithidae, Ibises and Spoonbills.
        Family Phoenicopteridae, Flamingos.
Order Anseriformes, Lamellate-billed Swimmers.
        Family Anatidae, Swans, Geese, and Ducks.
            (Subfamily Cygninae, Swans.)
            (Subfamily Anserinae, Geese.)
            (Subfamily Dendrocygninae, Tree Ducks.)
            (Subfamily Anatinae, Dabbling Ducks.)
            (Subfamily Aythyinae, Diving Ducks.)
            (Subfamily Oxyurinae, Ruddy and Masked Ducks.)
            (Subfamily Merginae, Mergansers.)
Order Falconiformes, Vultures and Diurnal Birds of Prey.
        Family Cathartidae, New World Vultures.
        Family Accipitridae, Kites, Eagles, and Hawks.
        Family Pandionidae, Ospreys.
        Family Falconidae, Caracaras and Falcons.
Order Galliformes, Gallinaceous Birds and Hoatzins.
        Family Cracidae, Curassows, Guans, and Chachalacas
        Family Tetraonidae, Grouse and Ptarmigans.
        Family Phasianidae, Quails and Pheasants.
        Family Meleagrididae, Turkeys.
Order Gruiformes, Marsh Birds.
        Family Gruidae, Cranes.
        Family Aramidae, Limpkins.
        Family Rallidae, Rails, Gallinules, and Coots.
Order Diatrymiformes, *Diatryma* (fossil).
Order Charadriiformes, Shore Birds, Gulls, and Auks.
        Family Jacanidae, Jacanas.
        Family Haematopodidae, Oystercatchers.
        Family Charadriidae, Plovers and Turnstones.
        Family Scolopacidae, Snipe, Woodcock, and Sandpipers.
        Family Recurvirostridae, Avocets and Stilts.
        Family Phalaropodidae, Phalaropes.
        Family Stercorariidae, Skuas and Jaegers.
        Family Laridae, Gulls and Terns.
        Family Rynchopidae, Skimmers.
        Family Alcidae, Auks, Murres, and Puffins.
Order Columbiformes, Pigeon-like Birds.
        Family Columbidae, Pigeons and Doves.
Order Psittaciformes, Parrots, Parakeets, and Macaws.
Order Cuculiformes, Cuckoo-like Birds.
        Family Cuculidae, Cuckoos, Roadrunners, and Anis.
Order Strigiformes, Nocturnal Birds of Prey.
        Family Tytonidae, Barn Owls.
        Family Strigidae, Typical Owls.
Order Caprimulgiformes, Oil-birds and Goatsuckers.
        Family Caprimulgidae, Goatsuckers.

Order Apodiformes, Swifts and Hummingbirds.
    Family Apodidae, Swifts.
    Family Trochilidae, Hummingbirds.
Order Coliiformes, Colies.
Order Trogoniformes, Trogons.
Order Coraciiformes, Kingfishers, Motmots, Rollers, Bee-eaters, and Hornbills.
    Family Alcedinidae, Kingfishers.
Order Piciformes, Woodpeckers, Jacamars. Toucans, and Barbets:
    Family Picidae, Woodpeckers.
Order Passeriformes, Perching or Passerine Birds.
    Family Cotingidae, Cotingas.
    Family Tyrannidae, New World Flycatchers.
    Family Alaudidae, Larks.
    Family Hirundinidae, Swallows.
    Family Corvidae, Crows, Magpies, and Jays.
    Family Paridae, Titmice.
    Family Sittidae, Nuthatches.
    Family Certhiidae, Creepers.
    Family Chamaeidae, Wrentits.
    Family Cinclidae, Dippers.
    Family Troglodytidae, Wrens.
    Family Mimidae, Thrashers and Mockingbirds.
    Family Turdidae, Thrushes.
    Family Sylviidae, Gnatcatchers and Kinglets.
    Family Motacillidae, Wagtails and Pipits.
    Family Bombycillidae, Waxwings.
    Family Ptilogonatidae, Silky Flycatchers.
    Family Laniidae, Shrikes.
    Family Sturnidae, Starlings.
    Family Vireonidae, Vireos.
    Family Parulidae, Wood Warblers.
    Family Ploceidae, Weaver Finches.
    Family Icteridae, Blackbirds.
    Family Thraupidae, Tanagers.
    Family Fringillidae, Grosbeaks, Finches, Sparrows, and Buntings.

The above classification, or any such classification, though relying on phylogeny, does not attempt to show the phylogeny of birds. The fact that Struthioniformes heads the list of typical birds in no sense implies that all other birds stem from ostriches. One method for illustrating phylogeny is by means of a **tree** depicting the origin of different groups. For a tree that incorporates much of the latest thinking on avian phylogeny, see Welty (1962:16).

NOMENCLATURE

The system of giving technical names to birds and other animals in the United States and Canada conforms closely to the International Rules of Zoological Nomenclature, more familiarly known as the International Code. Power to act on the Rules is within the province of the International Commission on Zoological Nomenclature which obtains its authority from the International Congresses of Zoology.

The technical (i.e., "scientific") name is a combination of two, sometimes three, Latin or latinized words. The first is the name of the genus in which the bird is placed and is always written with a capital letter. The second and third names are the names of the species and subspecies, respectively, and are never capitalized. The name of the author of the name of the species, or subspecies, is placed directly after the technical name with no intervening punctuation. Parentheses are placed around the author's name if the genus is now different from the one in which the author placed the species originally.

When the technical name is written in long-hand, or is typewritten, it must always be underscored; when printed, it must be italicized. The author's name, however, is never underscored or italicized. Below are several complete technical names and their authors, selected to illustrate certain nomenclatural procedures.

Anas platyrhynchos platyrhynchos Linnaeus

The common subspecies of Mallard in Europe, Asia, and North America, is the **nominate subspecies**, because it shares the species name given by Linnaeus. That subspecies of every polytypic species which has the earliest valid name is called nominate.

Megaceryle alcyon alcyon (Linnaeus)

The subspecies of Belted Kingfisher in eastern North America is another nominate subspecies. In this case Linnaeus' name is in parentheses, because he named and described the species originally in *Alcedo*, a different genus.

Turdus migratorius propinquus Ridgway

The subspecies of Robin in western United States and parts of Mexico. Originally described and named by Ridgway as a species, *Turdus propinquus*, this form was later considered conspecific with *Turdus migratorius*; thus it was relegated to a subspecific status. Since the form still remains in the same genus, Ridgway's name is not placed in parentheses.

Hylocichla mustelina (Gmelin)

The Wood Thrush is a monotypic species. Gmelin originally assigned *mustelina* to the genus *Turdus*, but later systematists considered it a species of *Hylocichla*. Consequently Gmelin's name belongs in parentheses.

Dendroica pensylvanica (Linnaeus)

The Chestnut-sided Warbler is another monotypic species. The name *pensylvanica* exemplifies two nomenclatural procedures:

(1) The species name must remain uncapitalized even when based on a proper name.

(2) The spelling of a species name must be preserved as given by the author even when the spelling of the name is erroneous or does not otherwise conform to standard usage.

The established classification and nomenclature of birds occurring in North America north of Mexico is based on the *Check-list of North American Birds*, prepared by the Committee on Classification and Nomenclature of the American Ornithologists' Union. This is the standard work used by American ornithologists. Other works frequently consulted include Cory *et al.* (1918-49), Mayr and Amadon (1951), Peters (1931-64), Ridgway and Friedmann (1901-50), and Wetmore (1960).

The student should become fully acquainted with *"The A. O. U. Check-list,"* as it is familiarly called, and habitually rely upon it for the proper presentation of all technical and vernacular (i.e., "common" or "popular") names. Moreover, he should rely upon it when seeking the following information:

1. The categories into which birds are classified.
2. The sequence of categories.
3. The original authors and references to their original descriptions.
4. The **type localities**—the places of collection of the **type specimens** upon which the original descriptions were based.
5. The known ranges of species and subspecies.

For further information on nomenclatural procedures the student is referred to the book by Mayr (1969) in which he will find treated in detail the various rules governing the naming of birds and other animals, the significance of type specimens, the methods followed in naming new species and subspecies, and many related matters.

REFERENCES

Adams, C. T.
 1955 Comparative Osteology of the Night Herons. Condor, 57: 55-60. (A taxonomic evaluation of the skeletal characters of *Nycticorax nycticorax* and *Nyctanassa violacea*.)

Amadon, D.
 1943 Bird Weights as an Aid in Taxonomy. Wilson Bull., 55: 164-177.
 1949 The Seventy-five Per Cent Rule for Subspecies. Condor, 51: 250-258.
 1950 The Hawaiian Honeycreepers (Aves, Drepaniidae). Bull. Amer. Mus. Nat. Hist., 95: 151-262.
 1966 The Superspecies Concept. Systematic Zool., 15: 245-249.

American Ornithologists' Union
 1957 Check-list of North American Birds. Fifth edition. (For corrections in this edition, see *The Auk* for 1962, volume 79, pages 493-494.)

Arvey, M. D.
 1951 Phylogeny of the Waxwings and Allied Birds. Univ. Kansas Publ. Mus. Nat. Hist., 3: 473-530.

Ashmole, N. P.
 1968a Competition and Interspecific Territoriality in Empidonax Flycatchers. Systematic Zool., 17: 210-212.
 1968b Body Size, Prey Size, and Ecological Segregation in Five Sympatric Tropical Terns (Aves: Laridae). Systematic Zool., 17: 292-304.

Baker, R. H.
 1951 The Avifauna of Micronesia, Its Origin, Evolution, and Distribution. Univ. Kansas Publ. Mus. Nat. Hist., 3: 1-359.

Baldwin, P. H.
 1953 Annual Cycle, Environment and Evolution in the Hawaiian Honeycreepers (Aves: Drepaniidae). Univ. California Publ. in Zool., 52: 285-398. (An important contribution to an understanding of the evolution of an insular group of birds.)

Beecher, W. J.
 1953 A Phylogeny of the Oscines. Auk, 70: 270-333.

Behle, W. H.
 1950 Clines in the Yellow-throats of Western North America. Condor, 52: 193-219. (A model study of clines in one species.)

Bowman, R. I.
 1961 Morphological Differentiation and Adaptations in the Galapagos Finches. Univ. California Publ. in Zool., 58: i-viii; 1-326.

Brodkorb, P.
 1968 Birds. In *Vertebrates of the United States*. By W. F. Blair, A. P. Blair, P. Brodkorb, F. R. Cagle, and G. A. Moore. Second edition. McGraw-Hill Book Company, New York.

Brown, W. L., Jr., and E. O. Wilson
 1956 Character Displacement. Systematic Zool., 5: 49-64. (The authors propose the term for the phenomenon and illustrate it.)

Clay, T.
 1951 The Mallophaga as an Aid to the Classification of Birds, with Special Reference to the Structure of Feathers. Proc. Xth Internatl. Ornith. Congr., pp. 207-215.

Coble, M. F.
 1954 Introduction to Ornithological Nomenclature. American Book Institute, Los Angeles.

Cockrum, E. L.
 1952 A Check-list and Bibliography of Hybrid Birds in North America North of Mexico. Wilson Bull., 64: 140-159.

Cory, C. B., C. E. Hellmayr, and B. Conover
 1918- Catalogue of Birds of the Americas.
 49 Field Mus. Nat. Hist. Zool. Ser. 13.

Delacour, J., and E. Mayr
 1945 The Family Anatidae. Wilson Bull., 57: 3-55. (A revision based on both morphological and behavioral characters.)
 1946 Supplementary Notes on the Family Anatidae. Wilson Bull., 58: 104-110.

Ficken, M. S.
 1965 Mouth Color of Nestling Passerines and Its Use in Taxonomy. Wilson Bull., 77: 71-75. (With a few exceptions usually a good family character.)

Fisher, J.
 1954 A History of Birds. Houghton Mifflin Company, Boston. (Chapter 4, "Systematics"; Chapter 7, "Bird Speciation.")

Friedmann, H.
 1955 Recent Revisions in Classification and Their Biological Significance. In *Recent Studies in Avian Biology*.

Edited by A. Wolfson. University of Illinois Press, Urbana.

Huntington, C. E.
1952 Hybridization in the Purple Grackle, *Quiscalus quiscula*. Systematic Zool., 1: 149-170.

Johnsgard, P. A.
1961 The Taxonomy of the Anatidae—a Behavioural Analysis. Ibis, 103a: 71-85.

Johnson, N. K.
1963 Biosystematics of Sibling Species of Flycatchers in the *Empidonax hammondii-oberholseri-wrightii* Complex. Univ. California Publ. in Zool., 66: 79-237. (In these three sympatric species, reproductive isolation is maintained "solely through ecologic means." Individuals of all three species come into contact at the time of pair formation and then "segregate by behavioral means into conspecific pairs for reproduction.")
1966 Bill Size and the Question of Competition in Allopatric and Sympatric Populations of Dusky and Gray Flycatchers. Systematic Zool., 15: 70-87.

Johnston, R. F., and R. K. Selander
1964 House Sparrows: Rapid Evolution of Races in North America. Science, 144 (3618): 548-550.

Lack, D.
1945 The Galapagos Finches (Geospizinae): A Study in Variation. California Acad. Sci. Occas. Papers No. 21.
1947 Darwin's Finches. University Press, Cambridge, England.

Lanyon, W. E.
1957 The Comparative Biology of the Meadowlarks *(Sturnella)* in Wisconsin. Publ. Nuttall Ornith. Club No. 1.
1962 Specific Limits and Distribution of Meadowlarks of the Desert Grassland. Auk, 79: 183-207.

Mayr, E.
1942 Systematics and the Origin of Species. Revised edition. Columbia University Press, New York.
1946 The Number of Species of Birds. Auk, 63: 64-69.
1963 Animal Species and Evolution. Balknap Press of Harvard University Press, Cambridge, Massachusetts.
1969 Principles of Systematic Zoology. McGraw-Hill Book Company, New York.

Mayr, E., and D. Amadon
1951 A Classification of Recent Birds. Amer. Mus. Novitates No. 1496: 1-42.

Mengel, R. M.

1964 The Probable History of Species Formation in Some Northern Wood Warblers (Parulidae). Living Bird, 3: 9-43.

Miller, A. H.
1941 Speciation in the Avian Genus Junco. Univ. California Publ. in Zool., 44: 173-434.
1955 Concepts and Problems of Avian Systematics in Relation to Evolutionary Processes. In *Recent Studies in Avian Biology*. Edited by A. Wolfson. University of Illinois Press, Urbana. (Highly recommended reading for any student interested in modern systematics.)

Norris, R. A.
1958 Comparative Biosystematics and Life History of the Nuthatches *Sitta pygmaea* and *Sitta pusilla*. Univ. California Publ. in Zool., 56: 119-300.

Norris, R. A., and G. L. Hight, Jr.
1957 Subspecific Variation in Winter Populations of Savannah Sparrows: A Study in Field Taxonomy. Condor, 59: 40-52.

Packard, G. C.
1967 House Sparrows: Evolution of Populations from the Great Plains and Colorado Rockies. Systematic Zool., 16: 73-89.

Parkes, K. C.
1951 The Genetics of the Golden-winged × Blue-winged Warbler Complex. Wilson Bull., 63: 5-15.
1965 Character Displacement in Some Philippine Cuckoos. Living Bird, 4: 89-98.

Peakall, D. B.
1962 Electrophoresis of Egg-white Proteins as a Taxonomic Tool: A Critical Note. Ibis, 104: 567-568.

Peters, J. L.
1931- Check-list of Birds of the World. Harvard University Press, Cambridge, Massachusetts. (Volumes 1-7, 1931-51, by Peters. Remaining volumes by other authors are appearing at irregular intervals.)

Pitelka, F. A.
1950 Geographic Variation and the Species Problem in the Shore-bird Genus Limnodromus. Univ. California Publ. in Zool., 50: 1-108.
1951 Speciation and Ecologic Distribution in American Jays of the Genus Aphelocoma. Univ. California Publ. in Zool., 50: 195-464.

Rand, A. L.
1959 Tarsal Scutellation of Song Birds as a Taxonomic Character. Wilson Bull.,

71: 274-277. (An instance in classifying shrikes where tarsal scutellation fails as a key taxonomic character.)

Ridgway, R., and H. Friedmann
1901- The Birds of North and Middle Amer-
50   ica: A Descriptive Catalogue of the Higher Groups, Genera, Species and Subspecies of Birds Known to Occur in North America, from the Arctic Lands to the Isthmus of Panama, the West Indies and Other Islands of the Caribbean Sea, and the Galapagos Archipelago. Bull. U. S. Natl. Mus. No. 50.

Salomonsen, F.
1965 The Geographical Variation of the Fulmar *(Fulmarus glacialis)* and the Zones of Marine Environment in the North Atlantic. Auk, 82: 327-355.

Selander, R. K., and D. R. Giller
1959 Interspecific Relations of Woodpeckers in Texas. Wilson Bull., 71: 107-124.
1961 Analysis of Sympatry of Great-tailed and Boat-tailed Grackles. Condor, 63: 29-86.

Sharpe, R. B.
1899- A Hand-List of the Genera and Spe-
1909 cies of Birds. British Museum, London.

Short, L. L., Jr.
1963 Hybridization in the Wood Warblers *Vermivora pinus* and *V. chrysoptera.* Proc. XIIIth Internatl. Ornith. Congr., pp. 147-160.
1965 Hybridization in the Flickers *(Colaptes)* of North America. Bull. Amer. Mus. Nat. Hist., 129: 307-428.

Sibley, C. G.
1950 Species Formation in the Red-eyed Towhees of Mexico. Univ. California Publ. in Zool., 50: 109-194. (The author demonstrates that the Eastern Towhee, *Pipilo erythrophthalmus,* Swainson's Towhee, *P. macronyx,* and Spotted Towhee, *P. maculatus,* are conspecific and, therefore, subspecies.)
1954 Hybridization in the Red-eyed Towhees of Mexico. Evolution, 8: 252-290.
1957 The Evolutionary and Taxonomic Significance of Sexual Dimorphism and Hybridization in Birds. Condor, 59: 166-191.
1960 The Electrophoretic Patterns of Avian Egg-white Proteins as Taxonomic Characters. Ibis, 102: 215-284.
1961 Hybridization and Isolating Mechanisms. In *Vertebrate Speciation.* Edited by W. F. Blair. University of Texas Press, Austin.
1962 The Comparative Morphology of Protein Molecules as Data for Classification. Systematic Zool., 11: 108-118.
1967 Proteins: History Books of Evolution. Discovery, 3: 5-20.

Sibley, C. G., and O. S. Pettingill, Jr.
1955 A Hybrid Longspur from Saskatchewan. Auk, 72: 423-425.

Sibley, C. G., and L. L. Short, Jr.
1959 Hybridization in the Buntings *(Passerina)* of the Great Plains. Auk, 76: 443-463.

Sibley, C. G., and D. A. West
1958 Hybridization in the Red-eyed Towhees of Mexico: The Eastern Plateau Populations. Condor, 60: 85-104.
1959 Hybridization in the Rufous-sided Towhees of the Great Plains. Auk, 76: 326-338.

Stein, R. C.
1963 Isolating Mechanisms between Populations of Traill's Flycatchers. Proc. Amer. Phil. Soc., 107: 21-50. (Song is the main isolating mechanism.)

Storer, R. W.
1952 A Comparison of Variation, Behavior and Evolution in the Sea Bird Genera Uria and Cepphus. Univ. California Publ. in Zool., 52: 121-222.
1960 The Classification of Birds. In *Biology and Comparative Physiology of Birds.* Volume 1. Edited by A. J. Marshall. Academic Press, New York.

Tordoff, H. B.
1954a Relationships in the New World Nine-primaried Oscines. Auk, 71: 273-284.
1954b A Systematic Study of the Avian Family Fringillidae Based on the Structure of the Skull. Univ. Michigan Mus. Zool. Misc. Publ. No. 81.

Welty, J. C.
1962 The Life of Birds. W. B. Saunders Company, Philadelphia.

West, D. A.
1962 Hybridization in Grosbeaks *(Pheucticus)* of the Great Plains. Auk, 79: 399-424.

Wetmore, A.
1960 A Classification for the Birds of the World. Smithsonian Misc. Coll., 139 (11): 1-37.

Woods, R. S.
1944 The Naturalist's Lexicon. A List of Classical Greek and Latin Words Used or Suitable for Use in Biological Nomenclature. Abbey Garden Press, Pasadena, California.

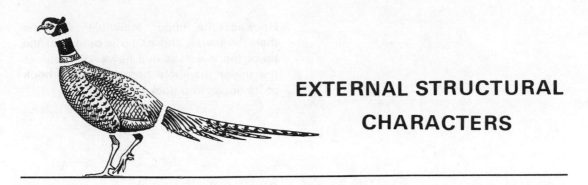

EXTERNAL STRUCTURAL
CHARACTERS

The preceding section of this book indicated the methods used in classifying birds into categories and pointed out that combinations of taxonomic characters determined these categories.

The laboratory identification of bird specimens to orders, families, genera, and sometimes species is commonly based on those taxonomic characters which are morphological and at the same time visible externally. They are usually called **external structural characters.** All keys and synopses employ them. Before undertaking laboratory identification, therefore, the student must know the external structural characters in order to be able to use keys and synopses effectively.

In the following pages are outlined the common external structural characters. Definitions of the terms used in describing them are included, together with the names of certain birds in which the characters are exemplified. Study these characters from actual specimens, and when spaces are available on the left-hand side of the pages, make a series of sketches similar to those already drawn.

Although "popular" manuals, keys, and synopses do not always employ technical terms in describing characters, a thorough student should be familiar with such terms and be able to use them when the occasion demands.

CHARACTERS OF THE BILL

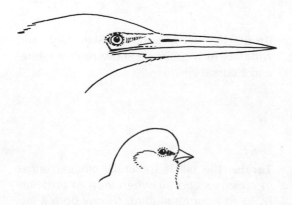

Long: the bill is decidedly longer than the head, as in a bittern.

Short: the bill is decidedly shorter than the head, as in a redpoll.

Hooked: the upper mandible is longer than the lower, and its tip is bent over the tip of the lower, as in a hawk. (Sometimes the upper mandible has a **nail-like hook** at its tip, as in a duck.)

Crossed: the tips of the mandibles cross each other, as in a crossbill.

Compressed: the bill for a good part of its length is higher than wide, as in a puffin or a kingfisher. (Show both a lateral and a frontal view.)

Depressed: the bill is wider than high, as in a duck. (Show both a lateral and a frontal view.)

Stout: the bill is conspicuously high and wide, as in a grouse. (Show both a lateral and a dorsal view.)

Terete: the bill is generally circular either in cross section, or when viewed anteriorly, as in a hummingbird. (Show both a lateral and a frontal view.)

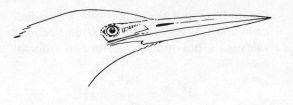

Straight: the line along which the mandibles close (i.e., the commissure) is in line with the axis of the head, as in a bittern.

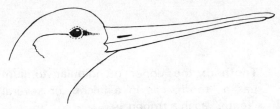

Recurved: the bill curves upward, as in a godwit.

Decurved: the bill curves downward, as in the Brown Creeper *(Certhia familiaris)*, or a curlew.

Bent: the bill is deflected at an angle (usually deflected downward at the middle), as in a flamingo.

Swollen: the sides of the mandibles are convex, as in a tanager *(Piranga)*. (Show a dorsal view.)

Acute: the bill tapers to a sharp point, as in the Yellow Warbler *(Dendroica petechia)*.

Chisel-like: the tip of the bill is beveled, as in the Hairy Woodpecker *(Dendrocopos villosus)*. (Show both a lateral and a dorsal view.)

Toothed: the upper mandibular tomium has a "tooth," as in a falcon, or several "teeth," as in a trogon.

Serrate: the bill has saw-like tomia, as in a merganser.

Gibbous: the bill has a pronounced hump, as in a scoter.

Spatulate, or **spoon-shaped**: the bill is much widened, or depressed, toward its tip, as in the Shoveler *(Spatula clypeata)*.

Notched: the bill has a slight nick in the tomia of one or both mandibles. Most frequently the notch occurs near the tip of the upper mandible, as in a thrush.

Conical: the bill has the shape of a cone, as in a redpoll.

Lamellate, or **sieve-billed**: the mandibles have just within their tomia a series of transverse tooth-like ridges, as in swans, geese, and ducks.

With **angulated commissure**: the commissure forms an angle at the point where the tomium proper meets the rictus, as in a grosbeak, finch, sparrow, or bunting. (Show the mouth closed. See Plate III, Figure 1, for the character in the House Sparrow with mouth open.)

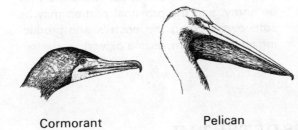

With **gular sac:** the chin, gular region, and jugulum are distended. In the pelican, the gular sac is conspicuous, outwardly membranous, and featherless; in the cormorant it is inconspicuous and partially feathered.

Cormorant   Pelican

The nostrils are generally separated from each other by a complete wall, or septum; they are, therefore, **imperforate.** A few groups of birds, such as the vultures, have nostrils without a medial septum; they communicate with each other and are, therefore, **perforate.** (See Plate XXVIII, Figure 6.) Nostrils show other characters:

Tubular: the nostrils are in the ends of short prolongations of the base of the upper mandible, as in an albatross, a shearwater, or a petrel.

Linear, oval, or **circular**: the nostril openings are thus shaped, as in a gull, an accipitrid hawk, and a falcon, respectively. The nostrils in the falcon possess **bony tubercles.**

The covering of the bill is generally horny throughout. Sometimes, as in the shore birds, it is **soft.** The covering may show two special modifications which constitute important characters.

The distal end of the upper mandible may be horny, and the proximal portion may be thick and soft, producing a **cere**, as in a hawk.

The distal end of the upper mandible may be horny, and the proximal portion may be soft, over-arching the nostrils and producing an **operculum,** as in a pigeon.

CHARACTERS OF THE TAIL

A tail is said to be **long** when it is decidedly longer than the trunk, as in a pheasant, or a cuckoo, and **short,** when it is either approximately the length of, or shorter than, the trunk, as in shore birds.

Due to the different relative lengths of the rectrices, the posterior margin of the tail assumes various shapes which are distinguishing characters.

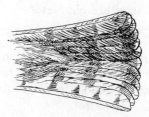

Square: the rectrices are all of the same length, as in the Sharp-shinned Hawk *(Accipiter striatus).*

Rounded: the rectrices shorten successively from the inside to the outside, in slight gradations, as in a crow.

Graduated: the rectrices shorten successively from the inside to the outside, in abrupt gradations, as in a cuckoo.

Pointed, or **acute**: the middle rectrices are much longer than the others, as in a Ring-necked Pheasant *(Phasianus colchicus)*.

Emarginate: the rectrices increase in length successively from the middle to the outermost pair, in slight gradations, as in a finch.

Forked: the rectrices increase in length successively from the middle to the outermost pair, in abrupt gradations, as in a tern.

CHARACTERS OF THE WINGS

A wing is said to be **long** when the distance from the bend to the tip is decidedly longer than the trunk, as in a tern, and **short** when the distance is either approximately the length of, or shorter than, the trunk, as in a grebe.

Spurred: the bend of the wing has a peculiar horny structure in the shape of a spur, as in a jacana.

The varying length of the primaries in different species causes the wing to assume different shapes.

Rounded: the middle primaries are the longest, and the remaining primaries are graduated, as in the Sharp-shinned Hawk *(Accipiter striatus)*.

Pointed: the outermost primaries are the longest, as in a gull.

The varying length of both primaries and secondaries in different species causes wings to show differences in width. A wing is **narrow** when the primaries, and particularly the secondaries, are relatively short throughout, as in a gull. A wing is **broad** when both the primaries and the secondaries are very long throughout, as in the Sharp-shinned Hawk. (See above drawings.)

The surface of the spread wing may vary in curvature. Although it is somewhat convex above and concave below, the curvature may sometimes be extreme, or it may sometimes be very slight. If the curvature is extreme, the wing is said to be **concave,** as in a grouse. If it is very slight, the wing is said to be **flat**, as in a swift or a hummingbird. (See drawings below.)

Wing of Grouse   Wing of Swift

CHARACTERS OF THE FEET

The tibia, when featherless, and the tarsus and toes usually have a horny investment. In different birds this investment is variously cut up.

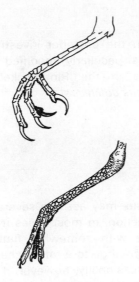

Scutellate: the investment is cut up into more or less imbricated (overlapping) scales, as the tarsus of a grosbeak, finch, sparrow, or bunting.

Reticulate: the investment is cut up into small irregular plates, as the tarsus of a plover.

Serrate: the investment plates have serrations, as on the posterior edge of the tarsus in a grebe.

Scutellate-reticulate: the investment is scutellate in front and reticulate behind, as in a pigeon.

Booted: the investment of the tarsus is continuously horny without scales or plates, as in a thrush.

Scutellate-booted: the tarsus is scutellate in front and booted behind, as in the Catbird *(Dumetella carolinensis).*

Spurred: the posterior investment of the tarsus is peculiarly modified to form a spur, as in the Ring-necked Pheasant *(Phasianus colchicus).*

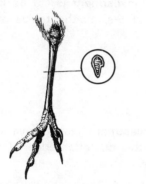

The tarsus may assume several shapes in cross section. In most cases it is **rounded in front** with somewhat flattened sides which converge to a rather sharp ridge behind. Occasionally, however, it is **rounded in front and behind,** as in a tyrannid flycatcher, and in the Horned Lark *(Eremophila alpestris).* (See the drawings for a comparison.) The tarsus of loons, grebes, and a few other aquatic birds is **compressed,** being very flat from side to side with rather sharp edges in front and behind.

Tarsus with Sharp   Tarsus Rounded
Ridge Behind   Behind

The position of the toes is important. In all birds the front toes are inserted on the metatarsus at the same level. But the hind toe, or hallux, varies in position.

Incumbent: the hallux is inserted on the metatarsus at the level of the other toes, as in a meadowlark.

Elevated: the hallux is inserted so high on the metatarsus that its tip does not reach the ground, as in a rail.

The nails of birds are generally curved and sharp-pointed. They are rounded above, flattened from side to side, and somewhat concave below. In certain birds these nails vary from the ordinary.

Acute: the nails are extremely curved and sharp-pointed, as in a woodpecker.

Obtuse: the nails are less curved and have rather blunt points, as in a grouse.

Lengthened: the nails are rather straight and elongated but sharp-pointed, as in the hallux nail of the Horned Lark *(Eremophila alpestris)*.

Pectinate: the nails have serrated edges, as the middle nail of a heron.

Flattened: the nails are so extremely flattened and broadened as to resemble a human finger nail, as in a grebe.

Birds' feet are of several types, depending on the arrangement of the toes and/or the particular functions which the feet perform. The ten types below are commonly used as characters for distinguishing groups of birds. Each type is subject to variation.

Anisodactyl: the hallux is behind and the other three toes are in front, as in a thrush.

Syndactyl: the third and fourth toes (outer and middle) are united for most of their length and have a broad sole in common, as in the Belted Kingfisher *(Megaceryle alcyon)*.

Zygodactyl: the toes are arranged in pairs, the second and third toes in front, the fourth and hallux behind, as in a woodpecker.

Heterodactyl: the toes are arranged in pairs, in this case, the third and fourth toes in front, the second and hallux behind, as in a trogon.

Pamprodactyl: all four toes are in front, the hallux being turned forward, as in a swift.

Raptorial: the toes are deeply cleft, with large, strong, sharply curved nails (talons), as in a hawk.

Semipalmate, or half-webbed: the anterior toes are joined part way by a small webbing, as in the Semipalmated Plover *(Charadrius semipalmatus)*, Willet *(Catoptrophorus semipalmatus)*, or Semipalmated Sandpiper *(Ereunetes pusillus)*.

Totipalmate, or fully webbed: all four toes are united by ample webs, as in a cormorant.

Palmate, or webbed: the front toes are united by ample webs, as in ducks and gulls.

Lobate, or lobed: a swimming foot with a series of lateral lobes on the toes, as in a grebe. Sometimes the foot may be palmate, but the hallux may bear a lobe, as in a diving duck.

CHARACTERS OF THE PLUMAGE

The distribution of plumage. Certain parts usually covered by feathers may be without well-developed feathers, or **bare,** as the lores of herons and the entire heads and upper necks of New World vultures (for example, the Turkey Vulture, *Cathartes aura;*

see Plate XXVIII, Figure 6). Certain parts usually uncovered may be **feathered,** as the tarsi of some owls.

The texture of plumage. The feathers of the goatsuckers and owls are generally **soft;** the rectrices of woodpeckers are **stiffened;** the tufts of feathers covering the nostrils of crows are **tough** and **bristle-like;** the plumage of wrentits is **lax;** certain rectrices of anhingas possess **flutings;** and the barbs of the outer vanes of the outermost primaries of the Rough-winged Swallow, *Stelgidopteryx ruficollis*, have stiffly hooked tips, which give the bird its name (see Plate XXVIII, Figure 9).

Some of the contour feathers may be modified to form **"horns,"** as in the Horned Lark, *Eremophila alpestris* (Plate XXVIII, Figure 1); **crests,** as in the Cedar Waxwing, *Bombycilla cedrorum* (Plate XXVIII, Figure 2); **ruffs,** as in the Ruffed Grouse, *Bonasa umbellus*, and **pinnae,** as in the Greater Prairie Chicken, *Tympanuchus cupido* (Plate XXVIII, Figures 3 and 4); **"ears"** and **facial discs,** as in the Great Horned Owl, *Bubo virginianus* (Plate XXVIII, Figure 5); the highly colored area, the **speculum,** on the secondaries of several ducks (Plate XXVIII, Figure 10); the **rictal bristles,** as in many kinds of birds such as the Brown Thrasher, *Toxostoma rufum*, and Whip-poor-will, *Caprimulgus vociferus* (Plate XXVIII, Figures 7 and 8).

Contour feathers are commonly **lanceolate,** like the primaries of pigeons. They may, however, be peculiarly modified in shape and structure.

Notched: a vane of the contour feather is incised toward the end, as the proximal vanes of the outer primaries of the Broad-winged Hawk *(Buteo platypterus)*.

Spinose: the shaft of the contour feather is prolonged distally without barbs, as in the rectrices of the Chimney Swift *(Chaetura pelagica)*.

Acuminate: the contour feather ends in a sharp point, as the rectrices of woodpeckers.

Plate XXVIII

Figure 1
Horned Lark

Figure 2
Cedar Waxwing

Figure 3
Ruffed Grouse

Figure 4
Greater Prairie Chicken

Figure 5
Great Horned Owl

Figure 6
Turkey Vulture

Figure 7
Brown Thrasher

Figure 8
Whip-poor-will

Figure 9
Rough-winged Swallow Wing

Figure 10
Duck Wing

CHARACTERS OF THE PLUMAGE

Filamentous, or **attenuate**: the contour feather is long and extremely narrow, as the outer rectrices of the Barn Swallow *(Hirundo rustica),* or the outermost primaries of the American Woodcock *(Philohela minor).*

Broad: the contour feather is extremely wide, as the rectrices of a trogon.

MISCELLANEOUS CHARACTERS

Numerous integumentary outgrowths may occur elsewhere than on the bill, wings, and feet, thus providing distinguishing characters. In the head region there may be small **eye scales** above and below the eyes, as in puffins; a **frontal plate** at the base of the upper mandible, as in gallinules; and **wattles** and **caruncles,** as in turkeys. (See the drawings below.)

| Puffin | Gallinule | Turkey |

SYNOPSIS OF NORTH AMERICAN ORDERS AND FAMILIES OF BIRDS

The orders and families of birds (except family Cotingidae) found in North America north of Mexico are presented below, together with their distinguishing external structural characters.

The methods employed in preparing the synopsis require brief explanation. Characters given are limited to those of bill, tail, wings, feet (including legs), and plumage, as outlined in the preceding pages. "Negative" characters are usually not given (i.e., absence of a character is not mentioned). Unless otherwise indicated, the characters listed for an order, or a family, apply not only to the North American representatives but to the entire group. Measurements given are the extremes in length among the species in each group treated. They are taken from bird skins and are at best only approximate.

The student should use the synopsis not only as a reference but also as a direct means of learning the distinguishing characters of each bird group. He can do this by first selecting conspicuous characters such as hooked bill, forked tail, or palmate feet, then listing after each character the bird groups possessing it. Later, as time permits, he may select the less conspicuous characters and proceed in the same manner. By learning these lists he will gain a knowledge that will greatly facilitate the work in laboratory identification of birds that follows in the next section of this manual.

The student desiring additional characters distinguishing the orders and families of North American Birds should consult Part 5 on birds by P. Brodkorb in *Vertebrates of the United States* (McGraw-Hill Book Company, New York, 1968). For characters distinguishing all the bird families of the world, the student should consult Chapter 13, "The Classification of World Birds by Families," in *Fundamentals of Ornithology* by J. Van Tyne and A. J. Berger (John Wiley and Sons, New York, 1959).

Order GAVIIFORMES. Family **Gaviidae.** Loons. (Length: 24-38 inches.)
Bill: straight; acute; compressed. **Tail:** short; rectrices stiff. **Wings:** well-developed but short; somewhat pointed. **Feet:** tarsi compressed and reticulate; toes four, and palmate.

Order PODICIPEDIFORMES. Family **Podicipedidae.** Grebes. (Length: 7.5-26 inches.)
Bill: straight; acute (one exception); compressed. **Tail:** rudimentary. **Wings:** poorly developed and short; somewhat pointed. **Feet:** tarsi compressed and scutellate with serrate posterior edges; toes four, and lobed; nails flattened.

Order PROCELLARIIFORMES. Tube-nosed Swimmers.
Bill: hooked; nostrils tubular. **Tail:** short to moderately long. **Wings:** long; narrow. **Feet:** palmate; hallux rudimentary or absent. **Plumage:** predominantly black and gray.

Family **Diomedeidae.** Albatrosses. (Length: 28-40 inches.)
Bill: nostril tubes lateral, separated by culmen.

Family **Procellariidae.** Shearwaters and Fulmars. (Length: 10-34 inches.)
Bill: nostril tubes on culmen; nostrils imperforate.

Family **Hydrobatidae.** Storm Petrels. (Length: 6-10 inches.)
Bill: nostrils on culmen, united in one tube.

Order PELECANIFORMES. Totipalmate Swimmers.
Bill: with gular sac. **Feet:** toes four, and totipalmate.

Family **Phaëthontidae.** Tropicbirds. (Length: 24-40 inches.)
Bill: as long as head; straight; compressed; acute; gular sac rudimentary; nostrils small and linear. **Tail:** pointed; middle rectrices filamentous. **Wings:** long; pointed. **Feet:** relatively small; hallux more elevated than in following families of order.

Family **Pelecanidae.** Pelicans. (Length: 52-72 inches.)
Bill: very long; straight; hooked; nostrils absent; gular sac large. **Tail:** very short. **Wings:** very long; rounded. **Feet:** legs short and stout; tarsi compressed and reticulate.

Family **Sulidae.** Gannets. (Length: 28-36 inches.)
Bill: slightly longer than head; straight, slightly decurved at tip; bluntly acute; exterior nostrils absent; gular sac very small. **Tail:** long; pointed. **Wings:** long; pointed. **Feet:** legs short and stout.

Family **Phalacrocoracidae.** Cormorants. (Length: 22-36 inches.)
Bill: long as head; straight; hooked; exterior nostrils absent; gular sac very small. **Tail:** long; rounded. **Wings:** short; rounded.

Family **Anhingidae.** Anhingas. (Length: 28-36 inches.)
Bill: long; straight; slender; acute; nostrils minute; gular sac moderate. **Tail:** long; rounded; middle pair of rectrices with flutings. **Wings:** moderately long.

Family **Fregatidae.** Frigatebirds. (Length: 30-40 inches.)
Bill: long; straight; hooked; nostrils small and linear. **Tail:** long; deeply forked. **Wings:** long; pointed. **Feet:** legs short; feet small; tarsi partly feathered; middle nail pectinate.

Order CICONIIFORMES. Deep-water Waders.
Bill: long. **Tail:** short. **Wings:** long; broad; rounded. **Feet:** legs very long; toes four. **Plumage:** lores usually bare.

Family **Ardeidae.** Herons and Bitterns. (Length: 10-56 inches.)
Bill: straight; acute. **Feet:** tarsi usually scutellate in front; toes long and on same level; middle nail pectinate. **Plumage:** lax; frequently with modified plumes; powder-down tracts.

Family **Ciconiidae.** Storks and Wood Ibises. (Length: 28-60 inches.)
Bill: straight; stout at base; sometimes recurved at tip; sometimes decurved at tip (American form). **Feet:** tarsi reticulate; toes long; hallux slightly elevated.

Family **Threskiornithidae.** Ibises and Spoonbills. (Length: 19-38 inches.)
Bill: either decurved, slender throughout, and somewhat terete; or straight, broad, and spatulate. **Feet:** tarsi usually reticulate; hallux slightly elevated.

Family **Phoenicopteridae.** Flamingos. (Length: 36-53 inches.)
Bill: bent in middle; lamellate. **Feet:** tarsi scutellate; hallux elevated; palmate.

Order ANSERIFORMES. Family **Anatidae.** Lamellate-billed Swimmers. (Length: 12-72 inches.)
Bill: either lamellate, or serrate (one subfamily); broad and depressed (except two subfamilies); with nail-like hook. **Tail:** short (except one subfamily); usually rounded. **Wings:** either pointed, or rounded (one subfamily). **Feet:** toes four, and palmate; hallux elevated and either lobate (three subfamilies), or not lobate. **Plumage:** lores either feathered, or bare (one subfamily).

Subfamily **Cygninae.** Swans. (Length: 47-72 inches.)
Feet: tarsi completely reticulate and shorter than middle toe with nail. **Plumage:** lores bare.

Subfamily **Anserinae.** Geese. (Length: 21-39 inches.)
Bill: compressed at base; narrowing toward tip. **Feet:** tarsi completely reticulate and longer than middle toe without nail.

Subfamily **Dendrocygninae.** Tree Ducks. (Length: 13-20 inches.)
Wings: rounded. **Feet:** tarsi completely reticulate and longer than middle toe with nail; legs, feet, and nails exceptionally long.

Subfamily **Anatinae.** Dabbling Ducks. (Length: 12-29 inches.)
Wings: usually with iridescent speculum. **Feet:** tarsi scutellate in front and shorter than middle toe without nail.

Subfamily **Aythyinae.** Diving Ducks. (Length: 12-27 inches.)
Wings: usually with non-iridescent speculum. **Feet:** tarsi scutellate in front and shorter than middle toe without nail; hallux lobate.

Subfamily **Oxyurinae.** Ruddy and Masked Ducks. (Length: 13-16 inches.)
Tail: long; rectrices narrow and stiffened; coverts extremely short. **Feet:** tarsi scutellate in front and shorter than middle toe with nail; hallux lobate.

Subfamily **Merginae.** Mergansers. (Length: 16-27 inches.)
Bill: narrow; terete; serrate. **Feet:** tarsi compressed, scutellate in front, and shorter than middle toe with nail; hallux lobate.

Order FALCONIFORMES. New World Vultures and Diurnal Birds of Prey.
Bill: hooked; with cere in which the nostrils open centrally. **Feet:** anisodactyl and raptorial. **Plumage:** lores with bristle-like feathers only.

Family **Cathartidae.** New World Vultures. (Length: 24-54 inches.)
Bill: moderately hooked; nostrils large, oval, and perforate. **Wings:** long, broad, and

rounded. **Feet:** weakly raptorial; hallux less than half the length of the middle toe; tarsi reticulate. **Plumage:** head in adults bare.

Family **Accipitridae.** Kites, Eagles, and Hawks. (Length: 10-48 inches.)
Bill: strongly hooked; nostrils small, usually oval, or slit-like with anterior end upper most, and imperforate. **Wings:** long, broad, and rounded (except in kites, in which they are long, narrow, and pointed). **Feet:** strongly raptorial; hallux usually the same length as, or slightly longer than, the shortest front toe; hallux nail larger than other nails; under surfaces of all nails grooved; tarsi more often scutellate than reticulate, feathered, or booted.

Family **Pandionidae.** Ospreys. (Length: 21-24.5 inches.)
Bill: strongly hooked; nostrils obliquely oval, with anterior end uppermost, and imperforate. **Wings:** long and pointed. **Feet:** strongly raptorial; hallux about the same length as the middle toe; hallux nail approximately the same size as other nails; under surfaces of nails rounded; under surfaces of toes with spiny scales; tarsi reticulate.

Family **Falconidae.** Caracaras and Falcons. (Length: 10-25 inches.)
Bill: strongly hooked; upper mandible toothed near tip (faintly so in caracaras); nostrils either circular with prominent central bony tubercle, or (as in caracaras) slit-like with posterior end uppermost, and imperforate. **Wings:** long, narrow, and pointed (except in caracaras, in which they are long, broad, and rounded). **Feet:** strongly raptorial; hallux usually the same length as, or slightly longer than, the shortest front toe; under surfaces of nails grooved; tarsi reticulate (front distinctly scutellate in caracaras).

Order GALLIFORMES. Gallinaceous Birds.
Bill: short, stout, culmen decurved; tip of upper mandible bent slightly over tip of lower mandible. **Wings:** short; concave; rounded; primaries stiff and usually curved. **Feet:** strong, tarsi scutellate when not feathered; nails obtuse.

Family **Cracidae.** (Genus *Ortalis* only.) Chachalacas. (Length: 20-24 inches.)
Bill: nostrils exposed. **Tail:** long; rounded. **Feet:** hallux incumbent. **Plumage:** area around the eyes and sides of the throat bare.

Family **Tetraonidae.** Grouse and Ptarmigans. (Length: 12-30 inches.)
Bill: nostrils feathered. **Tail:** variable in shape. **Feet:** tarsi feathered wholly or in part; hallux elevated; toes somewhat pectinate, especially in winter. **Plumage:** sides of neck often with inflatable air sacs and/or modified, erectile feathers.

Family **Phasianidae.** Quails and Pheasants. (Length: 8-36 inches.)
Bill: nostrils partly feathered or exposed. **Tail:** either extremely long and usually pointed, or extremely short and usually rounded. **Feet:** hallux elevated and sometimes spurred. **Plumage:** areas on head sometimes bare and wattled.

Family **Meleagrididae.** Turkeys. (Length: 30-40 inches.)
Bill: nostrils exposed. **Tail:** wide and rounded. **Feet:** hallux elevated and spurred in males. **Plumage:** with evident luster; head and neck bare, wattled and carunculated; individual body feathers and rectrices wide and somewhat square at ends.

Order GRUIFORMES. Suborder Grues. Marsh Birds.
Bill: variable in shape; nostrils perforate. **Tail:** short. **Wings:** rounded; tertiaries often as long as the primaries. **Plumage:** lores feathered, or with bristles.

Family **Gruidae.** Cranes. (Length: 33-54 inches.)
Bill: as long as, or slightly longer than, the head; straight; compressed. **Wings:** broad and long. **Feet:** legs very long; hallux elevated and short, being equal to the length of nail of middle toe. **Plumage:** lores and crown with bristles.

Family **Aramidae.** Limpkins. (Length: 24-28 inches.)
Bill: long; compressed; decurved at tip. **Wings:** short; outer primary stiff, attenuate, and incurved. **Feet:** legs moderately long; hallux elevated and twice the length of nail of middle toe. **Plumage:** predominantly brownish.

Family **Rallidae.** Rails, Gallinules, and Coots. (Length: 5-20 inches.)
Bill: variable in length and shape. **Wings:** short. **Feet:** legs moderately long, hallux elevated and longer than nail of middle toe.

Order CHARADRIIFORMES. Suborder Charadrii. Shore Birds.
Bill: as long as, or longer than, the head; usually slender and soft. **Tail:** short. **Wings:** long; pointed (except Jacanidae and a few aberrant species); tertiaries greatly lengthened. **Feet:** legs usually long and attached near middle of body; tibiae often bare; hallux short and elevated, or absent; anterior toes cleft to base, or sometimes semipalmate, or palmate.

Family **Jacanidae.** Jacanas. (Length: 6-12 inches.)
Bill: as long as head; straight; compressed; frontal plate above bill on forehead. **Wings:** spurred. **Feet:** toes very long; nails long, slender, acute; hallux nail much longer than hallux and somewhat recurved.

Family **Haematopodidae.** Oystercatchers. (Length: 16-21 inches.)
Bill: twice the length of the head; straight; compressed; constricted near base; tip chisel-like. **Tail:** square to slightly rounded. **Feet:** legs and toes stout; tarsi shorter than bill and reticulate; hallux absent; small webbing between toes. **Plumage:** black and white.

Family **Charadriidae.** Plovers, Surfbirds, and Turnstones. (Length: 5-15 inches.)
Bill: moderate in length, usually under one inch; lateral profile either slightly constricted at base, then tapering to acute tip (turnstones), or constricted in middle and swollen toward tip (plovers and surfbirds). **Feet:** tarsi either scutellate in front and reticulate behind (turnstones), or entirely reticulate (surfbirds and plovers); toes three (four in Black-bellied Plover, *Squatarola squatarola;* also in surfbirds and turnstones).

Family **Scolopacidae.** Woodcock, Snipe, and Sandpipers. (Length: 5-24 inches.)
Bill: variable in length; usually slender, pliable in life, and soft, with tip somewhat depressed. **Feet:** tarsi always scutellate in front and usually behind; toes four (three in Sanderling, *Crocethia alba*).

Family **Recurvirostridae.** Avocets and Stilts. (Length: 13-18 inches.)
Bill: long; slender; recurved. **Feet:** legs extremely long; tarsi reticulate; hallux either absent (stilts), or rudimentary (avocets); toes either palmate (avocets), or with webbing cleft nearly to base of toes (stilts).

Family **Phalaropodidae.** Phalaropes. (Length: 7-9 inches.)
Bill: less than an inch and a half in length; either slender and somewhat acute, or rather stout and appearing flattened near tip when viewed dorsally. **Feet:** tarsi compressed and scutellate in front; toes with lateral membranes, developed into lobes (lobes are indistinct in Wilson's Phalarope, *Steganopus tricolor*). **Plumage:** dense on under parts.

Order CHARADRIIFORMES. Suborder Lari. Gulls.
Bill: relatively short and stout; variable in shape; nostrils perforate. **Tail:** variable in length and shape. **Wings:** very long and relatively narrow; pointed. **Feet:** legs relatively short and stout and attached near center of body; tibiae partly bare; tarsi scutellate in front and reticulate elsewhere; hallux small and elevated (though sometimes rudimentary) or absent; anterior toes palmate. **Plumage:** compact on under parts; predominant coloration black, white, and gray.

Family **Stercorariidae.** Skuas and Jaegers. (Length: 20-24 inches.)
Bill: with nail-like hook; cere present. **Tail:** shorter than wing, except when middle rectrices are elongated; rounded to pointed. **Feet:** strong; tarsi longer than middle toe without nail; nails long, strongly hooked, and acute.

Family **Laridae.** Gulls and Terns. (Length: 9-30 inches.)
Bill: either hooked, but without nail-like hook (gulls), or straight throughout and somewhat acute (terns). **Tail:** usually either square to slightly rounded (gulls), or forked (terns). **Feet:** either extremely small and short, the tarsus being less than one-tenth as long as the wings (terns), or of moderate size and length, the tarsus being more than one-tenth as long as the wings (gulls).

Family **Rynchopidae.** Skimmers. (Length: 16-20 inches.)
Bill: straight; compressed to thinness of knife blade; lower mandible notably longer than the upper, and blunt at tip; upper mandible less blunt at tip. **Tail:** forked. **Feet:** small; tarsus short, approximately one-eleventh as long as the wing.

Order CHARADRIIFORMES. Suborder Alcae. Auks.
Family **Alcidae.** Auks, Murres, and Puffins. (Length: 8-18 inches.)

Bill: variable in shape and length; nostrils imperforate. **Tail:** short. **Wings:** moderately long to short; pointed. **Feet:** legs of moderate length and size and attached far back; tibiae bare near heel; tarsi compressed and either wholly reticulate or partly scutellate; hallux absent; palmate; nails curved and acute but not large. **Plumage:** compact throughout; head sometimes crested.

Order COLUMBIFORMES. Family **Columbidae.** Doves and Pigeons. (Length: 6-17 inches.)

Bill: relatively small and slender; basal part soft; terminal part horny with decurved culmen; middle part constricted; nostrils usually slit-like and overhung by operculum. **Tail:** long; in North American forms may be either square, rounded, or pointed. **Wings:** long and flat but variable in shape. **Feet:** tarsi scutellate in front (sometimes feathered on proximal part) and reticulate elsewhere; toes cleft to base; sometimes with slight webbing; hallux incumbent to slightly elevated. **Plumage:** dense; region in vicinity of eyes often quite bare.

Order PSITTACIFORMES. Family **Psittacidae.** Parrots, Parakeets, and Macaws. (Length: 5-36 inches.)

Bill: short and stout; culmen greatly decurved; strongly hooked; cere in which nostrils open. **Tail:** variable in form. **Wings:** variable in form. **Feet:** tarsi shorter than longest toe and reticulate (sometimes peculiarly granulated); zygodactyl; fourth toe reversible.

Order CUCULIFORMES. Family **Cuculidae.** Cuckoos, Roadrunners, and Anis. (Length: 5-24 inches.)

Bill: variable in size and shape; usually compressed; more or less decurved. **Tail:** usually long and graduated. **Wings:** variable in form. **Feet:** variable in size; zygodactyl; fourth toe permanently reversed. **Plumage:** predominantly brown and gray but occasionally black.

Order STRIGIFORMES. Nocturnal Birds of Prey.

Bill: hooked; culmen strongly decurved; cere present; nostrils opening at edge of cere except in genus *Speotyto*. **Tail:** variable in form. **Wings:** variable in form; inner webs of certain primaries may be either smooth or notched. **Feet:** strongly raptorial; tibiae and tarsi usually feathered; toes frequently feathered; zygodactyl, the fourth toe reversible. **Plumage:** soft and lax; facial disc present; "ears" in many species; lores with dense feathers which cover base of bill and hide nostrils; sexes alike in coloration.

Family **Tytonidae.** (Genus *Tyto* only.) Barn Owls. (Length: 15-21 inches.)

Tail: short; square to emarginate. **Wings:** long; pointed; inner webs of all primaries without notching. **Feet:** tarsi twice as long as middle toe without nail; feathers on back of tarsi pointed upward; middle toe as long as inner; middle nail with inner edge pectinate. **Plumage:** facial disc triangular.

Family **Strigidae.** Typical Owls. (Length: 6-30 inches.)

Tail: variable in length; usually somewhat rounded, rarely square. **Wings:** variable in form; from one to six of the outer primaries with notched edges. **Feet:** feathers on back of tarsi pointed downward; middle toe shorter than the inner. **Plumage:** facial disc circular; "ears" commonly present.

Order CAPRIMULGIFORMES. Family **Caprimulgidae.** Goatsuckers. (Length: 7-12 inches.)

Bill: short (small); weak; depressed; slightly hooked; gape very wide with commissural point below eyes; nostrils circular (sometimes tubular) and exposed. **Tail:** variable in form. **Wings:** long; pointed. **Feet:** small; weak; tarsi partly feathered and twice the length of hallux; hallux short and elevated; outer toe noticeably shorter than middle toe, having four phalanges only; nail of middle toe pectinate. **Plumage:** soft; lax; feathers with aftershafts; rictal bristles usually evident; coloration dull and usually streaked, mottled, or barred.

Order APODIFORMES. Suborder Apodi. Swifts.

Family **Apodidae.** Swifts. (Length: 4-9 inches.)

Bill: very short (small); culmen decurved; depressed; gape extremely wide; commissural point below eyes. **Tail:** variable in form, usually forked or emarginate; rectrices in genus *Chaetura* stiffened and spinose. **Wings:** long; flat; pointed; secondaries extremely short. **Feet:** small; weak; tarsi unfeathered or feathered; pamprodactyl, the small hallux frequently directed inward, but reversible; three anterior toes about equal in length; nails strongly curved and acute. **Plumage:** compact; feathers of forehead may partially conceal nostrils; coloration of sexes alike, uniformly dull but sometimes with white areas.

Order APODIFORMES. Suborder Trochili. Hummingbirds.
  Family **Trochilidae.** Hummingbirds. (Length: 2-9 inches.)
    Bill: variable in length; slender; straight (sometimes decurved; rarely recurved); terete (sometimes compressed). **Tail:** variable in form. **Wings:** long; flat; pointed; secondaries extremely short. **Feet:** small; weak; tarsi unfeathered or feathered and not longer than middle toe with claw; hallux large and incumbent; three anterior toes of different lengths; nails strongly curved and acute. **Plumage:** compact; coloration brilliantly metallic in both sexes but less so in females.

Order TROGONIFORMES. Family **Trogonidae.** (Genus *Trogon* only.) Trogons. (Length: 11-12 inches.)
    Bill: short; wide at base; culmen decurved; somewhat hooked; upper mandible with several "teeth." **Tail:** long, with broad rectrices; shape variable, usually graduated. **Wings:** short; concave; rounded. **Feet:** small and weak; tarsi shorter than longest toe and usually feathered; heterodactyl, the inner toe reversed; approximately half of anterior toes united. **Plumage:** soft; lax; feathers with afterfeathers; bristle-like feathers covering base of bill; coloration of males brilliantly metallic, of females less so.

Order CORACIIFORMES. Family **Alcedinidae.** Kingfishers. (Length: 5-17 inches.)
    Bill: long; straight; compressed; acute; nostrils linear. **Tail:** moderately long, being one-half to two-thirds as long as wing; slightly rounded. **Wings:** moderately long; pointed. **Feet:** small and weak; tibiae partly bare; tarsi extremely short and irregularly scutellate in front; syndactyl; hallux notably shorter than inner toe and partly connected with it; nails very acute, the middle one somewhat flattened. **Plumage:** head frequently crested.

Order PICIFORMES. Family **Picidae.** Woodpeckers. (Length: 5-21 inches.)
    Bill: strong; usually straight; usually chisel-like, but sometimes acute. **Tail:** either pointed, rounded, or graduated; rectrices acuminate and with stiffened tips. **Wings:** moderately long; more or less pointed but with outermost primary very short or rudimentary. **Feet:** strong; tarsi scutellate in front and reticulate behind; zygodactyl; fourth toe permanently reversed; hallux occasionally absent; nails strong, decurved, and very acute. **Plumage:** nostrils concealed by bristle-like feathers.

Order PASSERIFORMES. Perching or Passerine Birds.
    Bill: variable in form; covering horny; nostrils imperforate. **Tail:** variable in form; with 12 rectrices. **Wings:** variable in form; with 10 primaries, the outermost frequently rudimentary; with 6 or more secondaries. **Feet:** anisodactyl; hallux incumbent and as long as middle toe; hallux nail often as long as, or longer than, nail of middle toe.

  Family **Tyrannidae.** New World Flycatchers. (Length: 2.5-10 inches.)
    Bill: variable in size; straight; wide and depressed at base (triangular in outline when viewed from above); slightly hooked; culmen somewhat decurved toward tip; commissural point almost below nasal canthus of eye; nostrils circular. **Tail:** usually square; sometimes forked. **Wings:** obvious primaries 10, the outermost usually longer than the secondaries. **Feet:** small and weak; tarsi short (seldom longer than middle toe), irregularly scutellate, and rounded behind. **Plumage:** rictal bristles evident; coloration of sexes usually similar.

  Family **Alaudidae.** (Genus *Eremophila* only.) Horned Larks. (Length: 5-9.5 inches.)
    Bill: short; conical; acute. **Tail:** shorter than wing; nearly square. **Wings:** long; pointed; nine primaries. **Feet:** moderate size; tarsi longer than middle toe, rather stout, rounded behind, and scutellate; hallux nail very long, equaling hallux in length. **Plumage:** nostrils concealed by feather tufts; head frequently crested or "horned"; coloration predominantly brown, white, and black; conspicuous black patches on head and neck.

  Family **Hirundinidae.** Swallows. (Length: 4-8 inches.)
    Bill: short; wide at base (triangular in outline when viewed from above); depressed; slightly hooked; culmen somewhat decurved toward tip; gape wide and twice the length of culmen; commissural point below nasal canthus of eye. **Tail:** not longer than wing; emarginate or forked; lateral rectrices sometimes filamentous. **Wings:** long; pointed; nine obvious primaries; secondaries generally very short. **Feet:** small and weak; tarsi usually shorter than middle toe with nail, and scutellate. **Plumage:** compact; coloration often partly metallic.

Family Corvidae. (North American species only.) Crows, Magpies, and Jays. (Length: 8-25 inches.)

Bill: usually long; stout; culmen decurved toward tip; somewhat acute. **Tail:** rounded, sometimes graduated. **Wings:** either long and pointed (crows and ravens), or short and rounded (magpies and jays); 10 primaries. **Feet:** large and strong; tarsi longer than middle toe with nail, and scutellate. **Plumage:** nostrils concealed by dense tufts of stiff feathers; rictal bristles also evident.

Family Paridae. Titmice. (Length: 3.5-8 inches.)

Bill: short; straight; stout; compressed. **Tail:** either as long as, or longer than, the wings; slightly rounded, sometimes graduated. **Wings:** rounded; 10 primaries. **Feet:** strong; tarsi longer than middle toe with nail, and scutellate. **Plumage:** nostrils concealed by dense tufts of stiff feathers; rictal bristles sometimes evident; coloration without bars, streaks, or spots.

Family Sittidae. Nuthatches. (Length: 4-7 inches.)

Bill: as long as head; straight; slender; compressed; acute; gonys somewhat recurved toward tip. **Tail:** much shorter than wings; nearly square; rectrices broad with rounded tips. **Wings:** long; pointed; 10 primaries. **Feet:** strong; tarsi usually as long as middle toe with nail and scutellate in front; nails very curved and compressed. **Plumage:** compact; nostrils more or less covered with stiff feathers; rictal bristles evident; coloration in American species plain blue or gray above.

Family Certhiidae. (Genus *Certhia* only.) Creepers. (Length: 5-7 inches.)

Bill: variable in length, sometimes much shorter or longer than head; decurved; slender; compressed; nostrils entirely exposed. **Tail:** about as long as, or slightly longer than, the wings; rounded; rectrices stiff and acuminate. **Wings:** somewhat long; rounded; 10 primaries. **Feet:** strong; tarsi shorter than middle toe with nail, and scutellate; nails long and very curved. **Plumage:** coloration of upper parts brownish and streaked.

Family Chamaeidae. Wrentits. (Length: approximately 6 inches.)

Bill: short; straight; stout; compressed; conical; culmen very decurved; nostrils entirely exposed. **Tail:** much longer than wings; graduated; rectrices narrow, but rather broad toward rounded tips. **Wings:** short; rounded; 10 primaries. **Feet:** strong; tarsi much longer than middle toe with nail, and faintly scutellate (sometimes scutellation not evident). **Plumage:** soft; lax; rictal bristles evident; slightly crested; lores with bristly feathers; coloration plain olive-brown above.

Family Cinclidae. Dippers. (Length: 5-8 inches.)

Bill: short; straight; slender; compressed; culmen decurved toward tip; gonys recurved toward tip; upper mandible notched near tip. **Tail:** short, more than half as long as wings; square to slightly rounded; rectrices broad with rounded tips. **Wings:** short; concave; rounded; 10 primaries. **Feet:** strong; tarsi longer than middle toe with nail, and booted; nails very curved. **Plumage:** soft; compact; coloration with brown, or gray, predominating.

Family Troglodytidae. Wrens. (Length: 3-9 inches.)

Bill: varying in length from half as long to about as long as head; usually decurved; slender; compressed. **Tail:** varying in length from slightly longer than wings to two-thirds as long; rounded; rectrices soft, with rounded tips. **Wings:** short; concave; rounded; 10 primaries. **Feet:** strong; tarsi longer than middle toe with nail, and scutellate in front as well as (sometimes) behind. **Plumage:** brown coloration predominating; wings and tail usually barred.

Family Mimidae. Thrashers and Mockingbirds. (Length: 8-12 inches.)

Bill: variable in length; usually rather slender, terete, and decurved toward tip; upper mandible notched toward tip (except in genus *Toxostoma*). **Tail:** variable in length, usually somewhat longer than wings; rounded, sometimes graduated. **Wings:** variable in length, usually short; rounded; 10 primaries. **Feet:** strong; tarsi distinctly longer than middle toe without nail and scutellate in front (often booted behind). **Plumage:** rictal bristles evident.

Family Turdidae. Thrushes. (Length: 5-12 inches.)

Bill: variable in length, usually short; straight; slender; compressed; culmen decurved toward tip; upper mandible notched near tip. **Tail:** usually shorter than wings; square to slightly rounded. **Wings:** long; pointed; 10 primaries. **Feet:** strong; tarsi usually longer than middle toe with nail, and booted. **Plumage:** rictal bristles evident; juvenal plumage always spotted above and below.

Family **Sylviidae.** (Subfamilies Polioptilinae and Regulinae only.) Gnatcatchers and King-lets. (Length: 3-6 inches.)
   Bill: short; straight; slender; somewhat depressed at base; culmen decurved toward tip; upper mandible notched near tip. **Tail:** shorter than wing (except in gnatcatchers, in which it is longer than wings); variable in shape, usually square to rounded (except in kinglets, in which it is emarginate); rectrices broad and either acuminate at tips (kinglets) or rounded at tips (gnatcatchers). **Wings:** long; rounded; 10 primaries. **Feet:** moderately strong; tarsi longer than middle toe with nail, and either booted (kinglets) or scutellate (gnatcatchers). **Plumage:** rictal bristles evident; nostrils either exposed or partly covered by bristle-like feathers.

Family **Motacillidae.** Wagtails and Pipits. (Length: 4.5-8 inches.)
   Bill: short; straight; slender; acute; culmen decurved toward tip; upper mandible notched near tip. **Tail:** variable in length, never shorter than wings; variable in shape; rectrices narrow and acuminate (except possibly middle pair). **Wings:** long; pointed; tertiaries elongated, nearly equaling primaries in length; nine primaries. **Feet:** strong; tarsi usual-ly longer than middle toe with nail, and scutellate; hallux nail elongated, equaling hallux in length. **Plumage:** rictal bristles present but not evident.

Family **Bombycillidae.** Waxwings. (Length: 6-7.5 inches.)
   Bill: short; stout; straight; upper mandible slightly hooked, with notch near tip; culmen decurved; gape deeply cleft and wide, nearly equal to length of exposed culmen. **Tail:** shorter than wing; square to slightly rounded; upper tail coverts greatly elongated. **Wings:** long; pointed; 10 primaries, the outermost very short, being less than half as long as the primary coverts. **Feet:** strong; tarsi shorter than middle toe without nail, and scutellate. **Plumage:** soft and dense; nostrils nearly concealed by small dense feathers; head crested; coloration predominantly brownish, with black band from bill through eye, and with yellow-tipped tail.

Family **Ptilogonatidae.** Silky Flycatchers. (Length: 7-9 inches.)
   Bill: short; stout (less so than in Bombycillidae); straight; upper mandible slightly hooked, with notch near tip; culmen decurved; gape deeply cleft and wide, much less than length of exposed culmen. **Tail:** usually equal to, or longer than, the wings; variable in shape. **Wings:** short; rounded; 10 primaries, the outermost much longer than the primary coverts. **Feet:** strong; tarsi usually shorter than middle toe with nail, and scutellate; hallux very short. **Plumage:** soft; rictal bristles evident; head usually crested; coloration plain with-out markings.

Family **Laniidae.** (Subfamily Laniinae only.) American Shrikes. (Length: 8-10 inches.)
   Bill: short; compressed; hooked; upper mandible toothed near tip. **Tail:** variable in length, being nearly as long as, and sometimes much longer than, the wings; variable in shape, being either square, rounded, or graduated. **Wings:** short; rounded, 10 primaries. **Feet:** strong; tarsi longer than middle toe with nail, and scutellate; nails very curved and acute. **Plumage:** soft; rictal bristles evident; nostrils fringed by bristle-like feathers; coloration of sexes alike; with plain gray and brown, mixed with black and white, predominating.

Family **Sturnidae.** (Genus *Sturnus* only.) Starlings. (Length: 8-9 inches.)
   Bill: as long as head; straight; slightly depressed toward tip; commissure somewhat angu-lated; feathers of forehead partially divided by culmen. **Tail:** short, being half the length of the wings; nearly square to slightly emarginate. **Wings:** long; pointed; 10 primaries, the outermost rudimentary and acuminate. **Feet:** strong; tarsi longer than middle toe without nail, and scutellate. **Plumage:** feathers of head, neck, and breast long and narrow; coloration metallic and somewhat iridescent.

Family **Vireonidae.** Vireos. (Length: 5-7 inches.)
   Bill: usually short; rather straight; somewhat compressed at base; hooked; upper man-dible notched near tip. **Tail:** usually much shorter than wings; slightly rounded or emar-ginate; rectrices narrow. **Wings:** long; variable in shape, usually somewhat rounded; 10 primaries but the outermost rudimentary. **Feet:** strong; tarsi longer than middle toe with

nail, and scutellate. **Plumage:** rictal bristles present but not evident; nostrils somewhat concealed by bristle-like feathers; coloration of sexes nearly alike, with plain olive, olive-green, or gray above.

Family **Parulidae.** Wood Warblers. (Length: 4-7.5 inches.)
  Bill: variable in length, usually short; usually slender, straight, compressed, and acute, but with variations too numerous to itemize. **Tail:** generally shorter than wings; varying from square to slightly rounded. **Wings:** variable in length and shape but long and some-what pointed; nine primaries. **Feet:** strong; tarsi usually less than twice as long as middle toe without nail, and scutellate. **Plumage:** rictal bristles present (sometimes conspicuous, sometimes not evident) or absent.

Family **Ploceidae.** (Genus *Passer* only.) Weaver Finches. (Length: 6-7 inches.)
  Bill: short; conical; culmen slightly decurved; commissure angulated. **Tail:** shorter than wings; somewhat square. **Wings:** long; pointed; 10 primaries, the outermost rudimentary. **Feet:** strong; tarsi shorter than middle toe without nail, and scutellate. **Plumage:** rictal bristles present but not conspicuous.

Family **Icteridae.** Blackbirds. (Length: 7-19 inches.)
  Bill: variable in length, seldom conspicuously longer than head; somewhat conical; acute; culmen sometimes slightly decurved, elevated toward base, and extending far back to part the feathers of the forehead; commissure somewhat angulated. **Tail:** variable in length and shape, usually rather short (always more than half as long as wings, never conspicuously longer) and rounded. **Wings:** variable in length and shape, usually long and pointed; nine primaries. **Feet:** very strong; tarsi usually equal to, or slightly longer than, middle toe with nail, and scutellate.

Family **Thraupidae.** (Genus *Piranga* only.) Tanagers. (Length: 6.5-8 inches.)
  Bill: usually as long as head; somewhat conical, stout, and swollen; slightly hooked; upper mandible notched near tip and its tomium toothed near middle. **Tail:** shorter than wings; square to slightly rounded, sometimes emarginate. **Wings:** moderately long; more or less pointed; nine primaries. **Feet:** strong; tarsi longer than middle toe with nail, and scutel-late. **Plumage:** rictal bristles present but not conspicuous; coloration in adult males more or less red, replaced by olive-green in females.

Family **Fringillidae.** Grosbeaks, Finches, Sparrows, and Buntings. (Length: 3.5-9 inches.)
  Bill: short; stout; conical; culmen slightly decurved; commissure abruptly angulated. **Tail:** extremely variable in length and shape. **Wings:** extremely variable in length and shape; nine primaries. **Feet:** strong; tarsi variable in length but always scutellate. **Plum-age:** rictal bristles usually present, sometimes conspicuous; nostrils sometimes partially concealed by bristle-like feathers.

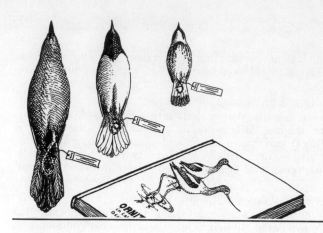

LABORATORY IDENTIFICATION

Having learned the methods used in classifying and naming birds, and having also acquired a preliminary knowledge of certain characters used in distinguishing birds, the student is now prepared to determine for himself the categories to which given birds belong. This is most quickly and satisfactorily accomplished by the use of **keys** based on external structural characters.

The student needs considerable practice before he can use the keys effectively. Therefore, he should identify, with the guidance of the instructor, a series of specimens representing a diversity of orders, families, and species.

IDENTIFICATION OF ORDERS AND FAMILIES

Seventy skins of birds native to the region of the conterminous United States, Alaska, or Canada, will be made available. Each skin bears a number from 1 through 70. In the following keys to orders and families, "run down" these specimens to their proper categories. On the ruled pages (184 and 186) entitled "Identification of Bird Skins to Orders and Families," place on the numbered lines, that correspond to the numbers of the specimens, the technical names of the orders and families as identified. When all the specimens have been run down, have the identifications checked by the instructor.

IDENTIFICATION OF SPECIES

Having identified the orders and families of the seventy specimens, run them down to species. Use the keys to species in Part 5 on birds by P. Brodkorb in *Vertebrates of the United States* (McGraw-Hill Book Company, New York, 1968). Or obtain keys to species based on the birds of the region. If the region is in North America east of the Mississippi River, use the keys in F. M. Chapman's *Handbook of Birds of Eastern North America* (D. Appleton and Company, 1932; Dover reprint available). If the region is in southeastern United States, use *A Key to Florida Birds* by H. M. Stevenson (Peninsular Publishing Company, Tallahassee, 1960). If the region is in north-central North America, including the northern prairie states and prairie provinces, use T. S. Roberts' *A Manual for the Identification of Birds of Minnesota and Neighboring States* (University of Minnesota Press, Minneapolis, Second Edition, 1955). If the specimens are from parts of North America west of the Mississippi River, use F. M. Bailey's *Handbook of Birds of the Western United States* (Houghton Mifflin Company, 1902). Before using any one of these keys, read the introduction in which the author explains his methods of obtaining measurements and his interpretation of terms. A clear understanding of such terms is quite necessary because the keys are based largely upon them.

On the accompanying ruled pages entitled "Identifications of Bird Skins to Species" (opposite those pages used for recording the identifications to orders and families), place the identifications to species, using the vernacular names. When the sexes and plumages (i.e., juvenal plumage, winter plumage, nuptial plumage) are readily recognizable, fill in the spaces provided. Then have the identifications checked by the instructor.

Once the student has satisfactorily identified the seventy specimens, he is ready to identify "on his own" an array of specimens representing all of the species found in the region. To save time, the instructor will group the various specimens by orders, or families. The label on each specimen may, or may not, give the name of the species, depending on the judgment of the instructor, but it should give the sex of the specimen and the locality and date where the specimen was collected. The keys already used will serve as adequate guides. (If specimens are identified as to species, the student is nevertheless advised to use the keys, thereby gaining a necessary knowledge of distinctions.) The methods, employed by the student in learning and remembering the characteristics of the various species, may be determined by him and the instructor.

IDENTIFICATION OF SUBSPECIES

No attempt will be made in this study to identify the subspecies of birds. Therefore, the student is requested to ignore both the technical and vernacular names of subspecies even though given in the keys.

KEY TO THE ORDERS OF BIRDS OF NORTH AMERICA NORTH OF MEXICO

(Roman numerals after the names of orders refer to the position
of the same orders in the key to families. See page 177.)

A. Feet webbed; neither semipalmated nor lobed.
   I. All four toes joined in web (totipalmate) ....**Pelecaniformes** (IV)
   II. Only front toes joined in web (palmate).
     a. Nostrils tubular ....**Procellariiformes** (III)
     b. Nostrils not tubular.
       1. Bill lamellate or serrate.
         (a) Bill decurved abruptly ....**Ciconiiformes** (V)
         (b) Bill more or less straight with nail-like hook
           at tip of upper mandible ....**Anseriformes** (VI)
       2. Bill not lamellate or serrate.
         (a) Legs inserted far behind middle of body; tarsi compressed.
           (1) Hallux present ....**Gaviiformes** (I)
           (2) Hallux absent ....**Charadriiformes** (X)
         (b) Legs not inserted far behind middle
           of body; tarsi rounded ....**Charadriiformes** (X)
B. Feet not webbed; may be semipalmated or lobed.
   I. Toes four.
     a. Toes three in front and one behind.
       1. Toes with flattened nails ....**Podicipediformes** (II)
       2. Toes without flattened nails.
         (a) Feet syndactyl ....**Coraciiformes** (XVIII)
         (b) Feet not syndactyl.
           (1) Feet raptorial ....**Falconiformes** (VII)
           (2) Feet not raptorial.

(c) Middle toe pectinate.

    (3) Bill long and acute; legs long;
        lores bare; plumage ordinary .... **Ciconiiformes** (V)

    (4) Bill very short and slightly hooked; feet small
        and weak; lores feathered; plumage soft ...**Caprimulgiformes** (XV)

(d) Middle toe not pectinate.

    (3) Small birds abundantly colored metallic green
        above, or uniformly sooty brown throughout, or
        sooty brown except for white on chin, throat, or
        rump; tarsus without obvious scales;
        feet small and weak .... **Apodiformes** (XVI)

    (4) Birds without the above combination of characters.

        (e) Lores entirely bare; rest of head
            feathered entirely or in part .... **Ciconiiformes** (V)

        (f) Lores feathered; or bare when rest of head is bare.

            (5) Entire head bare.

                (g) Hallux incumbent .... **Ciconiiformes** (V)
                (h) Hallux elevated.

                    (7) Nostrils perforate .... **Falconiformes** (VII)
                    (8) Nostrils imperforate .... **Galliformes** (VIII)

            (6) Entire head not bare.

                (g) Hallux incumbent.

                    (7) Nails acute .... **Passeriformes** (XX)
                    (8) Nails obtuse .... **Galliformes** (VIII)

                (h) Hallux elevated.

                  (7) Bill with operculum .... **Columbiformes** (XI)
                  (8) Bill without operculum.

                      (i) Bill short and stout; tip
                        of upper mandible noticeably
                        curved over lower one; all
                        primaries stiff and curved .. **Galliformes** (VIII)

                    (j) Bill may or may not be short and
                      stout; tip of upper mandible
                      not noticeably curved over lower one;
                      all primaries not stiff and curved.

                      (9) Wings pointed (in one species
                        the three outermost primaries
                        are attentuate and short, giving
                        rounded effect) ..... **Charadriiformes** (X)

                    (10) Wings rounded .... **Gruiformes** (IX)

b. Toes two in front and two behind.

  1. Eyes directed forward in facial
    discs; plumage soft .... **Strigiformes** (XIV)

  2. Eyes not in facial discs; plumage ordinary.

    (a) Bill with several "teeth" .... **Trogoniformes** (XVII)
    (b) Bill without "teeth."

        (1) Bill with cere and conspicuously hooked .... **Psittaciformes** (XII)
        (2) Bill without cere and not hooked.

            (c) Rectrices acuminate, with stiffened tips .... **Piciformes** (XIX)
            (d) Rectrices not acuminate, without stiffened tips .. **Cuculiformes** (XIII)

II. Toes three.

  a. Two toes in front and one behind .... **Piciformes** (XIX)
  b. Three toes in front .... **Charadriiformes** (X)

KEY TO THE FAMILIES OF BIRDS OF NORTH AMERICA NORTH OF MEXICO

(Family Cotingidae excluded)

I. Order GAVIIFORMES. Family *Gaviidae:* Loons

II. Order PODICIPEDIFORMES. Family *Podicipedidae:* Grebes

III. Order PROCELLARIIFORMES: Tube-nosed Swimmers

Key to Families

A. Large birds 28 to 40 inches long; nostrils opening on each
   side of the culmen in independent tubes ....**Diomedeidae:** Albatrosses
B. Medium-sized to small birds usually less than 28 inches
   long; nostrils not opening in independent tubes.
   I. Nostrils opening on top of the culmen
      in tubes separated only by a septum ....**Procellariidae:** Shearwaters and Fulmars
   II. Nostrils opening on top of the culmen in one tube ....**Hydrobatidae:** Storm Petrels

IV. Order PELECANIFORMES: Totipalmate Swimmers

Key to Families

A. Bill hooked.
   I. Tarsus partly feathered, tail deeply forked ....**Fregatidae:** Frigatebirds
   II. Tarsus not feathered, tail not forked.
      a. Bill over 10 inches long with huge gular sac,
         bare and suspended from the lower mandible
         and extending its entire length ....**Pelecanidae:** Pelicans
      b. Bill less than 4 inches long with
         inconspicuous gular sac ....**Phalacrocoracidae:** Cormorants
B. Bill unhooked.
   I. Middle rectrices filamentous; gular sac
      rudimentary and feathered ....**Phaëthontidae:** Tropicbirds
   II. Middle rectrices not filamentous; gular sac small and bare.
      a. Tail rounded; middle rectrices with flutings ....**Anhingidae:** Anhingas
      b. Tail pointed; rectrices without flutings ....**Sulidae:** Gannets

V. Order CICONIIFORMES: Deep-water Waders

Key to Families

A. Middle toe nail pectinate; bill long,
   straight, and acute ....**Ardeidae:** Herons and Bitters
B. Middle toe nail not pectinate; bill not acute.
   I. Bill bent abruptly down and lamellate ....**Phoenicopteridae:** Flamingos
   II. Bill not bent abruptly down and not lamellate.
      a. Large birds, never under 28 inches long; bill
         stout at base, tapering to tip (bill
         decurved toward tip in Wood Ibis) ....**Ciconiidae:** Storks and Wood Ibises
      b. Smaller birds, never over 28 inches long;
         bill either slender, decurved, and somewhat
         terete, or straight, very depressed,
         and spatulate ....**Threskiornithidae:** Ibises and Spoonbills

VI. Order ANSERIFORMES: Lamellate-billed Swimmers
Key to Subfamilies of the Anatidae

A. Lores bare; neck longer than body ....**Cygninae:** Swans
B. Lores feathered; neck not longer than body.
   I. Tarsus completely reticulate.
      a. Small birds not longer than 20 inches;
         wings rounded ....**Dendrocygninae:** Tree Ducks
      b. Large birds never shorter than 21
         inches; wings pointed ....**Anserinae:** Geese
   II. Tarsus scutellate in front.
      a. Rectrices narrow and stiff, with coverts
         extremely short ....**Oxyurinae:** Ruddy and Masked Ducks
      b. Rectrices of ordinary shape and texture,
         with coverts of ordinary length.
         1. Bill terete and serrate ....**Merginae:** Mergansers
         2. Bill broadly depressed and lamellate.
            (a) Hallux lobed; no iridescent speculum ....**Aythyinae:** Diving Ducks
            (b) Hallux not lobed; an iridescent
                speculum usually present ....**Anatinae:** Dabbling Ducks

VII. Order FALCONIFORMES. Vultures and Diurnal Birds of Prey
Key to Families

A. Head bare; nostrils perforate; feet
    weakly raptorial ....**Cathartidae:** New World Vultures
B. Head feathered; nostrils imperforate; feet strongly raptorial.
   I. Nostrils circular with central bony tubercle or (as in
      caracaras) slit-like with posterior end uppermost ....**Falconidae:** Falcons and Caracaras
   II. Nostrils oval or slit-like with anterior end uppermost.
      a. Under surfaces of toes smooth; hallux nail
         larger than other nails; under surfaces
         of nails grooved ....**Accipitridae:** Kites, Eagles, and Hawks
      b. Under surfaces of toes spiny; hallux
         approximately the same size of the other
         nails; under surfaces of nails rounded ....**Pandionidae:** Ospreys

VIII. Order GALLIFORMES: Gallinaceous Birds
Key to Families

A. Hallux incumbent ....Chachalacas of **Cracidae**
B. Hallux elevated.
   I. Tarsus either partly or wholly feathered;
      nostrils feathered ....**Tetraonidae:** Grouse and Ptarmigans
   II. Tarsus and nostrils not feathered.
      a. Head bare; rectrices somewhat square at ends ....**Meleagrididae:** Turkeys
      b. Head mostly feathered; rectrices
         ordinary, not square at ends ....**Phasianidae:** Quails and Pheasants

IX. Order GRUIFORMES: Marsh Birds
Key to Families

A. Crown more or less bare ....**Gruidae:** Cranes
B. Crown feathered.
   I. Length of bird approximately 26 inches; outer
      primary stiff, attenuate, and incurved ....**Aramidae:** Limpkins

II. Length of bird not over 20 inches;
outer primary of ordinary shape .... **Rallidae:** Rails, Gallinules, and Coots

X. Order CHARADRIIFORMES: Shore Birds, Gulls, and Auks
Key to Families

A. Wings spurred ....**Jacanidae:** Jacanas
B. Wings not spurred.

    I. Toes more or less completely webbed (palmate).

        a. Toes three ....**Alcidae:** Auks, Murres, and Puffins
        b. Toes four.

            1. Lower mandible blade-like, compressed, and
decidedly longer than upper mandible .... **Rynchopidae:** Skimmers
            2. Lower mandible not blade-like, not compressed,
and not longer than upper mandible.

                (a) Bill and legs long; bill strongly recurved .... Avocets of **Recurvirostridae**
                (b) Bill and legs not long; bill not recurved.

                    (1) Bill with a cere and a nail-like
hook; tail rounded to pointed
(middle rectrices often
greatly elongated) .... **Stercorariidae:** Skuas and Jaegers
                    (2) Bill without a cere and without a nail-like
hook; bill either plainly hooked, or sharply
pointed; tail either square to slightly
rounded, or forked .... **Laridae:** Gulls and Terns

    II. Toes not completely webbed.

        a. Toes either lobed or with lateral membranes .... **Phalaropodidae:** Phalaropes
        b. Toes neither lobed nor with lateral membranes.

            1. Toes four but hallux sometimes inconspicuous.

                (a) Bill slender, pliable in life, and slightly
depressed toward tip; bill frequently
straight, but sometimes either quite
recurved, or decurved .... **Scolopacidae:** Snipe, Woodcock,
and Sandpipers
                (b) Bill in lateral profile either constricted
in middle and swollen toward tip, or
constricted near base and tapering to
an acute tip (turnstones) .... Black-bellied Plover and
Turnstones of **Charadriidae.**

Toes with
Lateral
Membranes

Bill of
Plover

Bill of
Turnstone

            2. Toes three.

                (a) Tarsus reticulate.

                    (1) Bill longer than head.

                      (c) Bill slender; legs extremely
long and slender .... Stilts of **Recurvirostridae**

(d) Bill compressed, constricted
toward base and chisel-like at
tip; legs short and stout . . . . **Haematopodidae:** Oystercatchers
(2) Bill never longer, usually shorter, than head . . . . **Charadriidae:** Plovers
(b) Tarsus scutellate . . . . Sanderling of **Scolopacidae**

XI. Order COLUMBIFORMES. Family *Columbidae:* Pigeons and Doves

XII. Order PSITTACIFORMES. Family *Psittacidae:* Parrots, Parakeets, and Macaws

XIII. Order CUCULIFORMES. Family *Cuculidae:* Cuckoos, Roadrunners, and Anis

XIV. Order STRIGIFORMES: Nocturnal Birds of Prey

Key to Families

A. Middle toe nail pectinate . . . . **Tytonidae:** Barn Owls
B. Middle toe nail not pectinate . . . . **Strigidae:** Typical Owls

XV. Order CAPRIMULGIFORMES. Family *Caprimulgidae:* Goatsuckers

XVI. Order APODIFORMES: Swifts and Hummingbirds

Key to Families

A. Bill very short and depressed; gape extremely wide . . . . **Apodidae:** Swifts
B. Bill variable in length (usually long)
and terete; gape narrow . . . . **Trochilidae:** Hummingbirds

XVII. Order TROGONIFORMES. Family *Trogonidae:* Trogons.

XVIII. Order CORACIIFORMES. Family *Alcedinidae:* Kingfishers

XIX. Order PICIFORMES. Family *Picidae:* Woodpeckers

XX. Order PASSERIFORMES: Perching or Passerine Birds

Key to Families

A. Tarsus rounded in front and behind.
  I. Bill wider than high; bill also hooked; rictal bristles present;
  hallux nail ordinary length . . . . **Tyrannidae:** New World Flycatchers
  II. Bill not wider than high; bill not hooked; rictal bristles not
  present; hallux nail extraordinarily long . . . . **Alaudidae:** Larks
B. Tarsus rounded in front and ridged behind.
  I. Tarsus booted.
    a. Rictal bristles present.
      1. Small birds, not longer than 6 inches, wings rounded.
        (a) Wings shorter than tail;
        tail graduated; plumage lax . . . . **Chamaeidae:** Wrentits
        (b) Wings longer than tail; tail
        emarginate; plumage ordinary . . . . Kinglets of **Sylviidae**
      2. Birds of moderate size, not longer
      than 12 inches; wings pointed . . . . **Turdidae:** Thrushes
    b. Rictal bristles not present . . . . **Cinclidae:** Dippers

II. Tarsus not booted.
   a. Brownish birds with a combination of
      yellow-tipped tail and black band from
      bill through eye; head crested . . . .**Bombycillidae:** Waxwings
   b. Birds without the above color combination; head may or may not be crested.

     1. Bill *strongly* hooked and notched.
       (a) Small birds, between 5 and 7 inches in length,
          with olive-green (occasionally olive or
          gray) upper parts . . . .**Vireonidae:** Vireos
       (b) Birds of moderate size, between 8 and
          10 inches in length, with black, white,
          and gray plumage . . . .**Laniidae:** North American Shrikes

     2. Bill not strongly hooked, though it may or may not be slightly notched.
       (a) Bill slender and decidedly decurved;
          rectrices stiff and acuminate . . . .**Certhiidae:** Creepers
       (b) Bill and rectrices without the above combination of characters.
          (1) Bill slender, straight, with gonys
              somewhat recurved toward tip . . . .**Sittidae:** Nuthatches
          (2) Bill without the above combination of characters.
             (c) Nostrils entirely covered by feathers.
                (3) Small birds, between 3.5
                   and 8 inches in length . . . .**Paridae:** Titmice
                (4) Large birds, between 8 and 25
                   inches in length . . . .**Corvidae:** Crows, Magpies, and Jays
             (d) Nostrils uncovered.
                (3) Commissure of bill more or less angulated toward base.
                   (e) Bill somewhat depressed
                       toward tip (when viewed
                       dorsally); in adult plumage,
                       feathers of head, neck, and
                       breast long and narrow with
                       metallic coloration; in immature
                       plumage, upper parts grayish
                       brown, under parts streaked with
                       grayish white . . . .**Sturnidae:** Starlings of Genus *Sturnus*
                 (f) Combination of characters not as above.
                   (5) Bill conical, or crossed (in one genus);
                      commissure *abruptly* angulated; rictal
                      bristles usually evident; culmen not
                      parting feathers of forehead.
                     (g) Wings with 9 primaries . . . .**Fringillidae:** Grosbeaks,
                                   Finches, Sparrows, and Buntings
                     (h) Wings with 10 primaries, the
                       outermost rudimentary . . . . .**Ploceidae:** Weaver Finches
                                      of Genus *Passer*

                   (6) Bill less conical; commissure less
                      abruptly angulated; rictal bristles
                      absent; in some species culmen
                      extends far backward and parts the
                      feathers of forehead . . . .**Icteridae:** Meadowlarks,
Head of Meadowlark   Blackbirds, and Troupials
         (4) Commissure of bill not angulated toward base.
           (e) Wings rounded.
             (5) Rictal bristles evident.

(g) Head crested . . . .**Ptilogonatidae:** Silky Flycatchers

(h) Head not crested.

    (7) Small birds, between 4 and 5 inches in length; upper parts bluish gray; under parts white; outer rectrices white . . . . .Gnatcatchers of **Sylviidae**

    (8) Birds of moderate size, between 8 and 12 inches in length; coloration not always as above . . . .**Mimidae:** Thrashers and Mockingbirds

(6) Rictal bristles not evident.

    (g) Small brownish birds, under 9 inches in length, with bill slender and usually decurved; wings and tail indistinctly barred . . . .**Troglodytidae:** Wrens

    (h) Small variously colored birds, under 6 inches in length, with bill slender and usually straight, except one group (chats) which is approximately 7 inches in length with bill stout and decurved, wings and tail not barred . . . .**Parulidae:** Wood Warblers

(f) Wings pointed

    (5) Hallux nail as long as hallux itself . . . .**Motacillidae:** Wagtails and Pipits

    (6) Hallux nail shorter than hallux itself.

        (g) Bill somewhat swollen and conical; upper mandibular tomium toothed near middle . . . .**Thraupidae:** Tanagers of Genus *Piranga*

        (h) Bill not swollen and not conical; upper mandibular tomium not toothed.

            (7) Legs relatively short and weak; longest primary more than twice as long as the longest secondary . . . .**Hirundinidae:** Swallows

            (8) Legs of ordinary size and strength; primaries and secondaries in ordinary proportion to each other . . . .**Parulidae:** Wood Warblers

BRECKENRIDGE

IDENTIFICATIONS OF BIRD SKINS TO ORDERS AND FAMILIES

No.	Order	Family
1		
2		
3		
4		
5		
6		
7		
8		
9		
10		
11		
12		
13		
14		
15		
16		
17		
18		
19		
20		
21		
22		
23		
24		
25		
26		
27		
28		
29		
30		
31		
32		
33		
34		
35		
36		
37		
38		
39		

IDENTIFICATIONS OF BIRD SKINS TO SPECIES

No.	Species	Sex	Plumage
1			
2			
3			
4			
5			
6			
7			
8			
9			
10			
11			
12			
13			
14			
15			
16			
17			
18			
19			
20			
21			
22			
23			
24			
25			
26			
27			
28			
29			
30			
31			
32			
33			
34			
35			
36			
37			
38			
39			

IDENTIFICATIONS OF BIRD SKINS TO ORDERS AND FAMILIES

No.	Order	Family
40		
41		
42		
43		
44		
45		
46		
47		
48		
49		
50		
51		
52		
53		
54		
55		
56		
57		
58		
59		
60		
61		
62		
63		
64		
65		
66		
67		
68		
69		
70		

IDENTIFICATIONS OF BIRD SKINS TO SPECIES

No.	Species	Sex	Plumage
40			
41			
42			
43			
44			
45			
46			
47			
48			
49			
50			
51			
52			
53			
54			
55			
56			
57			
58			
59			
60			
61			
62			
63			
64			
65			
66			
67			
68			
69			
70			

PLUMAGES AND PLUMAGE COLORATION

Birds periodically shed and renew their **plumage** or feather covering by a process known as **molting.** Every species normally exhibits a definite sequence of plumages and molts.

SEQUENCE OF PLUMAGES AND MOLTS

The sequence of plumages and molts in each species proceeds in a relatively consistent manner, with the result that the successive plumages and intervening molts can be identified and named.

Natal Down. This plumage consists of down feathers. In birds belonging to such orders as the Anseriformes, Galliformes, and Charadriiformes natal down covers the entire body; in birds belonging to the Passeriformes down occurs only on the pterylae of the upper part of the body; in birds belonging to such orders as the Piciformes the body is naked at hatching, entirely without feathers of any sort. The natal down is lost by the **postnatal molt** which is always complete. The postnatal molt differs from later molts in that the feathers—the natal down feathers—are pushed out from their follicles by the tips of the feathers comprising the next generation—the juvenal feathers. In due course the down feathers, which are structurally continuous with the juvenal feathers, become dislodged, drop off, and disappear.

Juvenal Plumage. Consisting of juvenal feathers which are the first true contour feathers. In passerine species and some others the juvenal feathers are noticeably loose-textured, giving the plumage a soft, fluffy appearance. Species that remain in the nest long after hatching acquire the juvenal plumage before they leave; species that leave the nest soon after hatching acquire it later and often more slowly. The juvenal plumage is lost during the late summer and fall of the year by the **postjuvenal molt** which may or may not be complete.

First Winter Plumage. Retained throughout the first winter. It has the same texture as the adult plumage but is frequently different in coloration. In the late winter and early spring it may be lost wholly or in part by a **first prenuptial molt** or it may be retained as the plumage named below, the molt being entirely suppressed.

First Nuptial Plumage. Retained during the first breeding (i.e., nesting) season. In certain species the males are less brilliantly colored in their first nuptial plumages than they are in later years. The nuptial plumage is lost immediately after the nesting season by the **first postnuptial molt,** which is always complete.

Second Winter Plumage. Retained throughout the second winter of the bird's life and usually indistinguishable from winter plumages of later years. Depending on the species, it may be lost, wholly or in part, by the **second prenuptial molt,** or it may be retained as the second nuptial plumage, the molt being entirely suppressed.

Second Nuptial Plumage. Retained throughout the second breeding season and only rarely distinguishable from later nuptial plumages. It is lost immediately after the nesting season by the **second postnuptial molt.**

Third and Fourth Winter and Nuptial Plumages. In a few species the plumages acquired in the third and fourth year of life are distinguishable from later plumages. The plumages and molts are, therefore, designated accordingly.

Adult Winter and Adult Nuptial Plumages. When plumages become indistinguishable from those of adults the numerical designation (first, second, etc.) is dropped before the names of the plumages, and the term "adult" is substituted. In the majority of species, plumages become indistinguishable from adult plumages after the first year. In those species, the plumages following the First Winter Plumage are commonly referred to as the Adult Nuptial Plumage and Adult Winter Plumage. The names of the intervening molts are changed correspondingly.

In the chart outlined below, indicate, in one column, the sequence of plumages and, in the adjacent column, the molts by which they are lost. The student should thoroughly memorize this chart because there will be many occasions in later studies when a knowledge of this sequence will be essential.

The names given above for the plumages and their intervening molts apply satisfactorily to most birds in the temperate regions and have been used extensively, with some variation (e.g., summer plumage instead of nuptial plumage, non-nuptial plumage instead of winter plumage), in most works on North American and European birds. They are used in this book. The names, however, are not entirely suitable for tropical, oceanic, and antarctic birds. The difficulty arises from a nomenclature based on seasonal and annual cycles. In tropical species, the so-called "winter plumage" is correlated with the wet season and therefore becomes meaningless. Oceanic and antarctic species do not always breed annually; some breed less, others more often, than once a year. And some birds do not breed in their first "nuptial" plumage.

In an effort to derive a uniform series of terms for plumages and molts that express homologies between different groups of birds anywhere in the world, Humphrey and Parkes (1959) proposed a new terminology, which has already been adopted in recent works on birds. In order that the student may be cognizant of the Humphrey-Parkes terminology, a comparison is given below.

Traditional Terminology	Humphrey-Parkes Terminology
Natal Down	Natal Down
Postnatal Molt	Prejuvenal Molt
Juvenal Plumage	Juvenal Plumage
Postjuvenal Molt	First Prebasic Molt
First Winter Plumage	First Basic Plumage
First Prenuptial Molt	First Prealternate Molt
First Nuptial Plumage	First Alternate Plumage
First Postnuptial Molt	Second Prebasic Molt
Second Winter Plumage	Second Basic Plumage
Second Prenuptial Molt	Second Prealternate Molt
Second Nuptial Plumage	Second Alternate Plumage
Second Postnuptial Molt	Third Prebasic Molt
Etc.	Etc.

Humphrey and Parkes name the molt from the incoming plumage and confine the word "plumage" to a single generation of feathers acquired by a specific molt. In place of the "plumage" in the traditional sense, they use the term **feather coat** which constitutes all the feathers that a bird is wearing at a given time and may include not just one, but two or more plumages or generations of feathers. Humphrey and Parkes choose the word "basic" in place of winter plumage because in adult birds, which have only one plumage per cycle, this is the one almost invariably lost and renewed by a complete molt; and they consider "alternate" the logical name for the nuptial plumage because it is the second or alternate plumage in birds which have two plumages per cycle. Humphrey and Parkes also suggest a useful term, **definitive,** for any plumage that will not change further with age. Thus, the plumage of the Bald Eagle *(Haliaeetus leucocephalus)* is definitive when, after four or more years, the head and tail feathers are completely white.

ECLIPSE PLUMAGE

The so-called **eclipse plumage** occurs in certain birds, notably in most male ducks. In this plumage, acquired after the nesting season is under way and retained for only about two months, the males very closely resemble the females. Actually the eclipse plumage is homologous to the winter plumage of other birds, being acquired by a postnuptial molt and lost by a hastening of the prenuptial molt. As a result, male ducks wear their bright nuptial plumage during the fall, winter, and spring.

MOLTING IN ADULTS

All fully adult birds renew their feathers at least once a year by the postnuptial molt and many birds twice a year by the additional prenuptial molt. The postnuptial molt is complete in most all species. The prenuptial molt, on the other hand, is complete in very few species; if not suppressed altogether, it involves a few pterylae only. A third molt of certain pterylae, giving rise to a so-called **supplementary plumage,** occurs in the Old-squaw *(Clangula hyemalis)* during the winter, the Ruff *(Philomachus pugnax)* and possibly some other scolopacids during the late winter-early spring, and ptarmigans *(Lagopus* spp.) during the summer.

In general, the prenuptial molt immediately precedes the nesting season, and the postnuptial molt immediately follows. Thus the times of the year during which a species molts depends on when its nesting season begins and when it ends. Usually in migratory species the prenuptial molt is completed before they arrive on the nesting grounds, and the postnuptial molt is completed before they start southern migration. In the northern United States and Canada the majority of species undergo the postnuptial molt from late July through August to early September, although a few species begin as early as June and a few others are still molting as late as October. Swallows and some tyrannid flycatchers do not even begin their postnuptial molt until they are on their wintering grounds.

Molting is a gradual process; the feathers of a particular pteryla are normally not shed all at once, and thus the bird is assured continuance of adequate plumage protection. Among groups of birds there are marked differences with respect to the molting of the flight feathers (remiges and rectrices).

In the majority of flying birds the remiges are shed and replaced in an orderly sequence so that flight is still possible during molt. In a passerine bird, for instance, the primaries, secondaries, and tertiaries are shed one by one. The innermost primary is usually the first remex to drop out. As soon as the feather replacing it is partly grown, the adjoining feather drops out. The remaining primaries are molted successively until all are replaced. Coincident with the dropping out of the fifth or sixth primary is the molt of the outermost tertiary, followed by the adjoining tertiary, and then the third tertiary, which in most passerine birds is the last or innermost. The molt of the secondaries, beginning with the outermost, follows the molt of the last tertiary, and proceeds in series, the innermost secondary being the last secondary (and also the last remex) to drop out. The greater primary coverts are shed one at a time; each one is molted at about the same time as the primary which it overlies. The greater secondary coverts, though molted successively and in the same direction as the secondaries, are dropped out and replaced much more rapidly than the secondaries; as a consequence, a new set of greater secondary coverts is fully developed before the molt of the secondaries is well under way.

In a few groups of flying birds—notably, the loons, grebes, swans, geese, and ducks —all the remiges are shed simultaneously, with the result that the birds are temporarily incapable of flight. This occurs usually before fall migration, except in loons which delay the molt until they have reached southern coastal waters.

As a rule, the rectrices of flying birds are molted and replaced, beginning with the innermost pair, followed by the second innermost pair, and so on, until the outermost pair is shed. Thus the birds, even during molt, have ample tail surface for flight purposes. In tree-climbing birds, such as woodpeckers and creepers, the tail molt is similar, except that it begins with the second innermost pair, leaving the innermost pair to be shed last. In this way the birds during molt retain a tail that gives adequate support in climbing. Swans, geese, and ducks molt the rectrices simultaneously, as they do the remiges. The smaller owls also molt the rectrices simultaneously, but this peculiarity probably does not impair their flight, as they use their tails very little in flying.

Examine the wings and tails of a series of specimens of a passerine bird arranged in a sequence to show the progress of molt.

PLUMAGE-CHANGE WITHOUT MOLT

A number of species change the appearance of their winter plumage before the breeding season without molting. This is accomplished in two ways: (1) By **wear,** or abrasion; that is, contrastingly colored tips of feathers freshly acquired by the postnuptial molt wear away during the winter, revealing, before the breeding season, the basic color of the feathers. (2) By **fading;** that is, the color of the winter plumage gradually fades until, by the time of the breeding season, it has become noticeably lighter.

Examine bird skins of one or more species arranged in sequence to show plumage-change by wear and by fading.

A STUDY OF PLUMAGE-CHANGES

Study a series of male specimens of each of the following six species: Mallard (Anas platyrhynchos), Ring-billed Gull (Larus delawarensis), Black-capped Chickadee (Parus atricapillus), American Redstart (Setophaga ruticilla), Bobolink (Dolichonyx oryzivorus), and Rose-breasted Grosbeak (Pheucticus ludovicianus). These show the sequence of plumages and molts, and plumage-change by wear, and demonstrate some of the differences between species as to completeness or suppression of certain molts and the modification of molts by age. Note for each specimen the date of collection. On the basis of observations, fill in the chart, "Sequence of Plumages in the Males of Six Species," on the inside of the back cover of this book. In the columns captioned "How Changed," give the name of the molt and state whether it is complete or incomplete. If the molt is incomplete, state which pterylae show molt. If no molt occurs, mark "none"; if plumage-change by wear occurs, write "plumage-change by wear."

ABNORMAL PLUMAGE COLORATION

Birds sometimes vary from their normal coloring because of the lack or excess of pigments in one or more of the feathers. There are four variations.

Albinism. This is a variation caused by the reduction or absence of pigments in feathers. In the complete absence of pigments, the feathers' structure refracts light, giving white. The pigments may also be absent from the irises and from the normally featherless, colored skin of the lores, forehead, crown, and other topographical parts of certain species of birds. The colors of these areas will then have a light red or pinkish hue, owing to the blood showing through from the capillaries. There are four degrees of albinism in birds: **Total albinism,** when all pigments are completely absent from the plumage, irises, and skin. **Incomplete albinism,** when the pigments are completely absent from the plumage, or irises, or skin, but not from all three. **Imperfect albinism,** when all the pigments are reduced ("diluted"), or at least one of the pigments is absent, in any or all three areas. **Partial albinism,** when the pigments are reduced, or one or more is absent, from *the parts* of any or all three areas. Of the four degrees of albinism, partial albinism is the commonest. It frequently involves certain feathers only, and it is often symmetrical, each side of the bird being affected in the same way.

Albinism may have a genetic basis and be inherited. It may also be spontaneous, developing in an individual as a result of some physiological disturbance. An instance is reported (Frazier, 1952) of a Robin *(Turdus migratorius)* that was in normal plumage when banded, but was partially albinistic when recovered two years later.

Melanism, Erythrism, and **Xanthochroism.** These are variations in plumage color caused by either excessive amounts of pigments or the absence of certain pigments. (See the section, "The Coloration of Feathers," in this book, pages 47-50, for information on the different pigments.) Melanism results from large amounts of melanin, giving colors ranging from brown to black; erythrism and xanthochroism usually result from the absence of melanin and the retention of red or yellow carotenoid pigments. Xanthochroism may develop in captive birds, due to a dietary deficiency. Thus green or blue feathers of parrots and red feathers of tanagers and finches are sometimes replaced by yellow feathers.

COLOR PHASES

In certain wild species, two or more color phases, which presumably have a genetic basis, normally occur. When a species has only two color phases, the condition is known as **dichromatism.** In Atlantic populations of the Common Murre *(Uria aalge)* many individuals show a "ringed" or "bridled" phase distinguished by a white line encircling the eye and then running backward and downward for a short distance. The Screech Owl *(Otus asio)* and Ruffed Grouse *(Bonasa umbellus)* have "red" (erythristic) and "gray" phases. When a species has two or more phases often intergrading, the condition is **polychromatism.** The Swainson's Hawk *(Buteo swainsoni)* and Rough-legged Hawk *(B. lagopus)* have "light" and "dark" (melanistic) phases; the Ferruginous Hawk *(B. regalis)* has "normal" and "rufous" phases. The Gyrfalcon *(Falco rusticolus)* features three phases designated as "white," "gray," and "black." Color phases are sometimes clearly correlated with geographical location. Both red and gray phases, for example, occur in populations of the Screech Owl in eastern United States, but only the gray phase in western populations.

Examine several bird skins which show some of the color abnormalities and phases described above. List in the following chart the names of the species and state the kinds of abnormalities or phases observed.

Species	Color Abnormality or Phase

SEXUAL, AGE, AND SEASONAL DIFFERENCES IN PLUMAGE COLORATION

Within a species, plumage coloration is often subject to differences, sometimes extreme, dependent on sex, age, and season. The differences within the majority of species may be summarized as follows:

1. *The nuptial plumage of the adult male is more colorful, or more strikingly marked, than the plumage of the adult female, which is characteristically dull and inconspicuously marked the year-round.*
2. *The colorful, or strikingly marked, nuptial plumage of the male is replaced, following the breeding season, by a winter plumage closely resembling the plumage of the adult female.*
3. *The juvenal plumage of the immature bird, regardless of sex, is characteristically dull, or inconspicuously marked, and closely resembles the plumage of the adult female.*

While the majority of species may be said to conform to the above statements, a great number of species present exceptions to one or more of them. Every conceivable type of exception may be found:

On the chart on page 195, "Sexual, Age, and Seasonal Differences in Plumage Coloration," the main differences and several well-known exceptions are itemized in the left-hand column. Complete the chart in the following manner: (1) Fill in the spaces across the top of the chart with the names of species selected by the instructor to illustrate the differences and exceptions. (2) Examine skins of these species. Then, in the columns under the species names, put a check in the space opposite the difference or exception that each species represents. *Note:* when attempting to describe quality of coloration, use these criteria: A plumage is *colorful* when possessing one or more brilliant colors such as red, yellow, green, blue, etc.; *strikingly marked* when possessing conspicuous, contrasting dark or white markings with or without brilliant colors; *dull* when possessing colors predominantly brown or gray; *inconspicuously marked* when the markings are not sharply contrasted and are not bright in color.

USES OF PLUMAGE COLORATION

Plumage coloration has numerous uses, or functions, some of which are roughly classified as cryptic, deflective, and epigamic.

Cryptic Coloration. When coloration serves to conceal the bird in its environment, it is cryptic. Cryptic coloration functions in two ways: (1) It may conceal by **mimicking** the normal background of the bird. A bird thus colored is frequently **countershaded**, having light under parts that counteract dark shadows cast by its body, making it appear flat against the background. The bird has a tendency to remain motionless when approached by an enemy and to flush only as a last resort. (2) It may conceal by having **ruptive patterns**—designs with irregular, sharply defined areas of contrasting colors that "break up" the general contour of the bird into apparently unrelated, shapeless parts. When a bird with ruptive pattern is seen at a distance, these separate areas are more conspicuous than the bird itself.

Deflective Coloration. Certain birds have feathered areas colored in such a way as to become conspicuous when the birds are in flight. Examples are the white areas on the wings, rump, and tail which are sometimes called **banner-marks.** These areas are

SEXUAL, AGE, AND SEASONAL DIFFERENCES IN PLUMAGE COLORATION

Adult Plumage												
General rule: *♂ in breeding season only: more colorful, or more strikingly marked, than ♀*												
Exceptions: ♂ the year-round: more colorful, or more strikingly marked, than ♀												
♂ and ♀ similar: both colorful, or strikingly marked, in breeding season only												
♂ and ♀ similar: both colorful, or strikingly marked, the year-round												
♂ and ♀ similar: neither colorful nor strikingly marked the year-round												
♀ in breeding season only: more colorful, or more strikingly marked, than ♂												
♀ the year-round: more colorful, or more strikingly marked, than ♂												
Juvenal Plumage												
General rule: *Immature birds dull, or inconspicuously marked, and similar to adult ♀ ; unlike adult ♂*												
Exceptions: Immature birds dull, or inconspicuously marked, but different from both adult ♂ and ♀												
Immature birds similar to, and as colorful as, both adult ♂ and ♀												

USES OF PLUMAGE COLORATION

	Cryptic Coloration			Deflective Coloration			Epigamic Coloration	
	Mimicking	Countershading	Ruptive Patterns	Banner-marks on Tail	Banner-marks on Wings	Banner-marks on Rump	Sex-recognition Marks	Display-colors

considered to have a protective function in that they deflect a pursuing enemy's attack from a vital to a less vital part of the body. These areas are sometimes considered to have another protective function in that they serve to confuse an enemy and, in a sense, deflect his attention. When, for instance, a bird suddenly flushes before an enemy, the conspicuous areas unexpectedly appear and momentarily startle him, thus delaying (deflecting) the attack.

Epigamic Coloration. When coloration is used to bring the sexes together in any manner during the breeding season, it is epigamic. Plumage colors found in males, but not in females, are a means of intraspecific sex recognition. Facial markings, masks, and throat patches are examples. Brilliant colors in the feathers of many males enhance the role of plumage display prior to and during mating.

Examine a group of bird skins selected by the instructor to illustrate the above types of coloration. Opposite this page on the chart, "Uses of Plumage Coloration," list the species examined in the left-hand column. Check the spaces opposite the species that illustrate the types of coloration indicated across the top.

The student must bear in mind that plumage coloration serves many uses other than those classified and described above. Certain plumage colors in the same species may have more than one function. Banner-marks, while serving as a protective feature in one way or another, may act in intraspecific recognition as signals to keep members of a flock together. Epigamic coloration, besides bringing the opposite sexes together, may assist in threat displays against rivals of the same sex.

REFERENCES

Allen, A. A.
1961　The Book of Bird Life. Second edition. D. Van Nostrand Company, New York. (Chapter 12, "Plumage Coloration and Plumage Change.")

Allen, G. M.
1925　Birds and Their Attributes. Marshall Jones Company, Boston. (Dover reprint available; Chapter 3, "Birds' Colors and Their Uses.")

Beddard, F. E.
1892　Animal Coloration: An Account of the Principal Facts and Theories Relating to the Colors and Marking of Animals. Macmillan and Company, New York.

Behle, W. H., and R. K. Selander
1953　The Plumage Cycle of the California Gull (*Larus californicus*) with Notes on Color Changes of Soft Parts. Auk, 70: 239-260.

Chapman, F. M.
1932　Handbook of Birds of Eastern North America. Second revised edition. D. Appleton and Company, New York. (Dover reprint available; "The Plumage of Birds," pp. 91-104.)

Clay, W. M.
1953　Protective Coloration in the American Sparrow Hawk. Wilson Bull., 65: 129-134. (Shows how countershading, ruptive patterns, and deflective coloration in *Falco sparverius* assist the species in deceiving prey and/or enemies.)

Collins, C. T.
1961　Tail Molt of the Saw-whet Owl. Auk, 78: 634. (The author found the tail molt to be simultaneous in this species as did Mayr and Mayr, 1954, in other species of small owls.)

Cott, H. B.
1940　Adaptive Coloration in Animals. Reprinted with minor corrections, 1957. Methuen and Company, London.

Dixon, K. L.
1962　Notes on the Molt Schedule of the Plain Titmouse. Condor, 64: 134-139. (This species apparently begins the postnuptial molt in mid-June, while still feeding fledglings, thus contradicting the principle that the molt does not begin until the breeding season is over.)

Dwight, J., Jr.

1900a The Sequence of Plumages and Moults of the Passerine Birds of New York. Annals New York Acad. Sci., 13: 73-360.

1900b The Moult of the North American Shore Birds (Limicolae). Auk, 17: 368-385.

1903 How Birds Molt. Bird-Lore, 5: 156-160.

1920 The Plumages of Gulls in Relation to Age as Illustrated by the Herring Gull *(Larus argentatus)* and Other Species. Auk, 37: 262-268.

Frazier, F. P.

1952 Depigmentation of a Robin. Bird-Banding, 23: 114.

Friedmann, H.

1944 The Natural-history Background of Camouflage. Smithsonian Rept. for 1943, pp. 259-274.

Gross, A. O.

1965a The Incidence of Albinism in North American Birds. Bird-Banding, 36: 67-71.

1965b Melanism in North American Birds. Bird-Banding, 36: 240-242.

Gullion, G. W., and W. H. Marshall

1968 Survival of Ruffed Grouse in a Boreal Forest. Living Bird, 7: 117-167. (In Minnesota the gray color phase predominates in grouse populations occupying northern coniferous forests, the red phase in southern hardwoods. There appears to be a correlation between color phase and rate of survival.)

Holmes, R. T.

1966 Molt Cycle of the Red-backed Sandpiper *(Calidris alpina)* in Western North America. Auk, 83: 517-533.

Humphrey, P. S., and G. A. Clark, Jr.

1964 The Anatomy of Waterfowl. In *The Waterfowl of the World*, Volume 4, by J. Delacour. Country Life, London. ("Plumage Succession," pp. 175-180.)

Humphrey, P. S., and K. C. Parkes

1959 An Approach to the Study of Molts and Plumages. Auk, 76: 1-31. (The authors propose a new terminology for plumages and molts.)

1963 Comments on the Study of Plumage Succession. Auk, 80: 496-503. (The authors' response to the criticism by Stresemann, 1963, of the new terminology in their paper listed above.)

Huxley, J. S.

1938 Threat and Warning Coloration in Birds with a General Discussion on the Biological Functions of Colour.

Proc. VIIIth Internatl. Ornith. Congr., pp. 430-455.

Johnson, N. K.

1963 Comparative Molt Cycles in the Tyrannid Genus *Empidonax.* Proc. XIIIth Internatl. Ornith. Congr., pp. 870-883. (A great interspecific variation in the timing of molts, suggesting that it is "a reflection of adaptation at the species level.")

Johnston, D. W.

1961 Timing of Annual Molt in the Glaucous Gulls of Northern Alaska. Condor, 63: 474-478. (The adults begin their annual—i.e., postnuptial—molt "before or soon after the eggs are laid" by the end of May—a possible adaptation whereby the species is able to complete both breeding and molt during the abbreviated summer.)

Jones, L.

1930 The Sequence of the Molt. Wilson Bull., 42: 97-102.

Lesher, S. W., and S. C. Kendeigh

1941 Effect of Photoperiod on Molting of Feathers. Wilson Bull., 53: 169-180. (Experimental procedures show that captive birds, *Zonotrichia albicollis* and *Colinus virginianus,* which normally go through a prenuptial molt in spring as the days are increasing in length, will molt out of season when exposed to light-periods artificially increased to 15 hours.)

Martin, N. D.

1950 Colour Phase Investigations on the Screech Owl in Ontario. Canadian Field-Nat., 64: 208-211.

Mayr, E., and M. Mayr

1954 The Tail Molt of Small Owls. Auk, 71: 172-178. (The authors suggest that simultaneous tail molt may be due to a relaxation of selection pressure.)

Miller, A. H.

1928 The Molts of the Loggerhead Shrike, *Lanius ludovicianus* Linnaeus. Univ. California Publ. in Zool., 30: 393-417.

Mueller, C. D., and F. B. Hutt

1941 Genetics of the Fowl. 12. Sex-linked, Imperfect Albinism. Jour. Heredity, 32: 71-80. (Contains the classification of albinism used, with slight modification, in this manual.)

Nero, R. W.

1954 Plumage Aberrations of the Redwing *(Agelaius phoeniceus).* Auk, 71: 137-155. (The author found a surprisingly high percentage of aberrations, classified as: total albinism; imperfect albinsim; partial albinism—random and specific. Practically all 219 males collected near Madison, Wisconsin, showed some deviation from the wholly black plumage.)

1960   Additional Notes on the Plumage of the Redwinged Blackbird. Auk, 77: 298-305. (The author found a similarly high percentage of aberrations among 100 males near Regina, Saskatchewan, leading him to suggest that such aberrations are characteristic of the species.)

Noble, G. K.
1936   Courtship and Sexual Selection of the Flicker *(Colaptes auratus luteus)*. Auk, 53: 269-282.

Noble, G. K., and W. Vogt
1935   An Experimental Study of Sex Recognition in Birds. Auk, 52: 278-286.

Oring, L. W.
1968   Growth, Molts, and Plumages of the Gadwall. Auk, 85: 355-380. (An excellent study of an individual species.)

Owen, D. F.
1963   Polymorphism in the Screech Owl in Eastern North America. Wilson Bull., 75: 183-190.

Pitelka, F. A.
1945   Pterylography, Molt, and Age Determination of American Jays of the Genus Aphelocoma. Condor, 47: 229-260.

Poulton, E. B.
1890   The Colours of Animals, Their Meaning and Use. D. Appleton Company, New York.

Ralph, C. L.
1969   The Control of Color in Birds. Amer. Zool., 9: 521-530.

Salomonsen, F.
1939   Moults and Sequence of Plumages in the Rock Ptarmigan *(Lagopus mutus* (Montin)). Vid. Medd. Dansk Naturh. Foren., 103: 1-491. Reprinted by P. Haase and Son, Copenhagen. (A thorough study based on an enormous amount of data.)

Stone, W.
1896   The Molting of Birds with Special Reference to the Plumages of the Smaller Land Birds of Eastern North America. Proc. Acad. Nat. Sci. Philadelphia, 1896, pp. 108-167.

Stresemann, E.
1963   The Nomenclature of Plumages and Molts. Auk, 80: 1-8. (A strongly critical review of the new nomenclature proposed by Humphrey and Parkes, 1959.)

Stresemann, E., and V. Stresemann
1966   Die Mauser der Vögel. Jour. f. Ornith., 107 Sonderheft. (Although titled "The Molting of Birds," this monograph of 445 pages is concerned almost exclusively with the molting of the remiges and rectrices in selected species.)

Sutton, G. M.
1935   The Juvenal Plumage and Postjuvenal Molt in Several Species of Michigan Sparrows. Cranbrook Inst. Sci. Bull. No. 3, Bloomfield Hills, Michigan.

Test, F. H.
1945   Molt in Flight Feathers of Flickers. Condor, 47: 63-72.

Thayer, G. H.
1909   Concealing-coloration in the Animal Kingdom: An Exposition of the Laws of Disguise through Color and Pattern. Macmillan Company, New York. (Although much criticized, the work is nonetheless an important contribution to the subject.)

Thompson, D. R., and C. Kabat
1950   The Wing Molt of the Bob-white. Wilson Bull., 62: 20-31. (How age may be determined by the postjuvenal molt of the primaries.)

Tuck, L. M.
1960   The Murres: Their Distribution, Populations and Biology. Canadian Wildlife Series, 1. Canadian Wildlife Service, Ottawa. (Genetic Variations, pp. 32-33.)

Watson, G. E.
1963   Feather Replacement in Birds. Science, 139: 50-51. (Report of "passive" pushing out of old generation of pennaceous feathers; supports the Humphrey-Parkes system for naming plumages.)

Woolfenden, G. E.
1967   Selection for a Delayed Simultaneous Wing Molt in Loons (Gaviidae). Wilson Bull., 79: 416-420. (The author suggests selective factors that may account for the simultaneous molt following fall migration.)

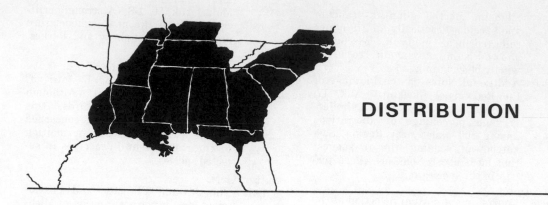

DISTRIBUTION

Despite their remarkable flying abilities, most birds are limited, as are other animals, to specific areas called **ranges.** Only a few bird species have ranges that are cosmopolitan. The Osprey *(Pandion haliaetus)* that occurs in all the continents except Antarctica is one example of a cosmopolitan species. Some birds represent the opposite extreme by having highly restricted ranges. The Kirtland's Warbler *(Dendroica kirtlandii),* for instance, nests solely in an area of Michigan, 100 by 60 miles, and winters in the Bahamas. Other species, confined the year round to oceanic islands or archipelagos, may have still smaller ranges.

Most birds occupy only parts of the ranges to which they are particularly suited and therefore are not evenly distributed. In some species the ranges are clearly **disjunct** or **discontinuous** with great distances between segments. The Scrub Jay *(Aphelocoma coerulescens),* occurs in central Flordia, separated by hundreds of miles from the rest of its range in the far western contiguous states and Mexico.

Numerous factors account for the present ranges of species. Some are past or historical factors such as the shaping, separating, and rejoining of continents, the invasion and recession of glaciers, the shifts in climate, and the changes in dominant vegetation. Others are present-day factors. They include:

Geographical barriers. Stretches of open water may limit the ranges of land species; large land masses, the ranges of marine species. Vast areas of continuous forest, grassland, or desert may be barriers to birds that are in no way adapted to them. High mountain ranges are often effective barriers. No general rule can explain why a barrier limits the range of one species and not another, and no correlation exists between the size and/or performance of the species and the limiting effect of a barrier.

Air temperature. High and low temperatures, sometimes in conjunction with humidity and precipitation, may be a factor in restricting a bird's range. In the Northern Hemisphere, low minimum temperature, usually reached at night, may limit a species' range northward; or a high maximum temperature, usually attained in the late afternoon, may prevent its expansion southward.

Sunlight. Seasonal variations in periods of light may be a limiting factor. For example, certain species cannot withstand the long cool nights and the loss of feeding time resulting from the long absence of light at northern latitudes in winter.

Winds. Some oceanic birds, such as the larger albatrosses, require winds for sustaining flight and consequently range only in areas where winds blow constantly.

GEOGRAPHICAL DISTRIBUTION

The arrangement of organisms with relation to areas is **geographical distribution.** The aforementioned ranges of the Osprey and Kirtland's Warbler are geographical— the general areas where the species occur. When a bird species is migratory, its range is customarily divided into **breeding range,** where the species nests, and **winter range,** where the species occurs between fall and spring migrations. For authoritative descriptions of the breeding and winter ranges of North American birds north of Mexico, consult *The A. O. U. Check-list* (American Ornithologists' Union, 1957). For more abbreviated but nonetheless useful descriptions, based largely on the Check-list, consult Reilly (1968) and the several field guides (see the section "Field Identification" in this book, page 230). The range maps in *Birds of North America: A Guide to Field Identification* by C. S. Robbins, B. Bruun, and H. S. Zim (Golden Press, New York, 1966) are especially helpful.

The ranges of birds, except for species confined to oceanic islands, are seldom static; indeed they change so constantly one should think of them as fluid.

Birds have an inherent tendency to disperse from their place of birth and invade new areas. If the areas harbor favorable conditions, birds can soon establish themselves and thereby expand their ranges.

Various factors aid in the constant dispersal of birds and may lead to the expansion of their ranges. Prevailing winds and cyclonic storms move individuals over seas, even to different continents where, provided the environment is suitable, they survive and reproduce. The Cattle Egret *(Bubulcus ibis)* may have arrived in South America from its native Africa by way of a storm. A failure in food supply may cause an irruption, or irregular migration, of a population of a somewhat sedentary species and hasten its expansion. In certain years and for some reason, perhaps in search of food, a number of northern species such as Hawk Owls *(Surnia ulula)* and Boreal Chickadees *(Parus hudsonicus)* suddenly move, in fall and winter, far south of their normal ranges. Although they usually retreat northward with the coming of spring, the possibility prevails that some individuals will find suitable breeding areas and establish themselves south of their present ranges. (For other instances of irruptions that could lead to range expansions, see "Migration" in this book, page 275.)

Civilization has changed the ranges of impressive numbers of North American birds. Some species, their populations reduced and natural environments depleted by man, have retreated to very limited ranges where favorable conditions still hold. The Whooping Crane *(Grus americana),* which once occurred widely through the interior of the continent, steadily declined in numbers through excessive shooting and loss of suitable marshes for feeding, nesting, and loitering. Today its breeding range is confined to a small remote area in northwestern Canada and its winter range to a peninsula on the coast of Texas.

On the other hand, some species have benefited so greatly by man's alteration of environments that they have extended their ranges and probably increased in numbers. The eastern subspecies of the Cardinal *(Richmondena cardinalis)* has gradually moved northwestward along the bottomlands of the Mississippi and Missouri Rivers into Minnesota and the Dakotas and northeastward into New York State and southern New England where it is now well established. The Robin *(Turdus migratorius),* House Wren *(Troglodytes aedon),* and Song Sparrow *(Melospiza melodia)* have extended their breeding ranges southward into Georgia. All these species show a marked attraction to open places—e.g., dooryards or shrubby situations—following the removal of forests. Apparently their earlier ranges were restricted less by climate than by available places for feeding and nesting. In the mid-continent, forest and forest-edge birds such as the Yellow-shafted Flicker *(Colaptes auratus)*, Baltimore Oriole *(Icterus galbula)*, Rose-breasted Grosbeak *(Pheucticus ludovicianus)*, and Indigo Bunting *(Passerina cyanea)* have extended their breeding ranges steadily across the Great Plains into western North and South Dakota, Nebraska, and Kansas, encouraged, no doubt, by the addition of tree and shrub plantations ("shelter belts") westward across the prairie where heretofore the only natural woodlands were along the rivers. At the same time, the Scissor-tailed Flycatcher *(Muscivora forficata)* has moved eastward into western Missouri and the Brewer's Blackbird *(Euphagus cyanocephalus)* into Michigan, both finding, in the place of formerly extensive forests, the open country to which they are adapted.

The rate at which a species may expand its range is sometimes astonishing. The most dramatic example is perhaps the movement of the Cattle Egret northward following its arrival in South America about 1930. In 1941 or 1942 it made its first appearance in the United States at Clewiston, Florida. By 1952 it was numerous in Florida and noted in New Jersey, Massachusetts, and northern Illinois. By 1954 it was breeding in Florida where it now numbers in the thousands. Since that time it has occurred in all the northeastern states and southeastern Canada and has nested in many places as far north as Ontario.

The extension of the range of one species may spell the decline of another species, into whose range it intrudes, by intensely and successfully competing with it. The spread of the Starling *(Sturnus vulgaris)* westward through conterminous United States very likely reduced the population of the Red-headed Woodpecker *(Melanerpes erythrocephalus)*, whose nesting sites the Starling often usurps. So far there is no evidence that the Cattle Egret has had any harmful effect on other herons with which it commonly associates and nests.

In any area, large or small, the ranges of different species overlap, forming an aggregation of organisms. All the animals in the aggregation are collectively called the **fauna.** Part of the study of geographical distribution concerns the species composition and development of faunas.

Continental Distribution

Each continent has its own fauna with certain forms that are **endemic,** or **indigenous,** meaning that they are restricted to, or native to, the area, or parts thereof, and occur nowhere else. The same forms, if they actually evolved in the area where they are found, may also be **autochthonous,** the term being merely a refinement of "endemic" and "indigenous."

It has long been the custom to divide the land areas of the world into **zoogeographical regions** based on the presence or absence of distinctive forms of animals. The exact number of these major units varies, depending on whether certain regions are considered as full regions or subdivisions, but six is the usually accepted number, as follows:

Nearctic: Greenland and North America south to and including the Mexican highlands.

Palaearctic: Europe and Africa south to and including most of the Sahara Desert; Asia north from the Himalayas and the Yangtze River.

Neotropical: South America, Central America, the Mexican lowlands, and the West Indies.

Ethiopian: Africa, south of the Sahara Desert, and southern Arabia; Madagascar.

Oriental: Asia south of the Himalayas and the Yangtze River; Ceylon, Sumatra, Java, Borneo, Celebes, Taiwan, and the Philippines.

Australian: Australia, New Zealand, New Guinea, all the islands of the southwest Pacific, and the Hawaiian Islands.

Some summations of the bird faunas, or **avifaunas,** of the above regions are given below. Figures for the numbers of families and breeding species are approximate since authorities differ in awarding full familial rank to certain groups of genera and full specific rank to many of the breeding populations. Marine birds are not included in the figures. For further data, see Darlington (1957), Serventy (1960), G. J. Wallace (1963), and Welty (1962).

The Nearctic and Palaearctic Regions, sometimes combined as the Holarctic Region because of their faunal similarities, have 650 and 750 species, respectively. Although these two regions have fewer species than any of the other regions, the number of individuals in each species is generally greater because of less interspecific competition and more extensive areas suitable for breeding. The Nearctic has 63 families, the Palaearctic 69; many are shared. Only one family—the Chamaeidae (wrentits) are endemic to the Nearctic; only one—the Prunellidae (accentors)—to the Palaearctic. Read Udvardy (1958) for an illuminating and detailed comparison of the Nearctic avifauna with the Palaearctic.

The Neotropical Region that embraces all of South America ("The Bird Continent") is by far the richest for birds with about 2,900 species classified into 97 families, 32 of which are endemic. A peculiarity of the avifauna, besides its great degree of endemism, is the absence of numerous forms widespread elsewhere in the world. Many of the species have exceedingly restricted ranges, some amounting to little more than a lake, valley, or mountain slope. Mountainous Colombia, with some 1,600 species within its borders, boasts more different kinds of birds than any other country in the world.

The Ethiopian Region is next to the Neotropical in variety of birds with 1,900 species, representing 67 families, eight of which are endemic to Africa and five to Madagascar.

The Oriental Region has perhaps 1,500 species among some 83 families, only one of which is endemic. All the other families are shared with one or more of the neighboring regions.

The Australian Region has about 1,200 species among 83 families and is next to the Neotropical Region in the number of endemic families—14 altogether.

History of a Continental Avifauna. The Zoogeographical Regions, as just demonstrated, indicate that forms of land life are not uniformly distributed over the earth, but tend to be grouped on different continents or parts of continents, and that between one such group and another there are at least a few recognizable differences. The most serious criticism of this procedure is that it implies a fixed grouping of organisms within geographical boundaries, as if all the animals of one region developed independently of the faunas of neighboring regions and became a sharply demarked assemblage. The truth is that the fauna of a region, continent, or smaller area has been changing and is still changing, ever modified by the disappearance of some forms and the invasion of forms from elsewhere.

In any continent as a whole the avifauna constitutes a complex of forms from various sources. Some forms originated within the area; others arrived from neighboring areas at different periods of geologic time, including recent centuries. In order to determine how any avifaunal complex has evolved, one must consider the geological history of the area, investigate the fossil record of bird forms, and analyze its present-day avifauna for clues as to where different forms came from and how they reached their present ranges. By combining these three approaches one can, possibly, derive a working hypothesis.

Mayr (1946) carefully analyzed the bird fauna of the North American continent—the Nearctic Region plus Central America and West Indian parts of the Neotropical Region. Early in his report, he pointed out two well-established facts relating to the continent's history during the Tertiary (see the Geologic Time Scale in this book, page 414): (1) North America was separated from South America when the isthmus between Colombia and central Mexico was broken up into islands with wide, intervening oceanic channels; (2) meanwhile North America was repeatedly connected to Asia by a land "bridge" across Bering Strait. Mayr then went on to show that the bird fauna of North America contains seven elements as follows:

Panboreal Element. Forms equally well represented in the northern latitudes of North America and the Old World, but possibly originating in the Old World. Examples: Loons (Gaviidae), phalaropes (Phalaropodidae), and auks (Alcidae).

Old World Element. Three groups of forms originating in the Old World and arriving in North America by way of the Bering Strait "bridge": (1) Very recent arrivals which have undergone no change. Examples: *Motacilla flava tschutschensis,* a subspecies of the Yellow Wagtail, and *Oenanthe oenanthe oenanthe,* a subspecies of Wheatear, which occur in both Alaska and the Eurasian continent to the west. (2) Fairly recent arrivals that have had time to attain subspecific distinction from Old World forms. Examples: *Certhia familiaris americana,* the North American boreal race of the Brown Creeper; *Lanius excubitor borealis* and *L. e. invictus,* the North American races of the Northern Shrike. (3) Much earlier arrivals that have had time to evolve into species, genera, or even subfamilies distinct from their Old World relatives. Examples: a variety of well-known North American birds, such as cranes (Gruidae), pigeons (Columbidae), cuckoos (Cuculidae), typical owls (Strigidae), kingfishers (Alcedinidae), crows and jays (Corvidae), titmice (Paridae), and nuthatches (Sittidae).

North American Element. Forms, developed in North America during the Tertiary, while the continent was partially isolated and much more of its southern area had a tropical or subtropical climate. Examples: New World vultures (Cathartidae), turkeys (Meleagrididae), limpkins (Aramidae), dippers (Cinclidae), wrens (Troglodytidae), thrashers and mockingbirds (Mimidae), waxwings (Bombycillidae), silky flycatchers (Ptilogonatidae), vireos (Vireonidae), and wood warblers (Parulidae).

Pan-American Element. Forms of South American origin which "island hopped," during the Tertiary, from South to North America and evolved endemic genera either on the islands, now Central America, or in southern North America. Examples: hummingbirds (Trochilidae), New World flycatchers (Tyrannidae), and blackbirds (Icteridae).

South American Element. A few forms of some families, well developed in South America and clearly originating there. Example: cotingas (Cotingidae). Of the 31 genera in this family, 12 have reached Central America; one, the Rose-throated Becard *(Platypsaris),* has reached the United States; none, however, has come to be restricted to any area outside South America.

Pantropical Element. Forms common to both the New and Old World tropics. Examples: parrots (Psittacidae) and trogons (Trogonidae) which differ at the generic level. Just where the parent stock originated and how the New and Old World forms came to be so widely separated geographically provide fascinating lines of speculation (see Mayr's paper for a review).

Unanalyzed Element. Forms so cosmopolitan in their distribution as to make any positive determination of their origin difficult, if not impossible. Examples: most oceanic birds, fresh-water birds, shore birds, and such land birds as the Osprey, hawks, eagles, goatsuckers, woodpeckers, and swallows.

Mayr concluded that most North American families and subfamilies of birds are clearly either of Old World origin, of South American origin, or members of an indigenous North American element which developed while the continent was partially isolated during the Tertiary.

Insular Distribution

Sea islands hold breeding populations of land birds and provide nesting sites for the great majority of marine birds. Since the main concern with distribution in this book is with land birds, including fresh-water and shore birds, the distribution of insular land birds is discussed first.

Practically all sea islands that are habitable—i.e., adequate in size with suitable environment—support, or have supported in recent times, land birds originating from the continents. In general, the more remote these islands are from the mainland, the fewer kinds of birds that colonize them, since the chances of stragglers, or "pioneers," reaching them in sufficient numbers to found colonies decrease with increased distance. Nevertheless, lists of birds reported over the years from sea islands include stragglers, representing a great array of species, far more than ever established residence.

Why have some birds successfully colonized remote islands and others have not? The size of the bird is not a determining factor. Certain passerines, despite their small size, have reached and founded populations on many of the distant islands. And flying ability is not necessarily a factor. Swallows, surely among the strongest flyers, have never colonized distant islands, while rails and gallinules, seemingly among the weakest flyers, have been some of the most successful colonists. The answer seems to lie not so much in the physical peculiarities of the birds as in their ability to adapt to the limited resources of an insular environment and to compete with other birds already established there.

The non-migratory forms among the populations on sea islands will probably show taxonomic distinctions, and the farther they are from their parent populations, the greater will be these distinctions.

Forms on islands close to mainlands, such as those on the continental shelf, may be expected to develop no more than minor peculiarities. The avifauna on the Queen Charlotte Islands, separated from the coast of British Columbia by a channel only 25 miles wide, contains essentially the same birds as that of the nearby mainland, except for the population of several permanent-resident species—e.g., the Saw-whet Owl *(Aegolius acadicus)* and Hairy Woodpecker *(Dendrocopos villosus)*—whose Queen Charlotte Island forms are peculiar enough to be considered endemic subspecies. Presumably, the bird fauna has not been isolated long enough or completely enough to evolve more than subspecific distinctions.

By contrast, the avifaunas of both the Hawaiian Islands in the north-central Pacific and the islands of Tristan da Cunha in the middle of the South Atlantic have developed marked peculiarities.

Besides two endemic subspecies of ducks and one endemic subspecies of a gallinule and stilt, the avifauna of the Hawaiian Islands includes 38 endemic forms ranked as full species: two ducks and two rails, a goose, hawk, and crow, one Old World flycatcher (Muscicapidae), two thrushes, five honeyeaters (Meliphagidae), and 23 honeycreepers grouped together as an endemic family, the Drepanididae. The two rails, three of the honeyeaters, and nine of the honeycreepers are now extinct. (See the section of this book, "Systematics," page 139, for a discussion of the origin of, and speciation in, the Drepanididae.)

On the three main islands of the Tristan da Cunhas, all very small with grass, ferns, shrubs, and (formerly) trees as the principal cover, the avifauna is remarkably distinguished by a monotypic genus of flightless rail on one island; a monotypic genus of thrush whose single species has developed three subspecies, one on each island; and two species of sparrows that constitute one genus and are divided into subspecies, each occupying one or two but not all three islands. Also included in the avifauna is a monotypic genus of flightless gallinule whose single species developed two subspecies, one (now extinct) on one of the Tristans and the other on Gough Island, some 217 miles to the southeast. Consult Rand (1955) for further details. Most of the evidence suggests that these endemic forms are of American origin, descended from stragglers that reached the islands accidentally, aided perhaps by the prevailing west winds.

The flightless condition of the Tristan rails is in keeping with rallid fauna endemic to the islands of the Pacific and Indian Oceans—and in line with the condition in a few other island forms of birds as well. Carlquist (1965) lists altogether 20 species of island Rallidae, all flightless. Nine are extinct and six are in danger of extinction—undoubtedly because man and the dogs, cats, rats, and other mammals that accompany him can kill them with ease.

Marine Distribution

True marine birds—also called oceanic birds or sea birds—are those species regularly inhabiting the sea and consistently obtaining their food from the sea the year round. All nest in colonies, usually on islands, or, more rarely, in continental situations adjacent to the sea. All feature salt glands (see the section of this book, "Anatomy and Physiology," page 121).

Marine birds vary in their habitual distance from land and general feeding habits. Albatrosses, shearwaters, and petrels—indeed, all members of the Procellariiformes—are pelagic, occurring on the open sea, ordinarily out of sight of land, and feeding largely on plankton. Most all the other marine species are confined to the shallow waters above the continental shelf, or waters adjacent to islands rising from the deep sea, and feed primarily on fish. Penguins, tropicbirds, boobies and gannets, frigatebirds, and the auks, auklets, murres and puffins commonly stay offshore; cormorants, skuas and jaegers, skimmers, and the marine species of pelicans, gulls, and terns stay inshore, seldom straying beyond sight of land. But there are exceptions. For example, small gulls, called kittiwakes, are at times pelagic as are some of the terns during their migrations. And some species of inshore birds partly forsake the marine environment altogether and move inland.

The seas of the world, like the land areas, may be divided into regions, distinguished by the presence or absence of animal forms. Ornithologists (e.g., see Serventy, 1960) recognize three major regions as follows:

Northern Marine Region: The frigid waters of the Arctic south to the Subtropical Convergence of the Northern Hemisphere where, between Latitudes 35 and 40 degrees, the water temperature rises rapidly.

Southern Marine Region: The frigid waters around Antarctica north to the Subtropical Convergence of the Southern Hemisphere where, between the same latitudes, the water temperature rises similarly.

Tropical Marine Region: The warm equatorial waters between the Subtropical Convergences of both hemispheres.

The Northern Marine Region is distinguished by the Alcidae (auks, etc., totaling 22 species), the only family wholly confined to the northern seas. More alcid species, probably with a higher average of individuals per species, occur in the North Pacific than in the North Atlantic.

The Southern Marine Region is the richest in its variety of bird species if not in numbers of individuals. Its principal distinctions are 16 of the 17 species of penguins, the other species reaching north to the Galapagos Islands at the Equator; all but three of the dozen species of albatrosses; and perhaps the majority of the 75 or more species of shearwaters and petrels. The penguins seem to fill the position in the Southern Region that the alcids occupy in the Northern. The high density of plankton and the strong, almost constant winds account for the presence of so many albatrosses; the plankton alone attracts the shearwaters and petrels.

The Tropical Marine Region is distinguished particularly for its wide variety of pelecaniform birds—tropicbirds, boobies, and frigatebirds—and terns. Compared to the open seas of the Southern Region, those of the Tropical seem devoid of bird life. This is due partly to a low supply of plankton, the food of pelagic birds, and to the habit of other avian residents of staying close to land where their food, small fish, is more abundant.

No doubt the Tropical Marine Region, with its warm water and low food content in the open sea, is an effective barrier to Northern Marine birds entering into the Southern Marine Region and *vice versa*. Almost certainly it has kept the poor-flying alcids and the flightless penguins in their respective cool, food-rich seas. But it seems to have in no way impeded the dispersal of many marine species. One, the Great Skua *(Catharacta skua)*, is comprised of two widely separate populations, one residing in the Northern Region, the other in the Southern. While it is true that there are more species of shearwaters and petrels in the Southern Region than elsewhere, some species are nevertheless cosmopolitan, frequenting nearly all the seas in one season or another.

SEASONAL DISTRIBUTION

The natural occurrence of animals in an area with relation to the seasons is **seasonal distribution.** In any part of North America the bird population is subject to changes during the course of the year as a result of seasonal migrations. In one season of the year certain species occur that are not ordinarily present at another time. Accordingly, in any area species may be grouped into seasonal categories as follows:

Summer Residents: Species in an area during the summer, coming from the South in the spring to breed and returning in the fall.

Transients: Species stopping temporarily in an area during their northward migration in the spring and during their southward migration in the fall.

Winter Visitants: Species in an area during the winter, having come from their northern nesting grounds to pass the winter in less rigorous climate and departing north in the spring.

Permanent Residents: Species not undergoing a regular periodical migration and consequently staying in one area the year round.

Determine the seasonal categories of the species found in the area where this study is being made. Obtain an up-to-date local check-list of species. A state or county list will suffice although a more local list will be better. On the check-list mark the seasonal categories beside the names of the more common species, using S for summer resident, T for transient, W for winter visitant, and P for permanent resident. If the information necessary for determining the seasonal categories is not available, secure data on local birds from the ranges in *The A. O. U. Check-list.*

ECOLOGICAL DISTRIBUTION

The natural arrangement of organisms with relation to environment is called **ecological distribution.** Whereas geographical distribution deals with organisms in areas, ecological distribution has to do with organisms in environments.

The environment that a species normally occupies in its geographical range is its **habitat.** For example, the deciduous forest is the habitat of a number of bird species whose breeding ranges cover wide areas in eastern North America. Generally, a number of different organisms have the same habitat and together constitute a biotic community.

A **biotic community** is an aggregation of organisms, both plant and animal, living in a given habitat. Any such community is characterized by the physical features of its habitat, by the complex of organisms occurring in it, and by the relationships of the organisms to the physical features of the habitat and to one another. Among the three principal biotic communities of the world—marine, fresh-water, and terrestrial—birds occupy essentially the terrestrial, even though many species have become secondarily adapted to either one of the others. It is beyond the scope of this section to consider birds with relation to the marine and fresh-water communities.

Major Biotic Communities

In North America north of Mexico, the bottomlands of the lower Rio Grande, and southern Florida, there are nine Major Biotic Communities (some are termed Biomes by a number of authorities). Each Major Community is made up of a series of communities, any one of which may be recognized by certain plant forms, or types of vegetation, that *dominate* the aggregation of organisms and certain other organisms (plants and animals) that *influence* the aggregation in varying degrees, depending on their abundance, size, and mode of life. The communities are named for their dominant plant forms.

One community in the Major Biotic Community is the **climax community;** the other communities in the Major Biotic Community are successive stages of development, or a **sere,** leading toward the climax community and are termed **seral communities.** The climax community is the final stage of development over a prolonged period of time and represents the highest possible development (from an ecological viewpoint) that can occur in an area under the prevailing conditions of soil and climate. A beech-maple forest is an example of a climax community in certain areas. The seral communities are transitory communities and eventually will be replaced. Thus a shrub community in certain areas will be succeeded by a beech-maple forest because the environmental conditions favor the development of the beech-maple forest as the climax community. Occasionally, the term **subclimax community** is used to designate a seral community which lasts for a very long time.

The meaning of "dominant plant forms" requires explanation. The controlling factor in plants for birds, as well as for other animals, is the form or type of plant rather than the species of plant. Thus certain birds are limited in habitat by their adaptations to coniferous trees, regardless of whether the trees are red spruces, eastern hemlocks, or alpine firs.

The nine Major Biotic Communities of North America are listed below. Nearly all are named for their dominant plant forms:

Tundra	Coniferous Forest
Deciduous Forest	Grassland

Southwestern Oak Woodland   Sagebrush
Pinyon-Juniper Woodland   Scrub Desert
Chaparral

The map on page 210, entitled "Major Biotic Communities of North America," shows the general extent of the climax communities in eight of the nine Major Biotic Communities. (The Southwestern Oak Woodland is not shown.) It also shows certain ecotones, to be defined and discussed later. The student should bear in mind that this map, or any attempt to depict distribution of communities over a wide area, has definite limitations. (1) It does not reveal seral communities, because they are usually too small. (2) It does not give the full extent of climax communities and ecotones, but only those parts broad enough to be mapped; in the western part of the conterminous United States, many parts of climax communities and ecotones are merely narrow belts on steep mountain slopes and consequently cannot be indicated.

In the following pages each Major Biotic Community is characterized by a condensed description of the climax community (see Kendeigh, 1961, for further details) and a list of some of the bird species either confined to it, or showing a marked preference for it, during the breeding season. An asterisk before the species name indicates that there is one or more subspecies with an affinity in the breeding season for that particular climax community. Following each list are useful references pertaining to the community.

Tundra

Arctic Tundra. Northern Alaska and northern Canada, south along the coasts of Hudson Bay and through most of Labrador to Newfoundland; **Alpine Tundra,** in discontinuous areas above 10,000 feet on the Rocky Mountains and the Sierra Nevada-Cascade system. Winter temperatures extremely cold; rainfall moderate to heavy. General aspect: treeless terrain (level to undulating in the Arctic Tundra). Dominant plants: lichens, grasses, sedges, and occasionally dwarf willows *(Salix)* and other small woody plants.

Arctic Tundra
Willow Ptarmigan *(Lagopus lagopus)*
Rock Ptarmigan *(Lagopus mutus)*
Snowy Owl *(Nyctea scandiaca)*
*Horned Lark *(Eremophila alpestris)*
*Water Pipit *(Anthus spinoletta)*
Lapland Longspur *(Calcarius lapponicus)*
Smith's Longspur *(Calcarius pictus)*
Snow Bunting *(Plectrophenax nivalis)*

Alpine Tundra
White-tailed Ptarmigan *(Lagopus leucurus)*
*Water Pipit *(Anthus spinoletta)*
Rosy Finch *(Leucosticte* spp.)

References: French and French (1965); Hayward (1952); Irving (1960); Shelford and Twomey (1941); Verbeek (1967).

Coniferous Forest

Transcontinental Coniferous Forest (also called **Taiga**), southern Canada, paralleling the Arctic Tundra on the north; **Eastern Montane Coniferous Forest,** northeastern United States southward at higher elevations of the Appalachians to North

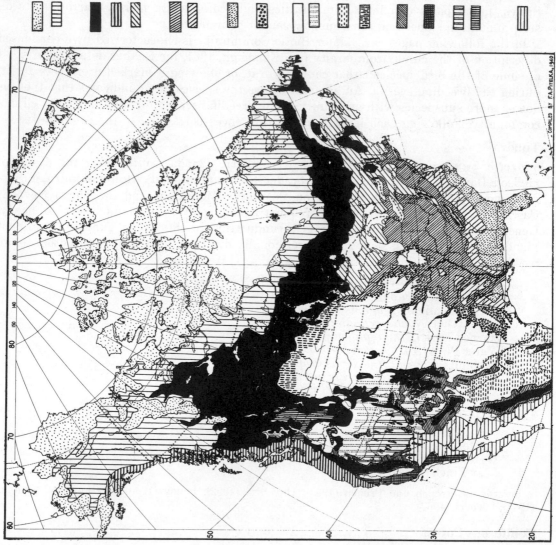

MAJOR BIOTIC COMMUNITIES OF NORTH AMERICA

ARCTIC TUNDRA [ALPINE TUNDRA]

Arctic Tundra—Coniferous Forest Ecotone

CONIFEROUS FOREST

Moist Coniferous Forest

Coniferous-Deciduous Forests Ecotone

DECIDUOUS FOREST

Deciduous Forest—Southeastern Pine Woodland Ecotone

Southeastern Pine Woodland Subclimax

Deciduous Forest—Grassland Ecotone

GRASSLAND

Grassland—Sagebrush Ecotone

Mesquite Subclimax

Grassland-Coniferous Forest Ecotone

PINYON-JUNIPER WOODLAND

CHAPARRAL

SAGEBRUSH

Sagebrush-Grassland—Coniferous Forest Ecotone

SCRUB DESERT

Carolina; **Western Montane Coniferous Forest,** southward on the Rocky Mountains to New Mexico and on the Sierra Nevada-Cascade system to California, below the Alpine Tundra. Climates subject to seasonal changes, summers always cool and winters very cold; rainfall moderate to heavy. General aspect: dense forest with deeply shaded floors supporting few small plants to open forest supporting shrubs and herbaceous plants. Dominant plants: coniferous, evergreen trees belonging to several genera—pines *(Pinus),* spruces *(Picea),* hemlocks *(Tsuga),* firs *(Abies),* cedars *(Thuja),* and, in the Western Montane division only, Douglas fir *(Pseudotsuga).* See the chart, "Birds of the Coniferous Forest," on page 212.

References: Farb *et al.,* (1961); Kendeigh (1947, 1948); Martin (1960); McCormick (1966); Rasmussen (1941); Richards (1965); Shelford and Olson (1935); Snyder (1950); Stewart and Aldrich (1952).

Deciduous Forest

Eastern United States, from southern New England and the Appalachians, at elevations below the coniferous forests, west to and including the bottomlands and bluffs along the Mississippi River and its tributaries. Climate moderate, with summers warm and winters cool; rainfall moderate to heavy. General aspect: forest of varying density, usually with partially shaded floors supporting small shrubs and herbaceous plants. Dominant plants: broad-leaved deciduous trees of several genera—beech *(Fagus),* maples *(Acer),* basswood *(Tilia),* oaks *(Quercus),* hickory *(Carya),* walnut *(Juglans),* and others.

*Red-shouldered Hawk *(Buteo lineatus)*
*Broad-winged Hawk *(Buteo platypterus)*
*Barred Owl *(Strix varia)*
*Whip-poor-will *(Caprimulgus vociferus)*
  Red-bellied Woodpecker *(Centurus carolinus)*
*Hairy Woodpecker *(Dendrocopos villosus)*
*Downy Woodpecker *(Dendrocopos pubescens)*
*Great Crested Flycatcher *(Myiarchus crinitus)*
  Acadian Flycatcher *(Empidonax virescens)*
  Eastern Wood Pewee *(Contopus virens)*
*Carolina Chickadee *(Parus carolinensis)*
  Tufted Titmouse *(Parus bicolor)*
*White-breasted Nuthatch *(Sitta carolinensis)*
  Wood Thrush *(Hylocichla mustelina)*
*Blue-gray Gnatcatcher *(Polioptila caerulea)*
  Yellow-throated Vireo *(Vireo flavifrons)*
  Worm-eating Warbler *(Helmitheros vermivorus)*
  Cerulean Warbler *(Dendroica caerulea)*
  Kentucky Warbler *(Oporornis formosus)*
  Hooded Warbler *(Wilsonia citrina)*

References: Bond (1957); Braun (1950); Farb *et al.* (1961); D. W. Johnston and Odum (1956); Kendeigh (1946, 1948); McCormick (1966); Williams (1936).

Grassland

Primarily the interior plains ("prairies") from the forested bottomlands and bluffs along the Mississippi River and its tributaries west to the Rocky Mountains, and from south-central Texas north into Canada. Climate subject to abrupt seasonal changes, winters being cold (southern division) to very cold (northern division); rainfall moderate to low. General aspect: flat to rolling, treeless, grass-covered terrain. Dominant grasses in the eastern section, where elevation is lower and rainfall moderate, of the "tall" type

BIRDS OF THE CONIFEROUS FOREST	Eastern Montane Coniferous Forest	Transcontinental Coniferous Forest	Western Montane Coniferous Forest
Goshawk (Accipiter gentilis)	X	X	X
Blue Grouse (Dendragapus obscurus)			X
Spruce Grouse (Canachites canadensis)		X	
Hawk Owl (Surnia ulula)		X	
Williamson's Sapsucker (Sphyrapicus thyroideus)			X
Black-backed Three-toed Woodpecker (Picoides arcticus)		X	X
Northern Three-toed Woodpecker (Picoides tridactylus)		X	X
Yellow-bellied Flycatcher (Empidonax flaviventris)		X	
Hammond's Flycatcher (Empidonax hammondi)			X
Olive-sided Flycatcher (Nuttallornis borealis)	X	X	X
Gray Jay (Perisoreus canadensis)		X	X
Clark's Nutcracker (Nucifraga columbiana)			X
Mountain Chickadee (Parus gambeli)			X
Boreal Chickadee (Parus hudsonicus)		X	
Red-breasted Nuthatch (Sitta canadensis)	X	X	X
Brown Creeper (Certhia familiaris)	X	X	X
Winter Wren (Troglodytes troglodytes)	X	X	X
Hermit Thrush (Hylocichla guttata)	X	X	X
Swainson's Thrush (Hylocichla ustulata)		X	X
Townsend's Solitaire (Myadestes townsendi)			X
Golden-crowned Kinglet (Regulus satrapa)	X	X	X
Ruby-crowned Kinglet (Regulus calendula)		X	X
Tennessee Warbler (Vermivora peregrina)		X	
Magnolia Warbler (Dendroica magnolia)	X	X	
Cape May Warbler (Dendroica tigrina)		X	
Myrtle Warbler (Dendroica coronata)		X	
Audubon's Warbler (Dendroica auduboni)			X
Townsend's Warbler (Dendroica townsendi)			X
Hermit Warbler (Dendroica occidentalis)			X
Blackburnian Warbler (Dendroica fusca)	X	X	
Bay-breasted Warbler (Dendroica castanea)		X	
Palm Warbler (Dendroica palmarum)		X	
Canada Warbler (Wilsonia canadensis)	X	X	
*Purple Finch (Carpodacus purpureus)	X	X	
Cassin's Finch (Carpodacus cassinii)			X
Pine Grosbeak (Pinicola enucleator)		X	X
Pine Siskin (Spinus pinus)	X	X	X
Red Crossbill (Loxia curvirostra)	X	X	X
White-winged Crossbill (Loxia leucoptera)		X	
*Slate-colored Junco (Junco hyemalis)	X	X	
Gray-headed Junco (Junco caniceps)			X
White-throated Sparrow (Zonotrichia albicollis)	X	X	

—bluestem *(Andropogon)*, Indian grass *(Sorghastrum)*, switch grass *(Panicum virgatum)*, and others; dominant grasses in the western section, on the Great Plains, where the elevation is higher and the rainfall low, of the "short" type—chiefly buffalo *(Buchloë)* and grama *(Bouteloua)*. Bird species marked by N breed only in the far northern sections.

> Swainson's Hawk *(Buteo swainsoni)*
> Ferruginous Hawk *(Buteo regalis)*
> Greater Prairie Chicken *(Tympanuchus cupido)*
> Sharp-tailed Grouse *(Pedioecetes phasianellus)*
> Long-billed Curlew *(Numenius americanus)*
> Burrowing Owl *(Speotyto cunicularia)*
> N Sprague's Pipit *(Anthus spragueii)*
> Western Meadowlark *(Sturnella neglecta)*
> Lark Bunting *(Calamospiza melanocorys)*
> Grasshopper Sparrow *(Ammodramus savannarum)*
> N Baird's Sparrow *(Ammodramus bairdii)*
> N McCown's Longspur *(Rhynchophanes mccownii)*
> N Chestnut-collared Longspur *(Calcarius ornatus)*

References: D. L. Allen (1967); Breckenridge (1965); Carpenter (1940); Gammell (1965); Kendeigh (1941); Mitchell (1961); Tester and Marshall (1961); Wiens (1969).

Southwestern Oak Woodland (not shown on map)

Southwestern United States—mainly Utah, Nevada, California, New Mexico, Arizona, and parts of Colorado and Oregon; usually on hills and mountain slopes. Summers warm, contrasting sharply with cool winters; rainfall low to moderate. General aspect: partially open woodland of oaks, 20 to 50 feet high; trees ordinarily close enough for branches to touch, but there may be wide spaces between trees which are covered with grasses and shrubs. Dominant plants: practically all oaks of one genus *(Quercus)*.

> Nuttall's Woodpecker *(Dendrocopos nuttallii)*
> Arizona Woodpecker *(Dendrocopos arizonae)*
> Bridled Titmouse *(Parus wollweberi)*
> Hutton's Vireo *(Vireo huttoni)*
> Virginia's Warbler *(Vermivora virginiae)*
> Black-throated Gray Warbler *(Dendroica nigrescens)*

References: Marshall (1957); Miller (1951).

Pinyon-Juniper Woodland

Primarily the Great Basin and the Colorado River region in Colorado, Utah, Nevada, Arizona, New Mexico, and the east side of the Sierra Nevada-Cascade system in California; situated on hills and mountain slopes at elevations above deserts or grasslands and below the coniferous forests. Summers warm, contrasting sharply with cool winters; low rainfall. General aspect: open, park-like woodland of small trees ("pigmy conifers"), 15 to 35 feet high. Dominant plants: pinyon pines *(Pinus edulis* or *P. monophylla)* and several species of juniper *(Juniperus)*; yuccas often prevalent.

> Gray Flycatcher *(Empidonax wrightii)*
> Piñon Jay *(Gymnorhinus cyanocephalus)*
> *Plain Titmouse *(Parus inornatus)*
> Common Bushtit *(Psaltriparus minimus)*
> *Bewick's wren *(Thryomanes bewickii)*

References: Hardy (1945); Rasmussen (1941); Woodbury (1947); Woodin and Lindsey (1954).

Chaparral

California, chiefly on hills and mountains of southwestern part of state, also on inner Coast Ranges and hills bordering the Great Valley. Mild climate; summers very dry and winters with abundant rainfall. General aspect: extensive tracts of densely growing, heavily branched shrubs and stunted trees, 2 to 8 feet high, interrupted now and then by grassy areas and rocky outcrops. Kinds of plants vary according to elevation above sea level and available moisture, but growth forms similar. Among dominant plants in one situation or another: snowbush *(Ceanothus)*, chamise *(Adenstoma)*, scrub oak *(Quercus)*, mountain mahogany *(Cercocarpus)*, coffeeberry *(Rhamnus)*, manzanita *(Arctostaphylos)*, poison oak *(Rhus)*, and baccharis *(Baccharis)*.

*Wrentit *(Chamaea fasciata)*
California Thrasher *(Toxostoma redivivum)*
Gray Vireo *(Vireo vicinior)*
*Orange-crowned Warbler *(Vermivora celata)*
*Sage Sparrow *(Amphispiza belli)*
Black-chinned Sparrow *(Spizella atrogularis)*
*White-crowned Sparrow *(Zonotrichia leucophrys)*

References: Cogswell (1947, 1965); Hayward (1948).

Sagebrush

At elevations above lower deserts and valley floors of Great Basin Plateau between the Rocky Mountains and the Sierra Nevada-Cascade system. Relatively dry climate with slight rainfall; summer temperatures high; winter temperatures cool. General aspect: densely foliaged shrubs, 2 to 5 feet high; plants not so widely spaced as in Scrub Desert, not so close together as in Chaparral. Dominant plants: sagebrush *(Artemisia)*, shadscale *(Atriplex)*, rabbitbrush *(Chrysothamnus)*, greasewood *(Sarcobatus)*, winterfat *(Eurotia)*, and occasional clumps of grasses.

Sage Grouse *(Centrocercus urophasianus)*
Sage Thrasher *(Oreoscoptes montanus)*
Sage Sparrow *(Amphispiza belli)*
Brewer's Sparrow *(Spizella breweri)*

References: Fautin (1946); Miller (1951).

Scrub Desert

Lowlands and valley floors from western Texas west to southwestern California. Semi-arid climate with scant rainfall; summer temperatures very high. General aspect: bare ground with widely spaced plants, 3 to 6 feet high. The dominant plant is creosote bush *(Larrea)*; prominent plants include mesquite *(Prosopis)*, paloverde *(Cercidium)*, catclaw *(Acacia)*, ironwood *(Olneya)*, ocotillo *(Fouquieria)*, agaves, cactuses, yuccas, and (occasionally) grasses.

Gambel's Quail *(Lophortyx gambelii)*
Roadrunner *(Geococcyx californianus)*
Elf Owl *(Micrathene whitneyi)*
Lesser Nighthawk *(Chordeiles acutipennis)*
Costa's Hummingbird *(Calypte costae)*
Gilded Flicker *(Colaptes chrysoides)*
Gila Woodpecker *(Centurus uropygialis)*
Vermilion Flycatcher *(Pyrocephalus rubinus)*

Verdin *(Auriparus flaviceps)*
Cactus Wren *(Campylorhynchus brunneicapillus)*
Bendire's Thrasher *(Toxostoma bendirei)*
Le Conte's Thrasher *(Toxostoma lecontei)*
Crissal Thrasher *(Toxostoma dorsale)*
Black-tailed Gnatcatcher *(Polioptila melanura)*
Phainopepla *(Phainopepla nitens)*

References: Dixon (1959); Hensley (1954); Jaeger (1957); Leopold *et al.* (1961); Miller (1963); Miller and Stebbins (1964); Monson (1965); Schmidt-Nielson (1964).

Several communities within the Major Biotic Communities, besides those with climax rating already treated, are great enough in extent and importance to warrant special mention.

The **Southeastern Pine Woodland Subclimax** (see map), on the Coastal Plain of the southeastern states, is characterized by the presence of open pine forests which, because of the poor soils and many fires, have never been succeeded by deciduous stands. Perhaps the most typical breeding birds confined to the Pine Woodland are the Red-cockaded Woodpecker *(Dendrocopos borealis)*, Brown-headed Nuthatch *(Sitta pusilla)*, and a subspecies of the Bachman's Sparrow *(Aimophila aestivalis)*. Reference: D. W. Johnston and Odum (1956).

In southwestern Texas and southern New Mexico the wide areas of mesquite *(Prosopis)* are considered by some authorities as **Mesquite Subclimax** (see map). In this community one finds such birds as the Golden-fronted Woodpecker *(Centurus aurifrons)* and the Black-crested Titmouse *(Parus atricristatus)*. Reference: Hamilton (1962).

The **Moist Coniferous Forest** (see map), sometimes called the **Coast Forest,** extending as a narrow belt along the western coast of North America from Alaska south into California to approximately the San Francisco area, is regarded by most ecologists as a division of the Coniferous Forest Climax. Owing to the great humidity in the region, the community has a distinctive vegetational aspect and, in turn, a few species of birds —for example, the Chestnut-backed Chickadee *(Parus rufescens)* and Varied Thrush *(Ixoreus naevius)*—which are generally limited to this particular biotic situation. References: Macnab (1958); Miller (1951).

Ecotones

When one community gives way to another, there is an area of overlap, or **ecotone,** where their respective plant and animal associations are intermixed or blended. In some of the broader ecotones, resulting from the overlapping of two Major Biotic Communities, one may find not only bird species from both communities, but occasionally species restricted more or less to the ecotone.

An exceptionally good example of the broader ecotone is the **(Arctic) Tundra-Coniferous Forest Ecotone** (see map), characterized by the interdigitation of Arctic Tundra and Coniferous Forest extensions. This area, loosely referred to as "timberline," has numerous low tundra shrubs and stands of stunted conifers. Here may be found such breeding birds as the Gray-cheeked Thrush *(Hylocichla minima)*, Northern Shrike *(Lanius excubitor)*, Blackpoll Warbler *(Dendroica striata)*, Common Redpoll *(Acanthis flammea)*, Tree Sparrow *(Spizella arborea)*, and Harris' Sparrow *(Zonotrichia querula)*, none of them characteristic of the open, uninterrupted tundra or the deep spruce forest. Reference: Taverner and Sutton (1934).

Another good example is the **Coniferous-Deciduous Forests Ecotone** (see map) where conifers (especially hemlocks) and maples, beech, and other hardwoods intermingle. The eastern subspecies of the Solitary Vireo *(Vireo solitarius)* and the Black-throated Blue Warbler *(Dendroica caerulescens)* are two of the several birds with affinities for this kind of forest during the breeding season. References: Kendeigh (1946, 1948).

None of the broader ecotones represent a perfect blending of neighboring communities; one finds numerous "islands" of these communities intermixed. Thus in a Coniferous-Deciduous Forests Ecotone there are isolated, pure stands of conifers and groves of hardwoods where the conditions of soil and moisture are especially suitable to one plant form. If such an island is large enough, it will attract the bird life usually associated with its particular plant form.

Seral Communities

Within the Major Biotic Community are the seral communities showing stages of development toward the climax communities. Some of the more common, natural communities are described below. For extensive information about them, consult Farb (1963), Kendeigh (1961), and Smith (1966).

Beaches: Bare areas of rock and sand, with little or no vegetation, bordering large bodies of water.

Lakes: Large bodies of water of varying depth from a few to many feet, with little or no rooted vegetation.

Ponds: Small, shallow bodies of water with plants, floating and submerged, rooted over most of the bottom. Reference: Amos (1967).

Marshes: Innundated land with emergent, herbaceous plants—bulrushes, cattails, sedges, grasses, etc. References: A. A. Allen (1914), Beecher (1942), Errington (1957, 1965), Hochbaum (1965), Niering (1966), Provost (1947), Robertson (1965), Weller and Spatcher (1965).

Swamps and *Bogs:* Undrained wetlands with wooded vegetation. Swamps are common in warm southern regions and feature deciduous growth. Bogs, more common in the cooler northern regions, feature coniferous growth and may have at their margins a semifloating mat of sphagnum moss and other cushiony plants. References: Aldrich (1943), Brewer (1967), Cottrille (1965), Murray (1965), Robertson (1965).

Shrublands: Areas with low, thickly growing woody plants predominating. Reference: D. W. Johnston and Odum (1956).

Riparian Woodlands: Mostly of deciduous trees, near streams, or bottomlands where there is an adequate supply of subsurface water; especially prominent in the more arid sections of western conterminous United States. References: Ingles (1950); Usinger (1967).

Any one of the above communities may be a stage in a succession leading to a particular climax community. For example, in eastern United States a pond may be succeeded by a marsh, a marsh by a shrubland, and a shrubland by a climax community. During the course of many years the seral communities will disappear as the climax community takes over. While the succession is in progress, many birds are associated with the seral stages. Pied-billed Grebes *Podilymbus podiceps)* and Black Terns *(Chlidonias niger)* feed in the open water of the pond and nest in the adjacent fringe of the marsh. American and Least Bitterns *(Botaurus lentiginosus* and *Ixobrychus exilis)*, Common Gallinules *(Gallinula chloropus)*, Virginia Rails *(Rallus limicola)* and Soras *(Porzana carolina)*, Long-billed and Short-billed Marsh Wrens *(Telmatodytes palustris* and

Cistothorus platensis), and Red-winged Blackbirds *(Agelaius phoeniceus)* nest in the bulrushes, cattails, or sedges in accordance with their preference. In the shrublands are nesting Yellowthroats *(Geothlypis trichas)* as well as many other species typical of shrublands. See Beecher (1942) for a classic study of marsh birds with relation to vegetation.

Primary and Secondary Communities

Seral communities are stages of primary succession, or **primary communities.** Any primary community is presumably the most highly developed association that has ever occurred in a given area during the present geological age.

In addition to primary communities there are stages in a secondary succession called **secondary communities.** Any secondary community is a biotic association that has developed in an area where a natural, primary community has been removed. The most common secondary communities are "artificial" in that man's activities have produced or influenced them directly or indirectly. Secondary communities include croplands, fields, pastures, shrub and tree plantations around dwellings, marshes and swamps resulting from dammed waterways, and deserts caused by overgrazing. Most woodlands in conterminous United States and southern Canada are secondary communities, replacing the primeval forests that have long since been lumbered or burned.

All secondary communities are more transitory than primary communities because areas previously occupied by primary communities are more receptive to replacement, and many plants, already present in neighboring areas, rapidly spread into them. As an illustration, a plot of cleared ground in a Deciduous Forest may, in the course of a man's lifetime, be covered first by grasses and other succulent plants, then shrubs, and finally young trees of the Deciduous Forest Climax.

Probably the majority of birds the student meets in the field are associated with secondary communities because, in the first place, man-made environments are predominant in the outskirts of cities and towns and, in the second place, secondary communities harbor a rich variety of birds. See Woolfenden and Rohwer (1969) for a study showing the remarkable density of birds in a suburban area.

Edge Effect

Some of the highest concentrations of birds anywhere are in the ecotones of seral communities. Here are many species characteristic of the communities involved plus those which are adapted more or less to the ecotonal conditions. Actually the number of species and their respective populations are often much greater in an ecotone than in the communities contributing to it. This phenomenon of increased variety and density of species where communities meet is spoken of as **edge effect.**

The importance of ecotones and edge effect is soon apparent to students looking for large numbers of birds. Frequently where such communities adjoin as woods and fields, woods and shores, or riparian woodlands and grasslands, there are shrubby areas which hold a variety of breeding birds that will include, depending on the part of the country, the Mockingbird *(Mimus polyglottos),* Catbird *(Dumetella carolinensis),* Yellow Warbler *(Dendroica petechia),* Chestnut-sided Warbler *(Dendroica pensylvanica),* Prairie Warbler *(Dendroica discolor),* Chat *(Icteria virens),* Cardinal, Blue Grosbeak *(Guiraca caerulea),* Indigo Bunting, Lazuli Bunting *(Passerina amoena),* Painted Bunting *(Passerina ciris),* Clay-colored Sparrow *(Spizella pallida),* Field Sparrow *(Spizella pusilla),* and Song Sparrow. Here, in addition to shrub birds, are open-country birds which seek trees and shrubs for nests and perches—e.g., Eastern Kingbird *(Tyrannus tyrannus),* Western Kingbird *(Tyrannus verticalis),* and Scissor-tailed Flycatcher—and woodland birds which often prefer trees or shrubs near openings for nests—Western Flycatcher *(Empidonax difficilis)*, Baltimore Oriole, and Bullock's Oriole *(Icterus bullockii)*.

In the Piedmont of Georgia, D. W. Johnston and Odum (1956) found that nearly 40 to 50 percent of the common breeding birds in the region were forest edge birds in their habitat requirements. In Illinois, V. R. Johnston (1947) concluded that the forest edge is actually a distinct community with characteristic species not found either in the forest or open-country communities.

As a rule, birds of seral communities have more extensive geographical ranges than those limited to climax communities because their communities are more widely distributed. One may find marsh birds—e.g., American Bittern, Marsh Hawk *(Circus cyaneus)*, Virginia Rail, Long-billed Marsh Wren, and Red-winged Blackbird—in several Major Biotic Communities for the reason that marsh seral communities occur in them. The same situation is true with regard to many "edge" birds and birds of artificial communities. If one takes a trip by car from the Atlantic Coast west to the Great Plains or Rocky Mountains, the species he will see along the roadsides and agricultural lands—e.g., Mourning Dove *(Zenaidura macroura)*, Robin, Eastern Bluebird *(Sialia sialis)*, Common Grackle *(Quiscalus quiscula)*, Chipping Sparrow *(Spizella passerina)* —will be monotonously the same despite the fact that he will pass through Deciduous Forest and Grassland where restricted climax species exist.

A comparatively small number of birds, usually those which nest on cliffs, in holes, cavities, or crevices, or in man-made "substitutes" such as buildings or birdboxes, are not restricted to any one biotic community and range widely in accordance with conditions of climate and other controlling factors. Among these birds are the Turkey Vulture *(Cathartes aura)*, Osprey, Peregrine Falcon *(Falco peregrinus)*, Barn Owl *(Tyto alba)*, Chimney Swift *(Chaetura pelagica)*, Belted Kingfisher *(Megaceryle alcyon)*, Common Raven *(Corvus corax)*, Cañon Wren *(Catherpes mexicanus)*, and Rock Wren *(Salpinctes obsoletus)*, together with eagles, phoebes, swallows, and several hawks.

The beginning student of ornithology should become aware of the distribution of birds in biotic communities of the area where the study is being made. By way of introducing him to the subject, the following procedure is suggested. Make a chart titled "Distribution of Birds in Biotic Communities" as shown on page 219. Place the name of the major biotic community, ecotone, or subclimax community in which, according to the map (page 210), the study is being conducted. In the left-hand column, labeled "species," list species (given by the instructor) that will presumably be seen on forthcoming field trips. List the species in the order of *The A. O. U. Check-list*. At the top of the adjoining columns place the names (also given by the instructor) of the climax and seral communities and their ecotones that are to be visited. At the conclusion of the first field trip, check the spaces in the columns opposite the species that represent the communities and ecotones in which they were observed. After each succeeding field trip, check any additional spaces. Meanwhile learn to associate different species with their communities. The far right-hand column of the chart, labeled "unrestricted," is for species apparently not confined to any one biotic community.

In becoming acquainted with biotic communities the student will find helpful the guides by Pettingill (1951, 1953) which include descriptions of the more important communities in each state and name some of their characteristic bird species. The climax communities are usually treated at greater length and include lists of species to be expected. For more extensive information on biotic communities, the student should consult Shelford (1963).

DISTRIBUTION OF BIRDS IN BIOTIC COMMUNITIES

Major Biotic Community:	Communities and Ecotones										Unrestricted
Species											

Niches

Within biotic communities, organisms are further distributed according to the niches they occupy. The **niche** of a bird species is its position or role in the community resulting from its structural, physiological, and behavioral adaptations. Do not confuse it with habitat which is the environment or place. Think of the habitat of a species as its "address" and the niche as its "occupation."

If one is to determine the niche of a species, he must not only discover where the species carries on its activities (e.g., feeding, singing, nesting, and roosting), but find out the kinds of food it requires, the relationships between the species and other organisms of the community, and the part played by the species in the general functioning of the community as a whole.

The problem of determining the niche of a species is invariably complicated by the natural variation of the community in different geographical areas. Thus the niche relationships in the northern part of its range may be very different from those in the southern, owing to the differences in vegetation, population densities of associated organisms, and other factors in the community. While a number of bird species in any one community may show similarities in their niche relationships, they seldom, if ever, share identical niches.

Altitudinal Distribution

In mountains, biotic communities succeed one another altitudinally with corresponding climatic changes just as they do latitudinally elsewhere on the continent. But due to the steepness of the slopes, mountain communities have a different configuration, which in turn sometimes creates puzzling ecological problems.

About 250 feet of elevation is equivalent to one degree of latitude—roughly 69 miles in length northward. In a general way, then, one may expect the same rate of change in climate when ascending a mountain for 250 feet that he will find when going north on the continent 69 miles. In the high Rocky Mountains, where many peaks reach well above 12,000 feet, a person will experience, during a climb, the same changes as when taking a trip north from the base of the Rockies to the Arctic Tundra. Furthermore, just as he will experience changes in climate, he will also observe essentially the same changes in communities. The one great difference between his climb and the trip, besides distance, is the abruptness of change on the mountain compared with the gradualness of change during the trip overland.

Instead of being spread over wide areas, the biotic communities on mountains are usually distinct belts corresponding to the same climatic features. Their width is a matter of footage rather than of mileage. Mountain ecotones, instead of being wide transitions between communities, are exceedingly narrow, the two communities standing in close proximity. Because of this distinct belting of mountain community units, their identification on the basis of plant formations is simpler and more readily seen.

The accompanying diagram (page 221), originally suggested by Woodbury (1947), illustrates the altitudinal succession of the Major Biotic Communities on a hypothetical peak of the south-central Rocky Mountains, approximately in eastern Utah. Note that on the warmer southern slope all the belts reach a greater altitude than on the northern.

As indicated in the preceding description of North American Major Biotic Communities, a number of bird species are characteristic of both the latitudinal and mountain communities—for example, species characteristic of both the Transcontinental and Western Montane Coniferous Forests. At the same time, populations of at least the higher communities on a mountain range are often sufficiently isolated from the corresponding communities on another range, or from corresponding latitudinal communities, as to become racially distinct. A study of *The A. O. U. Check-list* will show a conspicuous number of subspecies limited to certain mountain systems in western conterminous United States. In a few cases the populations of a bird form occupying widely separated mountain communities may develop sufficient distinctiveness to be accorded specific rank. For instance, there are three species of rosy finches (genus *Leucosticte*), each one limited to the Alpine Tundra of a separate group of western mountains. In effect, mountain biotic communities are like islands in that they hold certain bird populations in isolation where each population may develop peculiarities of its own.

For instructive reading on the relationships of birds and other organisms, animal and plant, to mountains in North America, read Brooks (1965, 1967).

Life Zones

Some authorities, in dealing with the distribution of birds in North America, follow the concept of life zones instead of communities. In place of the Major Biotic Communities they recognize six life zones.

Life zones are temperature zones mapped out in accordance with certain lines (isotherms) across the continent that have the same temperature at the same period of the year. (For a map of life zones in color, see the inside of the front cover in Chapman, 1932.) Life zones consequently appear on the map as broad transcontinental bands, except in the mountain regions, where they are extremely irregular owing to the effect of elevation on temperature. For each life zone there are certain characteristic species of birds and other organisms which authorities have designated as zonal "indicators."

The six life zones are divided into a boreal group containing, from north to south, the Arctic, Hudsonian, and Canadian; and an austral group, comprising from north to south,

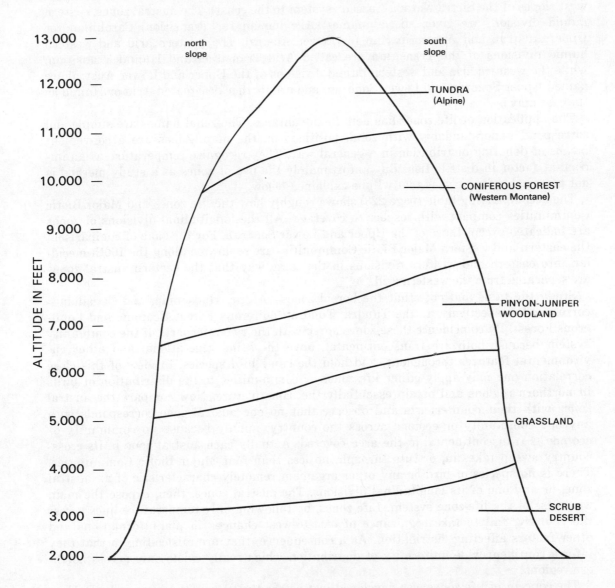

north slope

south slope

TUNDRA
(Alpine)

CONIFEROUS FOREST
(Western Montane)

PINYON-JUNIPER
WOODLAND

GRASSLAND

SCRUB
DESERT

ALTITUDE IN FEET

13,000 —
12,000 —
11,000 —
10,000 —
9,000 —
8,000 —
7,000 —
6,000 —
5,000 —
4,000 —
3,000 —
2,000 —

ALTITUDINAL SUCCESSION OF MAJOR BIOTIC COMMUNITIES

*(Modified from Woodbury, 1947; redrawn from Odum, "Fundamentals
of Ecology," W. B. Saunders Company, 1953.)*

the Transition, Upper Austral, and Lower Austral. The austral group is further divided longitudinally, on the basis of prevailing moisture, into three groups: eastern humid (from the Atlantic Coast west to the 100th meridian, or, approximately, the eastern edge of the Great Plains), the western arid division (from the 100th meridian west to the crests of the Sierra Nevada-Cascade system), and the western humid division (from the west slopes of the Sierra Nevada-Cascade system to the coast). The austral zones, eastern humid division, are given special names: Alleghanian for Transition, Carolinian for Upper Austral, and Austroriparian for Lower Austral. The western arid and western humid divisions of the Transition are called Arid Transition and Humid Transition, while the western arid and western humid divisions of the Upper and Lower Austral are named Upper Sonoran and Lower Sonoran, and are further designated Arid or Humid as the case may be.

The application of life zones has definite advantages. The zonal names are simple and correspond understandably with zonal positions on the map. Zones are a convenient means of denoting distribution in a general way. They recognize temperature as an important factor in distribution. But unfortunately the use of zones as a study method is not practicable for reasons that will be explained below.

The accompanying table (page 223) shows roughly how the life zones and Major Biotic Communities compare with respect to coverage. All the longitudinal divisions of zones are indicated except those of the Upper and Lower Sonoran. For the sake of comparison, the eastern and western Major Biotic Communities are separated along the 100th meridian into eastern and western divisions in the same way that the eastern austral zones are separated from the western arid.

Study the table and note that the boreal zones—Arctic, Hudsonian, and Canadian—correspond, respectively, to the Tundra, Tundra-Coniferous Forest Ecotone, and Coniferous Forest. By coincidence these zones agree with their counterparts of the community system because both are transcontinental, have the same vegetational and other environmental features throughout, and hold the same bird species. In view of this close correlation one may apply either life zones or communities to the distribution of birds in northern regions and obtain essentially the same picture. Now compare the austral zones with their counterparts and observe that no one austral zone corresponds with any one community or ecotone across the country, mainly because no community or ecotone is transcontinental in the area covered. Actually each austral zone in its cross-country sweep takes in, or cuts through, no less than four Major Biotic Communities. There is no species of bird or any other organism remotely characteristic of an austral zone, or any one of its longitudinal divisions. The austral zones, then, expose the main weakness of the life-zone system. Life zones, by following only temperature lines across the country, fail to take cognizance of east-to-west changes in plant formations and other factors affecting distribution. As a consequence, they are misleading in that they infer a continent-wide uniformity of distribution which really exists only across northern regions.

The concept of life zones was developed just before the turn of the century by C. Hart Merriam and has been followed for a long period in this country; the concept of communities, though of earlier origin, has since gained a wide following. (For a history of these concepts, see Kendeigh, 1954.) Most ornithologists now recognize the inadequacy of life zones in mapping distribution on a continental basis, but many are reluctant to abandon the concept entirely in studies of a local nature. The reason is understandable and often justified. In a mountainous area, for example, belts or zones of life very definitely succeed one another altitudinally as the temperature changes; they are usually obvious and readily described. The practice of these ornithologists is to apply the original life-zone terminology while at the same time stressing the importance of plant formations and other factors in holding bird species to each zone. An outstanding advantage of this procedure is that, by using zonal names, one may avoid committing himself on the seral status of communities and naming the communities accordingly.

For an illustration of how life zones with a modern interpretation may be used successfully in a state, the student is referred to the work by Gabrielson and Jewett (1940, pages 28-41). The student is also referred to Miller (1951), who related plant formations to life zones in his study of the distribution of California birds, and to V. R. Johnston (1965) and Small (1965), who applied life zones in describing the distribution of birds in the Sierra Nevadas.

MAJOR BIOTIC COMMUNITIES AND ECOTONES Eastern North America	LIFE ZONES		MAJOR BIOTIC COMMUNITIES AND ECOTONES Western North America
Tundra →	Arctic		← Tundra
Tundra-Coniferous Forest Ecotone →	Hudsonian		← Tundra-Coniferous Forest Ecotone
Coniferous Forest →	Canadian		← Coniferous Forest
Grassland (part) → Deciduous-Coniferous Forests Ecotone →	Transition — Alleghanian	Transition	← Grassland (part) ← Moist Coniferous Forest ← Western Oak Woodland (part) ← Sagebrush (part)
Grassland (part) → Deciduous Forest-Southwestern Pine Woodland Ecotone →	Upper Austral — Carolinian	Upper Sonoran	← Western Oak Woodland (part) ← Sagebrush (part) ← Grassland (part) ← Pinyon-Juniper Woodland ← Chaparral
Southeastern Pine Woodland Subclimax →	Lower Austral — Austroriparian	Lower Sonoran	← Grassland (part) ← Mesquite Subclimax ← Scrub Desert

REFERENCES

Aldrich, J. W.
  1943 Biological Survey of the Bogs and Swamps of Northeastern Ohio. Amer. Midland Nat., 30: 346-402.

Allee, W. C., and K. P. Schmidt
  1951 Ecological Animal Geography. Revised second edition. John Wiley and Sons, New York.

Allen, A. A.
  1914 The Red-winged Blackbird: A Study in the Ecology of a Cat-tail Marsh. Proc. Linnaean Soc. New York, Nos. 24-25: 43-128.

Allen, D. L.
  1967 The Life of Prairies and Plains. McGraw-Hill Book Company, New York.

Allen, G. M.
  1925 Birds and Their Attributes. Marshall Jones Company, Boston. (Dover reprint available. Chapter 6, "Origin and Distribution of Birds.")

Allen, J. A.
  1893 The Geographical Origin and Distribution of North American Birds, Considered in Relation to Faunal Areas of North America. Auk, 10: 97-150.

Amadon, D.
1966 Birds around the World: A Geographical Look at Evolution and Birds. Natural History Press, Garden City, New York. (Brief, highly readable coverage of both geographical and ecological distribution.)

American Ornithologists' Union
1957 Check-list of North American Birds. Fifth edition.

Amos, W. H.
1967 The Life of the Pond. McGraw-Hill Book Company, New York.

Barick, F. B.
1950 The Edge effect of the Lesser Vegetation of Certain Adirondack Forest Types with Particular Reference to Deer and Grouse. Roosevelt Wildlife Bull., 9: 1-146.

Beckwith, S. L.
1954 Ecological Succession on Abandoned Farm Lands and Its Relation to Wildlife Management. Ecol. Monogr., 24: 349-375.

Beddall, B. G.
1963 Range Expansion of the Cardinal and Other Birds in the Northeastern States. Wilson Bull., 75: 140-158.

Beecher, W. J.
1942 Nesting Birds and the Vegetation Substrate. Chicago Ornithological Society. (Includes good illustrations of the effect of edge.)

Bond, R. R.
1957 Ecological Distribution of Breeding Birds in the Upland Forests of Southern Wisconsin. Ecol. Monogr., 27: 351-384.

Braun, E. L.
1950 Deciduous Forests of Eastern North America. Blakiston Company, Philadelphia.

Breckenridge, W. J.
1965 A Virgin Prairie in Minnesota. In *The Bird Watcher's America*. Edited by O. S. Pettingill, Jr. McGraw-Hill Book Company, New York. (The birds one may expect in a prairie whose sod has never been broken.)

Brewer, R.
1967 Bird Populations of Bogs. Wilson Bull., 79: 371-396.

Brooks, M.
1952 The Allegheny Mountains as a Barrier to Bird Movement. Auk, 69: 192-198. (The author shows that this mountain range exerts a profound influence on bird movement.)

1965 The Appalachians. Houghton Mifflin Company, Boston.
1967 The Life of the Mountains. McGraw-Hill Book Company, New York.

Buchsbaum, R., and M. Buchsbaum
1957 Basic Ecology. Boxwood Press, Pittsburgh, Pennsylvania. (A book "for the general education of students without special interests in biology." Covers all the main principles of ecological distribution.)

Carlquist, S.
1965 Island Life: A Natural History of the Islands of the World. Natural History Press, Garden City, New York.

Carpenter, J. R.
1940 The Grassland Biome. Ecol. Monogr., 10: 617-684.

Chapman, F. M.
1932 Handbook of Birds of Eastern North America. Second revised edition. D. Appleton and company, New York. (Dover reprint available. "The Distribution of birds," pp. 29-35.)

Cogswell, H. L.
1947 Chaparral Country. Audubon Mag., 49: 74-81.
1965 The California Chaparral. In *The Bird Watcher's America*. Edited by O. S. Pettingill, Jr. McGraw-Hill Book Company, New York.

Cottrille, B. D.
1965 Northern Spruce Bogs. In *The Bird Watcher's America*. Edited by O. S. Pettingill, Jr. McGraw-Hill Book Company, New York. (Description of a typical spruce bog and some of the problems of studying birds in such a habitat.)

Darlington, P. J., Jr.
1957 Zoogeography: The Geographical Distribution of Animals. John Wiley and Sons, New York. (Probably the most useful and comprehensive treatise on the subject.)

Dice, L. R.
1943 The Biotic Provinces of North America. University of Michigan Press, Ann Arbor. (Contains descriptions of the provinces and a map. Biotic provinces are not to be confused with biotic communities. Each biotic province, according to Dice's definition, "covers a considerable and continuous geographic area and is characterized by the occurrence of one or more important ecologic associations that differ, at least in proportional area covered,

from the associations of adjacent provinces.")

1952 Natural Communities. University of Michigan Press, Ann Arbor. (An extensive summary of what is known about ecological communities.)

Dixon, K. L.
1959 Ecological and Distributional Relations of Desert Scrub Birds of Western Texas. Condor, 61: 397-409.

Ekman, S.
1953 Zoogeography of the Sea. Sidgwick and Jackson, London.

Errington, P. L.
1957 Of Men and Marshes. Macmillan Company, New York.
1965 An Iowa Marsh. In *The Bird Watcher's America*. Edited by O. S. Pettingill, Jr. McGraw-Hill Book Company, New York. (The seasonal succession of birds and other life in a large glacial marsh.)

Farb, P.
1963 Face of North America: The Natural History of a Continent. Harper and Row, New York.

Farb, P., and the Editors of *Life*
1961 The Forest. Time Incorporated, New York.
1963 Ecology. Time Incorporated, New York. (A treatise for the layman, handsomely illustrated, with good coverage of the subject including distribution.)

Fautin, R. W.
1946 Biotic Communities of the Northern Desert Shrub Biome in Western Utah. Ecol. Monogr., 16: 251-310.

Fisher, J.
1954 A History of Birds. Houghton Mifflin Company, Boston. (Chapter 5, "Geographical Distribution.")

French, N. R., and J. B. French
1965 Rosy Finches of the High Rockies. In *The Bird Watcher's America*. Edited by O. S. Pettingill, Jr. McGraw-Hill Book Company, New York. (A vivid description of habitat conditions in the Alpine Tundra and how different birds cope with them.)

Gabrielson, I. N., and S. G. Jewett
1940 Birds of Oregon. Oregon State College, Corvallis.

Gammell, A. M.
1965 North Dakota Prairie. In *The Bird Watcher's America*. Edited by O. S. Pettingill, Jr. McGraw-Hill Book Company, New York. (The large assortment of birds, found in prairie country, and the habitats they occupy.)

Hamilton, T. H.
1962 The Habitats of the Avifauna of the Mesquite Plains of Texas. Amer. Midland Nat., 67: 85-105.

Hardy, R.
1945 Breeding Birds of the Pigmy Conifers in the Book Cliff Region of Eastern Utah. Auk, 62: 523-542.

Haverschmidt, F.
1953 The Cattle Egret in South America. Audubon Mag., 55: 202-204, 236.

Hayward, C. L.
1948 Biotic Communities of the Wasatch Chaparral, Utah. Ecol. Monogr., 18: 473-506.
1952 Alpine Biotic Communities of the Uinta Mountains, Utah. Ecol. Monogr., 22: 93-120.

Hensley, M.
1954 Ecological Relations of the Breeding Bird Population of the Desert Biome of Arizona. Ecol. Monogr., 24: 185-207.

Hochbaum, H. A.
1965 The Delta Marshes of Manitoba. In *The Bird Watcher's America*. Edited by O. S. Pettingill, Jr. McGraw-Hill Book Company, New York. (One of North America's finest marshes for waterfowl.)

Hylander, C. J.
1966 Wildlife Communities: From the Tundra to the Tropics in North America. Houghton Mifflin Company, Boston. (A non-technical approach to the subject, aimed at the general reader; well illustrated.)

Ingles, L. G.
1950 Nesting Birds of the Willow-Cottonwood Community in California. Auk, 67: 325-332. (A good example of a riparian woodland community.)

Irving, L.
1960 Birds of Anaktuvuk Pass, Kobuk, and Old Crow: A Study in Arctic Adaptation. U. S. Natl. Mus. Bull. 217.

Jaeger, E. C.
1957 The North American Deserts. Stanford University Press, Stanford, California.

Johnston, D. W., and E. P. Odum
1956 Breeding Bird Populations in Relation to Plant Succession on the Piedmont of Georgia. Ecology, 37: 50-62.

Johnston, V. R.
1947 Breeding Birds of the Forest Edge in Illinois. Condor, 49: 45-53.

1965   The Sierra Nevada. In *The Bird Watcher's America.* Edited by O. S. Pettingill, Jr. McGraw-Hill Book Company, New York. (A colorful description of zonal distribution of birds in one of North America's highest mountain ranges.)

Kendeigh, S. C.
1932   A Study of Merriam's Temperature Laws. Wilson Bull., 44: 129-143.
1941   Birds of a Prairie Community. Condor, 43: 165-174.
1946   Breeding Birds of the Beech-Maple-Hemlock Community. Ecology, 27: 226-244.
1947   Bird Population Studies in the Coniferous Forest Biome during a Spruce Budworm Outbreak. Ontario Dept. Lands and Forests, Biol. Bull., 1: 1-100.
1948   Bird Populations and Biotic Communities in Northern Lower Michigan. Ecology, 29: 101-114.
1954   History and Evaluation of Various Concepts of Plant and Animal Communities in North America. Ecology, 35: 152-171. (An excellent review, with an important list of references.)
1961   Animal Ecology. Prentice-Hall, Englewood Cliffs, New Jersey. (Unquestionably the most comprehensive and scholarly work on the subject available.)

Krause, H., and S. G. Froiland
1956   Distribution of the Cardinal in South Dakota. Wilson Bull., 68: 111-117.

Lack, D.
1937   The Psychological Factor in Bird Distribution. Brit. Birds, 31: 130-136.

Leopold, A. S., and the Editors of *Life*
1961   The Desert. Time Incorporated, New York.

Lowry, W. P.
1967   Weather and Life: An Introduction to Biometeorology. Academic Press, New York. (A compendium of information on weather problems related to birds and other life.)

Macnab, J. A.
1958   Biotic Aspection in the Coast Range Mountains of Northwestern Oregon. Ecol. Monogr., 28: 21-54.

Marshall, J. T., Jr.
1957   Birds of the Pine-Oak Woodland in Southern Arizona and Adjacent Mexico. Pacific Coast Avifauna No. 32, Cooper Ornithological Society, Berkeley, California.

Martin, N. D.
1960   An Analysis of Bird Populations in Relation to Forest Succession in Algonquin Provincial Park, Ontario. Ecology, 41: 126-140.

Mayr, E.
1942   Systematics and the Origin of Species. Revised edition. Columbia University Press, New York.
1946   History of the North American Bird Fauna. Wilson Bull., 58: 3-41.

McCormick, J.
1966   The Life of the Forest. McGraw-Hill Book Company, New York.

Merriam, C. H.
1890   Results of a Biological Survey of the San Francisco Mountain Range and Desert of the Little Colorado, Arizona. N. Amer. Fauna No. 3: i-vii; 1-136, Division of Ornithology and Mammalogy, U. S. Department of Agriculture, Washington, D. C.
1892   The Geographic Distribution of Life in North America with Special Reference to the Mammalia. Proc. Biol. Soc. Washington, 7: 1-64.
1894   Laws of Temperature Control of the Geographic Distribution of Terrestrial Animals and Plants. Natl. Geog. Mag., 6: 229-238.

Miller, A. H.
1951   An Analysis of the Distribution of the Birds of California. Univ. California Publ. in Zool., 50: 531-644. (See S. C. Kendeigh's review of this paper in Auk, 69: 471-473, 1952.)
1963   Desert Adaptations in Birds. Proc. XIIIth Internatl. Ornith. Congr., pp. 666-674.

Miller, A. H., and R. C. Stebbins
1964   The Lives of Desert Animals in Joshua Tree National Monument. University of California Press, Berkeley.

Mitchell, M. J.
1961   Breeding Bird Populations in Relation to Grassland Succession on the Anoka Sand Plain. Flicker, 33: 102-108.

Monson, G.
1965   The Arizona Desert. In *The Bird Watcher's America.* Edited by O. S. Pettingill, Jr. McGraw-Hill Book Company, New York. (The seasonal succession of birds and other life in desert country.)

Murray, J. J.
1965   The Great Dismal Swamp. In *The Bird Watcher's America.* Edited by O. S. Pettingill, Jr. McGraw-Hill Book Company, New York. (Bird life in one

of the most famous southern wetlands.)

Niering, W. A.
1966   The Life of the Marsh: The North American Wetlands. McGraw-Hill Book Company, New York.

Odum, E. P.
1945   The Concept of the Biome as Applied to the Distribution of North American Birds. Wilson Bull., 57: 191-201. (An excellent presentation of the subject plus a noteworthy comparison with the life zone concept.)
1950   Bird Populations of the Highlands (North Carolina) Plateau in Relation to Plant Succession and Avian Invasion. Ecology, 31: 587-605.
1959   Fundamentals of Ecology. Second edition. W. B. Saunders Company, Philadelphia. (Highly recommended for general reading.)

Odum, E. P., and T. D. Burleigh
1946   Southward Invasion in Georgia. Auk, 63: 388-401.

Odum, E. P., and D. W. Johnston
1951   The House Wren Breeding in Georgia: An Analysis of a Range Extension. Auk, 68: 357-366.

Peterson, R. T.
1961   A Field Guide to Western Birds. Second edition. Houghton Mifflin Company, Boston. (Part 2: "The Hawaiian Islands.")

Pettingill, O. S., Jr.
1951   A Guide to Bird Finding East of the Mississippi. Oxford University Press, New York.
1953   A Guide to Bird Finding West of the Mississippi. Oxford University Press, New York.

Pitelka, F. A.
1941   Distribution of Birds in Relation to Major Biotic Communities. Amer. Midland Nat., 25: 113-137.

Provost, M. W.
1947   Nesting of Birds in the Marshes of Northwest Iowa. Amer. Midland Nat., 38: 485-503.

Rand, A. L.
1955   The Origin of the Land Birds of Tristan da Cunha. Fieldiana: Zoology, 37: 139-166.

Rasmussen, D. I.
1941   Biotic Communities of Kaibab Plateau, Arizona. Ecol. Monogr., 11: 229-275.

Reilly, E. M., Jr.
1968   The Audubon Illustrated Handbook of American Birds. Edited by O. S. Pettingill, Jr. McGraw-Hill Book Company, New York.

Rice, D. W.
1956   Dynamics of Range Expansion of Cattle Egrets in Florida. Auk, 73: 259-266.

Richards, T.
1965   In Northern New Hampshire. In *The Bird Watcher's America.* Edited by O. S. Pettingill, Jr. McGraw-Hill Book Company, New York. (Birds at the southern edge of the Coniferous Forest Major Biotic Community.)

Robertson, W. B., Jr.
1965   The Everglades. In *The Bird Watcher's America.* Edited by O. S. Pettingill, Jr. McGraw-Hill Book Company, New York. (Bird life in one of North America's most unique marsh-swamps.)

Root, R. B.
1967   The Niche Exploitation Pattern of the Blue-gray Gnatcatcher. Ecol. Monogr., 37: 317-350. (Shows how the exploitative behavior of one species is organized to achieve "optimal adaptation to the conflicting demands of a changing environment.")

Schmidt-Nielson, K.
1964   Desert Animals: Physiological Problems of Heat and Water. Oxford University Press, New York.

Serventy, D. L.
1960   Geographical Distribution of Living Birds. In *Biology and Comparative Physiology of Birds.* Volume 1. Edited by A. J. Marshall. Academic Press, New York.

Shelford, V. E.
1926   Naturalists' Guide to the Americas. Williams and Wilkins Company, Baltimore. (Contains very useful information on the location and distribution of natural areas.)
1963   The Ecology of North America. University of Illinois Press, Urbana. (Largely an up-dating of the above work.)

Shelford, V. E., and S. Olson
1935   Sere, Climax and Influent Animals with Special Reference to the Transcontinental Coniferous Forest of North America. Ecology, 16: 375-402.

Shelford, V. E., and A. C. Twomey
1941   Tundra Animal Communities in the Vicinity of Churchill. Ecology. 22: 47-69.

Small, A.
1965   From Monterey to Yosemite. In *The Bird Watcher's America.* Edited by

O. S. Pettingill, Jr. McGraw-Hill Book Company, New York. (A good description of zonal distribution of birds from sea to Alpine Tundra.)

Smith, R. L.
1966 Ecology and Field Biology. Harper and Row, New York. (A very readable and thorough text; includes a chapter-by-chapter treatment of communities.)

Snyder, D. P.
1950 Bird Communities in the Coniferous Forest Biome. Condor, 52: 17-27.

Sprunt, A., Jr.
1953 Newcomer [Cattle Egret] from the Old World. Audubon Mag., 55: 178-181.
1955 The Spread of the Cattle Egret (with Particular Reference to North America). Annual Rept. Smithsonian Inst. 1954: 259-276.

Stewart, R. E., and J. W. Aldrich
1952 Ecological Studies of Breeding Bird Populations in Northern Maine. Ecology, 33: 226-238.

Stewart, R. E., and others
1952 Seasonal Distribution of Bird Populations at the Patuxent Research Refuge. Amer. Midland Nat., 47: 257-363.

Tanner, J. T
1955 The Altitudinal Distribution of Birds in a Part of the Great Smoky Mountains. Migrant, 26: 37-40.

Taverner, P. A., and G. M. Sutton
1934 The Birds of Churchill, Manitoba. Annals Carnegie Mus., 23: 1-83.

Tester, J. R., and W. H. Marshall
1961 A Study of Certain Plant and Animal Interrelations on a Native Prairie in Northwestern Minnesota. Minnesota Mus. Nat. Hist., Occas. Papers No. 8.

Udvardy, M. D. F.
1958 Ecological and Distributional Analysis of North American Birds. Condor, 60: 50-66. (The author analyzes and compares the North American avifauna with the European. A significant paper requiring careful study by students interested in the origin of North American avifauna.)
1963 Bird Faunas of North America. Proc. XIIIth Internatl. Ornith. Congr., pp. 1147-1167. (Demonstrates with many maps how passerine species may be grouped as faunas on the basis of their geographical ranges and ecological preferences.)

Usinger, R. L.
1967 The Life of Rivers and Streams. McGraw-Hill Book Company, New York.

Van Tyne, J.
1951 The Distribution of the Kirtland Warbler (Dendroica kirtlandii). Proc. Xth Internatl. Ornith. Congr., pp. 537-544.

Verbeek, N. A. M.
1967 Breeding Biology and Ecology of the Horned Lark in Alpine Tundra. Wilson Bull., 79: 208-218.

Walkinshaw, L. H., and D. A. Zimmerman
1961 Range Expansion of the Brewer Blackbird in Eastern North America. Condor, 63: 162-177.

Wallace, A. R.
1876 The Geographical Distribution of Animals. Harper and Brothers, New York. (The classic work on the subject, containing the concept of the six zoogeographical regions, adapted from the earlier work of P. L. Sclater.)

Wallace, G. J.
1963 An Introduction to Ornithology. Second edition. Macmillan Company, New York. (Chapter 12, "The Distribution of Birds.")

Warner, A. C.
1966 Breeding-range Expansion of the Scissor-tailed Flycatcher into Missouri and in Other States. Wilson Bull., 78: 289-300.

Weller, M. W., and C. S. Spatcher
1965 Role of Habitat in the Distribution and Abundance of Marsh Birds. Agric. and Home Econ. Exp. Sta., Iowa State Univ., Spec. Rept. No. 43.

Welty, J. C.
1962 The Life of Birds. W. B. Saunders Company, Philadelphia. (Chapter 20, "The Geography of Birds.")

Wiens, J. A.
1969 An Approach to the Study of Ecological Relationships among Grassland Birds. Amer. Ornith. Union, Ornith. Monogr. No. 8.

Williams, A. B.
1936 The Composition and Dynamics of a Beech-Maple Climax Community. Ecol. Monogr., 6: 318-408.

Woodbury, A. M.
1947 Distribution of Pigmy Conifers in Utah and Northeastern Arizona. Ecology, 28: 113-126.

Woodin, H. E., and A. A. Lindsey
1954 Juniper-Pinyon East of the Continental Divide, as Analyzed by the Line-strip Method. Ecology, 35: 473-489.

Woolfenden, G. E., and S. A. Rohwer
1969 Breeding Birds in a Florida Suburb. Bull. Florida State Mus., 13(1): 1-83.

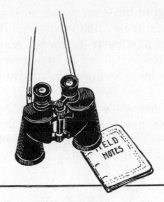

FIELD IDENTIFICATION

Ability to recognize birds in the field is one of the primary objectives of a beginning course in ornithology. A student should, at the first opportunity, start to identify birds and learn to recognize them by appearance and sound.

EQUIPMENT

The student should own or have access to the following:

Binocular. A 7x, 35 standard binocular, with coated optics and center focusing, is recommended for general use. The figures "7x, 35" signify that the binocular has a magnifying power of 7 times (making a bird at 70 feet seem only 10 feet away) and has objectives 35 mm in diameter. They also indicate that the binocular has a relative brightness value of 25, determined by dividing the diameter of the objective (35) by the magnifying power (7) and squaring the quotient (5). The term "standard" before binocular denotes a regular model as opposed to a wide-field or any other specially built model. The width of the field covered by the standard 7x, 35 model is 381 feet at 1,000 yards. A binocular with "coated optics" has both prisms and lenses treated to reduce light reflection. Thus more light passes through the binocular, giving greater brightness, when the instrument is used in poor light, and cutting down glare, when it is pointed in the direction of, but not into, the sun. A coating on the optics increases the relative brightness value by 50 percent; therefore, the brightness value of a 7x, 35 model with coated optics is actually 37 instead of 25. A binocular with "center focusing" is focused by turning a wheel on the central hinge-post after the right eyepiece has been adjusted to compensate for optical differences between the eyes.

Other models of binoculars may be useful for special purposes. All, however, should have coated optics. A 6x, 30 standard binocular, with its low power and consequent wider field, is advantageous when watching birds close at hand, as in woods or thickets. Standard binoculars with higher magnification, such as the 8x, 30 and 9x, 35, are particularly helpful when "reaching" for birds across water and canyons, or in mountains. They must be used in good light and clear atmosphere because their brightness value is lower, and they must be held steady, since their field of view is not increased in proportion to their magnification. For one doing bird work in dim light, as at dawn or dusk, or in shadowy places, the 7x, 50 special model is highly desirable. Though necessarily big and bulky to accommodate the large optics, it has the advantage of very great light-transmission qualities. If magnifying power above 9x is required, a telescope is preferable since a binocular above 9x cannot be held steady enough for accurate vision. Wide-field 7x, 35 and 8x, 35 models, which cover 525 and 445 feet, respectively, at 1,000 yards are available.

Their principal advantage is in enabling one to follow flying birds more easily. Unfortunately they are expensive. Nearly all standard binoculars are available with individual focusing—adjusting both eyepieces to one's eyes for a particular distance. Although models with individual focusing are inconvenient for looking at birds at varying distances because refocusing is slow, they are more sturdily built and more moisture-proof, hence less likely to be damaged in the field.

Telescope. A telescope, supported by a tripod, is indispensable for identification and observation of birds at great distances. Usually one telescope per class is sufficient for ordinary purposes. A recommended telescope is the so-called "spotting scope," 60 mm model, either with interchangeable eyepieces giving magnifications of 15x, 25x, 40x, and 60x, or with a lens that zooms from 15 to 60x.

Tape Recorder. A portable, fully transistorized recorder, one or more per class, for recording and playing back sounds on tape in cassettes, can be used effectively to bring singing birds into closer view. The technique is to pre-record (e.g., from a phonograph record) those songs of species most likely to inhabit the area under investigation, then play back the songs in suitable habitats. Most any male breeding bird, on hearing the song of his species, will mistake it for the song of a rival male, approach the source, and start singing, thereby revealing himself. This technique is especially effective in getting good views of birds in forests or dense shrubbery.

MATERIALS

Certain materials are necessary or desirable for identification work.

Field Guide. Among the available guides to the identification of bird species, those briefly described below are convenient in size to carry in the field and are especially useful for the reasons indicated. The selection of any one will depend on whether it covers the species in the region where the study is being undertaken and presents the information in the manner desired by the student or instructor.

For North America, north of Mexico: *Birds of North America: A Guide to Field Identification,* by C. S. Robbins, B. Bruun, and H. S. Zim; illustrated by A. Singer (Golden Press, New York, 1966; available in paperback or hard covers). Each species depicted in full color with pertinent text and range maps on opposite page; some similar species also compared in double-page, full-color spreads.

For North America, north of Mexico and east of the Rocky Mountains: *A Field Guide to the Birds,* by R. T. Peterson (Houghton Mifflin Company, Boston, 1947; available in paperback or hard covers). Many plates in color, comparing species of similar form; a few species illustrated in black and white only; concise text pointing out essential field characters and giving range. *Audubon Bird Guide: Eastern Land Birds* and *Audubon Water Bird Guide: Water, Game, and Large Land Birds,* by R. H. Pough (Doubleday and Company, Garden City, New York, 1946 and 1951). Plates in color by D. R. Eckelberry illustrating all species, including their color phases and many immature plumages; drawings in black and white by E. L. Poole showing many species in flight; text containing principal field characters and condensed information on habits, voice, nesting habits, and ranges of species.

For North America, north of Mexico and west of the Great Plains: *A Field Guide to Western Birds,* by R. T. Peterson (Houghton Mifflin Company, Boston, 1961). A companion to the volume for eastern North America; includes a section on the birds of the Hawaiian Islands and gives information on habitat and nest that is lacking in companion volume. *Audubon Western Bird Guide,* by R. H. Pough (Doubleday and Company,

Garden City, New York, 1957). A companion to the two other Audubon Guides (above), covering species not included in them.

Daily Field Check-list. This is a list of the birds regularly found in the region where the course is undertaken. Use new copy each day in the field. The format is small so that it folds and fits into one's pocket, field guide, or notebook, and contains spaces for checking each species seen or heard and for indicating the observer or observers, locality or localities, hours in the field, and weather conditions. Each copy, properly checked and filled out at the conclusion of the day in the field, serves as a permanent record. If a check-list is not already available for the region where the course is undertaken, a suitable one should be prepared. For a suggested format and arrangement, see the sample copy in the back pocket of this book.

Pocket Notebook. This notebook should have a durable cover (aluminum optional), be of a size to fit into one's pocket, have loose leaves so that they can be interchanged, and contain as many leaves as there are species likely to be observed.

PROCEDURE

The student's equipment and materials for each field trip should consist of a binocular, field guide, daily field check-list, pocket notebook, and pencil. As he identifies each species, he should check off its name on the field check-list and write the name on the top of a blank page in the pocket notebook. He should reserve this page thereafter for all information obtained on the species.

At the outset of this study the student must realize that identification of birds in the field is much more than matching their color patterns with pictures and descriptions in the field guide. Indeed, he will observe the majority of birds under such circumstances as to preclude the possibility of noting their colors at all. Many birds are seen in the distance as silhouettes; many appear only as forms dashing through the foliage; and many are merely heard. Successful identification involves knowing when to expect, what to notice, and how to observe.

Much of the student's initial difficulty in identification is his lack of observational acuity. Generally he is unused to detecting subtle differences in shape or motion of small forms, or noting distinctions in quality and pitch of sounds in the higher octaves. Thus, in the beginning, many birds look alike and sound alike to him. When the instructor readily recognizes one species after the other and all seem alike, the student may conclude that his own vision and hearing are impaired and that he has undertaken a task exceeding his capability. Any such conclusion, though understandable, is erroneous. He can acquire observational acuity, like a skill, only through conscientious practice and experience. As the diligent student proceeds with the study, he will discover that the recognition of the more common birds soon becomes second nature to him.

Identification Clues. The beginning student will find that he can make rapid progress in identification if he will pay special attention to the many kinds of identification clues. Some clues depend on a thorough knowledge of the distribution of birds by season and by habitat or biotic community. For instance, the winter season is a clue to certain species, a marsh community to others. An important group of clues is provided by the birds themselves even when they are not close enough for details of form and color pattern to appear. The general shape and posture of different birds, the way they fly or feed, and their flocking habits are among the many examples.

The drawings and comments on the following pages present a wide variety of clues. Before taking his first trip the student should study each drawing carefully, comparing and contrasting the birds depicted, so that he will watch for these clues himself and know what to look for when he meets the birds in the field.

The drawings of the four birds illustrate a situation where different species occur together in lighting conditions so poor as to prevent seeing either their colors or their markings. At first glance the birds appear much alike; closer inspection reveals several distinctions. They vary in size, the smallest being half the size of the largest. Their bills, though all acute, differ in length and thickness. The tail of one bird is exceedingly long; that of another quite stubby. These are clues that the student can obtain by comparing outlines of birds while they are together and that he can use in combination with other clues (e.g., call notes, peculiarities of behavior), noted at the same time.

Locomotion in birds is a source of many clues. When on the ground, many birds simply walk, but a number of species have the habit of hopping, keeping both feet together. In water, some birds swim while pumping their heads and necks back and forth. The mode of flight is an especially important clue to identification. The flight of a meadowlark (below) is an alternation between sailing with the wings spread and directed slightly downward and flying with rapid wing-beats. A flicker, like other woodpeckers, has an undulating flight, rising upward after several quick wing-strokes and then dipping downward. The Eastern Kingbird (next page) flies in a straight line, with a continuously quivering wing action.

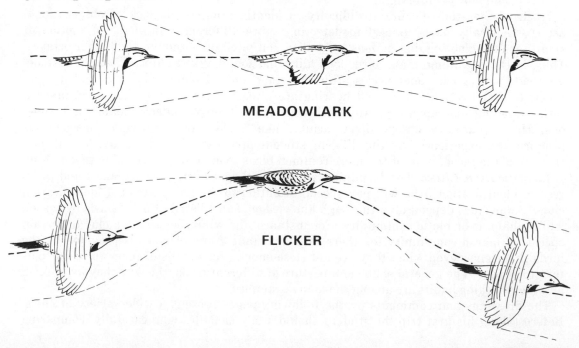

MEADOWLARK

FLICKER

EASTERN KINGBIRD

BIRDS OF A SEA COAST

The general color pattern of wings reveals clues: dark wing tips or white patches and stripes are excellent identification marks. The way a bird rests on the water gives other clues: pelicans and most ducks sit high on the water; loons and cormorants farther down in the water with less of their body showing. The tilt of the head may be a clue: a cormorant's head tips upward. Diving habits are clues: some species plunge into the water from the air; others dive from the surface; and a number of water birds seldom submerge their bodies completely.

In the air, left to right: Parasitic Jaeger, Black Skimmer, Common Tern, Gannet, Fulmar, Herring Gull. On the water, left to right: Brown Pelican, Cormorant, Common Merganser, Common Loon, Common Eider.

WIRE AND FENCEPOST SITTERS

The fact that these birds sit on wires and fenceposts is a clue—they are open-country birds. Forest birds, for instance, would not be seen in any such situation. Even at a distance these birds show a variety of recognition marks—stripes through the eyes, over the head, and across the breast; spots or streaks on the under parts; terminal bands on the tail. Their general color pattern provides clues: some are uniformly dark or light; some are dark above and light below. The birds on the wires demonstrate a variety of clues in tail shape, from square to extremely forked, while the birds on the posts illustrate clues in leg length and in posture, the owl standing almost vertically and the nighthawk with the body horizontal.

Left to right, top wire: Eastern Bluebird, Cedar Waxwing, Sparrow Hawk,
Scissor-tailed Flycatcher, Robin, Eastern Kingbird, Red-winged Blackbird.
Bottom wire: Loggerhead Shrike, Ruby-throated Hummingbird,
Purple Martin, Barn Swallow, Tree Swallow, Starling.

Left to right: Burrowing Owl, Common Nighthawk,
Meadowlark, Horned Lark, Upland Plover.

BIRDS OF A SHORE

The shore is a good habitat clue to all the birds shown. Birds of such a habitat are waders, characterized by long legs and often by long necks. A few have peculiar teetering and bobbing actions. Variations in bill length and shape are especially helpful clues to the identification of shore birds.

Left to right, front row: Dowitcher, Greater Yellowlegs, Semipalmated Sandpiper, Ruddy Turnstone. Middle row: Common Snipe, Killdeer, Semipalmated Plover, Spotted Sandpiper, Black-necked Stilt, Avocet, Whimbrel. Marbled Godwit, American Oystercatcher. In the background, Great Blue Heron, Black-crowned Night Heron, Glossy Ibis.

BIRDS OF A DECIDUOUS FOREST

Knowing what birds to expect in any given area the year round and during the different seasons is helpful. Furthermore, as birds spend much of their time feeding and have different adaptations for securing their food, their feeding habits are sometimes good clues to recognition. In a forest, woodpeckers, nuthatches, and creepers constantly search for food on tree trunks; tanagers and vireos in the forest canopy; ovenbirds on the forest floor; and other birds in a variety of places.

From the bottom, left side of trunk: Red-bellied Woodpecker, Ovenbird, Scarlet Tanager, Brown Creeper, Screech Owl, White-breasted Nuthatch. Right side of trunk, American Redstart, Least Flycatcher, Robin, Black-capped Chickadee, Red-eyed Vireo, Blue Jay, Ruffed Grouse.

BIRDS OF A MARSH

Becoming familiar with different biotic communities and learning what birds to expect in them is always an aid. Certain birds in a marsh community are particularly difficult to find because they are usually obscured by the vegetation. If one were not aware of the different kinds of marsh birds, some of them might be overlooked. Two marsh dwellers, the Red-winged Blackbird and Marsh Hawk, are readily identifiable by the banner-marks on wings and tail, respectively.

From the left, perching or standing: Red-winged Blackbird, Long-billed Marsh Wren, American Bittern, American Coot, Common Gallinule, Virginia Rail. In the water, Pied-billed Grebe. In the air, Marsh Hawk, Black Tern.

OUTLINES OF FLYING BIRDS

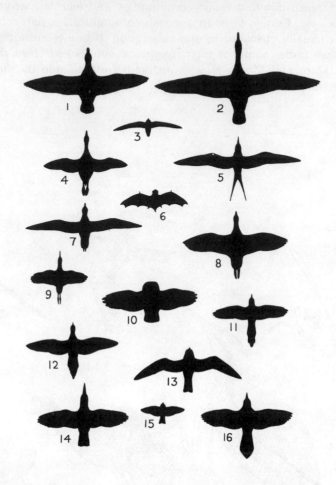

1. Redhead	9. Virginia Rail
2. Common Merganser	10. Screech Owl
3. Chimney Swift	11. Black-billed Cuckoo
4. Pied-billed Grebe	12. Mourning Dove
5. Forster's Tern	13. Common Nighthawk
6. Big Brown Bat	14. Belted Kingfisher
7. Black-bellied Plover	15. Purple Finch
8. American Coot	16. Common Grackle

Frequently all that can be seen of birds in flight are their outlines, which are, except for wing motions, the principal clues to identification. Note particularly some of the clues indicated in this plate and the one opposite: (1) the shape of the wings, whether pointed or rounded, narrow or broad, slotted or unslotted; (2) the length of the neck and tail in proportion to body length; (3) the shape of the tail; (4) the position of the feet, whether or not they are extended posteriorly beyond the body and tail. Observe how the outline of a bat compares with the outlines of the birds.

OUTLINES OF FLYING BIRDS

1. Whistling Swan	9. Sandhill Crane
2. Turkey Vulture	10. Common Raven
3. Brown Pelican	11. Common Loon
4. Bald Eagle	12. Cooper's Hawk
5. Canada Goose	13. Great Blue Heron
6. Osprey	14. Red-tailed Hawk
7. Peregrine Falcon	15. Double-crested Cormorant
8. Common Crow	16. Herring Gull

Songs and Call Notes. Learning to recognize birds by their songs and call notes is one of the most difficult tasks that confronts a student. This is usually because his ear is not trained to distinguish the extraordinarily fine tones given by so many birds. Nevertheless, learning to recognize songs and call notes is highly important because, on field trips, he will hear more birds than he sees. It is safe to say that the experienced ornithologist depends more on his ears than on his eyes and that, as a rule, over two-thirds of the birds that he records on a trip are only heard.

Conscientious and repeated listening to bird sounds will eventually lead to their ready identification. The following methods may aid in recording and remembering sounds:

1. Record the sounds phonetically—i.e., by the use of phrases that seem to resemble the sounds. Such phrases are easily remembered, particularly if they are coherent. Examples follow:

> SWEET SWEET SWEET I SWITCH YOU (Song of the Chestnut-sided Warbler)
> CHUCK BURR (Call of the Scarlet Tanager)
> O GURGLEEEE (Song of the Red-winged Blackbird)

2. Record and remember the songs by using diagrammatic symbols. The student may use any symbols that seem logical to him. The following are from *The Book of Bird Life* by Arthur A. Allen (D. Van Nostrand Company, New York, 1961).

Black-capped Chickadee ....:

Eastern Wood Pewee ....:

Eastern Meadowlark ....:

White-throated Sparrow ....:

Slate-colored Junco ....:

Field Sparrow ....:

Ruby-crowned Kinglet ....:

Veery ....:

The phonetical rendering of these same sounds would be as follows:

Black-capped Chickadee ....: PHE-BE-BE

Eastern Wood Pewee ....: PEE-A-WEE

Eastern Meadowlark ....: SPRING-IS-HERE

White-throated Sparrow ....: POOR-SAM-PEABODY, PEABODY, PEABODY

Slate-colored Junco ....: SWEET, SWEET, SWEET, SWEET, SWEET, SWEET, SWEET

Field Sparrow ....: HERE, HERE, HERE, HERE, SWEET, SWEET, SWEET

Ruby-crowned Kinglet ....: SEE-SEE-SEE, JUST LOOK AT ME, JUST LOOK AT ME, JUST LOOK AT ME, SEE-SEE-SEE

Veery ....: TUREE, AREE, AREE, AREE, AREE

According to Allen's method, a line indicates a whistle-like sound, the thickness of the line varying with the intensity of the sound. Thus, a thin line means a whispering whistle, a broad line is a clear whistle, and a series of small circles a trill. A small dash represents a clear note, a wavy line a tremulous note, etc. A line running upward points out a rise in pitch; a line running downward, a drop in pitch. In a continuous song the lines are connected; in a song interrupted by silent intervals the lines are correspondingly broken.

3. A. A. Saunders, in *A Guide to Bird Songs* (Doubleday and Company, Garden City, New York, 1951), has developed a somewhat different method of recording sounds. He diagrams a song as a series of lines, chiefly horizontal. In his words: "Each line represents one note of a song. Its horizontal length represents the period of time the note occupies. Its vertical height, in relation to other notes in the song, represents the note's pitch. Its heaviness represents its loudness or intensity." Saunders then explains in detail how his method is applied to various types of songs. The main text of Saunders' book gives a concise description, illustrated by diagrams when possible, of the songs of most of the land birds and a number of water birds in the United States east of the Great Plains. The student will find the book very instructive. Even if it does not cover the species in the region where the study is being made, the student may find it worthwhile to record the songs he hears by Saunders' method.

4. Remember the songs by using descriptive words. Among words commonly used are the following:

abrupt	broken	canary-like	cheeping	chinking
alarming	bell-like	carol-like	chucking	cat-like
ascending	babbling	chipping	churring	coarse
		chirping	choppy	
buzzing	continuous	chattering	cooing	discordant

double-toned	hiccuping	owl-like	short	twittering
			slow	tinkling
effervescent	instrumental	piping	shrill	tremulous
energetic	insect-like	pealing	soft	throaty
explosive	intoned	piercing		ticking
		puttering	screaming	tinking
flowing	jabbering	plaintive	strident	throbbing
faint		penetrating	sipping	
flute-like	long	peeping	sighing	ululating
forced	low		screeching	unmusical
fife-like	loud	quivering	sibilant	
	liquid	quacking	sweet	varied
grating	lively	quawking	sputtering	vibratory
gurgling	lispy		smacking	ventriloquial
guttural	laughing	rapid	squawking	voluble
grasshopper-		rich	squealing	
like	mewing	ringing	staccato	whining
gabbling	moderate	rattling	sonorous	weak
gibbering	melodious	rasping	stentorian	warbling
gushing	metallic	resonant	squeaking	whistling
	modulated	reedy		watery
	musical	raucous	strained	whirring
hurried	mellow	reverberant	slurring	whooping
harsh		repetitive		
hollow	nasal	rollicking	thin	yelping
husky	noisy		trilling	
hooting				

5. Remember bird songs and become familiar with them by recording them in the field with a portable tape recorder (see above) and then reviewing them, or reviewing them from phonograph records. Heard repeatedly, the songs will gradually become impressed upon the memory.

A number of phonograph records of actual bird songs are available. The following, all 33 1/3 RPM, obtainable from the Cornell Laboratory of Ornithology (159 Sapsucker Woods Road, Ithaca, New York 14850) and the Federation of Ontario Naturalists (Edwards Gardens, Don Mills, Ontario) are particularly recommended.

A Field Guide to Bird Songs (two 12-inch records), designed to accompany, page by page, *A Field Guide to the Birds*, by R. T. Peterson, and *A Field Guide to Western Bird Songs* (three 12-inch records), designed to accompany, page by page, *A Field Guide to Western Birds* (second edition), by Peterson. In order for so many species to be represented on the records, the selection for each species is necessarily brief.

American Bird Songs, Volume 1: game birds, southern birds, and birds of fields and prairies. *American Bird Songs*, Volume 2: birds of the lakes, marshes, and roadsides, concluding with 12 North American warblers. *Warblers:* About 400 songs from 150 birds of 38 species known to breed regularly in eastern North America. *Finches:* About 400 songs from 226 birds of 43 species of fringillids in eastern and central North America. *Thrushes, Wrens and Mockingbirds:* More than 100 birds of 17 species.

Other Data. In addition to recording the information suggested above, in the pocket notebook make detailed notes on anything concerned with habitat, general behavior, and the breeding cycle—displays, nests and nest-building, eggs, young, defense activities, and so on. Describe actions and related circumstances but do not attempt interpretations because they may bias later use of the information. For example, take careful note of the way a particular bird moves its wings and tail and the conditions occurring at the time of the action but do not state as a fact what appears to be the reason for the action since it may be incorrect. Careful studies under varying circumstances are necessary before one can attempt interpretation of a behavior.

RESULTS

Organize the information, set down in the Daily Field Check-list and pocket notebook, as indicated below:

1. Keep the "Seasonal Check-list" (in the back pocket of this book) up to date. After the first field trip, write in the left-hand column of this list the names of the birds checked off on the Daily Field Check-list. Indicate the date of the first trip at the top of the first vertical column corresponding to the correct day of the week; then check the spaces in the vertical column that are in line with the names of the birds. After each succeeding field trip, date the next correct vertical column, but write in the left-hand column only the names of birds seen for the first time during the season. Simply check, in the correct spaces, the birds previously seen.

2. After each field trip go over the notes in the pocket notebook and rearrange the pages so that the species will be in order—preferably taxonomic order, as in the field guide. Then add any observations or impressions that seem important or desirable for future reference. Do not recopy field notes unless absolutely necessary; mistakes may result. And never destroy the original record. Notes are more valuable in their original form, even though they are not always neatly written in the field. As pages are filled, they may be withdrawn and filed, to save increasing the bulk of the notebook. Add new pages in their place. The importance of note-taking cannot be too greatly emphasized. In the first place, putting down observations and impressions in writing is the most dependable means of remembering what one has seen or heard. In the second place, notes are the most reliable record of field work and will almost invariably be useful at some future date.

SUPPLEMENTARY GUIDES

Besides the regular guide to the identification of species, there are two other kinds of guides that may be useful to the student, either during the course or at a later date.

For the identification of birds' nests: *Birds' Nests: A Field Guide* and *Birds' Nests of the West: A Field Guide*, by Richard Headstrom (Ives Washburn, New York; 1949 and 1951, respectively). The first volume covers species east, the other covers species west, of the 100th meridian in the United States. Identification is by means of keys. The nest of each species is described and in some cases illustrated by photographs.

For locating birds: *A Guide to Bird Finding East of the Mississippi* and *A Guide to Bird Finding West of the Mississippi*, by O. S. Pettingill, Jr. (Oxford University Press, New York; 1951 and 1953, respectively). The volumes include detailed instructions for reaching the outstanding places for birds in the contiguous United States and tell what the places are particularly noted for and when they should be visited.

OTHER GUIDES AND REFERENCES

Allen, A. A.
  1961 The Book of Bird Life. Second edition. D. Van Nostrand Company, New York. (Chapter 18: "Learning Bird-Songs.")

Blachly, L., and R. Jenks
  1963 Naming the Birds at a Glance. Alfred A. Knopf, New York. (A system of identification based on color combinations.)

Booth, E. S.
  1950 Birds of the West. Stanford University Press, Stanford, California. (Approaches the problem of identification by keys to field characters; notations on nests; most species illustrated by black and white; several plates in color comparing numerous species of similar forms.)

Cruickshank, A. D.
  1953 Cruickshank's Pocket Guide to the Birds: Eastern and Central North America. Dodd, Mead and Company, New York. (Also available under another title, "The Pocket Guide to Birds," published in 1954 by Pocket Books, New York.)

Green, R.
  1947 Wing-Tips. The Identification of Birds in Flight. Adam and Charles Black, London. (Contains helpful suggestions on how to distinguish birds in flight. Based on British species.)

Hausman, L. A.
  1946 Field Book of Eastern Birds. G. P. Putnam's Sons, New York.

Hickey, J. J.
  1943 A Guide to Bird Watching. Oxford University Press, New York. (Chapter 1, "How to Begin Bird Study.")

Hunt, R.
  1923 The Phonetics of Bird-Sound. Condor, 25: 202-208.

Kortright, F. H.
  1942 The Ducks, Geese, and Swans of North America. American Wildlife Institute, Washington, D. C.

Peterson, R. T.
  1949 How to Know the Birds. Houghton Mifflin Company, Boston. (Also available as a Mentor Book, New American Library of World Literature, New York.)

Reichert, R. J., and E.
  1949 The Inside Story of Binoculars. Mirakel Optical Company, Mount Vernon, New York.
  1951 Know Your Binoculars. Audubon Mag., 53: 45-50; 105-109. (Revised and available in reprint form. The authors operate the Mirakel Optical Company, Mount Vernon, New York 10550. They are always glad to answer questions about binoculars.)

Scott, P.
  1951 A Coloured Key to the Wildfowl of the World. Revised and reprinted. Charles Scribner's Sons, New York.

BEHAVIOR

The **behavior** of a bird is what it does and how it responds to its surroundings. Like anatomy and physiology, behavior is adapted to the environment. How successfully a bird species survives and produces young depends as much on the way it reacts to other living things, takes care of itself, mates, and rears young, as it does on its structural and functional attributes.

INHERITED AND LEARNED BEHAVIORS

Many behaviors in birds are inherited and thus **innate** or **"instinctive"**—they are performed without preliminary experience or learning. Innate behaviors normally occur in patterns—**fixed action patterns**—so remarkably constant that they must be triggered by stimuli and can be discharged only after being activitated by **internal releasing mechanisms** selectively responsive to the stimuli.

The stimuli triggering a particular innate behavior are special features of the environment—form, color patterns, movements, sounds, etc.—that function as cues and are called **releasers.** Practically all bird species have social releasers frequently termed **signals** or **displays.** These may be visible or audible, and their principal and often exclusive function is to broadcast information to fellow members of the species.

Two near-classic examples of social releasers, proved by experiments, are the black malar strip ("mustache") of the male Yellow-shafted Flicker *(Colaptes auratus)* that releases courtship behavior in the female and attack by an intruding male (Noble, 1936), and the red patch on the lower mandible of the Herring Gull *(Larus argentatus)* that releases food-begging in the chick (Tinbergen and Perdeck, 1950). Both examples are derived from markings that stand out in contrast to the adjacent color and are thus conspicuous. Other examples of releasers that send out stimuli are particular movements and positioning of the head and/or other parts of the bird's body, vocalizations such as songs and alarm calls, and so on.

Through experimentation, Tinbergen (1951) discovered that he could produce **supernormal stimuli** that would elicit a more effective response. For instance, he found that an Eurasian Oystercatcher *(Haematopus ostralegus)*, which normally lays three, sometimes four, eggs, prefers five if given a choice between three and five, and if offered a giant model of an egg painted to match the natural colors of its own eggs, makes frantic attempts to sit on it rather than on its own eggs.

Theoretically, releasing stimuli will not trigger innate behaviors until the bird is motivated—i.e., has built up a **specific action potential** through hormonal actions and other factors to perform behaviors. This build-up toward discharge of most any innate behavior involves restlessness and exploratory maneuvers referred to as **appetitive activity.** Once the bird is sufficiently motivated, it then responds to the appropriate releasing stimuli with **appetitive behavior.** Provided the bird's searches are successful and the stimuli act on its internal releasing mechanisms, the build-up of action potential acquired by the bird is then discharged in the performance of the **consummatory act** which consumes the energy.

Two illustrations of appetitive behavior leading to the consummatory act are in feeding and nest-building. As the motivation ("drive") to feed or to nest develops, the bird shows appetitive activity by searching for food or for nest site and nesting materials. When, finally, by responding to stimuli, it finds and swallows the food, or it selects the nest site and builds the nest, it performs the consummatory act. For the immediate future the bird has no "desire" to eat or to build another nest.

Occasionally, a bird performs an act in the absence of a discernible stimulus. This suggests some sort of build-up of motivation to the point where, in the continued absence of a proper stimulus, the energy is discharged in a behavior called a **vacuum activity.** A good example is a male passerine bird bringing food to the eggs in his nest, "anticipating" the feeding of young. If the incubating female is absent at the time, he makes futile attempts to offer the food to the eggs and then departs, swallowing the food beforehand or taking it with him.

Often a certain behavior appears irrelevant, being "out of context." An incubating bird, disturbed and driven from its nest by an intruder, does not attack, perhaps from fear, but instead proceeds to feed when it is actually not hungry or to preen when its plumage is in no need of attention. Two shore birds suddenly stop fighting and one of them tucks its bill under the scapulars as though it were going to sleep. These are examples of what is commonly called a **displacement activity.**

Now and then a bird directs a certain behavior at an object to which it does not normally respond. Two Herring Gulls, in a territorial dispute, vent their hostility by pulling grass instead of fighting each other. An incubating bird, disturbed at its nest by a human being, may attack another bird in the vicinity when it would not ordinarily notice that bird, or it may peck vigorously at the branch of a tree as if in anger. Any such behavior is called a **redirected activity.**

Rather frequently a bird starts to perform a behavior but does not complete it. For example, a bird crouches as though intending to spring into flight, but it does not fly. A behavior of this sort, very low in intensity, is called an **intention movement.**

These behaviors—displacement activities, redirected activities, and intention movements which so often appear in conflict situations—are of special significance in that many seem to have become incorporated through evolution in more elaborate behaviors such as mating displays. They are then highly stereotyped or "formal" and are commonly referred to as **ritualized.** In this new role they serve as signals to elicit a response in another bird.

When any one of these conflict behaviors becomes ritualized for mating, the action is more exaggerated, more emphatic, and is consequently more conspicuous. Frequently, the conspicuousness is enhanced by bright colors in the plumage involved. Displacement preening in ducks such as the Mallard (*Anas platyrhynchos*) consists of lifting sharply and spreading the secondaries and adjacent scapulars of one wing and, while turning the head under the secondaries and scapulars as if intending to preen, revealing the iridescent speculum. Ritualized behavior, in addition to being more conspicuous, tends to be species-specific—i.e., peculiar to the species performing it. Thus displacement preening in the Mallard differs in minor ways from that in other ducks, thereby serving as an isolating mechanism that prevents cross-breeding.

Learned behavior is essentially adaptive behavior resulting from practice or experience.

Basic to learned behavior is, of course, inherited behavior which often predetermines the extent to which learned behavior may develop. Learned behavior, however, is frequently so fused with, and inseparable from, inherited behavior that most behaviors cannot be declared the result of inheritance or learning or both. Thorpe (1961) provided a good example from his research on the development of song in the Chaffinch *(Fringilla coelebs).* He discovered that Chaffinches reared in auditory isolation could give extremely simple songs (subsongs), not the full song of the species. The simple songs were obviously inherited, but why not the full song? He found the answer when he exposed isolated Chaffinches at the age of about six months to recordings of the full Chaffinch song. They could learn to sing it. Yet other Chaffinches which he reared in isolation and exposed, at the same age, to recordings of songs of different species could not learn them well, if at all. Thorpe concluded that Chaffinches must learn the full song from their own species. It is only the components of the simple songs—in a sense, the ability to sing their own songs—that they inherit. They are born with no appreciable ability to learn the songs of other species.

The study of learning in birds poses numerous problems in interpretation and identification. Certain behaviors are sometimes misconstrued as the product of learning when learning is actually not involved. This is the case with mainly neuromuscular activities (motor activities) such as flying. A young bird just out of nest does not fly as expertly as one a little older, presumably because it has to learn how to fly. Yet a pigeon raised in a tube, without being able to use its wings, to the age when it should fly expertly, will indeed fly expertly when first released. Flying ability is inherent; it shows up not with practice or experience but with the maturation of neuromuscular coordinations.

Learned behaviors resulting from practice or experience usually develop in the early stages of a bird's life. Thorpe (1961) found that young Chaffinches must hear the full adult song of the species during their first year of development, if they are to learn it and sing it normally thereafter. Not hearing it, they may never learn to sing it even though they may be exposed to it later. A learned response in birds sometimes shows up long after the process of learning has taken place. Thus Chaffinches exposed to normal full songs as wild juveniles, then captured in the fall of their first year and isolated singly so that they cannot hear any other Chaffinch sing, may not show the effect of the original exposure until they begin to sing at the start of the first breeding season.

Learning takes different forms, all of which are attributed to experience, not maturation, and all of which may intergrade.

Habituation, the declining in a response to repeated stimulation without reinforcement ("reward or punishment"), is one form of learning. Stated another way, habituation is learning to avoid commonplace stimuli that are irrelevant to the welfare of the individual. Stated still another way, habituation is becoming used to, and no longer reacting to, anything new that proves to be harmless or non-rewarding. Tameness in a bird is the result of habituation whereby the bird becomes accustomed to unnatural conditions such as a cage or the heretofore unfamiliar actions of human beings. Once the bird is tame it is likely to remain so.

Another form of learning is **trial and error,** the selective acquisition during appetitive behavior of a response resulting from reinforcement. A bird learns to swallow palatable objects and to reject objects with a bitter taste. In the same way, a bird when first flying learns to take off and land *in* the wind rather than with it and to select perches that can accommodate its feet and support its weight; a bird when first building a nest learns to select those materials that it can manipulate best. Playing (see below under Individual Behavior) in a young bird may be a means of learning through trial and error.

A third form of learning is **imprinting** which usually takes place during a brief, critical period in the developmental stage of a bird's social life. Once acquired, the learned habit may sometimes remain stable for the duration of its life. Occasionally the behavior manifests itself or reappears long after the process of learning. Imprinting is commonly observed in waterfowl and gallinaceous birds. Ducklings and goslings, for example, will learn to follow a substitute "parent" such as a human being, thereafter ignoring its natural parents (Lorenz, 1935, 1937). In the Mallard the period of "learning to follow" is between 13 and 16 hours (Hess, 1959). With age, or with the tendency to flee strange objects, the following response is apt to weaken, even disappear, though it may show up again under favorable circumstances. Imprinting is much less commonly observed in passerine and other birds reared in nests. Nevertheless, Thorpe (1958) has suggested a form of learning, probably imprinting, in Chaffinches. The young Chaffinch, hand-reared in isolation, sings simple songs that are innate; but the young Chaffinch, reared in the wild, on hearing the songs of its parents or other Chaffinches, acquires, in due course of time, characteristic features of the full song without apparently practicing them. Later, in its first breeding season, it learns further characteristics from competing males. After 13 months, however, it can no longer learn features of other songs because its period of learning—presumably its period of imprinting—has ceased.

There are a few species of birds which manipulate or actually use objects to solve simple problems. The Tailorbird *(Orthotomus sutorius)* of southeast Asia forms a receptacle for its cocoon-like nest by perforating the edges of broad leaves with its bill and then drawing them together with plant fibers (Wood, 1936). A number of passerine species have the ability to pull up string to which food is attached. From the air the Black-breasted Buzzard *(Hamirostra melanosternon)* of Australia drops stones onto emu eggs in order to break them open (Chisholm, 1954), and from a standing position the Egyptian Vulture *(Neophron percnopterus)* tosses stones onto ostrich eggs for the same purpose (the van Lawick-Goodalls, 1966). The Woodpecker Finch *(Cactospiza pallida)* of the Galapagos Islands has the singular habit of picking up or breaking off cactus spines or twigs and using them as tools for extracting insects from holes and narrow crevices (Lack, 1945, 1947; Bowman, 1961). Whether or not the bird acquires this habit by learning is open to question. Millikan and Bowman (1967) suggest that the habit may be a form of displacement activity "when another behavior performance has been thwarted in some way." Possibly the ability is simply an inherent behavior re-adapted through experience to perform a task more efficiently. Only a careful study of the way the habit develops among several individuals can prove whether or not learning is involved.

A few experiments have shown that birds possess quite remarkable abilities to learn, presumably by **insight**—the "sudden adaptive re-organization of experience" (Thorpe, 1963). For example, Pastore (1954) proved with a series of 21 experiments on Canaries *(Serinus canarius)* that they can learn to respond to a unique object among a group of nine, the other eight being the same in all respects. For his experiments he put before each bird a board with nine small depressions in a row. In the first trial of his first experiment he capped eight of the nine depressions with aspirin tablets and the other with the head of a wood screw, the unique object, under which he hid some food, the reward. In the next trial he hid the food under an aspirin tablet, this time the unique object. In later trials the food was placed under either an aspirin tablet or a wood screw in depressions selected at random. The purpose of all the trials was to train the birds to find the food by pushing off the unique object. Each of the four Canaries, after an average of nearly 160 trials, achieved the ability to make the right choice in 15 out of 20 attempts. In a subsequent series of 20 experiments, using differing pairs of objects, the average number of trials required by the Canaries to choose the unique object became steadily fewer, to under 30 by the 13th in the series of experiments. Obviously, the Canaries learned to respond to *uniqueness* although they must have perceived it as abstract quality.

As another example of what may be learning by insight, Köhler (1950) taught a Common Raven *(Corvus corax)* to "count" an "unnamed number" of objects. To do this he trained the bird to open the one box in five which had the same number of black spots on its lid as on the key card on the ground in front of the box. The raven eventually learned to distinguish between five groups of spots—two, three, four, five, and six. To prove that the bird was choosing the lid with the number of spots corresponding to the number on the card, Köhler changed the size, shape, and arrangement of the spots on both the lids and the card in a random manner. Even the position of the boxes in relation to the key card was changed. With this elimination of extraneous cues, the raven could still distinguish numbers of objects, but without consciously recognizing the numbers as such.

INDIVIDUAL BEHAVIOR

Many behaviors are directed toward the bird's own welfare. These include **maintenance behaviors**, concerned with the care and comfort of the body, and behaviors associated with selection of habitat and cover, with food and feeding, and with playing.

Maintenance Behaviors

A large number of movements relate to the care of the plumage and, to a lesser extent, the skin and soft parts. Since the feathers, especially those used in flight or as insulation for the body, are vital to the bird's survival, the movements directed toward their care have evolved through natural selection to an elaborate degree and become highly stereotyped. Some of the maintenance behaviors have served as fruitful sources for displacement activities and displays.

A brief survey of the more common maintenance activities is given below, drawn in part from the very thorough summary by Simmons (1964).

Preening. Probably the most important as well as the most frequent activity in plumage care; performed by the bill. Usually a bird deals with one feather at a time, seizing it at the base between the mandibles and working toward the tip, either nibbling it continuously or drawing it through the mandibles in one movement. This serves to clean the feather, to interlock barbs that have become separated, and to smooth down the plumage. Less often the bird moves the bill rapidly along the feather from tip to base near the shaft either in a stroking action or in a trembling action. When preening, the bird fluffs the feathers to make them more accessible and turns its head toward the plumage areas needing attention. Meanwhile, the bird contorts itself and assumes numerous postures, all quite stereotyped.

Head-scratching. Performed by the foot. Sometimes, along with preening, head-scratching is an activity for the care of the plumage, especially the feathers of the head and upper neck that are inaccessible to the bill; otherwise it is a simple reflex action in response to an irritation in the head region and is therefore a comfort movement (see below). Birds use two methods in reaching the head: **directly,** by lifting the foot straight up from under the wing; **indirectly,** by bringing the foot up and over the lowered wing. As a rule, different taxonomic groups of birds use one method or the other (see Simmons, 1957b). Neither is reliable as a taxonomic character since some closely related groups, even some individuals of the same species, use both methods.

Bathing. This activity differs in water birds and land birds. Water birds (e.g., loons, grebes, cormorants, and waterfowl), while floating, immerse their heads and shoulders in a scooping motion which sends the water over their backs while they rub their sides and flanks with their heads, ruffle their feathers, and lift and splash more water over their plumage by beating their wings. Wading birds stand in the water but bathe the same way. Land birds, going into shallow water, first dip their throats and breasts and flap their wings up and down, then raise their heads and foreparts and lower their tails and hindparts into the water and flap their wings in and out of the water and over the back, spraying their plumage which is ruffled so that water can reach all the feathers and even the skin. There are many exceptions to this method of bathing. Swifts, swallows, and other highly aerial birds, as well as hummingbirds, kingfishers, and some owls, drop repeatedly to the water's surface and perform a skimming action sufficient to wet their plumage. Swifts bathe in falling rain (Lack, 1956). Tropical species of many varieties bathe mainly in the rain or wet foliage. Almost all land birds will take advantage of rain or even lawn sprays. Bathing functions as a means of keeping the plumage in order, rarely for cleaning it. While bathing, a bird is liable to predation; hence it is innately apprehensive and habitually avoids drenching itself to the extent of impeding its escape by flight.

Oiling. A form of preening among birds with an oil gland (see page 105 for the structure and function). Typically and often in association with preening, the bird touches or pinches the oil gland with its bill, stimulating the flow of oil to the bill's surface. The bird then smears the oil on the plumage areas with rubbing movements of the bill. But there are many variations. Long-necked birds often first rub the head on the oil gland and then rub it immediately on the plumage areas. Short-necked birds, such as passerines, oil the head with the foot, first taking the oil with the bill, then scratching the bill with the foot.

Sunning. An activity among practically all birds, sometimes performed after preening, bathing, or dusting. Typically, in a moderate reaction to warm sunlight, passerine bird raises the feathers of the head, neck, back, and rump, crouches with the axis of the body perpendicular to the source of light, spreads the flight feathers of the wing nearer the sun, and fans the tail. It sometimes pants with opened mouth. It may draw the nictitating membrane over the eye and perhaps close the lower lid. Non-passerine birds react in different ways. Vultures and other large diurnal birds of prey may stand, facing the sun, with wings extended. Cormorants sun themselves similarly after coming out of the water or long after their plumage is dry. Birds being photographed and exposed too long under bright lamps that radiate considerable heat, give the normal sunning response. During moderate- to high-intensity sunning, some birds seem to be in a semi-hypnotic state and much less cautious than is normal. Whether sunning directly benefits the plumage or serves any other significant function is undetermined.

Dusting. A behavior in which dust—powdery dry surface soil—is forced through the plumage. In an average performance the bird proceeds to scratch and/or peck out a depression while crouching and usually turning its body. Once the hollow is deep enough and the soil loosened, the bird sits in the hollow, raises dust through the ruffled plumage with a flicking and scooping action of the wings alternating with a scratching action of one foot or the other. Intermittently, the bird compresses its feathers and pauses to peck dust against itself and under the body and, at least in the gallinaceous bird, to lower and rub its head and neck in the dust. On ceasing to dust, the bird stands, ruffles its plumage and vigorously shakes out all dust particles. Unlike bathing, dusting is characteristic of comparatively few birds. It appears to have developed commonly among birds of deserts and grasslands. Some birds—e.g., wrens, House Sparrows *(Passer domesticus)*— bathe as well as dust; others such as gallinaceous birds and goatsuckers dust only, rarely if ever taking a bath by deliberately standing in water. Dusting no doubt removes ectoparasites and may have other functions as yet undetermined.

Anting. A behavior in which the fluids from ants, notably formic acid, are applied to the plumage and possibly the skin. There are two basic types of the behavior: **active anting**, in which the bird anoints its plumage with the bill, and **passive anting**, in which the bird allows ants to invade and anoint the plumage. Whitaker (1957) has reviewed the subject and described active anting in the Orchard Oriole *(Icterus spurius)*. In a typical performance, the oriole crushed the ants and dabbed them, with rapidly vibrating head, to the ventral surface of the wing tips and especially the under tail coverts and bases of the rectrices. It anointed the feathers of the belly and tibiae only occasionally. A regular feature of the behavior was tripping and tumbling, the result of bringing the tail forward beneath the body. In passive anting, as described by Simmons (1957a), "birds squat or lie down among the ants, often while applying them directly and going through the motions of doing so with empty bill, while the wings and tail are partly or wholly spread out." Whitaker listed 148 species, 16 of them non-passerine, which have been observed anting. Some individuals of these species, in unnatural situations such as in captivity, perform active anting with burning matches and cigarette smoke, moth balls, hair tonic, mustard, citrus fruits, numerous insects other than ants, and other substitutes. What causes anting and what purpose or purposes it serves, if any, need investigation. V. B. Dubinin (see Kelso and Nice, 1963) discovered that the formic acid from ants may destroy feather mites and thus he believes that anting functions as a defense against them.

Comfort Movements. Any number of actions which presumably give the bird "comfort." Some common movements are **feather settling** by first raising the feathers, shaking the body, flapping the wings, and then depressing the feathers into proper position; **stretching**, often by extending the leg and wing of one side while fanning the tail on that side, then repeating the action on the other side; **yawning**, by opening and closing the free-moving mandible, in most birds the lower one; and **resting**, by standing on one or both feet, or by squatting down, the plumage relaxed, the head retracted and seeming to rest on the shoulders, sometimes turned with the bill tucked under the scapulars in a sleeping posture.

Sleeping. Birds ordinarily sleep with the head retracted and turned so that it rests on the back with the bill tucked under the scapulars. But there is both intraspecific and interspecific variation. Some individuals may sleep with the head retracted to the shoulders on which it seems to rest and the bill forward. Several familial groups, such as pigeons, normally sleep this way.

If not sleeping when floating on water, birds usually squat rather than stand. If squatting on a perch, the skeletal and muscular arrangement of the hindlimbs is such that the toes automatically grasp and hold fast to a twig or branch (see page 27). Goatsuckers, though lacking this arrangement, can sleep in trees by squatting with the body lengthwise on a wide branch.

Instead of squatting during sleep, a few groups of birds such as woodpeckers, swifts, and presumably creepers (Certhiidae) cling to vertical surfaces and a few others such as the mousebirds of Africa (Coliidae) and some parrots *(Loriculus)* hang upside down from perches.

Selecting Habitat and Shelter

Most species of birds select distinct habitats. This allows an efficient use of the natural environment and assists in reducing competition between species. Birds no doubt recognize their habitats by visible cues, but what are the cues and how do preferences become established in the first place? To answer the question, one may identify the components of an environment and then determine how the preference develops by comparing individual birds raised in captivity under different conditions with wild trapped birds. Klopfer (1963) did just that with Chipping Sparrows *(Spizella passerina)*, taken from mixed stands of pine and oak in the North Carolina Piedmont and kept in a chamber where they were exposed to both fresh pine and oak foliage. Since Klopfer designed the study to measure the birds' preference of foliage for perch sites, all other environmental variables except foliage and light were minimized as much as possible. Klopfer found that adults, trapped in the wild, and young, reared in isolation with no foliage, both preferred pine over oak; but the young, reared in isolation with oak leaves, preferred oak over pine. Their choices suggest that size and shape of leaves and to a less extent light-shadow patterns are the cues and that the preference of one foliage over the other, though innate, may be modified by experience.

Klopfer's study does not suggest that the selection of a habitat constitutes an innate response only to such simple features as size and shape of foliage. Almost certainly the selection depends on many cues, single or in combination. However, Klopfer's work does quite properly imply that birds, while innately preferring a particular habitat, are nonetheless genetically equipped to re-adapt to others and thus demonstrate considerable plasticity or **opportunism**. Field observations support this. For example, in northern Michigan the Wood Thrush *(Hylocichla mustelina)* prefers cedar-spruce bogs to the deep deciduous woods one would expect (Root, 1942). In West Virginia the Magnolia Warbler *(Dendroica magnolia)* appears in oak forests whereas northward in North America it is ordinarily partial to spruce and hemlock forests (Brooks, 1940).

Klopfer's study as well as field observations tend to support the concept that birds breeding in the temperate areas of North America show opportunism because their environment is unstable and they must cope with variability. Birds breeding in tropical areas are less opportunistic (postulated by Klopfer and MacArthur, 1961) because their environment is comparatively stable.

The habitats which birds select must meet requirements of shelter favorable to their survival, day and night, in all seasons of their occupancy. Besides shelter that provides escape and protection from predators, there must be shelter for roosting. Sometimes one shelter may serve two or more purposes—e.g., a shelter for escape may also be a nesting shelter or even a roosting shelter. But whether serving one or more than one purpose, birds are nevertheless as selective of these shelters as they are of the overall habitat.

A few remarks about roosting shelter are appropriate since most works on ornithology give relatively little attention to the subject.

Klimstra and Ziccardi (1963) found that covies of Bobwhites *(Colinus virginianus)* in southern Illinois much preferred to roost on well-drained ground at medium to low elevations facing south or southwest and protected by a low, sparse, canopy of vegetation.

Any shift in roosting ground was mostly from lower to higher elevations, except in windy and inclement weather when the shift was to poorly drained sites at lower elevation. Some cavity-nesting birds may use their own nests as roosts in the winter, or construct or excavate special nests for roosting. Berger (1961) watched a male Downy Woodpecker *(Dendrocopos pubescens)* in October excavate a roosting cavity in a dead maple, much as it would a nesting cavity, and use it regularly through the winter until late April. Frazier and Nolan (1959) noted five to 14 Eastern Bluebirds *(Sialia sialis)* roosting together in a birdbox during the winter in Indiana. Many other observers have reported instances of two or more birds roosting together in such enclosures or even natural cavities. After a heavy fall of snow the Ruffed Grouse *(Bonasa umbellus)* will plunge into the snow and pass the night.

The above illustrations bring out several important points. Besides providing protection from predators, the roosting shelter gives protection from unfavorable weather and helps conserve body heat. When two or more birds choose to roost together in a small cavity, they naturally conserve heat. Also, in selecting roosting shelter, birds show opportunism as, for example, when choosing a nestbox instead of a natural cavity or when taking advantage of a snowfall.

Selecting Food and Feeding

Most species of birds have the structure and motor coordinations for taking a wide variety of food although they usually learn to select particular items. Among closely allied species in the same habitat this reduces competition and assures each species its share of the total food supply.

This book does not treat the various means, both behavioral and structural, by which birds obtain their food. The concern here is with the question: How do birds learn to select their food?

Each species is innately equipped with fixed action patterns for seeking food and preparing it for consumption. In the young bird these movements may be elicited initially by somewhat generalized stimuli. Gallinaceous chicks, for example, begin seeking food by pecking at small objects with color values contrasting with the background. To test the food-pecking responses of young Indian Peafowl *(Pavo cristatus)* to objects of different sizes and colors against different background colors, Dilger and Wallen (1966) exposed four newly hatched chicks to circular objects ranging in size from two to 14 millimeters, painted in six different colors, and placed against backgrounds with the same six colors. The chicks preferred the smallest objects when given a choice of all sizes, presented in different colors in random combination with differently colored backgrounds. They also preferred the combinations of objects and backgrounds that offered medium rather than maximum or minimum contrast in both hue and brightness. In all probability the chicks were innately adapted to the medium rather than the maximum contrast that one might expect, because edible objects in the wild are more usually in medium contrast with the background. Dilger and Wallen suggest that maximum contrasts may be "somewhat inhibiting or possibly even somewhat frightening."

The size of the bill and its special adaptations are among the important factors in determining the food items a bird learns to select. In general, young birds select smaller items than older birds; as their bills harden and their motor coordinations mature, they take larger items though not beyond the point where they can handle them efficiently. Provided they have the necessary behavioral and structural equipment, which in many species includes the feet, birds eventually learn from their own trial-and-error responses to generalized stimuli, or from watching their parents and conspecific associates, to select particular items in various local situations.

In England since 1911, titmice and other small birds—at least 11 species altogether—have learned the habit of drinking from milk bottles by either puncturing metal foil caps, or stripping off or removing paper or cardboard caps (Hinde and Fisher, 1952). Some observers reported titmice in groups following milkcarts and attacking bottles immediately upon delivery. While many titmice no doubt learned the habit from other titmice, it is just as likely that it developed independently among others.

Playing

Birds perform playful activities though not to the extent that mammals do. Young birds still in the nest show **exploratory pecking**—gently pecking one another and the nesting materials. Chicks in a brood, after leaving the nest, similarly peck one another and peck at objects on the ground. Occasionally, they pick up morsels of food and run with them, eating them later. Gull chicks when scarcely half-grown pick up, run about with, and drop sticks, plant stalks, and other objects. All young birds, well in advance of flying, fan their wings from time to time. Such activities constitute what amounts to "practicing" and may be a form of learning to forage, escape, or fly.

Fledglings and chicks of land species commonly make short flights or runs, turning sharply—an activity called **play-fleeing**. Aquatic chicks give a comparable performance on the water by running, splashing, and turning on the surface, even attempting to dive.

Adults of some species perform activities that have all the appearances of play and seem to serve no function other than the release of pent-up energy. At the Falkland Islands in the South Atlantic, Pettingill (1964) observed groups of Magellanic Penguins *(Spheniscus magellanicus)* running from a beach into the sea, quickly submerging, torpedoing away from shore in a wide circle, and as quickly emerging and hustling up the beach. Parrots are notoriously playful. The Kea *(Nestor notabilis),* a large powerful mountain species in New Zealand, pulls windshield wipers off cars and tosses empty tin cans about for no apparent reason than to be doing something.

SOCIAL BEHAVIOR

Social behavior involves the many kinds of interactions or joint activities that have to do with two or more individuals in the same species or even individuals in different species.

Agonistic Behavior

One of the most common forms of social behavior is **agonistic behavior** which includes all manner of hostile activities or displays from overt attack to overt escape.

Perhaps most obvious hostile displays are those that have become ritualized as social releasers (Moynihan, 1955b). The most abundant of these are probably the **threat displays**, "designed" to produce a "frightening" effect, forcing an opponent to retreat. Threat displays are thought to be activated by drives to attack and escape. For example, a bird approaching the territorial boundary of another exhibits attack, but, as it crosses the boundary, the motivation to escape sets in, coming into balance with the motivation to attack. Meanwhile, the owner of the territory, instead of being intimidated, proceeds to attack the intruder. If, at this point, the attack drive of the intruder still continues more intense than its escape drive, an encounter or fight between the two birds results; if the escape drive of the intruder becomes strong enough to inhibit or outbalance its attack drive, the bird flees.

Threat displays are believed to be derived from a number of different sources (Moynihan, 1955a), among them the unritualized intention movements associated with locomotion. In certain passerine birds the threat displays may include lowering the body to the horizontal, facing and extending the head toward the opponent, sleeking the head feathers, holding the wings stiffly down or perhaps flitting them, depressing and fanning the tail or sometimes flipping it. All such actions suggest the movements of a bird about to take flight.

Almost as common as threat displays are the **appeasement displays** serving to prevent attack without provoking escape (Moynihan, 1955a). These displays serve to reduce the strength of the opponent's drive to attack and, to a lesser extent, the opponent's drive to escape.

As in threat displays, most of the movements have been derived from unritualized hostile sources, in this case, mostly from intention movements involving escape. These include, among certain passerine birds, sitting still, flexing the legs, "hunching" with the head held low and retracted to the shoulders, and turning away from the opponent.

Vocalizations frequently complement both threat and appeasement displays, giving them an auditory component. Most songs of male birds function in part as threat displays, warning and/or repelling rival males.

Threat displays serve to space out birds, thereby preventing over-population in any one area. Appeasement displays, on the other hand, bring birds closer together by inhibiting attack or aggression and promoting bonds between pairs for reproduction and between parents and young.

Defense Behavior

All bird species are subject to predation and, consequently, show behavioral adaptations that promote their survival. Like maintenance behaviors, the defense behavior of the individual is directed toward itself, in this case its own safety, but it may also serve to promote the survival of other individuals—e.g., the individual's own young, also the individuals of its own species and/or other species. Thus defense behavior, under given circumstances, can be a form of social behavior.

Birds respond to certain danger stimuli both auditory and visual. Unaccustomed sharp sounds and abrupt movements evoke defensive behavior of some sort, but if the same sounds are repeated again and again with further intensification, the defensive responses subside through habituation. Models of owls that have the right combination of shape, color pattern, and contour can elicit alarm reactions in small birds (Nice and ter Pelkwyk, 1941; Hartley, 1950). Cardboard models, cut in the shape of a short-necked, avian predator flying overhead and passed over the birds in the direction of the short neck, can induce escape reactions in waterfowl (Lorenz and Tinbergen, 1938). The exact shape of the wings and tail and the color are immaterial. Although it was once thought the responses to models were simply innate, more recent researches have demonstrated that at least some learning is involved.

The precise manner in which birds respond to danger signals depends on the physical and behavioral adaptations of the species responding and the type of predator, the intensity of signals, whether or not the responding individual is breeding, and the circumstances of the encounter. Some of the more common responses are the following:

Fleeing and *Freezing.* If the average passerine bird is exposed without benefit of cover and the danger signal, such as a hawk in flight, is strong, the bird flees to the nearest cover where it is likely to "freeze"—i.e., to keep completely motionless with the feathers sleeked, the legs flexed, the head down in line with the body but cocked so that one eye is fixed on the direction of the predator. If the bird is already in cover when receiving the danger signal, it freezes where it is. The freezing posture is concealing and, at the same time, practically identical to take-off posture; thus the bird is prepared to flee if necessary.

Birds such as grouse, snipe, woodcock, and goatsuckers with cryptic coloration respond to a strong danger signal by freezing even when not in a shelter. Most chicks, nestlings, and fledglings of all species normally freeze in response to a danger stimulus and often close their eyes.

Non-breeding sandpipers, plovers, small gulls, terns, swifts, and other shelter-free dwellers, capable of fast or erratic flight, on the appearance of a flying predator, may attempt to outdistance or outmaneuver it should it attack.

Communicating Danger. The vocal repertoires of practically all birds include special alarm or warning calls, given in response to danger signals, but not necessarily in all cases or at all times. A passerine bird at the sight of a cat invariably utters alarm calls, yet the same bird may freeze at the sight of a flying hawk and remain silent or give alarm calls only if it has young. Though presumably not consciously given to alarm or warn, the calls nonetheless alert flock associates and mates and induce freezing among young.

Threatening. Confronted with potential predators or animal forms that impose danger, birds may respond with actions that threaten or intimidate. In some instances the actions may appear identical to the threat displays against members of their own species, but often they are more exaggerated or more elaborate. Threatening behavior includes actions that make the performer seem more formidable. Thus most birds puff up their feathers, appearing larger than they are; and a great many birds open their mouths, as if threatening to bite, and lift their wings, as if intending to strike. Some birds utter infrequent sounds such as hisses that may function as an element of surprise and deter an approach. A Great Horned Owl *(Bubo virginianus)* not only hisses but also snaps its bill rapidly; a Least Bittern *(Ixobrychus exilis)* silently lifts its bill skyward and sways back and forth.

Birds with young most often employ threatening behavior and may vary the type and intensity with the kind of predator. A bird that performs vigorously at the approach of a small mammal may flee at the approach of a hawk or a human being.

Attacking. Many birds with nests or young, instead of threatening a predator, may attack it outright. If threatening behavior fails to intimidate a predator, attacking behavior may follow. In a sense, attacking behavior is a projection of threatening behavior.

Exploring. Occasionally, birds show a tendency to explore strange objects such as a blind suddenly placed near their nest. While examining it visually from different angles, they exhibit what appears to be a conflict between the urge to fly away (escape) and the urge to see more. Exploratory behavior of this sort may have survival value by leading to the discovery of incipient danger.

Mobbing. In any season, groups of small birds respond to a perched hawk or owl by "mobbing" it. During the breeding season, birds often leave their territories temporarily to join in the performance. The basic motivation may be a combination of two tendencies: to explore and to escape. In some instances, the tendency to escape may be supplanted by the tendency to attack outright. The function of mobbing is puzzling. Possibly it serves to put all birds in a given area on the alert and, during the breeding season, to distract a predator from discovering young birds in the vicinity.

Giving Distraction Displays. Parent birds, threatened by a predator, may perform movements that sometimes simulate incapacity from injury ("injury-feigning") or sometimes suggest the running away of a small mammal ("rodent-running"). The motivation is believed to be a combination of two drives—to escape and to attack—that has become ritualized and functions in diverting the attention of the predator from the nest or young to the parents.

Injury-feigning shows up in a great number of species representing many families. The principal movements consist of fanning, beating, or dragging one or both wings, spreading and depressing the tail, fluffing the back and rump feathers, giving "distress" calls, and alternately moving parallel with and away from the predator. The performance differs in detail from species to species depending to some extent on the habitat. In general, species nesting on the ground most fully perform injury-feigning, but species nesting in other situations, such as trees and marshes, at least use some of the movements and adapt them accordingly.

Rodent-running is characteristic of species nesting on the ground in thick grassy cover. Essentially it amounts to the bird sneaking off the nest, away from the predator, and running fast, occasionally hopping slightly and even fluttering its wings as it disappears from sight.

Full distraction displays at maximum intensity are given when the eggs are hatching and for one or two days thereafter. While some predators may elicit distraction displays in a species, others may instead elicit outright attack. If a displaying bird is harried by a predator or a human being for an extensive period, its performance gradually wanes in intensity and may cease altogether.

Flocking Behavior

The majority of birds tend to be gregarious at one time or another. During the breeding season, most sea birds (including pelicans, cormorants, gulls, and terns) and certain species of herons, swallows, and tropical icterids form **colonies** in which pairs establish and maintain nesting territories close to one another. Outside the breeding season a great many birds gather in **flocks** in which they remain until they disperse to nest. A few birds, such as the Monk Parakeet *(Myiopsitta monachus)* of Argentina and certain species of weaver finches (Ploceidae) of Africa, are gregarious at all times, nesting in colonies and staying in flocks when not breeding.

Flocks may be composed of one species or several species of similar size with approximately the same flight speeds but not necessarily the same food preferences and feeding adaptations. If a flock is of one species, as is more often the case, it may be comprised of one family or multiples of families from the preceding breeding season. This is believed true with small flocks of migrating geese (Elder and Elder, 1949). Indeed, at Delta, Manitoba, color-banded adult and young-of-the-year Canada Geese *(Branta canadensis)* were seen "to depart southward together in November and return, still as a united group, the next April" (Hochbaum, 1955). This same make-up may also be true of small flocks of permanent-resident chickadees *(Parus* spp.) during the winter. But there are numerous exceptions. For example, among shore birds migrating southward from the Far North, adult birds constitute the first flocks; birds-of-the-year come in later flocks. Early spring flocks of Red-winged Blackbirds *(Agelaius phoeniceus)* are likely to be made up entirely of males.

Each bird in a flock very rarely, except when roosting for the night, comes in physical contact with its fellows, but instead maintains its **individual distance**. Cliff Swallows *(Petrochelidon pyrrhonota)*, perched on a wire, keep about four inches apart (Emlen,

1952). Any individual attempting to sit closer is immediately threatened by the bird it is crowding.

Numerous signals such as peculiar vocalizations and color marks maintain the integrity of a flock. To these signals each individual responds. Should an individual become separated from the flock, it quickly shows appetitive behavior by calling and searching until it finds its flock members.

Within a closely integrated flock all the activities—feeding, preening, bathing, etc.—and movements tend to occur simultaneously. Usually one individual initiates the activities and the others soon perform them too—a procedure called **social facilitation** because the performance of one individual stimulates the behavior of its fellows to act similarly. Through social facilitation a bird will sometimes eat more than its normal capacity by seeing one or more of its associates feeding. The concerted, almost instantaneous movements in a flock are attributed to a "following" reaction, the instinctive urge among gregarious birds to stay close together. There is no better example of this trait in operation than in a compact group of shore birds suddenly rising as in a body from a mudflat, wheeling over the area in perfect unison, and settling again, still compact.

In small flocks of conspecific birds, whether caged or in the wild, a social arrangement may be established that is normally hierarchial with each individual having its own rank in a descending order of dominance. Schjelderup-Ebbe (1922) first brought this phenomenon to light in the Domestic Fowl *(Gallus gallus)*. Studying penned birds, he discovered the existence of a peck order wherein Bird A can peck all the others and is therefore dominant, Bird B can peck all but Bird A, Bird C can peck all but Birds A and B, and so on until there is one bird that can peck no other and is subordinate to all. Instead of a direct-line peck order, there can be a triangular or polygon arrangement in which Bird A pecks B, B pecks C, and C pecks A. This situation is apt to be only temporary and may revert back to the direct-line hierarchy.

The position, rank, or **peck right** of the bird in the hierarchy is determined by one or a combination of factors such as sex, size, vigor, stage of molt, color, and health. If something happens to the bird—for example, if it is injured and thus physically impaired—it may lose its peck right in the hierarchy and become the subordinate bird. Should the bird at the top of the hierarchy be removed, a scramble ensues in the flock for top position. A newcomer to a flock with an established hierarchy can work its way in social rank only by starting as the subordinate bird.

In some flocks of conspecific birds the peck order is less fixed. Instead of a bird having a peck right over another and making that bird an absolute subordinate, it may achieve a **peck dominance** over another by winning more encounters than it loses (Allee, 1936). This may be observed in flocks of chickadees or Tree Sparrows *(Spizella arborea)* at a feeding station. Among heterosexual flocks of birds there may be more than one hierarchy. In a caged flock of 12 Red Crossbills *(Loxia curvirostra)*, Tordoff (1954) noted three hierarchies, a peck order of males, a peck order of females, and a peck dominance of males over females. The lowest-ranking male was most active in dominating females, followed closely by the top-ranking male. In a wild flock of Ring-necked Pheasants *(Phasianus colchicus)*, Collias and Taber (1951) found a dominance order among both sexes, where all the cocks dominated all the hens until the onset of the breeding season when they stopped pecking them.

Heterospecific flocks of birds show a species hierarchy, generally established in accordance with the size and/or hostility of the species comprising it. Colquhoun (1942), for instance, observed that, when flocks of nuthatches and titmice were feeding, the

European Nuthatch *(Sitta europaea)* usually dominated the Great Tit *(Parus major)*, the Great Tit dominated the Blue Tit *(P. caeruleus)*, and the Blue Tit dominated the Marsh Tit *(P. palustris)*.

Whether or not a flock has leadership remains a moot question. A family flock of Canada Geese is commonly believed to be led by the adult male, and a migrating wedge of the same species made up of multiple families is thought to be led by one of the older males, but all this is supposition. Hanson (1965) found that the adult male Canada Goose will often initiate a local flight of his family but when the family is on the ground or in the water, the female often leads the flock while the male guards it at the rear. If there is a leadership in most flocks of adult birds, there is no evidence that it is taken by the top-ranking bird in the hierarchy. Quite possibly any persisting leadership of a flock of fully mature birds is non-existent since, as Edmund Selous (1931) once suggested, the flock is an individual, acting, moving, and behaving as a "collective bird."

Flocking behavior has survival value in at least two important respects: (1) Birds close together with integrated behavior are less vulnerable to predators than solitary individuals. With more eyes to watch, the flock has a greater awareness of the enemy's approach. When threatened, flock members acting in concert can take deterrent action. Either they can draw themselves together in a formidable mass as ducks resting on a pond converge to form a tight raft, or they can attack with a mobbing action as do gulls and terns in a nesting colony. (2) Birds foraging in a group increase the chances of success in finding and exploiting sources of food.

For further information on flocking, see under "Migration" in this book, page 294.

Reproductive Behavior

The interactions of birds that have to do with territory, mating, nesting, and parental care of the young come under the heading of reproductive behavior, treated in later sections of this book.

GENERAL STUDY PROCEDURES

The science of behavior, or **ethology**, is essentially the study of animal activities by direct observation, supplemented by experimental methods. While ethology is not exactly new, it does represent a recent "return-to-nature" approach in understanding the function of the live animal (see Tinbergen, 1962).

Although much is being learned rapidly about the behavior of birds, vast gaps in knowledge remain, even of the most common species. Avian ethology is still a wide open field for investigation.

In studying the behavior of any bird species, the student must describe and name its activities or behaviors. In other words, he must know thoroughly the behavioral make-up of his subject and be able to designate the different components.

Describing Behaviors. Describing a behavior is much more difficult than describing form, shape, color, and other morphological features. As Dilger (1962) stated, behavior is ephemeral—it takes place and is gone. It may last a fraction of a second, or, like sleeping, it may last for hours. Moreover, a behavior not only varies normally as does structure, but it also varies in the individual bird, depending on the subject's physiological state, motivation, and many other factors. An adequate description of a behavior must, therefore, be based on many more samples than would be required for structure.

Compounding the difficulty in describing behavior is the lack of suitable terms and the consequent recourse to words that describe human activities. This frequently results, however unintentionally, in ascribing to a bird certain human traits and motives.

In any scientific treatment of behavior always try to avoid anthropomorphisms—i.e., statements that suggest human thoughts, attitudes, and reactions—and concepts that involve teleology—the doctrine that gives a conscious purpose or objective to any activity undertaken.

Also compounding the difficulty in describing a behavior is its complexity, no matter how simple it may seem. Any behavior may often be comprised of two or more separate **acts**, each of which is a set of observable activities, seen in combination and not analyzable into separately occurring components (Russell *et al.*, 1954).

Dilger (1962) described a good example of a particular behavior and its component acts in the Peach-faced Lovebird *(Agapornis roseicollis)* which carries strips of nesting material in the feathers of its lower back and rump. When the bird puts a strip of paper in the feathers, first it *points* the strip either to the right or to the left. Next, the bird *tucks* the strip into the feathers of the lower back by turning its head over one shoulder and bringing its bill with the strip into the feathers, which are simultaneously *raised*. Then the bird *thrusts* the strip hard into the feathers with a *trembling* motion of its head, *releases* the strip, and *brings its head forward* while *compressing* the feathers. In all, Dilger noted at least eight separate acts performed in a characteristic manner and sequence. At the outset of his observations he considered thrusting and trembling a single act but, with later experiments, he found them to be separate because some birds trembled without thrusting.

Naming Behaviors. The established custom is to derive names that serve to describe or at least indicate the most prominent action. If the behavior is a display—i.e., a behavior that has evolved as a signal or releaser—the name is usually capitalized as, for example, Wings-up or Bill-up-and-open. But if the behavior is of any other sort, the name is not capitalized—e.g., bill-down or tremble-thrust. As a rule, it is inadvisable to include in the name of a behavior any word which designates its function or factors responsible for it—e.g., wings-up *threat* or *aggressive* bill-up-and-open. Such an interpretation is unwarranted, may eventually be proved wrong, and may bias future investigations.

Objectives of the Study

Having described the behaviors of a bird, the student may proceed to postulate their function, survival value, causes, and origin.

Function. What a behavior accomplishes. It may simply serve the individual performer, as in the case of a maintenance activity or in a flight from one place to another, or it may, in the case of a display, serve to elicit a response in another bird. One can usually postulate a function of a display by noting the kind of response elicited from the bird before which the display is performed.

Survival Value. Any normal behavior in an individual bird is presumed to enhance the welfare of its species, otherwise the behavior would not have evolved. The survival value or biological significance of many behaviors—e.g., flying to a food source, escaping from a predator, etc.—may be so obvious it needs no attention, or it may be obscure. In that case, numerous observations of the behavior, performed repeatedly under varying circumstances in the wild and/or in captivity, may provide the clue or at least the basis for making a reasonable guess. Why the occasional singing at night by certain species that regularly sing during the day? Why the teetering and tail-pumping action among many different species? Questions of this sort present a challenge to the avian ethologist.

Causes. A bird normally performs a behavior from both internal and external causes. Internal causes include drive, usually conditioned by experience or learning; external causes may be one or more sign stimuli arising from the environment. The student can often determine external causes by careful field observations. In the natural environment the situation in which a behavior occurs is rarely the same from one occasion to the next. Thus it is possible for the student to establish the cause by observing and comparing a long series of occasions, in which the behavior is performed, and noting the occasions when the behavior is *not* performed. In a sense, this is a "natural experiment" (N. Tinbergen, *in* Hutson, 1956). For example, what causes small birds to be alarmed by hawks on some occasions, to ignore them on others? Through observations, the student will discover that small birds generally show alarm when the hawk is in flight, but not when it is perched. The movement of the bird in flight is the principal cause of the alarm.

Another way of studying the external cause of behaviors is to compare similar behaviors apparently acquired independently by unrelated species—e.g., the habit developed by Cattle Egrets *(Bubulcus ibis)* and Brown-headed Cowbirds *(Molothrus ater)* of following large, grazing mammals in search of insects stirred up by their feet. There is always the chance that a comparative study of a habit common to very different birds will yield the clue to its common cause.

Establishing the internal causes of behavior and the role of experience in performing it requires sophisticated procedures under laboratory conditions. One method, to determine the role of experience (see Dilger, 1962, for further details), is the so-called deprivation experiment in which two groups of birds are used, one experimental, the other a control. The two groups are identical except that one group has been deprived of experience. By comparing subsequent changes in the performance of the experimental group to the "base line" of behavior of the control group, one can determine the extent to which experience plays a part in the behavior.

Origin. How behaviors have evolved is an important objective in ethological studies. After describing and observing all the behaviors in a species, the student may postulate that certain displays are derived from maintenance activities, others from displacement activities, intention movements, and so on. By comparing certain homologous behaviors in closely related bird species, the student may discern different stages in the evolution of these behaviors—information that may also shed light on the evolution of the species themselves.

THE STUDY OF BEHAVIOR

To give the beginning student in ornithology an introductory experience in observing behavior, the following two studies are outlined below. For other studies that may be undertaken, see Stokes (1968).

I. Behavior in a gull colony.

Procedure

Prior to the study, read *The Herring Gull's World* by Tinbergen (1953) with sufficient thoroughness to be aware of the various posturings and calls, the capacities of the adult bird's sense organs, the non-reproductive behaviors, and the more conspicuous behavioral features of territory establishment, pair-formation, incubation, and parent-chick relationships. Be equipped with notebook, stop watch, and (optional) a small portable tape recorder.

Erect a blind in the colony where several nests with eggs and/or small chicks are in view through one opening.

While waiting in the blind for the birds to settle down and become used to its presence, prepare a rough map of the immediate area in view, marking the positions of the nests and all adjacent features—boulders, clumps of vegetation, driftwood, and so on.

As soon as the birds return to the nearest nests, select for concentrated observation in the next three to four hours those pairs (no more than two or three) which will seem most likely to perform the greatest variety of normal activities. Attempt to recognize the adults individually by peculiarities of coloration, minor disfigurements, soiled areas on the plumage, etc. Give each adult either a numerical designation or a code number. Then follow all their behaviors, describing them either in the notebook or on the tape recorder. Make sketches to illustrate certain attitudes during the performance of the behaviors. Use the stop watch for timing the duration of behaviors.

Owing to the short time available for the study, pay attention mainly to social behavior. Look for instances of attack and appeasement; threat postures and reactions; displacement activities; courtship displays and responses; and defense behavior. If there are chicks, note their begging behavior and the adult's response.

Presentation of Results

Write a report on all observations. Consult Appendix B for directions on preparing the manuscript. Include in an introduction the location of the study, date, time involved, approximate size and general description of the colony, a map of the small area under study, and methods: how the adults were recognized and designated and how the observations were recorded. Name and give concise descriptions of all the behaviors and discuss their function. Include sketches that will enhance the descriptions.

II. Behavior in a single species.

Procedure

Select a bird species sufficiently conspicuous to enable direct observation without the use of a blind. Confine the study initially to the individual bird's behaviors—maintenance and feeding activities, displays, and defense actions. Use binocular, notebook, stop watch, and (optional) a portable tape recorder.

In studying each behavior, attempt when possible to identify its component acts. Give special attention to the following:

1. Movements of the head with relation to the horizontal plane of the body. Note whether the bill is opened or closed.

2. Extension or withdrawal of the neck.

3. Movements of the wings. When the bird is not in flight, observe whether the wings are spread, fanned, drooped, flicked, etc. When the bird is in flight, note in what manner, if any, the action of the wing is at variance with that in normal forward flight.

4. Movements of the tail. Look for: (1) Up and down actions together with the differences or similarities in the speed of the upstroke and downstroke. (2) Opening (spreading, fanning) and closing the tail together with the frequencies of the actions.

5. Movements of the legs. Note whether the legs are fully extended or flexed for a crouching position.

6. Movements of the body feathers. Note whether the feathers are sleeked (held tightly against the body) or fluffed (raised); whether only the feathers on parts of the body (e.g., crown or lower back) are raised.

In studying all movements, keep a record of the number of times the movements occur and the relative frequency of the different movements. With a stop watch, record the duration of the bouts (periods) during which the movements occur. A tape recorder will assist in recording observations quickly for later analysis. Sketch the attitudes assumed during movements as they may be helpful later in preparing the descriptions.

If time permits, the student may wish to extend the study to include the behaviors associated with reproduction—territory establishment, mating and courtship displays, song, incubation, etc. In that case he can refer to the special study directions in the succeeding sections of this book.

Presentation of Results

Write a report on all observations. Consult Appendix B for directions on preparing the manuscript. Describe in detail the different behaviors, grouping them according to function. Include dates and times of day observed and any pertinent ecological information; also include sketches when they will assist in describing the behaviors.

REFERENCES

Allee, W. C.
1936 Analytical Studies of Group Behavior in Birds. Wilson Bull., 48: 145-151.

Andrew, R. J.
1956 Intention Movements of Flight in Certain Passerines, and Their Use in Systematics. Behaviour, 10: 179-204.
1961 The Displays Given by Passerines in Courtship and Reproductive Fighting: A Review. Ibis, 103a: 315-348, 549-579.

Armstrong, E. A.
1947 Bird Display and Behaviour: An Introduction to the Study of Bird Psychology. Revised edition. Lindsay Drummond, London. (Contains an extensive coverage of the literature on reproductive behavior.)

Berger, A. J.
1961 Bird Study. John Wiley and Sons, New York. (Chapter 5 deals with behavior.)

Bowman, R. I.
1961 Morphological Differentiation and Adaptation in the Galápagos Finches. Univ. California Publ. in Zool., 58: 1-302.

Brooks, M.
1940 The Breeding Warblers of the Central Allegheny Mountain Region. Wilson Bull., 52: 249-266.

Brown, J. L.
1963 Ecogeographic Variation and Introgression in an Avian Visual Signal: The Crest of the Steller's Jay, *Cyanocitta stelleri*. Evolution, 17: 23-39.
1964 The Integration of Agonistic Behavior in the Steller's Jay *Cyanocitta stelleri* (Gmelin). Univ. California Publ. in Zool., 60: 223-328.

Chisholm, A. H.
1954 The Use by Birds of "Tools" or "Instruments." Ibis, 96: 380-383.

Collias, N. E., and R. D. Taber
1951 A Field Study of Some Groupings and Dominance Relations in Ring-necked Pheasants. Condor, 53: 265-275.

Colquhoun, M. K.
1942 Notes on the Social Behaviour of Blue Tits. Brit. Birds, 35: 234-240.

Craig, W.
1918 Appetites and Aversions as Constituents of Instincts. Biol. Bull, 34: 91-107.

Crook, J. H.
1964 The Evolution of Social Organisation and Visual Communication in the Weaver Birds (Ploceinae). Behaviour Supplement 10. E. J. Brill, Leiden.

Daanje, A.
1950 On Locomotory Movements in Birds and the Intention Movements Derived from Them. Behaviour, 3: 48-98.

Dilger, W. C.
1962 Methods and Objectives of Ethology. Living Bird, 1: 83-92.

Dilger, W. C., and J. C. Wallen
1966 The Pecking Responses of Peafowl Chicks. Living Bird, 5: 115-125.

Elder, W. H., and N. L. Elder
1949 Role of the Family in the Formation of Goose Flocks. Wilson Bull., 61: 133-140.

Emlen, J. T., Jr.
1952 Flocking Behavior in Birds. Auk, 69: 160-170.
1954 Territory, Nest Building, and Pair Formation in the Cliff Swallow. Auk, 71: 16-35.
1955 The Study of Behavior in Birds. In

Recent Studies in Avian Biology. Edited by A. Wolfson. University of Illinois Press, Urbana.

Etkin, W., Editor
1964 Social Behavior and Organization among Vertebrates. University of Chicago Press, Chicago. (Deals extensively with birds.)

Farner, D. S.
1960 Metabolic Adaptations in Migration. Proc. XIIth Internatl. Ornith. Congr., pp. 197-208.

Ficken, M. S., and W. C. Dilger
1960 Comments on Redirection with Examples of Avian Copulations with Substitute Objects. Animal Behaviour, 8: 219-222.

Ficken, M. S., and R. W. Ficken
1962 The Comparative Ethology of the Wood Warblers: *A Review.* Living Bird, 1: 103-122.

Ficken, R. W., and M. S. Ficken
1966 A Review of Some Aspects of Avian Field Ethology. Auk, 83: 637-661.

Frazier, A., and V. Nolan, Jr.
1959 Communal Roosting by the Eastern Bluebird in Winter. Bird-Banding, 30: 219-226.

Hailman, J. P.
1967 The Ontogeny of an Instinct: The Pecking Responses in Chicks of the Laughing Gull (*Larus atricilla* L.) and Related Species. Behaviour Supplement 15, E. J. Brill, Leiden. (The division of behavior patterns into either "instinctive" or "learned" is rejected; such concepts as "fixed action pattern" and "innate releasing mechanism" are criticized.)

Hanson, H. C.
1965 The Giant Canada Goose. Southern Illinois University Press, Carbondale.

Hartley, P. H. T.
1950 An Experimental Analysis of Interspecific Recognition. Symp. Soc. Exp. Biol., 4: 313-336.

Hauser, D. C.
1957 Some Observations on Sun-bathing in Birds. Wilson Bull., 69: 78-90.

Hess, E. H.
1959 Imprinting. Science, 130: 133-141.

Hinde, R. A.
1952 The Behaviour of the Great Tit (*Parus major*) and Some Other Related Species. Behaviour Supplement 5. E. J. Brill, Leiden. (Highly recommended as a model for the study of a single species.)

1961 Behavior. In *Biology and Comparative Physiology of Birds.* Volume 2. Edited by A. J. Marshall. Academic Press, New York. (A review concentrating on some of the more recent work in the field.)

1966 Animal Behaviour: A Synthesis of Ethology and Comparative Psychology. McGraw-Hill Book Company, New York. (For the advanced student of ethology.)

Hinde, R. A., and J. Fisher
1952 Further Observations on the Opening of Milk Bottle Tops by Birds. Brit. Birds, 44: 393-396.

Hochbaum, H. A.
1955 Travels and Traditions of Waterfowl. University of Minnesota Press, Minneapolis.

Hutson, H. P. W., Editor
1956 The Ornithologists' Guide. British Ornithologists' Union, London. (Contains many suggestions for beginners for the study of behavior in the field.)

Johnsgard, P. A.
1965 Handbook of Waterfowl Behavior. Cornell University Press, Ithaca, New York. (The first book ever to summarize comprehensively the behavior patterns in a major avian family.)

1967 Animal Behavior. Wm. C. Brown Company, Dubuque, Iowa. (A review of the subject from the ethological approach with the concept of evolution as the dominating issue.)

Kelso, L., and M. M. Nice
1963 A Russian Contribution to Anting and Feather Mites. Wilson Bull., 75: 23-26.

Klimstra, W. D., and V. C. Ziccardi
1963 Night-roosting Habit of Bobwhites. Jour. Wildlife Management, 27: 202-214.

Klopfer, P.
1963 Behavioral Aspects of Habitat Selection: The Role of Early Experience. Wilson Bull., 75: 15-22.

1965 Behavioral Aspects of Habitat Selection: A Preliminary Report on Stereotypy in Foliage Preferences of Birds. Wilson Bull., 77: 376-381.

Klopfer, P. H., and J. P. Hailman
1967 An Introduction to Animal Behavior: Ethology's First Century. Prentice-Hall, Englewood Cliffs, New Jersey. (A survey with emphasis on aspects of greatest current interest.)

Klopfer, P. H., and R. H. MacArthur

1961   On the Causes of Tropical Species Diversity and Niche Overlap. Amer. Midland Nat., 95: 223-236.

Köhler, O.
1950   The Ability of Birds to "Count." Bull. Animal Behaviour, No. 9: 41-45.

Kruijt, J. P.
1964   Ontogeny of Social Behaviour in Burmese Red Junglefowl *(Gallus gallus spadiceus)* Bonnaterre. Behaviour Supplement 12. E. J. Brill, Leiden.

Kruuk, H.
1964   Predators and Anti-predator Behaviour of the Black-headed Gull *(Larus ridibundus* L.). Behaviour Supplement 11. E. J. Brill, Leiden.

Lack, D.
1945   The Galápagos Finches (Geospizinae): A Study in Variation. Occas. Papers California Acad. Sci., 21.
1947   Darwin's Finches. University Press, Cambridge.
1956   Swifts in a Tower. Methuen and Company, London.

Lorenz, K.
1935   Der Kumpan in der Umwelt des Vögels. Jour. f. Ornith., 83: 137-213; 289-413.
1937   The Companion in the Bird's World. Auk, 54: 245-273. (Translation of the author's 1935 paper.)
1950   The Comparative Method in Studying Innate Behavior Patterns. Symp. Soc. Exp. Biol., 4: 221-268.

Lorenz, K., and N. Tinbergen
1938   Taxis und Instinkthandlung in der Eirollbewegung der Grangans: I. Zeits. Tierpsychol., 2: 1-29.

Marler, P.
1956   Behaviour of the Chaffinch *Fringilla coelebs.* Behaviour Supplement 5. E. J. Brill, Leiden. (Highly recommended as a model for the study of an individual species.)

Marler, P., and W. J. Hamilton III
1966   Mechanisms of Animal Behavior. John Wiley and Sons, New York. (Concerned to a large extent with the physiological processes of behavior; recommended to the advanced student.)

Millikan, G. C., and R. I. Bowman
1967   Observations on Galápagos Tool-using Finches in Captivity. Living Bird, 6: 23-41.

Moynihan, M.
1955a   Remarks on the Original Sources of Displays. Auk, 72: 240-246.

1955b   Types of Hostile Display. Auk, 72: 247-259.

Nice, M. M.
1953   Some Experiences in Imprinting Ducklings. Condor, 55: 33-37.
1962   Development of Behavior in Precocial Birds. Trans. Linnaean Soc. New York, 8: i-xii; 1-211.

Nice, M. M., and J. ter Pelkwyk
1941   Enemy Recognition by the Song Sparrow. Auk, 58: 195-214.

Noble, G. K.
1936   Courtship and Sexual Selection of the Flicker *(Colaptes auratus luteus).* Auk, 53: 269-282.

Orians, G. H., and G. M. Christman
1968   A Comparative Study of the Behavior of the Red-winged, Tricolored, and Yellow-headed Blackbirds. Univ. California Publ. in Zool., 84: i-vi; 1-85.

Pastore, N.
1954   Discrimination Learning in the Canary. Jour. Comp. Physiol. Psychol., 47: 288-289; 389-390.

Penney, R. L.
1968   Territorial and Social Behavior in the Adelie Penguin. American Geophysical Union. Antarctic Res. Ser., 12: 83-131.

Pettingill, O. S., Jr.
1964   Penguins Ashore at the Falkland Islands. Living Bird, 3: 45-64.

Root, O. M.
1942   Wood Thrush Nesting in the Coniferous Bogs of Canadian Zone. Auk, 59: 113-114.

Rowell, C. H. F.
1961   Displacement Grooming in the Chaffinch. Animal Behaviour, 9: 38-63.

Russell, E. S.
1943   Perceptual and Sensory Signs in Instinctive Behaviour. Proc. Linnaean Soc. London, 154: 195-216.

Russell, W. M. S., A. P. Means, and J. S. Hayes
1954   A Basis for the Quantitative Study of the Structure of Behaviour. Behaviour, 6: 153-205.

Schjelderup-Ebbe, T.
1922   Beiträge zur Sozialpsychologie des Haushühns. Zeits. Psychol., 88: 225-252.

Selous, E.
1931   Thought-transference (or What?) in Birds. Constable and Company, London.

Simmons, K. E. L.
1957a   A Review of Anting Behaviour of Passerine Birds. Brit. Birds, 50: 401-424.

1957b The Taxonomic Significance of the Head-scratching Methods of Birds. Ibis, 99: 178-181.

1964 Feather Maintenance. In *A New Dictionary of Birds*. Edited by A. L. Thomson. McGraw-Hill Book Company, New York.

Stokes, A. W., Editor

1968 Animal Behavior in Laboratory and Field. W. H. Freeman and Company, San Francisco.

Thomson, A. L., Editor

1964 A New Dictionary of Birds. McGraw-Hill Book Company, New York. (Contains separate articles, each by an authority, on Learning, Imprinting, Sign Stimulus, Releaser, Fixed Action Pattern, Redirection, Displacement Activity, Ritualization, Development of Behavior, Counting, Play, and Tameness.)

Thorpe, W. H.

1958 The Learning of Song Patterns by Birds, with Especial Reference to the Song of the Chaffinch *Fringilla coelebs*. Ibis, 100: 535-570.

1961 Bird-song: The Biology of Vocal Communication and Expression in Birds. University Press, Cambridge.

1963 Learning and Instinct in Animals. Second edition. Methuen and Company Limited, London. (Contains a thorough review of the literature on the subject.)

Tinbergen, N.

1948 Social Releasers and the Experimental Method Required for Their Study. Wilson Bull., 60: 6-51.

1951 The Study of Instinct. Oxford University Press, London. (A thorough review of the subject.)

1952 Derived Activities: Their Causation, Biological Significance, Origin and Emancipation during Evolution. Quart. Rev. Biol., 27: 1-32.

1953 The Herring Gull's World: A Study of the Social Behaviour of Birds. Collins, London.

1962 Behavioral Research at the Cornell Laboratory of Ornithology. Living Bird, 1: 79-82.

Tinbergen, N., and A. C. Perdeck

1950 On the Stimulus Situation Releasing the Begging Response in the Newly-hatched Herring Gull Chick (*Larus a. argentatus* Pontopp.). Behaviour, 3: 1-38.

Tordoff, H. B.

1954 Social Organization and Behavior in a Flock of Captive, Nonbreeding Red Crossbills. Condor, 56: 346-358.

van Iersel, J. J. A., and A. Bol

1958 Preening of Two Tern Species: A Study of Displacement Activities. Behaviour, 13: 1-88.

van Lawick-Goodall, J., and H. van Lawick-Goodall

1966 Use of Tools by the Egyptian Vulture *Neophron percnopterus*. Nature, 212: 1468-1469.

van Tets, G. F.

1965 A Comparative Study of Some Social Communication Patterns in the Pelecaniformes. Amer. Ornith. Union, Ornith. Monogr. No. 2.

Welty, J. C.

1962 The Life of Birds. W. B. Saunders Company, Philadelphia. (Chapter 9 deals with behavior.)

Whitaker, L. M.

1957 A Résumé of Anting, with Particular Reference to a Captive Orchard Oriole. Wilson Bull., 69: 195-262.

Wood, C. O.

1936 Some of the Commoner Birds of Ceylon. Rept. Smithsonian Inst. for 1936: 297-302.

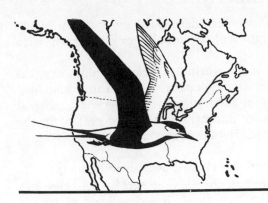

MIGRATION

Any movement between two areas is called **migration.** As a rule, it is a response of an animal population to changes in environmental conditions.

Birds are more uniformly migratory than any other group of animals. Nearly all orders of birds include species which perform migrations; in other vertebrates and in the lower groups of animals the migratory habit occurs in species scattered through a smaller proportion of orders.

Among birds there are two common kinds of migration, **daily** and **seasonal.** A daily migration is a movement to and from a familiar place such as a roosting area. A seasonal migration, on the other hand, involves a passage at one season from a place of hatching and a return at another season to the same general area. This section of the book is concerned entirely with seasonal migration.

For centuries the phenomenon of migration has been primarily associated with birds; indeed, judging by the references to the subject in the earliest literature, migration was first observed in birds. Since civilization developed in a temperate region of the world where migratory movements are especially pronounced, it is hardly surprising that bird migration has long received attention. Man could not help noticing the flocks of birds in the spring and fall, the seasonal disappearance of some species and reappearance of others; nor could he help being interested in what he saw and eager to investigate what he could not understand.

Today there is an enormous literature on bird migration, based on extensive studies in Europe and North America. And yet the causes and processes of migration are not fully known. The root of the problem is that bird migration, whether it occurs by day or night, is "an unseen movement" (Lowery, 1951). One must investigate its mechanisms indirectly through laboratory research on physiological and environmental influences; the analysis of migrant birds mist-netted or otherwise captured, of birds killed during migration by television towers and other man-made hazards, and of returns from banded birds; mathematical calculations based on kinds and numbers of birds observed through a telescope as they fly across the face of the moon; correlation of meteorological data with known migratory activity; deductions derived from direct field observations, radar surveillance, and tracking by radiotelemetry; and experiments on homing and direction-finding. The following pages present briefly some of the important facts and concepts of migration and suggest a few studies that students may undertake in conjunction with class work.

CAUSES OF MIGRATION

All modern birds—even those incapable of flight—are descended from volant stock. Early in their history birds had the power of flight and presumably could migrate with facility.

Today many species in the temperate regions of the world are strongly migratory, exhibiting mass movements away from both poles as day-length, temperature, and food supply diminish, then reversing the movements at the season when there is a general augmentation of these environmental factors. Such migrations are more evident in the temperate region of the Northern, or Continental, Hemisphere since more species are involved.

Because the north-south migrations of the Northern Hemisphere include many well-known and conspicuous birds which seem to move in the same general way, there has been a tendency to conclude that migration, like bird flight, has developed along the same line in all species, with deviations for adaptive purposes. A study of the migration phenomenon in all groups of birds soon shows the fallacy of this reasoning. One finds that:

1. Some species migrate in directions other than north-south.
2. Some species migrate irrespective of day-length, as in tropical lands.
3. Some species migrate when the temperature is mild and the food supply ample, and others when the opposite conditions are true.
4. Some species migrate as a result of seasonal alternation in rainfall and drought.
5. In certain species, some populations migrate while others do not.
6. In some populations, some individuals migrate while others do not.
7. Some individuals migrate in some years but not in others.

The only conclusion one can safely reach after considering the above peculiarities is that there is no one line along which migration in all birds developed and that there must be different causes of migration in different groups of birds.

A number of authorities have sought causes of migration in historical factors. Three of the several causes suggested bear mention. (1) Bird migration, at least in the Northern Hemisphere, was initiated by the effects of the Ice Age (Pleistocene). Prior to the coming of the great glaciers to the polar regions, birds lived the year round in the Northern Hemisphere where they originated. Though forced to retreat with the advance of the glaciers, they nevertheless continued to return to nest in the summer because of an innate attachment to their homeland. Objections to this suggestion are several. Many of the birds which are today typical migrants were in existence before the Pleistocene and there is no reason to suppose that they were not already migrating before the glaciers advanced. The retreat from the glaciers does not account for migrations in directions other than north-south, nor does it account for migrations of birds in tropical regions that were never glaciated. (2) Birds migrate in the fall because of the oncoming shortage of food during the winter in their breeding areas. Not that they "know" of the impending lack of food; it is the fact of the food shortage that has caused fall migration to evolve among species which, owing to the shortage, would fail to survive if they stayed for the winter. Birds return in the spring because the northern areas provide more favorable and less crowded conditions for nesting and rearing families. (3) Intraspecific and interspecific competition for food, territory, nest sites, and so on may have been important factors initiating migration. When, as suggested by Cox (1968), individuals among normally sedentary species could benefit by moving into adjacent areas where the season was favorable and competition reduced, they did so if the hazards of moving did not exceed the gains in their survival and reproduction.

Unquestionably, certain historical factors and various factors prevailing today such as day-length, air temperature, and food supply have influenced migration indirectly by affecting the environment, but no one factor can account for migration or has played the principal role in establishing migration. In view of the current diversity of migratory movements, it is evident that migration has evolved independently in different bird populations through selective pressure. When any resident population experienced an unfavorable situation during a seasonal period in the area it occupied and could gain an advantage by shifting to another area for that period, it gradually developed a migratory pattern, probably with a genetic basis.

PREPARATION AND STIMULUS FOR MIGRATION

Migration is synchronized with the annual seasonal changes, but it will not take place until the bird is internally prepared and outwardly responsive to a stimulus.

Before migrating, a bird must be ready to meet the energy requirements for prolonged flight. It accomplishes this by eating amounts of food in excess of its daily needs and thereby storing energy in the form of subcutaneous fat. At the same time, the bird must become predisposed to migrate by developing a condition commonly called **migratory restlessness.** In the spring, the physiological process leading toward this state is influenced by the pituitary gland whose activity at this time is stimulated by the total effects of day-length; in the fall, the bird reaches a similar metabolic state during a period when the pituitary is "refractory" (see page 123). When the bird has attained the necessary physiological and behavioral conditions, an outside stimulus is required to trigger migratory behavior. The stimulus is probably some meteorological factor, or combination of factors, such as a change in the temperature of the air, the direction and velocity of the wind, or the onset or passage of a cold front.

Normally an adult migratory bird goes through the special metabolic cycle twice a year, once before the journey to the nesting area and once before the journey away from it. If the bird does not reach the necessary physiological and behavioral condition, it cannot migrate. Furthermore, if the external stimulus for migration is absent, the bird will tend not to migrate even though physiologically and behaviorally capable.

For a summary of knowledge concerning the nature and mechanisms of periodic preparation and stimulus for migration, see Farner (1955, 1960).

DIURNAL AND NOCTURNAL MIGRATIONS

Migrations proceed by day or night. Many birds—e.g., loons, geese, ducks, gulls, terns, and shore birds—travel by day or night, apparently indifferent to daylight or darkness. But this is not the case with many other birds. Herons, hawks, eagles, falcons, crows, hummingbirds, swifts, and swallows migrate only during the day, while nearly all passerine birds (excepting crows, swallows, and a few others) migrate primarily during the night from after sunset until dawn.

At least two explanations have been advanced for the development of nocturnal migration. (1) Movement by night affords birds, which normally live in thick vegetational cover and rarely take long flights away from it, the protection of darkness against their diurnal predators. (2) Movement by night affords birds the opportunity of using all the daylight hours for feeding, thereby enabling them to build up sufficient energy resources for sustained long-distance flights.

EFFECTS OF WEATHER ON MIGRATION

Normal weather alternates between fair and inclement conditions. The movements of the vast majority of migrants show a close correlation with these conditions which are largely governed by barometric pressure patterns, temperature, and wind directions.

To understand how migration takes place with relation to weather, the student must first be acquainted with a few basic facts about weather elements and their sequence over a ground area.

Perpetually sweeping across the continent in an average easterly direction are air masses that vary in velocity, depending on the season and numerous other factors, from 500 to 700 miles a day. In these masses, which may be visualized as roughly circular, are centers of low barometric pressure ("lows"), with generally warmer, more moist air, and centers of high pressure ("highs") with cooler, drier air. Within the lows the air circulates counterclockwise; within the highs, air circulates clockwise. Where the lows and highs adjoin there are boundaries called "fronts." The front of an oncoming high is the cold front; of an oncoming low, the warm front. The area of generally low pressure between the cold and warm fronts is called the warm sector. Any weather map appearing in daily newspapers will show the lines of equal barometric pressure (isobars) as roughly concentric circles around lows and highs, and cold and warm fronts as heavy lines with marks indicating the direction they face. Areas where there has been precipitation are shaded.

The table on page 272, "Relation of Weather Elements to Migration," demonstrates in a general way the sequence of spring and fall weather elements when a warm and then a cold front pass over a given area in conterminous United States. The accompanying sketch, "Pattern of Fronts and Wind Directions," shows how the sequence indicated in the table might appear on a weather map. The movements of the elements in both the table and the sketch is to the right (east). Thus the set of weather elements characterizing "Ahead of Warm Front" will be followed by the elements of "Warm Front" and so on. The rapidity with which one set of elements—e.g., the elements of a warm front—arrive, prevail, and finally disappear in an area is dependent on the width of the front and the speed with which the front travels. The width and speed of the front are in turn dependent on the season of the year, the temperature discrepancy between fronts, the wind direction, and continent-wide weather conditions at the time.

Migration in the spring usually takes place with warm weather. Studies of spring migratory movements in eastern United States and Canada (see Bagg and others, 1950) show that movements begin at the onset of warm fronts, when barometric pressure is dropping and warm moist air from the Gulf of Mexico and Caribbean Sea is flowing in from a southerly direction. As each low (with mild temperature and southerly winds) passes, movements proceed in full force. On the approach of cold frints, the movements usually slow down—though they may occasionally be heavy, as when a cold front advances against a western flank of warm air. When cold fronts arrive, movements stop. Not until the highs have passed will movements begin again.

Migration in the fall usually takes place with cold weather. In the Chicago area (see Bennett, 1952), migratory movements in September and October start immediately after the passage of cold fronts when there is a flow of continental polar air from a northerly direction. Movements then proceed in full force until the first part of each low passes; thereafter the movements decrease somewhat in intensity and may even cease altogether as the next cold fronts approach.

On page 272, the table, "Relation of Weather Elements to Migration," relates spring and fall migration movements to the sequence of weather elements.

Many species are reluctant to initiate migration under an overcast (Bellrose and Graber, 1963). Air temperature is probably the principal factor in starting migration; wind direction or air flow is a decidedly critical factor in regulating migratory movement. Most movements take place when the wind is favorable—i.e., blowing in the direction of flight—and after migrants have been held back for long intervals by such weather conditions as fog, rain, and headwinds. In the spring the weather that ordinarily accompanies cold fronts is especially obstructive to movements, forcing migrants to take to the ground without delay.

The arrival of any migration movement in an area is more often controlled by the weather at the point of departure than during its course. Thus at the height of the migration season, when the weather appears suitable in his area, the student will not always observe migration movement because inclement weather at the point of departure or some intervening point has arrested the flight.

For further details and comments on the relation of weather to migration, see Lack (1960b, 1963).

The study of weather maps in conjunction with observations on movements is very useful. By means of weather maps one can frequently anticipate a migration movement in a given area. For instance, if during the height of spring migration the map indicates a cold front moving in from the northwest, one may expect its arrival to stop and hold numerous transients in the area. Weather maps also assist in explaining the failure of a migration movement to appear. If there is no pronounced migration movement in an area for a long period, even though the season and weather are favorable, maps may reveal that a pressure area along the migration route has become quasi-stationary, either "damming up" migrants (in case the pressure area is unfavorable to movement), or (in case it is favorable) allowing migrants to move along from day to day without massing in conspicuous waves.

Exceptional weather conditions with unusually high winds often deflect migrants from their usual routes and at the same time carry many non-migrating birds from their regular ranges. The hurricanes or heavy northeast storms that occasionally move north along the eastern Atlantic seaboard, by the counterclockwise direction of their winds, force many south-bound migrants and sea birds far inland. As a result, ducks, geese, and gulls are reported in great abundance in places where they do not ordinarily occur, and such sea birds as petrels show up far inland. Two very severe northeast storms in December, 1927, and January, 1966, bore spectacular numbers of Eurasian Lapwings *(Vanellus vanellus)* over the North Atlantic from their migration route in western Europe to the vicinity of Newfoundland. The student will find instructive the paper by Bagg (1967) showing in detail, with a series of weather maps, how the great storms brought the Lapwings to North America.

REGULARITY OF MIGRATORY TRAVEL

Despite the effects of weather on migration, migratory travel over a period of years is regular on the average. This is apparent to a student observing bird populations from year to year in any given area of North America north of Mexico.

If one keeps records for several years of the days when common summer-resident species first arrive in the spring, and computes average dates of arrival of each species, he can eventually predict within a few days when these species will appear. Similarly, by keeping records of departure of transient spring species—i.e., dates when species going through the area are last seen—he will know approximately when these species depart each year.

RELATION OF WEATHER ELEMENTS TO MIGRATION

→ Direction of air masses →
→ Sequence of weather elements in a given area →

	HIGH			LOW		
	BEHIND COLD FRONT	COLD FRONT	AHEAD OF COLD FRONT	BEHIND WARM FRONT	WARM FRONT	AHEAD OF WARM FRONT
Barometric Pressure	Rapidly rising	Low	Dropping steadily	Rising slightly	Low	Dropping rapidly
Wind Direction	From northwest and west; strong and steady	From west; strong and gusty	Easterly; slight	From southwest	From southeast and south	Easterly; increasing
Temperature	Dropping	Unchanged	Rising	Rising	Rising slowly	Rising slightly
Weather Aspect	Clearing	Low clouds and heavy rain	Cloudy; occasional fog or rain	Showers, but generally clearing	Cloudy; drizzle or showers	Cloudy; drizzle or showers

Sequence of migration in the same area →

Spring Migration	MIGRATION STOPPED	MIGRATION STOPPED	MIGRATION SLOWED BUT MAY BE HEAVY	MIGRATION IN FULL FORCE	MIGRATION BEGINS	NO MIGRATION
Fall Migration	MIGRATION BEGINS	MIGRATION STOPPED	MIGRATION SLOWED	MIGRATION SLOWED	MIGRATION IN FULL FORCE	MIGRATION IN FULL FORCE

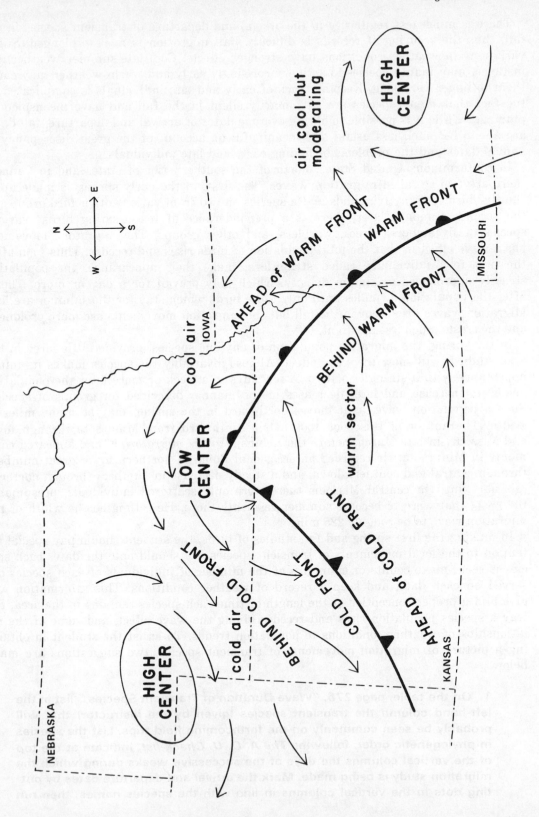

PATTERN OF FRONTS AND WIND DIRECTIONS

HIGH CENTER

air cool but moderating

WARM FRONT

AHEAD of WARM FRONT

WARM FRONT

BEHIND WARM FRONT

cool air

IOWA

MISSOURI

warm sector

LOW CENTER

COLD FRONT

AHEAD of COLD FRONT

BEHIND COLD FRONT

cold air

cold air

HIGH CENTER

NEBRASKA

KANSAS

N
E
W
S

There is much less regularity in the arrival and departure of transient species in the fall, thus the securing of records is difficult. Fall migration is more prolonged. Slight variations in weather conditions have stronger effects. Cool late-summer weather, for instance, may induce species to arrive surprisingly early and warm weather may cause them to linger very long. Keeping track of early and late individuals is complicated by the fact that many species are customarily silent in the fall and have inconspicuous plumage. While it is possible to obtain average dates of arrival and departure, fall dates are apt to be much less useful and meaningful on account of the great discrepancy in annual dates and the problems of finding early and late individuals.

In conterminous United States, north of the southern tier of states and in Canada, there are two so-called **migration waves**. The first, in the early spring, is made up of "hardy" birds—many fringillids and a species or two from various other bird groups; in the second, a month later, there is a preponderance of insect-eating birds—various species of flycatchers, vireos, warblers, and other groups. The migratory movement has a wave effect in that the total population of birds rises and recedes. Thus soon after the first few individuals, called **stragglers**, make their appearance, the population steadily approaches maximum density, which may prevail for a day or more. Thereafter the population dwindles until only those birds which stay for the summer are left. Migration waves also occur in the fall but the migration movements are more prolonged and the crests much less apparent.

In the spring, the migration population of any one species, provided it is large in the area studied, will show the wave effect. Almost invariably the species makes its initial appearance with a straggler or two. A few days to a week or more later the population begins to increase and later decreases in the manner described for groups of species. Such a population wave, as it moves northward in the spring, may be many miles in width. The author of this book took a trip southward from Minnesota through Iowa and Missouri in late March, before the Robins *(Turdus migratorius)* had appeared commonly in Minnesota. He recorded an occasional Robin in northern Iowa; great numbers through central and southern Iowa; and a steady decrease in numbers through northern Missouri until in central Missouri there were only scattered individuals, presumably the birds that were to become summer residents there. He estimated the width of the migration wave to be roughly 225 miles.

In making the first spring and fall studies of birds, the student should pay special attention to the local movements of transient species. He should note the dates each species is seen, make counts or estimates of the number of individuals of each species observed on each date, and keep a record of weather conditions. This information will give him a proper conception of the length of time each species remains in the area, the way a species population rises and recedes giving the wave effect, and some of the relationships of weather conditions to population trends. To assist the student in obtaining a picture of migration movements of transient species, two suggestions are made below.

1. On the table, page 276, "Wave Duration of Transient Species," list in the left-hand column the transient species (given by the instructor) that will probably be seen commonly on the forthcoming field trips. List the species in phylogenetic order, following *The A. O. U. Check-list.* Indicate at the top of the vertical columns the days of the successive weeks during which the migration study is being made. Mark the arrival and departure dates by putting dots in the vertical columns in line with the species names, then run

continuous lines between each set of dots. The lines will represent the duration of each species wave. Each line may be thickened or narrowed during its course to represent increase or decrease in abundance of the species during the wave.

2. On the graph, page 277, "Population Changes of Several Transient Species," plot the population curves of a number of transient species (to be designated by the instructor) that will be common enough to provide adequate figures. Three sets of population figures are given in the left-hand column. Use the set that will best accommodate the figures obtained. Indicate the dates of counts or estimates across the top of the graph. For the population counts of each species, put dots in the appropriate spaces, then connect them with a line, using a crayon of different color for each species. At the bottom of the graph, fill in the key, giving the name of each species and the color of the line that represents it. Though the various lines will be jerky, owing to daily fluctuations in the counts, the high points in each line will show a steady rise and subsequent fall as the waves of the species represented pass through the study area.

IRREGULAR MIGRATIONS

The many cases of movements among bird populations, which either do not conform to the usual seasonal migration pattern or are not sufficiently well understood to seem a part of the pattern, are loosely classified as **irregular migrations**.

In certain permanent-resident species there may be mass movements of particular populations with some periodicity. The Blue Jay *(Cyanocitta cristata)*, especially in the northern part of its range, shows some migratory movement. Each fall, small numbers usually move south past Hawk Mountain in Pennsylvania (Broun, 1941); in 1939 there was an exceptionally heavy migration (over 7,000 individuals counted), which may have been due to a shortage of beechnuts and acorns in northern forests. In northwestern Oklahoma, the Bobwhite *(Colinus virginianus)* shows a distinct seasonal population shift by moving from summer habitats in the uplands to pass the winter in bottomlands and dunes where there is better cover (Duck, 1943). Their movements, involving distances up to 26 miles, are apparently heavier during severe winters.

Populations of species which are permanent residents may on occasion show a mass movement—**invasion** or **irruption** without periodicity. Now and then there is a winter when considerable numbers of Snowy Owls *(Nyctea scandiaca)* leave the Arctic and invade southern Canada and northern conterminous United States. The cause of this behavior is sometimes attributed to a sharp reduction in the lemming on which the Snowy Owl preys to a large extent. Crossbills *(Loxia* spp.) and Evening Grosbeaks *(Hesperiphona vespertina)*, normally residents of the Coniferous Forest Biotic Community, occasionally appear in flocks as far south as Florida, presumably due to a failure of their food supply. Sometimes invasions involve mainly young birds. At Cedar Grove, Wisconsin, on the west shore of Lake Michigan, Mueller and Berger (1967) reported a southern invasion of Goshawks *(Accipiter gentilis)* in the years 1961 through 1963. Most of the birds that they trapped and examined were apparently hatched in 1961 shortly after a decrease in snowshoe hares and grouse, their principal prey. Mueller and Berger hypothesized that these young birds had come south after being displaced by adults already well established in a range that could not support large wintering populations of the species.

WAVE DURATION OF TRANSIENT SPECIES

Name of Species	S M T W T F S	S M T W T F S	S M T W T F S	S M T W T F S	S M T W T F S	S M T W T F S	S M T W T F S	S M T W T F S	S M T W T F S

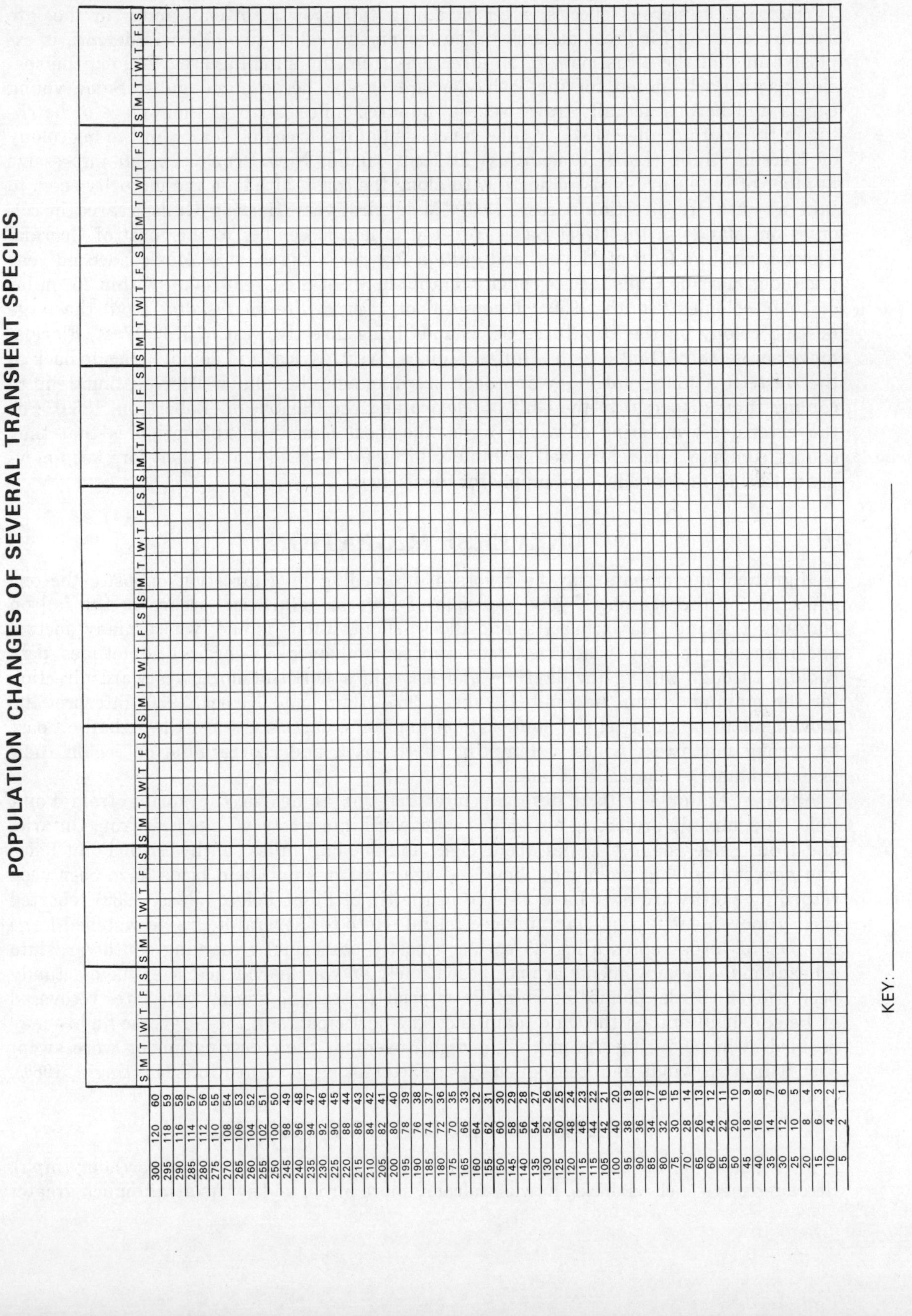
POPULATION CHANGES OF SEVERAL TRANSIENT SPECIES
S M T W T F S
S M T W T F S
S M T W T F S
S M T W T F S
S M T W T F S
S M T W T F S
S M T W T F S
300 295 290 285 280 275 270 265 260 255 250 245 240 235 230 225 220 215 210 205 200 195 190 185 180 175 170 165 160 155 150 145 140 135 130 125 120 115 110 105 100 95 90 85 80 75 70 65 60 55 50 45 40 35 30 25 20 15 10 5
120 118 116 114 112 110 108 106 104 102 100 98 96 94 92 90 88 86 84 82 80 78 76 74 72 70 66 64 62 60 58 56 54 52 50 48 46 44 42 40 38 36 34 32 30 28 26 24 22 20 18 16 14 12 10 8 6 4 2
60 59 58 57 56 55 54 53 52 51 50 49 48 47 46 45 44 43 42 41 40 39 38 37 36 35 33 32 31 30 29 28 27 26 25 24 23 22 21 20 19 18 17 16 15 14 13 12 11 10 9 8 7 6 5 4 3 2 1
KEY:

Young of numerous species, after attaining full growth, often wander in the late summer and fall for great distances. The movement, called **juvenile wandering,** is explosive in that the birds move in all directions from the hatching area. Among the species particularly noted for this behavior are egrets, herons, and gulls. Some young egrets and herons actually travel several hundred miles north of their place of hatching in the south (Coffey, 1943). At the conclusion of the breeding season in the big colony of Herring Gulls (*Larus argentatus*) at Kent Island, New Brunswick, an impressive number of immature birds go northward along the coast, although the majority seem to take a southerly direction (Gross, 1940). Many first-year Herring Gulls, reared in colonies on islands in the Great Lakes, proceed in late December to the coast of Georgia, Florida, and the Gulf of Mexico, and quite a few reach the coast of Mexico; second-year and older Herring Gulls tend to remain on the shores of the Great Lakes within 300 miles of their colonies (Smith, 1959). Juvenile Sooty Terns (*Sterna fuscata*) from the large nesting colony on the Dry Tortugas, islands lying directly west of Key West, Florida, move across the Atlantic to the Gulf of Guinea, West Africa, and do not straggle back to the western Atlantic until they approach breeding age; the adults, after nesting, tend to confine their dispersal to the Gulf of Mexico and the Caribbean (Robertson, 1969). For the extensive wandering of young birds, the most plausible explanation is that they cannot compete sufficiently well with older birds for food and must therefore keep moving until they find an adequate supply for themselves.

REVERSE MIGRATION

Migratory movements may be reversed, proceeding in a direction opposite the one expected for the season. A good example of **reverse migration** occurs in the fall on Nantucket Island, Massachusetts, and Block Island, Rhode Island, where many nocturnal migrants (mostly passerines representing well over 50 species) sometimes pass rapidly through during the daytime and leave in a north or northwestward direction for the mainland, into the wind (Hawkes, 1965). Baird and Nisbet (1960) interpret the movement to be the result of south-bound migrants, carried toward the Atlantic Coast by strong northwest winds, attempting to fly back and redetermine or regain their preferred lanes of passage overland.

Another example with a different interpretation comes in the spring from Point Pelee, a peninsula projecting nine miles southward into western Lake Erie from Ontario, and from Pelee Island that lies about eight and one-half miles southwest of Point Pelee and nearer the Ohio mainland. Time and again many small land birds have been seen returning southward over Lake Erie from the tip of Point Pelee. Lewis (1939) reported such a movement in mid-May at Pelee Island. Here for several hours he watched large number of birds representing 35 species (mostly passerine) streaming southward into a headwind. The movements are not based solely on visible evidence; they have actually been proven. Birds, banded at Point Pelee prior to starting south, were later recovered at Pelee Island and on the Ohio mainland. The participants in these reverse flights may be birds which, during the preceding night, overshot their destination or were swept past it in high winds and, consequently, are attempting to return to it (Pettingill, 1964).

RATE OF MIGRATORY TRAVEL

Most passerine birds fly at ground-speeds averaging 18 to 25 miles per hour (mph). Stronger fliers such as ducks, hawks, falcons, shore birds, and swifts attain much greater speeds. Any bird can accelerate its speed in special circumstances as when frightened or diving earthward. In general, the normal, unhurried, cruising speed of a bird is much slower than suggested by published records, most of which, until recently, were estimated by observers in automobiles or airplanes moving parallel to the bird's line of flight.

Schnell (1965), using Doppler radar equipment similar to that operated by law enforcement agencies in determining speed of automobiles, measured the ground flight-speeds of 17 species of birds in northern Michigan. He recorded on windless days two speeds of the Spotted Sandpiper *(Actitis macularia)* at 25 mph; four speeds of the Eastern Kingbird *(Tyrannus tyrannus)* at 21 mph and one at 13 mph; three speeds of the Cedar Waxwing *(Bombycilla cedrorum)* at 21, 23, and 29 mph; and three speeds of the Red-winged Blackbird *(Agelaius phoeniceus)*, one at 17 mph, and two at 23 mph. Had the wind been blowing, the speeds might well have been slower for birds flying into it and faster for birds flying with it. Strong winds can significantly affect flight-speed as Schnell proved with the 267 speeds of the Herring Gull *(Larus argentatus)* that he recorded in different wind velocities. Speeds, he found, averaged 25 mph in winds less than 6 mph, but averaged 18 mph (extremes 7 and 39 mph) *into* winds of 6 to 15 mph and 34 mph (extremes of 21 and 49 mph) *with* the same winds.

During migration, according to radar surveillance by Bellrose (1967a), birds appear to reduce their flight-speed somewhat proportionately to the increase in favorable wind-speed. Apparently they adjust their flight efforts in relation to the degree of wind assistance or resistance. Thus the ground-speeds of migrants tend to remain fairly constant despite variations in wind-speed whereas the ground-speeds of birds in the daily activity flights, as shown by Schnell, are definitely influenced by winds.

Many of the stronger flying birds show great ability for fast migratory travel. McCabe (1942), in an airplane going at an air-speed of 90 miles per hour, was overtaken by two flocks of sandpipers flying at an estimated air-speed of 110 mph. Speirs (1945) once estimated the average ground-speed of the Oldsquaw *(Clangula hyemalis)* at 61.5 mph and the air-speed at 50.5 mph.

Birds homing to their breeding sites, after being displaced at great distances away, demonstrate impressive ability for sustained speed for many hours. A female Purple Martin *(Progne subis)*, taken from her colony at the University of Michigan Biological Station in northern Lower Michigan and released in Ann Arbor, Michigan, 234 miles to the south, at 10:40 PM, was back feeding her young at 7:15 AM, having made the return flight in not more than 8.6 hours at an average speed of 27.2 miles per hour (Southern, 1959). A Manx Shearwater *(Puffinus puffinus)*, removed from its nesting burrow on Skokholm off the west coast of Wales and released in Boston, Massachusetts, reached its burrow after at least 3,200 miles in 12 days and some 13 hours, or an average of 250 miles a day (Mazzeo, 1953). Another sea bird, a Leach's Petrel *(Oceanodroma leucorhoa)*, averaged about 300 miles a day for nine days from its point of release at Prestwick, Scotland, back to its nesting burrow on New Brunswick's Kent Island in the Bay of Fundy (Huntington, 1967). Of the 18 Laysan Albatrosses *(Diomedea immutabilis)*, taken from their nests on Midway Island—one of the Hawaiian Leewards in the north-central Pacific—and released at widely scattered points in the northern Pacific, 14 returned, one from the Philippines, a distance of 4,120 miles in approximately 32 days, and one from Whidby Island off the coast of Washington, a distance of 3,200 miles in 10.1 days at an average speed of 317 miles a day (Kenyon and Rice, 1958). Presumably, not one of these birds homing to its breeding site had the fat reserves for energy that migrants acquire prior to their long journeys.

There is much additional evidence, obtained by other means, of the bird's ability for sustained speed during long distances of migration. Cochran, Montgomery, and Graber (1967), using radiotelemetry, tracked migrating *Hylocichla* thrushes nearly all night from Illinois northward into Michigan, Wisconsin, and Minnesota. Although they found considerable variation, most flights were at air-speeds — i.e., speeds with relation to the winds aloft — between 25 and 35 mph. These were usually less than ground-speeds — speeds with relation to the earth — and thus suggested that the birds were aided by favorable winds. One of the most remarkable records of a long, sustained flight is that of a banded Ruddy Turnstone (*Arenaria interpres*) released by Max C. Thompson at St. George Island, one of the Pribilofs in the Bering Sea, on August 27, 1965, and shot four days later, on August 31, at French Frigate Shoals in the Hawaiian Leeward Islands (data from Philip S. Humphrey). Assuming that this individual covered 2,300 miles between St. George Island and French Frigate Shoals in a steady bee-line flight, its average speed was 575 miles a day.

Recently, radar studies have provided many reliable estimates of the rate at which migratory birds travel. W. R. P. Bourne (*in* Lack, 1962) assessed the air-speed of Lapwings (*Vanellus vanellus*) in their flights during June over the southern North Sea to England at 35 knots (40 mph). Lee (1963), at the Isle of Lewis in the Hebrides, Scotland, showed that the air-speed of Wheatears (*Oenanthe oenanthe*) from Iceland approximated 20 knots (23 mph); Redwings (*Turdus iliacus*), from 30 to 35 knots (34.5 to 40 mph); and Grey Lag Geese (*Anser anser*), from 53 to 55 knots (61 to 63 mph). Bergman and Donner (1964) demonstrated that the still-air-speed for the Oldsquaw *(Clangula hyemalis)* at low altitudes over the Gulf of Finland was 40 knots (46 mph) and for the Common Scoter (*Oidemia nigra*), 45 knots (52 mph). Once the birds reached a higher altitude inland, their speed increased by about 10 percent. After recording air-speeds of passerine birds off the coast of Norfolk, England, for a whole year, Tedd and Lack (1958), on analyzing the results, found evidence of a seasonal difference: in the spring, the speed averaged 27 knots (31 mph), 4 knots faster than in the fall. At Cape Cod, Massachusetts, Nisbet and Drury (1967) found another seasonal difference in that the directions of migration in the spring were much less diverse than in the fall, thereby suggesting much less time lost in passage.

In the late fall, long-distance migrating ducks commonly pass from breeding area to winter quarters in a short series of mass movements, each of which carries them many hundreds of miles in one continuous flight. They start each flight immediately after the passage of a cold front when temperature has dropped and the sky is clear, but they may overtake bad weather as they proceed. Sometimes, owing to the triggering effect of extremely low temperatures resulting from a strong flow of polar air, the mass movements are spectacular both in numbers of birds involved and distances covered. Bellrose (1957) documented one such migration in 1955 that moved with unusual rapidity from the Great Plains of Canada to the marshes of southern Louisiana. The exodus began from Canada on October 31; early on November 1 the flight was in full force through the Dakotas; and on November 2 the vanguards had reached northern Tennessee and Arkansas shortly after sunrise and Louisiana later in the day. Many thousands of ducks made the flight from Canada to southern Louisiana, a distance of 1,200 to 2,000 miles, in two days, or roughly 35 to 50 hours, at an average speed of 40 miles per hour. No doubt some of the birds covered the distance without stopping, accomplishing their migration in one flight.

In undertaking long-distance migrations, many small land birds tend to begin with short flights and complete them with longer flights. Indirect evidence of this procedure was reported by Caldwell, Odum, and Marshall (1963), after comparing the fat reserves of six species of tropical-wintering North American passerines killed during fall migration by television towers, one near the Florida Gulf Coast and the other in central Michigan. All the birds killed by the Florida tower showed significantly greater amounts of fat, strongly suggesting that these migrants began with low to moderate fat reserves that allowed only short flights and then increased their reserves until they had acquired a maximum amount for the long, non-stop flights such as across the Gulf of Mexico. European birds migrating south across Africa build up fat reserves of 30 to 40 percent of their body weight by the time they set out across the Sahara (Ward, 1963).

By making longer flights as they near their destinations, birds gradually speed up their migrations. Cooke (1904) provided evidence for this acceleration when he analyzed migration dates of North American species, mostly passerines, approaching their northern nesting areas. "Sixteen species," he wrote, "maintain a daily average of 40 miles from southern Minnesota to southern Manitoba, and from this point 12 species travel to Lake Athabasca at an average of 72 miles a day, 5 others to Great Slave Lake at 116 miles a day, and 5 more to Alaska at 150 miles a day."

From all these studies and reports, several generalizations on the rate of migratory travel emerge. Strong winds can affect ground-speed of birds in their daily activity flights but not in migration. Birds make long, sustained flights, usually at increasingly higher speeds at higher altitudes. Spring migration proceeds at a greater rate with less time loss than fall migration. Larger birds such as ducks accomplish their migrations in a short series of a few mass movements, occasionally in one non-stop mass movement. Small land birds, however, tend to begin their migrations in many short flights, gradually building fat reserves for long, non-stop flights, thereby accelerating their migrations.

MORTALITY IN MIGRATION

Migrating is dangerous for all birds. In their long flights over land or water they are likely to meet disaster through vagaries of the weather. When forced to land by cold fronts, frequently they must accept environments where, because of inadequate cover, they are easy victims of predators.

While migrations are adjusted to the normal alternation of fair and inclement weather during spring and fall, sudden and unseasonable changes in the weather occasionally have serious effects. Once, during fall migration in the vicinity of Lake Huron, untold numbers of birds crossing this huge lake were forced into the water and drowned because of a very quick drop in temperature and an exceptionally heavy snowfall. After the storm, one observer (Saunders, 1907) reported an estimated 5,000 dead birds washed up on a one-mile stretch of shore. In their flights north in the spring, birds are sometimes caught in severe storms and killed by becoming first exhausted and then being exposed to excessively low temperature coupled with heavy rain or wet snow. After a blinding March snowstorm in Minnesota as many as 750,000 Lapland Longspurs (*Calcarius lapponicus*) were found dead on the ice of two lakes, each of which covered only a square mile (Roberts, 1907).

Man has created awesome hazards for migrating birds by erecting lighthouses with strong light beams and by illuminating various tall structures such as the Washington Monument in the District of Columbia and the Empire State Building in New York City. During nights in the spring and fall when migration is proceeding at a low elevation because of an unsurmountable cloud layer, passing birds are attracted by the brightness and, approaching it, soon become blinded and fly into its source, killing themselves.

Under certain circumstances, even street lights can be a hazard. Vast numbers of parulid warblers, driven ashore on the Texas coast by a northeast storm while migrating northward across the Gulf of Mexico during a night in early May, met their death by flying into street lights. In a park on Padre Island, James (1956) counted more than 900 dead birds under just one light pole plus an estimated 100 on the adjacent pavement. There were nine other light poles in the vicinity with similar tolls.

Airport ceilometers indirectly cause mortality among small, nocturnal migrants. These instruments, which are used to determine the cloud ceiling, direct a narrow, extremely brilliant beam of light straight upward. At night when the clouds are low, they produce a bright spot on the cloud ceiling that can be seen for a considerable distance. On mornings following a heavy, nocturnal migration, numbers of birds varying from three to over a thousand have been found dead near spots where ceilometers are used. Three ornithologists, Howell, Laskey, and Tanner (1954), who have investigated many of these accidents, explain the cause as follows: When there is a pronounced migration and a low ceiling, the beam attracts the migrants. After circling through the bright light, many circle back to fly in and about it. While in the beam their bodies reflect light attracting still other migrants. Blinded by the light, the birds die by collision, either with each other, with the ground, or (rarely) with a building.

A direct cause of mortality among small, nocturnal migrants are television towers erected to heights of 900 to 1,000 feet or more and, as is often the case, situated on hills or bluffs where they reach even greater heights above the local terrain. Each tower is supported by guy wires and has a system of steady, flashing red lights, mandatory on all tall structures that are potential hazards to airplanes.

When migrants are flying under a low ceiling, they are attracted to a television tower because of its lighted area and become reluctant to leave. Just as birds, released at night in a lighted room with doors and windows open, continue to fly about in the room rather than escape into the darkness, the migrants fly through the tower framework and circle out to the edge of the lighted area, then return toward the light. Mortality results when the birds strike the dark guy wires while circling. The above observations and explanations come from Graber (1968). For examples of the extent of avian mortality around television towers, see the papers by Tordoff and Mengel (1956) and Stoddard and Norris (1967).

ALTITUDES OF MIGRATORY FLIGHT

The recent analysis of migratory flight by means of radar shows not only that birds move at altitudes averaging higher than formerly believed, but also that their altitude varies widely depending on the circumstances.

Nisbet (1963), studying radar heights of nocturnal fall migrants above Cape Cod, Massachusetts, and the outlying ocean, found that the most frequent height was usually between 1,500 and 2,500 feet. About 90 percent of the birds, probably small passerines, were below 5,000 feet. On some nights they were lower than 2,500 feet, while on others they were up to 6,000 or even 8,000 feet.

The presence or absence of cloud cover may determine the altitude chosen by birds. Bellrose and Graber (1963) discovered, during their radar studies of nocturnal migrants in central Illinois, that birds are prone to migrate at higher altitudes when the skies are overcast than when they are clear. If the clouds are not too high, the birds apparently attempt to surmount them; but if the clouds are too high, they usually continue to fly, sometimes in the clouds, although usually immediately under them. When the birds are flying under an overcast and, consequently, at much lower altitude, they can usually be heard from the ground (Graber, 1968).

Birds fly higher by night than by day. Lack (1960a) first noted this tendency, by radar, among migrants crossing the southern North Sea between England and the Continent. Later, Eastwood and Rider (1965) at the Bushy Hill station in England proved that the tendency is significant. From their considerable data they were able to show that 80 percent of the birds fly below 5,000 feet at night and 80 percent below 3,500 during the day. They further demonstrated that migrating birds, in a 24-hour day, have a tendency to fly at the lowest altitudes in the afternoon and the highest just before midnight. From radar studies and tape recordings of call notes in Illinois, Graber (1968) has concluded that migrants reduce their altitude after midnight to 1,500 feet or less, although they continue their flight until daylight. After their descent to lower altitude, they increase their calling. This helps to explain why nocturnal migrants can be heard more frequently as dawn approaches than earlier in the night.

There are seasonal variations in altitudes. Bellrose and Graber (1963) found that migrating birds in Illinois fly higher during the fall than during the spring, possibly because the winds during the fall are more favorable for southward migration at higher altitudes. Eastwood and Rider (1965), on the other hand, found the reverse to be the case in England. They suggest as one reason for this seasonal difference that flocks of fall migrants include many young birds whose flight capabilities are inferior to those of adults and which are, consequently, unable to achieve the higher altitudes of the more mature spring migrants.

Birds migrate higher over land than sea. Common Scoters (*Oidemia nigra*) and Oldsquaws (*Clangula hyemalis*), in their spring passage over southern Finland and the Gulf of Finland, were noted by Bergman and Donner (1964) to fly at altitudes that averaged 3,400 feet over land and ranged from 300 to 1,000 feet above water. Passerine migrants, in passing from sea to land in England, were shown by Eastwood and Rider (1965) to make a similarly significant though not as great change in altitude, climbing from a median height of 1,700 feet to a median height of 2,200 feet.

Birds have long been known to reach very high altitudes during flight. Direct observations from aircraft proved that large birds can fly over the highest mountain ranges — for example, the Himalayas between central Russia and India. The Yellow-billed Chough (*Pyrrhocorax graculus*) was actually found on Mt. Everest at an elevation of 27,000 feet (Gilliard, 1958). The use of radar in determining heights of migration shows that while most birds rarely exceed 8,000 to 10,000 feet, a small proportion of migrants, particularly the stronger fliers, nonetheless attain great heights, in some cases astonishing. In his radar studies of migrants at Cape Cod, Massachusetts, Nisbet (1963) recorded a number of birds on several dates in September and October, usually before midnight, or before and after sunrise, at altitudes between 8,000 and 15,000 feet and a few birds as high as 20,000 feet. These may have been sandpipers and plovers, flying over the ocean, toward the Lesser Antilles and eastern South America. Using an especially powerful and accurate height-finder at Norfolk in southeast England, Lack (1960a) observed at sunrise on 16 dates in September a thin scattering of birds extending fairly uniformly up to at least 15,000 feet. On seven of these mornings he noted that the greatest height was 19,000 feet and on two mornings it was 21,000 feet. Probably the highest migrants were small shore birds such as the Dunlin (*Erolia alpina*) which had left Scandinavia the night before.

Bearing in mind that the oxygen content of the air at 18,000 feet is 50 percent less than the air at sea level, one wonders whether high-flying birds suffer "altitude sickness." When a man sets out to climb a lofty mountain he acclimates himself gradually over a period of days, but birds take off and reach a comparable elevation in a matter of a few hours. Do birds have special adaptations that enable them to avoid altitude sickness? The answer may come someday from experimental studies of bird flight in pressure chambers.

COURSE OF MIGRATION AND MIGRATION ROUTES

Many species breeding in North America north of Mexico have their winter ranges far south and southeast, in southern Mexico, Central America, and South America. To reach their winter ranges, most of these species proceed at night from their breeding ranges to southern United States in broad fronts without notable regard to topographical features. Radar and other observations confirm that their movements trend southeast—toward their winter ranges. This direction, coming as it does in the wake of a cold front, is with the wind and, therefore, beneficial to the migrants. In the spring, the direction trends northwest, in reverse, and again with the wind since the movements take place with the onset of a warm front. Many species, however, do not exactly retrace their course in the spring, but fly instead somewhat to the west of their fall passage. Evidence of this elliptical course—going to southern United States one way and coming back another—is borne out by kills at television towers. Certain species are well represented in the spring migration but seldom or not at all in the fall, and *vice versa.* For further details on this subject, see Bellrose and Graber (1963) and Graber (1968).

Owing to the narrowing of the North American continent southward and the intervention of the Gulf of Mexico and Caribbean Sea, all species moving southeastward through the United States toward their wintering ranges in southern Mexico, Central America, and South America converge on special routes. There are five altogether. Certain species use mainly one route, others two or three; no species is known to use more than three. The routes are as follows:

Route 1: From the coasts of Newfoundland, Nova Scotia, New England, and New Jersey southward over the Atlantic Ocean to the Lesser Antilles and the northeastern coast of South America. A few shore birds use this route.

Route 2: From Florida southward over the Bahamas, Hispaniola, Puerto Rico, and the Lesser Antilles to South America. The few birds which frequent this route are seldom far from land as there are many small islands along the way.

Route 3: From Florida southward over Cuba and Jamaica across 400 miles of the Caribbean Sea to South America.

Route 4: From the shores of the Gulf states across the Gulf of Mexico to the Yucatan Peninsula and southern Mexico. This is the route most frequently used by the many species of birds from eastern United States and Canada.

Route 5: From Texas, New Mexico, Arizona, and California through northern Mexico. The majority of birds from western conterminous United States, western Canada, and Alaska use the western side of this route.

Species vary greatly in their course of migration and use of routes. A few species move south over one route and return by another; a few species move eastward or westward before going south; and a few species have spectacularly long routes that take them as far south as southern South America. Among a few species which breed in western United States—e.g., the Western Kingbird *(Tyrannus verticalis)* and the Scissor-tailed Flycatcher *(Muscivora forficata)*—some individuals in the fall move eastward across the

Gulf states to winter in Florida (see McAtee and others, 1944) whereas most of the population moves directly south into Mexico.

To illustrate the wide variation in migratory movements and the choice of routes, the migrations of six species are described below.

The American Golden Plover *(Pluvialis dominica)* passes eastward from its breeding range to the Atlantic Coast, where it turns southward over Route 1. Once in South America, it flies directly across the continent to its winter range. It returns, however, by another route, coming up across northwestern South America and the Gulf of Mexico and reaching the United States along the coast of Texas and Louisiana. From there it continues up the Mississippi Valley and through central Canada to its breeding range.

The Blackpoll Warbler *(Dendroica striata)* shows a remarkable convergence in its southward migration. From its vast transcontinental breeding range it converges on the Atlantic coastal plain as far north as Virginia and from there proceeds to Florida where it leaves for South America over Routes 2 and 3. It returns to Florida by the same routes but from there fans out northward to its breeding range.

The Mourning Warbler *(Oporornis philadelphia)* shows a similar convergence. From its breeding range of more modest extent it converges on southern Texas, then goes southward along the eastern portion of Route 5 through eastern Mexico and Central America to its winter range. It returns via the same part of Route 5 and fans out from Texas northward.

The American Redstart *(Setophaga ruticilla)* shows little convergence. Instead, it passes southward more or less directly over a broad front of nearly 2,500 miles, eventually using Routes 2, 3, and 4. It returns the same way.

The Bobolink *(Dolichonyx oryzivorus)* migrates an exceptionally long distance, averaging farther than any other passerine species. From its transcontinental breeding range, it converges southward to leave the United States over Routes 3 and 4. Once in South America, it continues directly overland to its winter range in extreme southern Brazil, southeastern Bolivia, Paraguay, and northern Argentina. It returns the same way.

The Connecticut Warbler *(Oporornis agilis)* migrates in an eccentric manner. From its breeding range it flies directly eastward to New England, then to South America along the Atlantic coastal plain and eventually Route 3. From its winter range in South America it returns over Route 3 to Florida, but from there it passes diagonally to the Mississippi Valley and northward to its breeding range.

> **On the outline maps of North and South America, pages 286-291, indicate the ranges of the above six species, the breeding ranges by stippling and the winter ranges by cross-hatching. Use one map for each species. Refer to *The A. O. U. Check-list* for the description of these ranges and an atlas for the location of the places mentioned. Then by means of lines, indicate the courses and routes of fall migration in broken lines, of spring migration in solid lines. Use arrows to show the direction of migratory movements.**

ALTITUDINAL MIGRATION

Some bird populations of high mountains in both temperate and tropical regions move down to the lower slopes and valleys when winter sets in at higher elevations. In the descent of a few hundred feet they accomplish what many other populations do in their latitudinal migrations of many hundreds of miles. (For various migrational data in a typical section of the Rocky Mountains, see Packard, 1945).

OUTLINE MAP OF NORTH AND SOUTH AMERICA

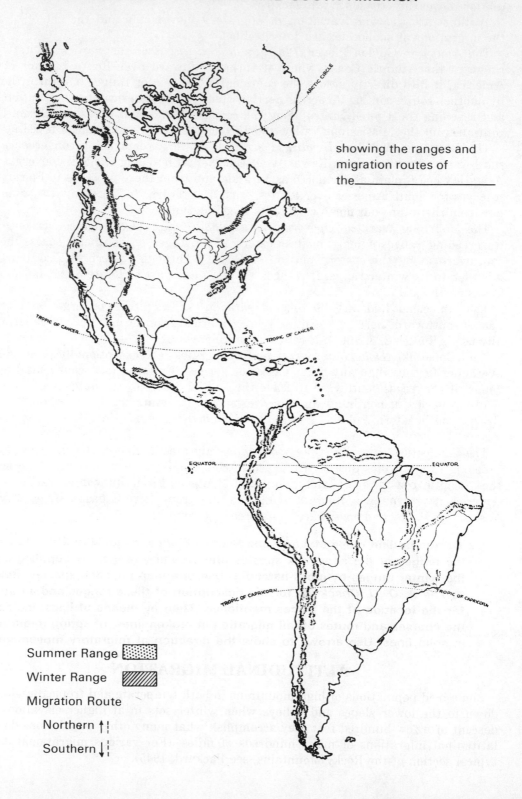

showing the ranges and
migration routes of
the_____

Summer Range

Winter Range

Migration Route

Northern ↑↑

Southern ↓↓

OUTLINE MAP OF NORTH AND SOUTH AMERICA

showing the ranges and
migration routes of
the_____

Summer Range
Winter Range
Migration Route
  Northern ↑⋮
  Southern ↓⋮

OUTLINE MAP OF NORTH AND SOUTH AMERICA

showing the ranges and
migration routes of
the_____

Summer Range

Winter Range

Migration Route

   Northern ↑⋮

   Southern ↓⋮

OUTLINE MAP OF NORTH AND SOUTH AMERICA

showing the ranges and
migration routes of
the_____

Summer Range

Winter Range

Migration Route

Northern ↑

Southern ↓

OUTLINE MAP OF NORTH AND SOUTH AMERICA

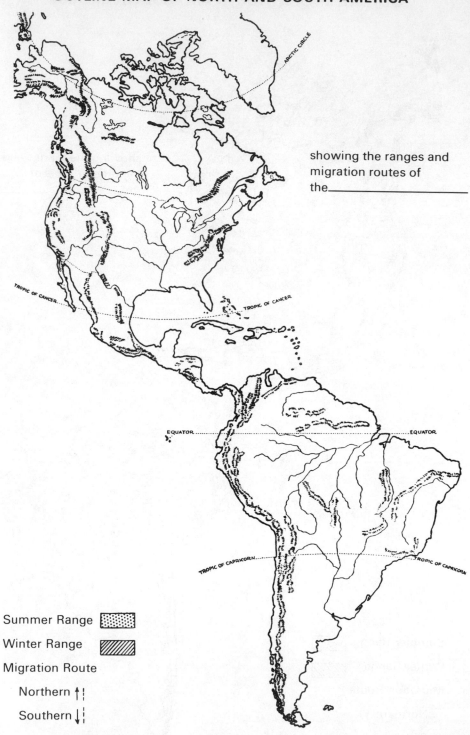

showing the ranges and
migration routes of
the_____

Summer Range

Winter Range

Migration Route

Northern ↑⁝

Southern ↓⁝

OUTLINE MAP OF NORTH AND SOUTH AMERICA

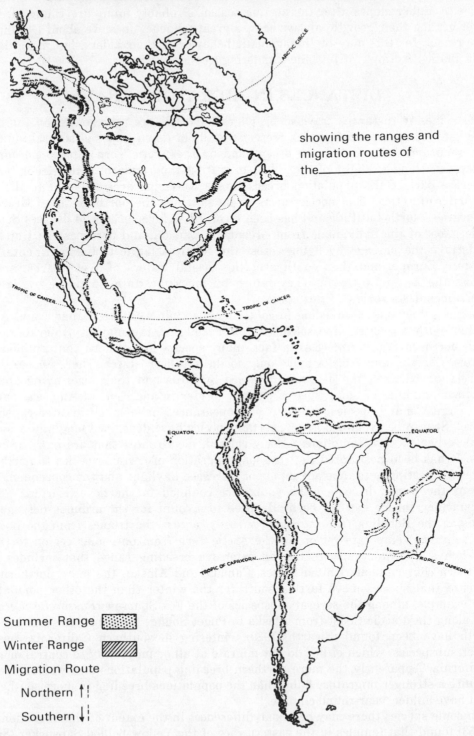

showing the ranges and
migration routes of
the_____

Summer Range

Winter Range

Migration Route

Northern ↑↑

Southern ↓↓

Several observers (e.g., see Packard, 1946) have noticed that mountain birds tend to move to higher slopes after the nesting season. Probably many are young birds which, after getting their growth, always show a great tendency to move about (Phillips, 1951). The reason for this may be the agonistic behavior of their elders, but more than likely it is another example of juvenile wandering (see page 278).

DISTANCES IN MIGRATORY TRAVEL

Extremes in distances traveled by migrating birds are represented on the one hand by high-mountain species, which merely pass up or down slopes of several hundred to a few thousand feet, and on the other hand by the Arctic Tern *(Sterna paradisaea)*. It makes the longest flight of any bird, migrating thousands of miles from the Arctic, where a part of the population breeds, to its wintering area adjacent to the pack ice around Antarctica. This species nests as far north as the northern tip of Greenland at 83 degrees North Latitude and has been recorded as far south as 74 degrees South Latitude. Most of the individuals from Greenland, Canada, and northeastern United States undertake the passage by flying across the North Atlantic to the continental shelf of western Europe, and then south over the coastal waters of West Africa and finally across the Antarctic Ocean. They return by the same route. A paper by Salomonsen (1967) includes a review of this subject.

While a few species breeding in conterminous United States, Canada, and Alaska go as far south as central and southern South America, the majority migrate no farther than northern South America. In fact, many species in northern conterminous United States, Canada, and Alaska move only to the southern states. There appears to be no correlation between the distances traveled by birds and their size, flying abilities, or habits. A number of small species journey to Mexico and Central America, outdistancing a great many species which to all appearances have much greater capacities for travel. Still to be explained is why certain species have developed long migration routes, while other species, sometimes closely related, have routes that are half the length or less. It has been postulated that the long migration of certain species to northern latitudes for nesting is to take advantage of increased daylight and the consequent shortening of the period in which the young are confined to the nest (Karplus, 1952). No satisfactory theory has yet been advanced to account for the arduous passage of such birds as the Bobolink *(Dolichonyx oryzivorus)* across the tropics from the northern to the southern temperate region, or the Arctic Tern from one polar region to the other.

When a migratory species has an extensive breeding range that includes parts of northern conterminous United States, Canada, and Alaska, the more northern populations of that species move farther south for the winter than the other populations do. For example, among the several subspecies of the Fox Sparrow *(Passerella iliaca)* breeding along the Pacific Coast from Alaska to Puget Sound, the subspecies nesting farthest north have been found (Swarth, 1920) wintering in southern California, passing by other subspecies, which either do not migrate at all or move only to central or northern California. Apparently the more northern breeding populations of a widespread species acquire a stronger migratory habit than the populations breeding in more southern areas that have milder, year-round climate.

In some species there may be sexual differences in the extent of the migration. Howell (1953) found that females in the eastern race of the Yellow-bellied Sapsucker *(Sphyrapicus varius)* outnumber the males in the southern part of the winter range by about three and one-half to one.

FLIGHT LANES AND CONCENTRATIONS

In North America migratory movement is continent-wide. There are probably no areas over which birds do not pass in their latitudinal migrations. Prairies, forests, mountains, lakes, and inland extensions of the oceans fail to stop or divert migration altogether. Even so, as diurnal migrants move northward or southward across Canada and the conterminous United States, many species tend to favor or be influenced by certain topographic features which trend in a north-south direction. Some species fly along ridges; others follow the coasts, large rivers, and chains of lakes. A great many species go through valleys, move along peninsulas, or pass from island to island across large bodies of water. In both Canada and the United States there are many places where the topography is such as to cause narrow flight lanes in which migratory movement is especially conspicuous.

Particularly in the fall, hawks and a few other large birds follow the crests of north-south ridges, riding on the updrafts as they proceed southward. The traffic in these lanes is unusually heavy on clear, windy days following the passage of a cold front, because there is considerable wind deflected upward, thus making these lanes advantageous to travel. One of the best known places is the Kittatinny Ridge in eastern Pennsylvania. At one point, Hawk Mountain (Broun, 1949, 1951, 1965), where the ridge becomes suddenly high and slender, the birds are brought together and closer to the ground. As a result of this narrowing of the flight lane, observers at Hawk Mountain have been able to count over 22,000 hawks moving by in a single season.

Large bodies of water constitute barriers to day-migrating birds and thus cause flight lanes to curve around them. The Great Lakes are a good example. In the fall, southbound hawks approaching their north shores from western Quebec and southern Ontario take the shortest courses around them that geography and air movements will allow (Pettingill, 1962). If the birds are migrating in great numbers, as on the second days after cold fronts when there are steady westerly winds and ample sunlight producing thermals, large numbers can be seen in continuous passage at such points as Port Credit on the northwest shore of Lake Ontario, Port Stanley on the north side of Lake Erie, Cedar Grove on the west side of Lake Michigan (see Mueller and Berger, 1961), and Duluth at the westernmost extension of Lake Superior (Hofslund, 1965, 1966).

Peninsulas projecting into large bodies of water that lie athwart the direction of migration may become funnels for land birds in diurnal passage. Hawks moving northward in the spring through Michigan to Canada converge in large numbers on the northern tip of Lower Michigan at the Straits of Mackinac (Sheldon, 1965). Here, if the weather is rainy and windless, they settle on trees and other perches. With the advent of the next clear day and a favoring wind, the hawks begin to spiral higher and higher until one bird peels out and heads northward over the Straits with the others following. Small numbers of north-bound Blue Jays (*Cyanocitta cristata*), in order to get across Lake Mendota at Madison, Wisconsin, converge first at Picnic Point and then spiral upward until "barely visible to the naked eye" before crossing the 1.7 miles of open water to Fox Bluff on the north shore (Schorger, 1961). Cape May, the southern tip of New Jersey between the Atlantic Ocean and Delaware Bay, is noted for its hordes of land migrants, large and small, which gather from August through November when northerly winds are strong (Stone, 1937). Some of the migrants held up here are those which regularly follow the Atlantic Coast southward, but many are birds which have been drifted by wind southeastward from their usual flight lines. Frequently, migrants at Cape May may be seen in the day flying north and northwestward into the wind as they skirt Delaware Bay before continuing their journey. Similar concentrations may be observed at Cape Charles, a south-pointing peninsula separating the waters of Chesapeake Bay from the Atlantic (Pettingill, 1951).

Night migrants do not follow topographically determined flight lanes to any significant degree (Lowery, 1951). Instead, as radar surveillance shows, birds migrate at night without regard to what lies below.

In the spring, land migrants returning to the United States from across the Gulf of Mexico vary in their manner of arrival in accordance with weather conditions. If the weather is mild and the wind favorable, birds bound for more northern destinations continue inland from the Gulf for considerable distances before coming to land. The coastal area thus appears to be an ornithological "hiatus" (Lowery, 1945). But if a cold front with strong northerly winds moves in over the area while migration is in progress, migrants are forced to come down to the first land reached, with the result that the coastal area is flooded with birds (see Pettingill, 1953) that linger here until the weather again becomes favorable. A somewhat similar situation occurs during spring migration at Point Pelee. Here, since it is the nearest land, north-bound small land birds, on meeting a cold front from the north while they are over Lake Erie are forced to descend (Pettingill, 1964; Livingston, 1965a) and remain until the cold front abates. At such times Point Pelee swarms with birds.

FLOCKING DURING MIGRATION

During migration bird species show wide differences in flocking habit. A number of diurnal migrants, notably many hawks and other predators, are little inclined to move in groups, preferring to travel solitarily. But the majority of migrants, diurnal or nocturnal, exhibit the flocking habit.

Certain species migrate in flocks strictly of their own kind. These are usually birds whose flight-speed, feeding habits, or roosting preferences are so individual as to make them incompatible traveling companions. The Common Nighthawk *(Chordeiles minor)* and Chimney Swift *(Chaetura pelagica)* are good examples of birds which migrate in their own company. Neither waxwings nor crossbills will migrate with other birds, but Cedar and Bohemian Waxwings *(Bombycilla cedrorum* and *B. garrulus)* have been seen in the same flocks and so have Red and White-winged Crossbills *(Loxia curvirostra* and *L. leucoptera)*. Some of the larger birds—e.g., pelicans, cormorants, storks, swans, geese, and cranes—noted for V-shaped or linear flock formations likewise tend to travel in unmixed groups.

The majority of species traveling in flocks, whether unmixed or mixed, give call notes. This is especially true of nocturnal migrants: some species such as *Hylocichla* thrushes utter calls heard only in night migration (Ball, 1952); other species such as the Bobolink and Dickcissel *(Spiza americana)* give calls that are the same as ones heard in the daytime on their breeding grounds. If the calls are distinctive as in the case of the Bobolink and Dickcissel, it is possible to identify the species, but the calls of most species are faint chips and lisps that sound more or less the same to the human ear. See Graber and Cochran (1960) for an analysis of bird sounds heard at night.

Unmixed or mixed flocks of some species of smaller birds—e.g., certain sandpipers and plovers—fly in compact formations, all individuals in the flock simultaneously performing almost identical maneuvers. Many more species of smaller birds, including the majority of passerines, travel in flocks that are loosely formed, though still cohesive, the individuals proceeding in the same direction. At night, the cohesion of flocks is probably maintained by call notes (Hamilton, 1962c). At the same time, the call notes serve to space out individuals in each flock so that they will not collide with one another.

No flock of migrants appears to have a persisting leader. The direction taken by a flock, as suggested by Hamilton (1967), represents a compromise by each individual to the directional preference of the other individuals of the flock. At night, call notes convey directional information from one individual to the other. When a nocturnal flock shifts its direction or lowers its altitude, or when it is disoriented for one reason or another, all its members greatly increase their rate of calling (see Graber and Cochran, 1960).

Whether or not flocks remain intact for the duration of migration has not been determined. Nor is it known for certain whether flocks remain together during the winter. In all probability, most flocks, if they are comprised of common, widely distributed species, change from day to day during migration and in the winter split up into smaller groups, or combine with other flocks to form larger groups.

Flocking during migration is undoubtedly advantageous to the individuals concerned. Just as flocking among resident birds provides group protection against predation or increases the success in finding and exploiting food sources (see page 259), flocking in migration greatly facilitates the attainment of destination. Younger birds traveling with more seasoned adults benefit from their experience. As another advantage, Hamilton (1962c, 1967) suggests that groups of birds on an average determine their direction with greater accuracy than single individuals. Thus flocking assists any migrant that goes a long distance, or any migrant required to pinpoint its destination on a small land area in mid-ocean. Hamilton cites as examples: (1) The Broad-winged and Swainson's Hawks *(Buteo platypterus* and *B. swainsoni)*. Both of these North American species travel in large flocks to winter in South America. By contrast, the Red-tailed Hawk *(B. jamaicensis)* and certain other buteos, which rarely leave the North American continent, seldom move in appreciable groups. (2) The Long-tailed Cuckoo *(Urodynamis taitensis)*. After breeding in New Zealand, this species gathers in large flocks for the exceedingly long, non-stop passage to winter quarters on tiny islands in the west-central Pacific. By contrast, the Yellow-billed and Black-billed Cuckoos *(Coccyzus americanus* and *C. erythropthalmus)* of North America, which have only to reach a broad tropical region of South America for the winter, migrate solitarily.

V-shaped formations help to conserve energy by creating favorable air currents for all individuals in the flock except the leader. When fatigued, the leader drops back and is replaced by another bird in the flock. An alternate advantage hypothesized by Hamilton (1967) is that the V-shaped structure serves as a means of communication, enabling the individuals to profit fully from the collective direction-finding of the group. By flying in parallel alignment, each individual moves in the same direction as the leader. If, as in flocks of geese traveling in poor visibility, difficulties in establishing direction arise, the leadership changes frequently in order that collective "judgment" may prevail in maintaining the proper course. Sometimes the V-structure gives way temporarily to a crescent form; forward flight remains in the same direction but is slowed until a new leader takes over and the V-shape is resumed.

Unmixed or mixed flocks may contain only immature individuals, only adults, or only adults of one sex; or they may contain individuals of all ages and both sexes. Flocks of geese and cranes may be comprised of one family or several families.

During the fall migration the adults may precede the immature birds, the birds-of-the-year. Hagar (1966) reported that adult Hudsonian Godwits *(Limosa haemastica)* withdraw from their nesting grounds in central and northwestern subarctic Canada in late July; the immatures follow a month later. Among passerine species there is convincing evidence assembled by Murray and Jehl (1964) from several thousand migrants, mist-netted in the fall at Island Beach, New Jersey, and analyzed as to age, that adults and immatures travel at approximately the same time. However, in the case of the Least Flycatcher *(Empidonax minimus)*, Hussell, Davis, and Montgomerie (1967) concluded from an analysis of 182 individuals trapped in the late summer at Long Point, Ontario, that most of the adults migrate in advance of the immatures: the adults during the second half of July and first half of August; the majority of immatures from the second week of August to the end of September.

In the spring migration many of the first flocks of certain species coming north have a preponderance of adult males which reach the breeding grounds and establish territories before the rest of the population arrives. A. A. Allen (1914) once carefully studied the spring migration of Red-winged Blackbirds *(Agelaius phoeniceus)* at Ithaca, New York. Since the males and females have distinct plumages and birds hatched the previous year (i.e., the immature birds) differ sufficiently in color from the adults, he was able to analyze flocks and work out a migration schedule according to age and sex. While his findings may not hold for all Red-wing populations, it is presented below as a useful guide and basis for comparison.

Migrant adult males	March 13-April 21
Resident adult males	March 25-April 10
Migrant females and immature males	March 29-April 24
Resident adult females	April 10-May 1
Resident immature males	May 6-June 1
Resident immature females	May 10-June 11

In watching spring migration in his own area, the student will find that not all species follow such a schedule. In fact, males and females of quite a few species migrate together. Only rarely, however, do females precede males.

DIRECTION-FINDING

How migrating birds determine their direction when migrating or homing by day or night over areas unfamiliar to them is one of the most fascinating aspects of migration. Before considering the subject of direction-finding in migration, it is worthwhile to review some of the problems associated with homing in birds. The basic means by which birds orient themselves, whether migrating or homing, are much the same.

A great many experiments have demonstrated the remarkable ability of wild birds to return to their eggs or young after being removed great distances and released. Years ago Watson (1909) tested the homing ability of Sooty and Noddy Terns *(Sterna fuscata* and *Anoüs stolidus)*, which nest on the Dry Tortugas, the islands in the Gulf of Mexico west of Florida. These species come to the Tortugas from tropical seas and are seldom seen farther north. Two nesting Sooty Terns and three Noddy Terns were captured, marked, and transported northward in a ship to a point off Cape Hatteras, about 1,000 miles by sea from the Tortugas. Here they were released. Just five days later the two Sooty Terns were back on their nests; one of the Noddy Terns showed up after a few more days. Some of the more recent experiments demonstrating the sustained speed of homing flight by the Purple Martin, Manx Shearwater, Leach's Petrel, and Laysan Albatross (see above) attest at the same time to their precision in direction-finding.

The inducement of birds to home is not necessarily provided by their eggs or young; it can be the breeding area or home range. Breeding Brown-headed Cowbirds *(Molothrus ater),* which are brood parasites, will home from maximum distances of 250 to 380 miles (Miller, 1953; Wharton, 1959; Manwell, 1962). An adult female Bobolink *(Dolichonyx oryzivorus),* escaping from captivity in September at Berkeley, California, was recaptured the following June at Kenmare, North Dakota, where it was originally trapped as a breeding bird (Hamilton, 1962b). Six hundred and sixty Golden-crowned and White-crowned Sparrows *(Zonotrichia atricapilla* and *Z. leucophrys),* captured while wintering in the San Jose area of California, were immediately carried by plane to Laurel, Maryland, and released. Fifteen were known to have come back the following winter. In the interim they had presumably found their way in the spring to the nesting grounds in northwestern Canada and Alaska, then returned to California in normal migration (Mewaldt, 1964).

Homing flights are more or less routine with homing pigeons which are Common Pigeons *(Columba livia)* specially bred for racing. Their precision and speed of return to the home loft is developed by training and experience. Sometimes the experience of only one flight to the home loft is sufficient to determine the proper direction for successive flights. In his work on homing pigeons, Matthews (1951, 1952b, 1963) found that certain individuals, trained to maintain a given direction, can adhere to that direction over unfamiliar terrain and that certain other individuals can fly straight toward the home loft from unfamiliar territory regardless of the direction of the home loft. In the course of their training, pigeons must be made familiar with the area around the loft so that they will have a broad or reasonable target. See Graue (1965) for later work demonstrating how experience plays a prominent role in the return flight of homing pigeons.

Although homing has been amply demonstrated in both wild birds and racing pigeons, the perplexing question remains: How do homing birds find their way? In seeking an answer, Griffin and Hock (1949) attempted to determine how displaced and released nesting birds find their way by following them in an airplane and watching their behavior. For their experiment they selected the Gannet *(Morus bassanus),* a marine bird which rarely occurs inland; being large and white, it was easy to follow. Taking 17 individuals from their nests on an island off the Gaspé Peninsula, Quebec, they carried them to a point in northern Maine, 100 miles from salt water and 215 miles from the home island. There they released nine of the birds (the others were used as controls) and, from an airplane, traced their flights from 25 to 230 miles. The investigators were careful to keep the plane 1,500 feet or more away from the birds so as not to frighten them. The experimental birds flew in all directions with no significant tendency to head directly toward their nests. Their flight paths were generally gradual curves. The first birds to reach the coast within the first few hours were the first to get back to their nests. Altogether 62.5 percent of these birds eventually reached their nests, in from one to four days.

The result of Griffin and Hock's work on the Gannets suggests that the ability of birds to home—that is, to return to a known or familiar site such as a nesting area—is dependent on random searching. In a strange territory birds keep circling and exploring by trial and error until they find familiar landmarks within familiar territory. However, the concept of random searching as a means by which all birds find their way cannot be reconciled with the rapid homing exemplified by Sooty Terns and other species already mentioned (see above). Many birds, if not all, obviously have an ability to find their way by orienting themselves—determining their position with respect to their environment—and by following directional cues or navigating.

In taking up the subject of the means by which birds orient themselves and navigate, it is well to consider first the question whether or not birds inherit at least part of their ability to find their way. Rowan (1946) at Edmonton, Alberta, caught young Common Crows *(Corvus brachyrhyncos)* in the late summer and kept them in captivity. In November, when winter conditions had begun to set in and all adult Common Crows had left for their winter range in Kansas and Oklahoma, he banded the captive birds and released them. Altogether 54 individuals were set free, and in the next few days he received reports of recoveries. Apparently some of the birds had not traveled very far, but those which had gone an appreciable distance were headed toward their winter range. Schüz (1949) at Rossitten on the Baltic Coast of eastern Prussia tried similar banding experiments with White Storks *(Ciconia ciconia)*. In the middle of September, after all the local population had departed, he released 73 young banded birds. Most of them traveled eastward toward the Black Sea, paralleling the normal flight line for the local population, though some of the birds flew in a more southerly direction and three went southwestward to Italy. Perdeck (1958), over a period of four years, caught and banded over 11,000 fall-migrating Starlings *(Sturnus vulgaris)* in Holland and released them in Switzerland. Recoveries totaling 354 later showed that the juveniles took their ancestral direction southwest, paralleling the normal route, to a new winter area, whereas the adults soon separated and veered westward toward their ancestral winter range. From these experiments alone, it seems clear that at least some birds must have an innate ability to follow the normal migratory route or one parallel to it. How the birds "know" when they have flown far enough is yet to be determined.

Granting that migrating birds have an innate ability to find their way does not deny that some birds acquire their ability by experience. Young Indigo Buntings *(Passerina cyanea)*, hand-raised in various conditions of isolation, apparently do not attain the accuracy of orientation typical of adult buntings and thus seem to depend, at least partly, on some kind of experience (S. T. Emlen, 1969b). Young geese and cranes, which migrate in families or groups of families, probably learn the migration routes by following their elders. They no doubt "memorize" features of the landscape when they migrate by day as they often do. But whether or not birds inherit or acquire their ability to find their way, the fact remains that they must depend on cues other than landscape features to orient themselves and navigate when traveling at night, over the sea, or above or in a heavy overcast. What are the cues?

The first break-through came in the early 1950's when it was proved that the sun is a cue to orientation and even a guide in navigation. At Wilhelmshaven, Germany, Kramer (1952) placed a small, circular cage high in the center of a circular pavilion that was completely enclosed except for six windows, each high enough to give only a view of the sky from the cage. Kramer put a hand-reared Starling in the cage at the time in the spring when it would normally migrate, and from below the cage he recorded the direction in which the bird, in its migratory restlessness, showed a tendency to flutter. The bird fluttered persistently in the normal direction of spring migration if the sun was shining but not if the sky was heavily overcast. When, by an ingenious use of mirrors at each of the windows, Kramer altered the sun's apparent position, the bird deflected its direction in order to maintain the same angle as before. Kramer and his associates soon demonstrated in subsequent experiments that birds have some sort of "internal clock" that enables them to compensate for the sun's daily "movement" across the sky and thus can hold the same direction of flight despite the sun's steadily changing position.

Hoffmann (1960) experimented further on the internal clock as a basis for orientation by the sun. He trained caged Starlings to seek a food reward at a particular time of day and always in the same direction. Then he tested them without rewards at the time they had come to expect and at other times. The birds, he found, could allow for the daily movement of the sun. Having proved this, he subjected the trained birds to a regime of light and darkness that shifted their internal clocks six hours ahead or six hours behind the natural day outside. Later he exposed the birds to the natural day. The birds responded by shifting their directions accordingly by 90 degrees, counterclockwise if the clocks were set ahead, clockwise if they were set behind. These and later tests showed that the birds did not orient themselves by the elevation or altitude of the sun in the sky but by its azimuth or position with relation to compass direction. The birds thus determined direction by what is called "sun compass orientation."

Soon after Kramer reported his initial experiments with Starlings, Matthews (1951, 1953, 1955b) found that homing pigeons, which are non-migratory, use the sun as a guide. When he released pigeons in unfamiliar territory under clear skies, they headed in the direction of home. If the sky was overcast, they failed to do so. The concept of the sun as a compass for homing pigeons has since been supported by other experimental investigators using different techniques. For example, Schmidt-Koenig (1961) shifted the internal clocks of pigeons by six hours in advance and six hours behind the natural day. On releasing the birds under a clear sky at varying distances and different directions from the home loft, he found that the birds deviated appropriately to the left or right of the correct flight direction taken by the control pigeons whose clocks had not been shifted. This suggests that the sun played a role in their orientation. Experience also plays a role in the orientation. Each time a pigeon is released, the direction it takes is more accurate. For a review of current knowledge on pigeon homing, the student is referred to the paper by Wallraff (1967).

Some birds may show one-directional orientation that is not necessarily related to migration or homing. Matthews (1961), experimenting with non-migratory Mallards *(Anas platyrhynchos)* at Slimbridge, England, discovered that when birds of any age were displaced during a clear day for any distance in any direction at any season, they demonstrated a strong tendency to fly in one direction, namely, northwest. The flights were always short and seemed to be guided by the sun, but since they had no discernible purpose other than possibly escape, Matthews called the behavior "nonsense orientation." He believed that it might be innate or perhaps developed at an early age. Other birds may show the same phenomenon. Migratory Mallards wintering in Illinois, on being trapped and released during a clear day or night at points far east and west of their winter-home lakes, consistently flew northward regardless of the direction of their home-lakes (Bellrose, 1958). Common Terns *(Sterna hirundo)*, displaced from their breeding colony on Penikese Island off Cape Cod, Massachusetts, on a clear day, flew southeast, irrespective of where they were released or the direction of Penikese Island (Griffin and Goldsmith, 1955; Goldsmith and Griffin, 1956).

Adélie Penguins *(Pygoscelis adeliae)*, when displaced from their breeding colonies at Cape Crozier on the coast of Antarctica to the featureless expanse of compacted snow in the interior, oriented themselves by the sun, departed in one direction, and ultimately reached their nest sites. The investigators, J. T. Emlen and Penney (1964), used breeding males, some of whose internal clocks had been artificially shifted in advance. Releasing them individually from three groups, each at widely separated points, they observed that the birds headed straight for the coast on courses which were essentially parallel without any convergence toward Cape Crozier. The birds seemed to have no information on their location at release with respect to their home colonies. They took their courses from the sun and, as shown by the breakdown in orientation among those birds whose clocks had been shifted, maintained their courses by referring to the sun's azimuth. If heavy clouds obscured the sun and eliminated all shadows, their orientation soon deteriorated. Emlen and Penney considered such one-directional courses to be fixed and to be maintained by an inherent time sense; they likened the procedure to nonsense orientation as described in waterfowl and other birds, but believed that it might be escape orientation with survival value by steering the penguins to off-coast feeding areas whence they would be guided, perhaps in part by familiar landmarks, to their home colonies. Later experiments by the two investigators (see Penney and Emlen, 1967) confirmed their original findings and conclusions.

A few years following Kramer's initial discovery that migratory restlessness in caged Starlings was directed toward the sun, Sauer and his wife (Sauer, 1957; Sauer and Sauer, 1960) undertook experiments in Germany to determine by what means nocturnal migrants find their way. The Sauers used three species of sylviid warblers, reared in captivity without ever having seen the natural sky. During the period when the species would normally migrate, the Sauers put each bird in a rotatable circular cage and exposed it to the night sky. Under a clear, starry sky, even when their cages were turned periodically, the test birds fluttered persistently in the direction that the species takes at the season of the experiments; under heavily overcast skies the birds fluttered randomly in various directions, obviously disoriented. Then the Sauers placed the cage in a small planetarium with a starry sky, adjusted to the local time and latitude. The birds fluttered in the direction appropriate to that season depicted. There seemed to be no doubt that these birds obtained their information for direction from the starry sky.

The Sauers carried their experiments further. Placing the birds under skies representing the non-migratory seasons, they noticed that some of the birds were completely disoriented. From these experiments and others, Sauer concluded that birds do not rely on the stars themselves for direction but on the azimuth and altitude of the starry sky. This bicoordinate system would give birds the necessary information on their location since azimuth or hour angle would denote longitude and altitude would indicate latitude.

Working with mist-netted Indigo Buntings in the United States, S. T. Emlen (1967) repeated many of the Sauers' experiments and made still other well-controlled tests under planetarium skies. The results confirmed some of the Sauers' studies but differed sharply from the others. Emlen found no evidence that the buntings relied on a bicoordinate celestial system for orientation but rather made use of the numerous stars, particularly "the constant, two-dimensional spatial relationships" existing between them. He believed that no single star, with the possible exception of Polaris (the North Star) could give the birds sufficient information for direction. It is the configuration or patterning of the bright stars in the constellations such as Ursa Major (the Big Dipper) that gives the cues. While Emlen could not determine with any certainty which patterns were of special importance or essential to orientation, the evidence seemed to point to the northern astral sky, especially the area within 35 degrees of Polaris.

What prompts birds to take the appropriate seasonal direction? Seeking an answer to this question, S. T. Emlen (1969a) induced physiological readiness for spring and fall migration in two groups of Indigo Buntings and then tested them simultaneously under the spring sky of a planetarium. The group conditioned for spring migration oriented northward; the group in condition for fall migration took the opposite direction. This clearly suggested to Emlen that the internal physiological changes in Indigo Buntings, rather than differences in external stimuli (e.g., in the night sky), are responsible for their direction in spring and fall.

Practically all the experiments on the responses of homing and migrating birds to the sun and stars have demonstrated that when skies are heavily overcast the birds show disorientation. The consequent implication is that birds cannot find their way when celestial cues are obscured. Very recent studies, however, indicate that this may be untrue. Keeton (1969) has shown that homing pigeons, while using the sun as a compass when available, can navigate accurately under total overcast and even without familiar landmarks. Radar studies reveal that birds migrate successfully when there is a heavy overcast. If the cloud layer is too high to surmount, birds will fly under it or even in it (Bellrose and Graber, 1963; Eastwood and Rider, 1965). This being the case, how do birds orient themselves without celestial cues and, if they are flying at night or in a cloud layer, without being able to see landmarks or topographical features familiar to them? The cue may be wind direction.

Nisbet (1955) once theorized that the birds might use wind—more specifically atmospheric or wind turbulence — in orientation. In central Illinois, Bellrose (1967b), with his associates, has gathered ample experimental evidence to show that Blue-winged Teal (*Anas discors*), a long-distance migrant that commonly winters in northwestern South America, does indeed use the turbulent structure of the wind for direction. The species migrates by day or night, and radar surveillance and visual sightings by Bellrose confirm its ability to migrate under overcast skies. In the fall and early winter, Bellrose used both hand-raised immature teal that he released at various points up to 100 miles distant from their home pens and wild immature teal, trapped during migration, that he held in pens and released from 10 to 40 days after other Blue-winged Teal had left the region. All his experimental birds were banded. The hand-reared birds had never flown over the area and none of the birds had ever viewed the landscape along the standard migration routes southward. When released under both clear and overcast skies, nearly all the birds tended to start in the same direction — with the wind. Analyzing his data from visual sightings and banding recoveries and drawing upon his knowledge of waterfowl movements from radar studies and airplane tracking, Bellrose concluded that Blue-winged Teal, and probably other species of waterfowl as well, prefer landscape features for orientation, but, if unavailable, they resort to celestial cues. Both being unavailable, they use wind direction, perhaps referring at the outset of flight to landscape cues for information on the sector of the compass from which the wind is blowing.

The present knowledge of direction-finding, the highlights of which are briefly reviewed in the foregoing paragraphs, points all too clearly to the danger of generalizing on the means of orientation and navigation in migration. Different groups of birds in their adaptations to different modes of existence may well have developed correspondingly different means of finding their way from one place to another in accordance with the prevailing ecological conditions. It is unlikely that any species orients itself entirely by one cue. A one-directional orientation may suffice for some birds while a bicoordinate system of navigation may be necessary for others. Direction-finding, though appearing to be basically inherent, may actually prove through eventual research to be acquired chiefly by experience. In all probability, direction-finding in strongly migratory birds is a highly complex procedure involving physiological and behavioral responses to not one but several environmental factors — celestial bodies, wind, and perhaps others presently considered as having no relationship to the problem.

AIDS AND ADDITIONAL SUGGESTIONS FOR THE STUDY OF MIGRATION

Books by the following authors contain good, highly readable information on such aspects of migration as probable origin, routes, distances traveled, relation to weather and season: G. M. Allen (1925), Chapman (1932), Dorst (1962), Griscom (1945), Hochbaum (1955), Lincoln (1950), Thomson (1949), Wetmore (1926).

For general information pertaining to migration movements in conterminous United States the student is referred to the guides by Pettingill (1951, 1953). The introduction to the chapters on each state contains a description of the principal migration concentrations in the state and, in most cases, a series of inclusive dates when one may expect to observe peak flights of the majority of waterfowl, shore birds, and land birds. If there are places in the state where one may observe hawk flights or especially heavy concentrations of migrating birds, they are indicated in the body of the chapter.

Of great value and interest to any student following migration in his own area of conterminous United States or southern Canada are the annual first and fourth numbers of *Audubon Field Notes* (see Appendix G). The first number each year, published in February, contains a country-wide summary and regional reports of the preceding fall migration; the fourth number, published in August, contains similar coverage of the preceding spring migration. The vast amount of information in these numbers is gathered through the cooperative effort of many observers. Any student who is able to identify birds is urged to send in reports of his observations to the particular regional editor handling data from his region.

The U.S. Department of Commerce (Environmental Science Services Administration) publishes a weekly series of *Daily Weather Maps*. These present a surface weather chart at 7:00 A.M. (EST) for each day of the seven-day period. The subscription price as of 1969 is $4.50 per year. To subscribe to this publication, make out an application and send remittance (Post Office Money Order or personal check) to: Superintendent of Documents, Government Printing Office, Washington, D.C. 20402. Most weather maps appearing in newspapers are based on these weather maps.

Study of Nocturnal Migration

There are at least five field methods of studying nocturnal migration. (1) By observations of movements across the face of the moon. The principal equipment needed is a telescope with a power of 15 or greater. For methods and procedures, consult Lowery (1951) and Nisbet (1959) and for the application of accumulated data, see Newman and Lowery (1964) and Lowery and Newman (1955, 1966). (2) By radio-tracking. The equipment needed includes a transmitter and receiver. For general information on the type of equipment required, consult Sclater (1963), Cochran and Lord (1963), and Southern (1965). For information on a transmitter adapted to small migrants, as well as an example of methods and procedures in radio-tracking small migrants, see Cochran, Montgomery, and Graber (1967). (3) By radar surveillance. This requires the use of a special facility operated by skilled technicians. Consult Eastwood (1967) and papers by Drury and Keith (1962), Bellrose and Graber (1963), Hassler, Graber, and Bellrose (1963), and Nisbet (1963). (4) By tape-recording. Besides a tape recorder, this requires a parabolic reflector, microphone, and amplifier. For equipment, methods, and procedures, consult Graber and Cochran (1959) and for an analysis of data recorded, see the same authors (1960). (5) By the study of migrants killed at television towers or migrants killed as a result of airport ceilometers or other man-made interferences with migration.

Of the five listed methods, the most practical for the beginning student is the study of migrants killed at television towers. It requires no special equipment. Practically every city or large community has on its outskirts one or more television towers that almost certainly take a toll of nocturnal migrants. Much as these kills are regretted, they nonetheless provide rewarding material for the study of many aspects of migration locally.

An early-morning search of the ground under a television tower and its guy wires following a night during which migration proceeded at low altitude will usually yield birds of many species. If the student chooses to make such a search as often as these accidents occur through the migration season and on successive seasons, keeping a careful record with dates of the number of individuals of each species killed, he will obtain a true sampling of the nocturnal migratory activity in the area of the tower. He may also discover the presence of species not heretofore reported, or even suspected, in the area because they have always passed through at night unseen.

Given the time necessary, the student may collect specimens for making one or more special studies such as the following:

1. Succession of sexes and age groups in different species: whether males precede females and adults precede immatures, or *vice versa.*

2. Geographic variation in a wide-ranging species migrating through a particular area: whether the color and measurements of individuals reveal one or more populations or subspecies.

3. Molt with relation to migration: whether certain species are in stages of molt and, if so, which feather tracts are involved. Normally birds do not molt the remiges and their primary coverts during migration, but the rest of the alar tract and all the other tracts may be in the process of molt, depending on the species or the sex and/or stage of maturity of individuals in each species.

4. Weight with relation to the length of night-flight: whether the birds are quite heavy, having been killed soon after take-off, or quite light, having been killed after a long flight.

5. Fat condition with relation to stage of migration and duration of night-flight: whether the birds are quite obese, indicating that their migration was well under way with long night-flights, or quite lean, indicating that their migration was just beginning with short night-flights.

Before collecting any specimens, the student must obtain both a federal and a state (or provincial) collecting permit (see Appendix A) authorizing him to take dead birds "for salvage."

Collect the specimens after a kill as early in the morning as possible since the carcasses not only decompose rapidly but are soon molested by insects or eaten by house cats, crows, gulls, and other creatures that inevitably find the ground around a television tower a promising source of food. Weigh each specimen without too much delay as the body begins losing weight shortly after death. Mark the weight on a tag and attach it to a leg. Seal all specimens collected in plastic bags to avoid their dehydration, mark with the date, time of day, and place of collection, and put at once in a deep-freezer for study and analysis later.

Prior to undertaking the study, the student is advised to peruse the following papers: For recording, analyzing, and summarizing data, the papers by Tordoff and Mengel (1956), Brewer and Ellis (1958), and Stoddard and Norris (1967). For studies of special aspects of nocturnal migration, the papers by Odum, Connell, and Stoddard (1961), Graber and Graber (1962), Caldwell, Odum, and Marshall (1963), Raveling (1965), Johnston (1966), and Raveling and LeFebvre (1967).

If the student is to use kills resulting from airport ceilometers or from other lethal interferences with nocturnal migration, he can apply the same study procedures.

Study of Diurnal Migration

Whenever opportunity permits, the student should watch the diurnal movements of migrants in his own area. He will soon discover how greatly birds are influenced in their daytime movements by the terrain, vegetation, and waterways. He is almost certain to note that passerine migrants, although continually searching for food, follow much the same courses and in the same direction. If the birds inhabit trees or shrubs, they choose paths in which these grow and avoid, when possible, crossing wide stretches of water and open country. For an idea of how much detailed information one can obtain in a local study and some of the procedures to follow, the student is referred to the classic work by Ball (1952), in which he reported on six fall migrations on a small point of land of the great Gaspé Peninsula, Quebec.

Given the time necessary, the student may capture migrants for undertaking studies of live birds similar to the first three special studies of specimens killed in nocturnal migration (see above). For methods of trapping and mist-netting migrants and for the procedure in obtaining permits, see Appendix A. Papers that the student may wish to consult for suggestions in recording, analyzing, and summarizing data are the following: Woodford and Lovesy (1958), Annan (1962), Murray and Jehl (1964), Murray (1966b), and Hussell, Davis, and Montgomerie (1967).

REFERENCES

Allen, A. A.
1914 The Red-winged Blackbird: A Study in the Ecology of a Cat-tail Marsh. Proc. Linnaean Soc. New York, Nos. 24-25: 43-128.

Allen, G. M.
1925 Birds and Their Attributes. Marshall Jones Company, Boston. (Dover reprint available. Chapter 13, "Bird Migration.")

Annan, O.
1962 Sequence of Migration, by Sex, Age, and Species, of Thrushes of the Genus Hylocichla, through Chicago. Bird-Banding, 33: 130-137.

Bagg, A. M.
1950 Reverse Warbler Migration in the Connecticut Valley. Auk, 67: 244-245.
1955 Airborne from Gulf to Gulf: April 16-18, 1954. Bull. Massachusetts Audubon Soc., 39: 106-110, 159-168. (A migration of the Indigo Bunting, *Passerina cyanea*, from the Gulf of Mexico to the Gulf of Maine. An illuminating account showing the relation of weather phenomena to migration.)
1967 Factors Affecting the Occurrence of the Eurasian Lapwing in Eastern North America. Living Bird, 6: 87-121.

Bagg, A. M., and others
1950 Barometric Pressure-patterns and Spring Bird Migration. Wilson Bull., 62: 5-19.

Baird, J., and I. C. T. Nisbet
1960 Northward Fall Migration on the Atlantic Coast and Its Relation to Offshore Drift. Auk, 77: 119-149.

Ball, S. C.
1952 Fall Bird Migration on the Gaspé Peninsula. Peabody Mus. Nat. Hist., Bull. 7.

Bellrose, F. C.
1957 A Spectacular Waterfowl Migration through Central North America. Illinois Dept. Registration and Education, Nat. Hist. Surv. Div., Biol. Notes No. 36.
1958 Celestial Orientation by Wild Mallards. Bird-Banding, 29: 75-90.
1963 Orientation Behavior of Four Species of Waterfowl. Auk, 80: 257-289.
1967a Radar in Orientation Research. Proc. XIVth Internatl. Ornith. Congr., pp. 281-309.
1967b Orientation in Waterfowl Migration. In *Animal Orientation and Navigation*. Edited by R. M. Storm. Oregon State University Press, Corvallis.

Bellrose, F. C., and R. R. Graber
1963 A Radar Study of the Flight Directions of Nocturnal Migrants. Proc. XIIIth Internatl. Ornith. Congr., pp. 362-389.

Bennett, H. R.
1952 Fall Migration of Birds at Chicago. Wilson Bull., 64: 197-220.

Bergman, G., and K. O. Donner
1964 An Analysis of the Spring Migration of the Common Scoter and Long-tailed Duck in Southern Finland. Acta Zool. Fennica 105.

Bergstrom, E. A., and W. H. Drury, Jr.
1956 Migration Sampling by Trapping: A Brief Review. Bird-Banding, 27: 107-120. (How to gather data and some suggestions for their use.)

Billings, S. M.
1968 Homing in Leach's Petrel. Auk, 85: 36-43. (Two breeding adults, released from England, returned to their nests at Kent Island, New Brunswick in 13.7 days, averaging 217 miles per day.)

Brackbill, H.
1959 Migration of Breeding American Robins at Baltimore. Bird-Banding, 30: 122. (Majority of males return before majority of females.)

Brewer, R., and J. A. Ellis
1958 An Analysis of Migrating Birds Killed at a Television Tower in East-central Illinois, September 1955-May 1957. Auk, 75: 400-414.

Broun, M.
1941 Migration of Blue Jays. Auk, 58: 262-263.
1949 Hawks Aloft: The Story of Hawk Mountain. Dodd, Mead, and Company, New York.
1951 Hawks and the Weather. Atlantic Nat., 6: 105-112.
1965 At Hawk Mountain Sanctuary. In *The Bird Watcher's America*. Edited by O. S. Pettingill, Jr. McGraw-Hill Book Company, New York.

Caldwell, L. D., E. P. Odum, and S. G. Marshall
1963 Comparison of Fat Levels in Migrating Birds Killed at Central Michigan and a Florida Gulf Coast Television Tower. Wilson Bull., 75: 428-434.

Carthy, J. D.
1956 Animal Navigation. George Allen and Unwin, London.

Chapman, F. M.
1932 Handbook of Birds of Eastern North America. Second revised edition. D. Appleton Company, New York. (Dover reprint available. "The Migration of Birds," pp. 35-66.)

Christian, G.
1961 Down the Long Wind: A Study of Bird Migration. Newnes, London.

Cochran, W. W., and R. R. Graber
1958 Attraction of Nocturnal Migrants by Lights on a Television Tower. Wilson Bull., 70: 378-380.

Cochran, W. W., and R. D. Lord, Jr.
1963 A Radio-tracking System for Wild Animals. Jour. Wildlife Management, 27: 9-24.

Cochran, W. W., G. G. Montgomery, and R. R. Graber
1967 Migratory Flights of *Hylocichla* Thrushes in Spring: A Radiotelemetry Study. Living Bird, 6: 213-225.

Coffey, B. B., Jr.
1943 Post-juvenal Migration of Herons. Bird-Banding, 14: 34-39.

Cooke, W. W.
1904 Some New Facts about the Migration of Birds. In: *Yearbook of the United States Department of Agriculture—1903.*
1905 Routes of Bird Migration. Auk, 22: 1-11.
1915 Bird Migration. U. S. Dept. Agric. Bull. No. 185.

Cox, G. W.
1968 The Role of Competition in the Evolution of Migration. Evolution, 22: 180-192.

Dorst, J.
1962 The Migrations of Birds. Houghton Mifflin Company, Boston. (Deals primarily with migratory movements in different parts of the world.)

Drury, W. H., and J. A. Keith
1962 Radar Studies of Songbird Migration in Coastal New England. Ibis, 104: 449-489.

Drury, W. H., Jr., and I. C. T. Nisbet
1964 Radar Studies of Orientation of Songbird Migrants in Southeastern New England. Bird-Banding, 35: 69-119.

Duck, L. G.
1943 Seasonal Movements of Bobwhite Quail in Northwestern Oklahoma. Jour. Wildlife Management, 7: 365-368.

Eastwood, E.
1967 Radar Ornithology. Methuen and Company, London.

Eastwood, E., and G. C. Rider
1965 Some Radar Measurements of the Altitude of Bird Flight. Brit. Birds, 58: 393-426.

Elder, W. H., and N. L. Elder
1949 Role of the Family in the Formation of Goose Flocks. Wilson Bull., 61: 133-140.

Emlen, J. T., and R. L. Penney
1964 Distance Navigation in the Adelie Penguin. Ibis, 106: 417-431.

Emlen, S. T.
1967 Migratory Orientation in the Indigo Bunting, *Passerina cyanea*. Part I: Evidence for Use of Celestial Cues. Auk, 84: 309-342. Part II: Mechanism of Celestial Orientation. Auk, 84: 463-489.

1969a Bird Migration: Influence of Physiological State upon Celestial Orientation. Science, 165: 716-718.

1969b The Development of Migratory Orientation in Young Indigo Buntings. Living Bird 8:113-126.

Emlen, S. T., and J. T. Emlen
1966 A Technique for Recording Migratory Orientation of Captive Birds. Auk, 83: 361-367.

Farner, D. S.
1955 The Annual Stimulus for Migration: Experimental and Physiologic Aspects. In *Recent Studies in Avian Biology*. Edited by A. Wolfson. University of Illinois Press, Urbana.

1960 Metabolic Adaptations in Migration. Proc. XIIth Internatl. Ornith. Congr., pp. 197-208.

Fisher, R. M.
1953 How to Know and Predict the Weather. (A Mentor Book.) The New American Library, New York. (A useful work for one wishing to learn about the weather.)

Gill, G.
1941 Notes on the Migration of Blue Jays. Bird-Banding, 12: 109-112.

Gilliard, E. T.
1958 Living Birds of the World. Doubleday and Company, Garden City, New York.

Goldsmith, T. H., and D. R. Griffin
1956 Further Observations on Homing Terns. Biol. Bull., 111: 235-239.

Graber, R. R.
1968 Nocturnal Migration in Illinois—Different Points of View. Wilson Bull., 80: 36-71.

Graber, R. R., and W. W. Cochran
1959 An Audio Technique for the Study of Nocturnal Migration of Birds. Wilson Bull., 71: 220-236. (With the use of a parabolic reflector, microphone, amplifier, and recorder, the authors show how they gather data on hours of nocturnal calling, flight call density, and identity of species calling.)

1960 Evaluation of an Aural Record of Nocturnal Migration. Wilson Bull., 72: 253-273. (A wealth of information gathered by applying the technique described in the earlier paper.)

Graber, R. R., and J. W. Graber
1962 Weight Characteristics of Birds Killed in Nocturnal Migration. Wilson Bull., 74: 74-88.

Graue, L. C.
1965 Experience Effect on Initial Orientation in Pigeon Homing. Animal Behaviour, 13: 149-153.

Griffin, D. R.
1955 Bird Navigation. In *Recent Studies in Avian Biology*. Edited by A. Wolfson. University of Illinois Press, Urbana.

1964 Bird Migration. Doubleday and Company, Garden City, New York.

Griffin, D. R., and T. H. Goldsmith
1955 Initial Flight Direction of Homing Birds. Biol. Bull., 108: 264-276.

Griffin, D. R., and R. J. Hock
1949 Airplane Observations of Homing Birds. Ecology, 30: 176-198.

Griscom, L.
1945 Modern Bird Study. Harvard University Press, Cambridge, Massachusetts. (Chapter 4, "Migration: Causes and Origin"; Chapter 5, "Migration: Factors and Routes.")

Gross, A. O.
1940 The Migration of Kent Island Herring Gulls. Bird-Banding, 11: 129-155.

Hagar, J. A.
1966 Nesting of the Hudsonian Godwit at Churchill, Manitoba. Living Bird, 5: 5-43.

Hamilton, W. J., III
1962a Bobolink Migratory Pathways and Their Experimental Analysis under Night Skies. Auk, 79: 208-233. (Orientational choice based on the stars and internal clock in phase with local time.)

1962b Does the Bobolink Navigate? Wilson Bull., 74: 357-366.

1962c Evidence Concerning the Function of Nocturnal Call Notes of Migratory Birds. Condor, 64: 390-401.

1967 Social Aspects of Bird Orientation Mechanisms. In *Animal Orientation and Navigation*. Edited by R. M. Storm. Oregon State University Press, Corvallis.

Hassler, S. S., R. R. Graber, and F. C. Bellrose
1963 Fall Migration and Weather, a Radar Study. Wilson Bull., 75: 56-77. (Departures of nocturnal migrants—particularly long-distance migrants—are released by a change in wind direction from south to north. Such wind shifts almost always accompany cold fronts, but the migrants react to the wind shift whether or not a cold front is involved. Following wind is important for long-distance migrants. Cloud cover can modify response by deterring some migrants, or postpone flights.)

Hawkes, A. L.
1965 At Block Island in the Fall. In *The*

Bird Watcher's America. Edited by O. S. Pettingill, Jr. McGraw-Hill Book Company, New York.

Hochbaum, H. A.
1955 Travels and Traditions of Waterfowl. University of Minnesota Press, Minmeapolis. (Paperback edition, 1967.)

Hoffmann, K.
1960 Experimental Manipulation of the Orientational Clock in Birds. Cold Spring Harbor Symp. Quant. Biol., 25: 379-387.

Hofslund, P. B.
1965 Hawks above Duluth. In *The Bird Watcher's America*. Edited by O. S. Pettingill, Jr. McGraw-Hill Book Company, New York.
1966 Hawk Migration over the Western Tip of Lake Superior. Wilson Bull., 78: 79-87.

Howell, J. C., A. R. Laskey, and J. T. Tanner
1954 Bird Mortality at Airport Ceilometers. Wilson Bull., 66: 207-215.

Howell, T. R.
1953 Racial and Sexual Differences in Migration in *Sphyrapicus varius*. Auk, 70: 118-126.

Huntington, C. E.
1967 Leach's Petrel. Bowdoin Alumnus, 42 (1): 6-10.

Hussell, D. J. T., T. Davis, and R. D. Montgomerie
1967 Differential Fall Migration of Adult and Immature Least Flycatchers. Bird-Banding, 38: 61-66.

Imhof, T. A.
1953 Effect of Weather on Spring Bird Migration in Northern Alabama. Wilson Bull., 65: 184-195.

James, P.
1956 Destruction of Warblers on Padre Island, Texas, in May, 1951. Wilson Bull., 68: 224-227.

Johnston, D. W.
1966 A Review of the Vernal Fat Deposition Picture in Overland Migrant Birds. Bird-Banding, 37: 172-183. (The premigratory bird contains enough fat to sustain a flight of about 90 miles. Premigrants initiate northward movement with short flights. As they migrate northward and as the season progresses, they become more and more obese. The farther north they move, the longer the flights and less frequent the stopovers.)

Johnston, D. W., and R. W. McFarlane
1967 Migration and Bioenergetics of Flight in the Pacific Golden Plover. Condor, 69: 156-168. (The authors raise doubts about the migration rate of the American Golden Plover.)

Karplus, M.
1952 Bird Activity in the Continuous Daylight of Arctic Summer. Ecology, 33: 129-134.

Keeton, W. T.
1969 Orientation by Pigeons: Is the Sun Necessary? Science, 165: 922-928.

Kenyon, K. W., and D. W. Rice
1958 Homing of Laysan Albatrosses. Condor, 60: 3-6.

Kramer, G.
1952 Experiments on Bird Orientation. Ibis, 94: 265-285.
1959 Recent Experiments on Bird Orientation. Ibis, 101: 399-416.
1961 Long-distance Orientation. In *Biology and Comparative Physiology of Birds*. Edited by A. J. Marshall. Academic Press, New York.

Krause, H.
1965 Geese along the Missouri. In *The Bird Watcher's America*. Edited by O. S. Pettingill, Jr. McGraw-Hill Book Company, New York. (The spectacular spring movements, colorfully described.)

Lack, D.
1960a The Height of Bird Migration. Brit. Birds., 53: 5-10.
1960b The Influence of Weather on Passerine Migration. A Review. Auk, 77: 171-209.
1962 Migration across the Southern North Sea Studied by Radar. Part 3. Movements in June and July. Ibis, 104: 74-85.
1963 Weather Factors Initiating Migration. Proc. XIIIth Internatl. Ornith. Congr., pp. 412-414.

Lee, S. L. B.
1963 Migration in the Outer Hebrides Studied by Radar. Ibis, 105: 493-515.

Lewis, H. F.
1939 Reverse Migration. Auk, 56: 13-27.

Lincoln, F. C.
1950 Migration of Birds. U. S. Dept. Interior, Fish and Wildlife Service Circ. 16. (Commercial edition published in 1952 by Doubleday and Company, Garden City, New York.)

Livingston, J. A.
1965a At Point Pelee in the Spring. In *The Bird Watcher's America*. Edited by O. S. Pettingill, Jr. McGraw-Hill Book Company, New York.
1965b The Cranes at Last Mountain Lake.

In *The Bird Watcher's America*. Edited by O. S Pettingill, Jr. McGraw-Hill Book Company, New York. (In late summer thousands of Sandhill Cranes gather in central Saskatchewan before moving southward. A vivid description of their behavior.)

Lowery, G. H., Jr.
1945 Trans-Gulf Spring Migration of Birds and the Coastal Hiatus. Wilson Bull., 57: 92-121.
1946 Evidence of Trans-Gulf Migration. Auk, 63: 175-211.
1951 A Quantitative Study of the Nocturnal Migration of Birds. Univ. Kans. Publ. Mus. Nat. Hist., 3: 361-472.

Lowery, G. H., Jr., and R. J. Newman
1955 Direct Studies of Nocturnal Bird Migration. In *Recent Studies in Avian Biology*. Edited by A. Wolfson. University of Illinois Press, Urbana.
1966 A Continentwide View of Bird Migration on Four Nights in October. Auk, 83: 547-586.

Manwell, R. D.
1962 The Homing of Cowbirds. Auk, 79: 649-654.

Marler, P., and W. J. Hamilton, III
1956 Mechanisms of Animal Behavior. John Wiley and Sons, New York. (Chapter 15: "Spatial Orientation"; Chapter 16: "Navigation and Homing.")

Matthews, G. V. T.
1951 The Experimental Investigation of Navigation in Homing Pigeons. Jour. Exp. Biol., 28: 508-536.
1952a An Investigation of Homing Ability in Two Species of Gulls. Ibis, 94: 243-264.
1952b The Relation of Learning and Memory to the Orientation and Homing of Pigeons. Behaviour, 4: 202-221.
1953 Sun Navigation in Homing Pigeons. Jour. Exp. Biol., 30: 243-267.
1955a An Investigation of the "Chronometer" Factor in Bird Navigation. Jour. Exp. Biol., 32: 39-58.
1955b Bird Navigation. University Press, Cambridge.
1961 "Nonsense" Orientation in Mallard *Anas platyrhynchos* and Its Relation to Experiments on Bird Navigation. Ibis, 103a: 211-230.
1963 The Orientation of Pigeons as Affected by Learning of Landmarks and by the Distance of Displacement. Animal Behaviour, 11: 310-317.

1964a Individual Experience as a Factor in the Navigation of Manx Shearwaters. Auk, 81: 132-146.
1964b Navigation. In *A New Dictionary of Birds*. Edited by A. L. Thomson. McGraw-Hill Book Company, New York.
1968 Bird Navigation. Second edition. University Press, Cambridge. (A complete revision of his 1955 book, including a thorough coverage of the literature on the subject.)

Mazzeo, R.
1953 Homing of the Manx Shearwater. Auk, 70: 200-201.

McAtee, W. L., and others
1944 Eastward Migration through the Gulf States. Wilson Bull., 56: 152-160.

McCabe, T. T.
1942 Types of Shorebird Flight. Auk, 59: 110-111.

Mewaldt, L. R.
1964 California Sparrows Return from Displacement to Maryland. Science, 146 (3646): 941-942.

Mewaldt, L. R., M. L. Morton, and I. L. Brown
1964 Orientation of Migratory Restlessness in Zonotrichia. Condor, 66: 377-417.

Miller, A. D.
1953 Further Evidence of the Homing Ability of the Cowbird. Wilson Bull., 65: 206-207.

Mueller, H. C., and D. D. Berger
1961 Weather and Fall Migration of Hawks at Cedar Grove, Wisconsin. Wilson Bull., 73: 171-192.
1967 Some Observations and Comments on the Periodic Invasions of Goshawks. Auk, 84: 183-191.

Murray, B. G., Jr.
1965 On the Autumn Migration of the Blackpoll Warbler. Wilson Bull., 77: 122-133.
1966a Blackpoll Warbler Migration in Michigan. Jack-Pine Warbler, 44: 23-29.
1966b Migration of Age and Sex Classes of Passerines on the Atlantic Coast in Autumn. Auk, 83: 352-360.

Murray, B. G., Jr., and J. R. Jehl, Jr.
1964 Weights of Autumn Migrants from Coastal New Jersey. Bird-Banding, 35: 253-263.

Newman, R. H., and G. H. Lowery, Jr.
1964 Selected Quantitative Data on Night Migration in Autumn. Louisiana State Univ. Mus. Zool. Special Publs., 3: 1-39.

Nisbet, I. C. T.
  1955 Atmospheric Turbulence and Bird Flight. Brit. Birds, 48: 557-559.
  1959 Calculation of Flight Directions of Birds Observed Crossing the Face of the Moon. Wilson Bull., 71: 237-243. ("Instructions...from which observers without mathematical training can calculate flight directions and approximate densities of migrating birds observed flying across the face of the moon.")
  1963 Measurements with Radar of the Height of Nocturnal Migration over Cape Cod, Massachusetts. Bird-Banding, 34: 57-67.

Nisbet, I. C. T., and W. H. Drury, Jr.
  1967 Orientation of Spring Migrants Studied by Radar. Bird-Banding, 38: 173-186.

Norris, R. A.
  1961 A New Method of Preserving Bird Specimens. Auk, 78: 436-440. (Mounting skins flat on cards on which there is room for data; a particularly economical method for handling large kills from television towers.)

Odum, E. P., C. E. Connell, and H. L. Stoddard
  1961 Flight Energy and Estimated Flight Ranges of Some Migratory Birds. Auk, 78: 515-527.

Packard, F. M.
  1945 The Birds of Rocky Mountain National Park, Colorado. Auk, 62: 371-394.
  1946 Midsummer Wandering of Certain Rocky Mountain Birds. Auk, 63: 152-158.

Penney, R. L., and J. T. Emlen
  1967 Further Experiments on Distance Navigation in the Adelie Penguin *Pygoscelis adeliae*. Ibis, 109: 99-109.

Perdeck, A. C.
  1958 Two Types of Orientation in Migrating Starlings, *Sturnus vulgaris* L., and Chaffinches, *Fringilla coelebs* L., as Revealed by Displacement Experiments. Ardea, 46: 1-37.

Pettingill, O. S., Jr.
  1951 A Guide to Bird Finding East of the Mississippi. Oxford University Press, New York.
  1953 A Guide to Bird Finding West of the Mississippi. Oxford University Press, New York.
  1962 Hawk Migrations around the Great Lakes. Audubon Mag., 64: 44-45, 49.
  1964 Spring Migration at Point Pelee. Audubon Mag., 66: 78-80.

Phillips, A. R.
  1951 Complexities of Migration: A Review. With Original Data from Arizona. Wilson Bull., 63: 129-136.

Raveling, D. G.
  1965 Geographic Variation and Measurements of Tennessee Warblers Killed at a TV Tower. Bird-Banding, 36: 89-101.

Raveling, D. G., and E. A. LeFebvre
  1967 Energy Metabolism and Theoretical Flight Range of Birds. Bird-Banding, 38: 97-113.

Roberts, T. S.
  1907 A Lapland Longspur Tragedy. Auk, 24: 369-377.

Robertson, W. B., Jr.
  1969 Transatlantic Migration of Juvenile Sooty Terns. Nature, 222: 632-634.

Rowan, W.
  1946 Experiments in Bird Migration. Trans. Royal Soc. Canada, 40: 123-135.

Salomonsen, F.
  1967 Migratory Movements of the Arctic Tern *(Sterna paradisaea* Pontoppidan) in the Southern Ocean. Det Kgl. Danske Vid. Selsk., Biol. Medd., 24, 1.

Sauer, E. G. F.
  1957 Die Sternenorientierung nächtlich ziehender Grasmücken *(Sylvia atricapilla, borin,* und *curruca* L.). Zeits. Tierpsychol., 14: 29-70.

Sauer, E. G. F., and E. M. Sauer
  1960 Star Navigation of Nocturnal Migrating Birds. Cold Spring Harbor Symp. Quant. Biol., 25: 463-473.

Saunders, W. E.
  1907 A Migration Disaster in Western Ontario. Auk, 24: 108-110.

Schmidt-Koenig, K.
  1961 Die Sonne als Kompass im Heim-Orientierungssytem der Brieftauben. Zeits. Tierpsychol., 18: 221-224.
  1965 Current Problems in Bird Orientation. In *Advances in the Study of Behavior*. Volume 1. Edited by D. S. Lehrman, R. A. Hinde, and E. Shaw. Academic Press, New York.

Schnell, G. D.
  1965 Recording the Flight-speed of Birds by Doppler Radar. Living Bird, 4: 79-87.

Schorger, A. W.
  1961 Migration of Blue Jays at Madison, Wisconsin. Wilson Bull., 73: 393-394.

Schüz, E.
  1949 Die Spät-Auflassung Ostpreussischer Jungstörche in West-Deutschland

durch die Vogelwarte Rossitten 1933. Vogelwarte, 14: 63-78.

Sclater, L., Editor
1963   Bio-Telemetry: The Use of Telemetry in Animal Behavior and Physiology in Relation to Ecological Problems. Pergamon Press, New York.

Shaub, M. S.
1963   Evening Grosbeak Winter Incursions —1958-59, 1959-60, 1960-61. Bird-Banding, 34: 1-22.

Sheldon, W.
1965   Hawk Migration in Michigan and the Straits of Mackinac. Jack-Pine Warbler, 43: 79-83.

Shelford, V. E.
1945   The Relation of Snowy Owl Migration to the Abundance of the Collared Lemming. Auk, 62: 592-596.

Smith, W. J.
1959   Movements of Michigan Herring Gulls. Bird-Banding, 30: 69-104.

Snyder, L. L.
1943   The Snowy Owl Migration of 1941-42. Wilson Bull., 55: 8-10.

Southern, W. E.
1959   Homing of Purple Martins. Wilson Bull., 71: 254-261.
1965   Biotelemetry: A New Technique for Wildlife Research. Living Bird, 4: 45-58.

Speirs, J. M.
1945   Flight Speed of the Old-squaw. Auk, 62: 135-136.

Stevenson, H. M.
1957   The Relative Magnitude of the Trans-Gulf and Circum-Gulf Spring Migrations. Wilson Bull., 69: 39-77.

Stoddard, H. L. Sr., and R. A. Norris
1967   Bird Casualties at a Leon County, Florida TV Tower: An Eleven-year Study. Bull. Tall Timbers Res. Sta. No. 8.

Stone, W.
1937   Bird Studies at Old Cape May. Two volumes. Academy of Natural Sciences of Philadelphia. (Dover reprint available.)

Swarth, H. S.
1920   Revision of the Avian Genus Passerella, with Special Reference to the Distribution and Migration of the Races in California. Univ. California Publ. in Zool., 21: 75-224.

Tedd, J. G., and D. Lack
1958   The Detection of Bird Migration by High Power Radar. Proc. Royal Soc., B, 149: 503-510.

Thomson, A. L.
1949   Bird Migration: A Short Account.

Third edition. H. F. and G. Witherby, London.

Tordoff, H. B., and R. M. Mengel
1956   Studies of Birds Killed in Nocturnal Migration. Univ. Kansas Publ. Mus. Nat. Hist., 10: 1-44.

Walcott, C., and M. Michener
1967   Analysis of Tracks of Single Homing Pigeons. Proc. XIVth Internatl. Ornith. Congr., pp. 311-329.

Wallraff, H. G.
1967   The Present Status of Our Knowledge about Pigeon Homing. Proc. XIVth Internatl. Ornith. Congr., pp. 331-358.

Ward, P.
1963   Lipid Levels in Birds Preparing to Cross the Sahara. Ibis, 105: 109-111.

Watson, J. B.
1909   Some Experiments on Distant Orientation. Papers Tortugas Lab. Carnegie Inst., 2: 227-230.

Wetmore, A.
1926   The Migrations of Birds. Harvard University Press, Cambridge, Massachusetts.

Wharton, W. P.
1959   Homing by a Female Cowbird. Bird-Banding, 30: 228.

Wolfson, A.
1945   The Role of the Pituitary, Fat Deposition, and Body Weight in Bird Migration. Condor, 47: 95-127.
1952   Day Length, Migration, and Breeding Cycles in Birds. Sci. Monthly, 74: 191-200.
1954a Body Weight and Fat Deposition in Captive White-throated Sparrows in Relation to the Mechanics of Migration. Wilson Bull., 66: 112-118.
1954b Weight and Fat Deposition in Relation to Spring Migration in Transient White-throated Sparrows. Auk, 71: 413-434.

Woodford, J., and F. T. Lovesy
1958   Weights and Measurements of Wood Warblers at Pelee Island. Bird-Banding, 29: 109-110.

Yeagley, H. L.
1947   A Preliminary Study of a Physical Basis of Bird Navigation. Jour. Applied Phys., 18: 1035-1063.
1951   A Preliminary Study of a Physical Basis of Bird Navigation. II. Jour. Applied Phys., 22: 746-760. (In these papers the author advances the hypothesis that homing pigeons are guided latitudinally by the Coriolis force and longitudinally by the earth's magnetic field.)

TERRITORY

Any area defended by a bird against individuals of its own species is **territory.** Most bird species show at least some type of territorialism.

CLASSIFICATION OF TERRITORY

Territory may be roughly classified into two main categories — **breeding territory** and **non-breeding territory;** each category includes several types. Indicated below are the types of territory. Enter names of species showing each type in the spaces provided.

I. Breeding Territory

    A. Mating, nesting, and feeding area for adults and young. The commonest type of territory and characteristic of most passerine birds. Once a pair of birds establishes a territory, they customarily remain on it until the young are independent.

    _____   _____

    B. Mating and nesting (but not feeding) area. Characteristic of a few species. Closely related to and sometimes not easily distinguished from Type A.

    _____   _____

    C. Mating area only. A "court" or "singing ground" apart from the nest, maintained by the male in certain species. "Arenas," "tournament grounds," or "leks," where two or more males of a particular species gather for mating purposes, are considered breeding territories of this type even though the males do not defend the areas in the strict sense. The nesting area, maintained by the female only, is not considered breeding territory.

    _____   _____

    D. Restricted mating and nesting area. Found in colonial species or in a few solitary nesting species. Different from Type B in that the territory is in the immediate vicinity of the nest.

    _____   _____

II. Non-Breeding Territory

    A. Feeding territories. Feeding areas outside the breeding territory but nevertheless defended.

    _____   _____

B. **Winter territories.** Areas defended through the winter months, particularly by permanent-resident birds that may or may not use the same territories during the breeding season.

_____   _____

_____   _____

C. **Roosting territories.** Specific areas used for night roosting. Evidence suggests that certain species defend such areas.

_____   _____

_____   _____

Not all species show the types of territorialism outlined above. Some show combinations of two or more types; a few show no evidence of any type. Furthermore, a species may show a **strong** territorialism, or a **weak** territorialism, or a species may vary as to its territorialism under different ecological conditions.

For a discussion of the different types of territory, consult the papers by Mayr (1935) and Nice (1941) from which the above classification has been adapted. The information below concerns breeding territory only.

ESTABLISHMENT AND MAINTENANCE OF TERRITORY

The male of the species usually establishes a breeding territory and defends it against other males of the same species. The female may or may not participate in territorial defense; if she does participate, her role is generally less active and is usually directed toward other females of the species. The primary purpose of the defense is the territory itself, not the sex-partner, nor the nest and young. Competition for territory is theoretically intraspecific, not interspecific, although interspecific competition sometimes occurs between closely related species.

The methods by which males establish their territories involve hostile behaviors — threat displays, physical encounters, appeasement displays, pursuit-flying, and singing or other vocalizations. (For further information on hostile behavior, see the section, "Behavior," page 254.) All such hostile activities continue intensively while the females are building nests and laying eggs. Generally older males show more aggression than younger and consequently succeed in establishing their territories in areas with more optimal habitat.

How the territories are maintained after the eggs are laid depends on the particular species concerned, type of territory, population density, and so on. As a rule, in passerine species with Breeding Territory Type A, any two neighboring males, once their mates are incubating full clutches of eggs, tend to moderate their aggressive behavior. Encounters between them become briefer and less frequent. More often when they meet at the common boundary, they jockey into position for encounters without going through with them. After the eggs in the nest have hatched, the males increasingly orient their attention toward the defense of their mates and young with the result that territory as a defended area begins to break down.

SIZE OF TERRITORY

A combination of factors dictates the size of a breeding bird's territory: (1) The size of the species — the larger it is, the larger the territory it requires. (2) Density of the species' population — the greater the population is, the smaller each individual territory. (3) Behavior of the species — whether it is strongly or mildly aggressive; whether it nests in colonies or solitarily. (4) Available habitat—whether it is extensive or restricted. (5) Food supply—whether it is ample or sparse.

Extremes in the size of territory are demonstrated on the one hand by the Bald Eagle *(Haliaeetus leucocephalus)* and on the other by the murres *(Uria* spp.). The eagle maintains a territory of about one square mile (Broley, 1947); the murre, nesting in a colony on a rocky ledge overlooking the sea, has an average territory of about one square foot, so small that occupant birds sometimes touch one another (Tuck, 1960). Territories of many other colonial sea birds, such as Gentoo Penguins *(Pygoscelis papua)*, are just large enough to prevent the incubating birds from reaching one another.

Occasionally, species of similar size in similar habitats may have vastly different space requirements. For example, 13 pairs of Least Flycatchers *(Empidonax minimus)* in a Michigan woods had territories averaging 0.18 acre (MacQueen, 1950), yet 15 pairs of Black-capped Chickadees in a New York woods had territories averaging 13.2 acres (Odum, 1941). Territory in a forest may involve only a particular vertical section. Thus the territory of a Red-eyed Vireo *(Vireo olivaceus)*, even though it averages 1.7 acres of forest, actually includes just the upper canopy down to the edge of the lower (Southern, 1958).

FUNCTIONS OF TERRITORY

Breeding territory may (1) guarantee essential cover, nesting materials, and food for developing young, (2) protect the nest, sex-partner, and young against despotism of other males, (3) limit populations in a bird community so that it will not exceed its carrying capacity, (4) serve as a means of instigating the sexual bond by attracting the female through the singing of the male, and (5) facilitate the re-mating of sex-partners from the previous breeding season.

The student should realize that all of the above functions, attributed to territory, are based largely on speculation rather than proven facts. According to the consensus of most authorities, different bird species are innately adapted to particular habitats (see the section, "Behavior," p. 252) and, within their respective habitats, each species through its aggressive behavior spaces itself out (Tinbergen, 1957). This control of population density by avoiding crowding allows for survival factors such as a reserved food supply. On a 40-acre tract of coniferous forest in northern Maine, Stewart and Aldrich (1951) killed 81 percent of the territorial passerine species, mostly parulid warblers, and during the succeeding days from June 15 to July 8 kept the numbers reduced to about that level. When they were through, they had eliminated twice as many males as were present when the experiment began. The same procedure was repeated the next season in the same area (Hensley and Cope, 1951) with similar results. Both experiments show that the potential for crowding exists in an area and suggest that only the hostile activities of territorial males keep an area from being overpopulated. Lack (1954, 1965) dissents from this view, believing that males space themselves out by *avoiding* occupied areas rather than by aggressive behavior. He also believes that food supply is not a factor in maintaining territory since, in his opinion, few territorial birds feed and gather food for their young solely on their own territories.

Tinbergen (1957) stresses the point that once a bird establishes its territory it becomes conditioned or closely attached to it with many resulting advantages besides reserving an adequate food supply. Close attachment, for example, enables the bird to become intimately acquainted with suitable cover for ready escape from predators, with sources of nest material, with places for bathing or dusting, and so on—all certain to enhance its survival. Many birds become so closely attached to their territories that they return to them and re-mate season after breeding season.

The student desiring further information on the subject of territory will do well to consult the July number of *The Ibis* for 1956 (Number 3, Volume 98,) in which there are 18 papers on territory in different species of Old and New World birds, mainly by British and American authors. He will also find profitable reading on territory in Lack (1965), Penney (1968), Snow (1958), and Tinbergen (1939, 1953, 1957). For a recent review and re-evaluation of territorial behavior as it relates to the limitation of populations, the student is referred to Brown (1969).

HOME RANGE

Home range is the total area that a bird habitually occupies. It may be the same as territory if the bird defends the whole area. It may be territory and the area outside that the bird frequents for food, water, bathing, dusting, etc. Or it may contain no territory—for instance, in cases where territory has broken down after the breeding season.

The home range of species with Breeding Territory Types B and D comprises vastly more area outside the territories than within, whereas the home range of species with Breeding Territory Type A is little larger than the territory itself. Any home range has indefinite boundaries that can be determined only by observing the extent and frequency of the individual bird's movements.

THE STUDY OF TERRITORY

Find several adjacent nests of one species in a given area and determine the associated territories. If possible, find the nests early in the breeding cycle—preferably at the time of nest-building or during the egg-laying period—when the birds defend their territories vigorously, making them more evident. Prepare a rough map of the area; indicate the topographical features, distribution of vegetational types, conspicuous landmarks (roads, fences, houses, large tree stubs, boulders, etc.), and the location of each nest.

Procedure

Follow the activities of each nesting male, marking on the map the points where singing, threat displays, pursuit-flying, and fighting take place. In determining the extent of the male's territory, play tape recordings of another singing male and also affix in full view either a stuffed specimen or a realistic model of a male in full breeding plumage. Presumably the points where the occupant male fails to show aggression toward the recordings or visual representation of another male will mark the boundary of his territory. (For other methods of determining territory, see the paper by Odum and Kuenzler, 1955.) Follow the activities of the female, looking for any evidence of intraspecific competition. If possible, capture both sexes and mark them for individual identification. (See Appendix A for methods of capturing and marking.)

Make a series of observations throughout the nesting cycle, taking full notes on all territorial activities, and recording the date, length, and time of day of each activity. After gaining experience in recognizing territories already established, attempt to find territories being established.

Presentation of Results

The marked areas on the map will show, at least roughly, the extent of each territory. Now prepare a finished map similar in content to the first one, but show only the boundary of each territory by drawing a line through the extreme points where the males appeared. Then prepare a paper based on a careful analysis of notes taken. (See Apendix B for directions on preparing the paper.) Give attention to some of the important problems suggested below:

Size of territory estimated in square feet or acres.

Physical and biological characteristics of territory. Illustrate with at least one photograph.

Establishment and defense of territory (with emphasis on the methods employed and the role of the sexes).

Location of nest (or nests, if the species is polygamous) in the territory and noteworthy relationships between nest location and territorial activity.

Length of time during the breeding season that the bird maintains territory.

Changes in extent of territory and in behavior of the adults during second and third nestings of a season.

Maintenance of territory at times other than the breeding season.

The student working on territories of the same species in the same area in subsequent years should attempt to find out whether or not the same individuals re-occupy the territories.

REFERENCES

Armstrong, J. T.
    1965 Breeding Home Range in the Nighthawk and Other Birds; Its Evolutionary and Ecological Significance. Ecology, 46: 619-629. (Neither refutes nor supports argument that food needs determine home range size in mammals. In birds, territories are not adjusted to food needs, but are in excess of them. Population increase may be "damped" by social process of Type A territoriality.)

Broley, C. L.
    1947 Migration and Nesting of Florida Bald Eagles. Wilson Bull., 59:3-20.

Brown, J. L.
    1969 Territorial Behavior and Population Regulation in Birds: A Review and Re-evaluation. Wilson Bull., 81:293-329.

Chapman, F. M.
    1935 The Courtship of Gould's Manakin (*Manacus vitellinus vitellinus*) on Barro Colorado Island, Canal Zone. Bull. Amer. Mus. Nat. Hist., 68:471-525. (Breeding Territory Type C.)

Davis, D. E.
    1941 The Belligerency of the Kingbird. Wilson Bull., 53:157-168.
    1959 Observations on Territorial Behavior of Least Flycatchers. Wilson Bull., 71: 73-85. (Special attention is given to the role of vocalizations.)

Dixon, J. B.
    1937 The Golden Eagle in San Diego County, California. Condor, 39:49-56. (The average area — i.e., home range —to support a pair of Golden Eagles comprises close to 36 square miles.)

Dixon, K. L.
    1956 Territoriality and Survival in the Plain Titmouse. Condor. 58:169-182.

Drum, M.
    1939 Territorial Studies on the Eastern Goldfinch. Wilson Bull., 51: 69-77. (Breeding Territory Type B.)

Emlen, J. T., Jr.
    1954 Territory, Nest Building, and Pair Formation in the Cliff Swallow. Auk, 71: 16-35. (Breeding Territory Type D.)

Erickson, M. A.
    1938 Territory, Annual Cycle, and Numbers in a Population of Wren-tits (*Chamaea fasciata*). Univ. California Publ. in Zool., 42: 247-334. (Breeding Territory Type A and Non-Breeding Territory Type B.)

Fautin, R. W.
    1940 The Establishment and Maintenance of Territories by the Yellow-headed Blackbird in Utah. Great Basin Nat., 1:75-91. (Breeding Territory Type B.)

Ficken, M. S.
    1962 Agonistic Behavior and Territory in the American Redstart. Auk, 79:607-632.

Fitch, H. S.
1958 Home Ranges, Territories, and Seasonal Movements of Vertebrates of the Natural History Reservation. Univ. Kansas Publ. Mus. Nat. Hist., 11: 63-326.

Gould, P. J.
1961 Territorial Relationships between Cardinals and Pyrrhuloxias. Condor, 63: 246-256.

Gullion, G. W.
1953 Territorial Behavior of the American Coot. Condor, 55: 169-186. (Breeding Territory Type A.)

Hann, H. W.
1937 Life History of the Oven-bird in Southern Michigan. Wilson Bull., 49: 145-237. (Breeding Territory Type A.)

Hensley, M. M., and J. B. Cope
1951 Further Data on Removal and Repopulation of the Breeding Birds in a Spruce-Fir Forest Community. Auk, 68: 483-493.

Hickey, J. J.
1940 Territorial Aspects of the American Redstart. Auk, 57: 255-256. (Breeding Territory Type A.)

Hinde, R. A.
1956 The Biological Significance of the Territories of Birds. Ibis, 98: 340-369.

Hochbaum, H. A.
1959 The Canvasback on a Prairie Marsh. Second edition. Stackpole Company, Harrisburg, Pennsylvania, and Wildlife Management Institute, Washington, D. C. (Breeding Territory Type C. Chapter 5, pages 56-87, contains an excellent account of territory and territorial problems in ducks.)

Howard, H. E.
1948 Territory in Bird Life. Collins, London. (A republication of Howard's 1920 work, with an introduction by Julian Huxley and James Fisher; a basic work on territory.)

Johnson, R. A.
1941 Nesting Behavior of the Atlantic Murre. Auk, 58: 153-163. (Breeding Territory Type D.)

Kendeigh, S. C.
1941 Territorial and Mating Behavior of the House Wren. Illinois Biol. Monogr., 18: 1-120. (Breeding Territory Type A.)

Kilham, L.
1958 Territorial Behavior of Wintering Red-headed Woodpeckers. Wilson Bull., 70: 347-358.

1960 Courtship and Territorial Behavior of Hairy Woodpeckers. Auk, 77: 259-270.

Kuerzi, R. G.
1941 Life History Studies of the Tree Swallow. Proc. Linnaean Soc. New York, Nos. 52-53: 1-52. (Breeding Territory Type D.)

Lack, D.
1954 The Natural Regulation of Animal Numbers. Oxford University Press, London.

1965 The Life of the Robin. Fourth edition. H. F. and G. Witherby, London. (Concerns *Erithacus rubecula;* Chapter 11 deals with the significance of territory.)

MacDonald, S. D.
1968 The Courtship and Territorial Behavior of Franklin's Race of the Spruce Grouse. Living Bird, 7: 5-25. (The male's home range "is too large to have distinct boundaries" and "firm territorial lines are established only in areas of interaction with an adjacent male.")

MacQueen, P. M.
1950 Territory and Song in the Least Flycatcher. Wilson Bull., 62: 194-205. (Breeding Territory Types A and B.)

Mayr, E.
1935 Bernard Altum and the Territory Theory. Proc. Linnaean Soc. New York, Nos. 45-46: 24-38.

Michael, C. W.
1935 Feeding Habits of the Black-bellied Plover in Winter. Condor, 37: 169. (Non-Breeding Territory Type A.)

Michener, H., and J. R. Michener
1935 Mockingbirds, Their Territories and Individualities. Condor, 37: 97-140. (Breeding Territory Type A and Non-Breeding Territory Type B.)

Michener, J. R.
1951 Territorial Behavior and Age Composition in a Population of Mockingbirds at a Feeding Station. Condor, 53: 276-283.

Nero, R. W.
1956 A Behavior Study of the Red-winged Blackbird. II. Territoriality. Wilson Bull., 68: 129-150. (Breeding Territory Type B.)

Nero, R. W., and J. T. Emlen, Jr.
1951 An Experimental Study of Territorial Behavior in Breeding Red-winged Blackbirds. Condor, 53: 105-116.

Nice, M. M.
1937   Studies in the Life History of the Song Sparrow, I. Trans. Linnaean Soc. New York, 4: i-vi; 1-247. (Breeding Territory Type A. Chapters 6-8 deal extensively with territory and related problems.)
1941   The Role of Territory in Bird Life. Amer. Midland Nat., 26: 441-487.
1943   Studies in the Life History of the Song Sparrow, II. Trans. Linnaean Soc. New York, 6: i-viii; 1-328. (Chapters 9-13.)

Odum, E. P.
1941   Annual Cycle of the Black-capped Chickadee-1. Auk, 58: 314-333.
Odum, E. P., and E. J. Kuenzler
1955   Measurement of Territory and Home Range Size in Birds. Auk, 72: 128-137. (The authors make a distinction between *maximum* territory (defended area) or home range (in case the area is not defended) and *utilized* territory; and they suggest two new methods of measuring and expressing maximum territory or home range.)

Orians, G. H., and M. F. Willson
1964   Interspecific Territories of Birds. Ecology, 45: 736-745. (An important paper on the subject.)

Penney, R. L.
1968   Territorial and Social Behavior in the Adélie Penguin. American Geophysical Union, Antarctic Res. Ser., 12: 83-131. (An excellent, detailed study of territorial behavior in a colonial nesting species.)

Pettingill, O. S., Jr.
1936   The American Woodcock *Philohela minor* (Gmelin). Mem. Boston Soc. Nat. Hist., 9: 167-391. (Breeding Territory Type C.)

Pitelka, F. A.
1942   Territoriality and Related Problems in North American Hummingbirds. Condor, 44: 189-204. (Breeding Territory Type C, but modified to include feeding area. The author suggests the possibility that Non-Breeding Territory Type A is also maintained.)
1943   Territoriality, Display, and Certain Ecological Relations of the American Woodcock. Wilson Bull., 55: 88-114.
1959   Numbers, Breeding Schedule, and Territoriality in Pectoral Sandpipers of Northern Alaska. Condor, 61: 233-264.

Rankin, M. N., and D. H. Rankin
1940   Additional Notes on the Roosting Habits of the Tree-creeper. Brit. Birds, 34: 46-60. (Non-Breeding Territory Type C.)
Schwartz, P.
1964   The Northern Waterthrush in Venezuela. Living Bird, 3: 169-184. (Non-Breeding Territory Type B.)

Simmons, K. E. L.
1951   Interspecific Territorialism. Ibis, 93: 407-413.

Snow, D. W.
1958   A Study of Blackbirds. George Allen and Unwin, London. (Concerns *Turdus merula*; Chapter 4 deals with territory.)
1968   The Singing Assemblies of Little Hermits. Living Bird, 7: 47-55. (The tropical hummingbird, *Phaethornis longuemareus*, has singing grounds—Breeding Territory Type C.)

Southern, W. E.
1958   Nesting of the Red-eyed Vireo in the Douglas Lake Region, Michigan. Jack-Pine Warbler, 36: 105-130, 185-207. (Breeding Territory Type A.)

Stewart, R. E., and J. W. Aldrich
1951   Removal and Repopulation of Breeding Birds in a Spruce-Fir Forest Community. Auk, 68: 471-482.

Stickel, D. W.
1965   Territorial and Breeding Habits of Red-bellied Woodpeckers. Amer. Midland Nat. 74: 110-118.

Suthers, R. A.
1960   Measurements of Some Lake-shore Territories of the Song Sparrow. Wilson Bull., 72: 232-237.

Swanberg, P. O.
1951   Food Storage, Territory and Song in the Thick-billed Nutcracker. Proc. Xth Internatl. Ornith. Congr., pp. 545-554. (In this species a pair of individuals will apparently hold a special area or territory for life.)

Tinbergen, N.
1939   The Behavior of the Snow Bunting in Spring. Trans. Linnaean Soc. New York, 5: 1-94. (Breeding Territory Type A.)
1953   The Herring Gull's World: A Study of the Social Behaviour of Birds. Collins, London. (Chapters 9 and 10 are particularly pertinent to territorial behavior.)
1957   The Functions of Territory. Bird Study, 4: 14-27.

Tompa, F. S.
1962 Territorial Behavior: The Main Controlling Factor of a Local Song Sparrow Population. Auk, 79:687-697.

Tuck, L. M.
1960 The Murres: Their Distribution, Populations and Biology: A Study of the Genus *Uria*. Canadian Wildlife Series, 1. Canadian Wildlife Service, Ottawa. (Breeding Territory Type D.)

Venables, L. S. V., and D. L. Lack
1934 Territory and the Great Crested Grebe. Brit. Birds, 28:191-198. (Breeding Territory Type B.)

Young, H.
1951 Territorial Behavior in the Eastern Robin. Proc. Linnaean Soc. New York, Nos. 58-62: 1-37.

SONG

Song is a vocal display in which one or more sounds are consistently repeated in a specific pattern. It is given mainly by males, usually during the breeding season. All other bird vocalization are collectively termed **call notes** or simply **calls**.

FUNCTIONS OF SONG

The chief functions of song are: (1) To proclaim territory and warn conspecific intruders. (2) To advertise the species and the sex of the singer, thereby inviting attention of the opposite sex of the same species, and later to maintain the bond between the singer and the sex-partner.

SUBSTITUTES FOR SONG

Like all vocalizations, songs originate in a part of the respiratory tract called the syrinx (see under "Anatomy and Physiology" in this book, pages 83 and 109).

Many birds, particularly among some of the lower orders, produce sounds by the wings, bill, tail, or other parts of the body. These sounds enhance the effects of songs or function as songs. For example, the Common Nighthawk (*Chordeiles minor*), while in the air, utters a rasping *peent* at wide intervals and then, as it checks a quick dive earthward, lets the air rush through its wing feathers, producing a loud, hollow *whoooom*. The Ruffed Grouse (*Bonasa umbellus*) on a log makes a solely mechanical sound by "drumming" — beating the air with its wings, the tempo increasing until the pulsating sounds come together in a roar. Woodpeckers drum by hammering with their bills, preferably on a hollow stump or some other object that provides resonance. Very few species of birds fail to produce sounds that are either songs in the strict sense or at least substitutes for songs.

MALE SONGS

Songs among different species vary tremendously, ranging from a repetition of one syllable to a highly complex series of sounds. The "best" songsters — i.e., those giving loud, extended songs with a strong musical quality — are passerine species in which the males are usually dull in color and closely resemble the females. Such males, it is believed, depend more on sounds than they do on appearance for the identification of their species.

Duration of Singing

Most species come into full song with the establishment of territory. While singing may decrease in intensity or cease altogether during the short mating period, the birds resume singing with considerable vigor during nest-building, egg-laying, and incubation. When the period of caring for the young is reached, singing almost invariably wanes. Unless the pair rears a second brood, singing usually stops altogether between the period of caring for the young and the end of the postnuptial molt. (A. A. Saunders, 1948a).

Some species resume singing in the fall after the postnuptial molt, though the songs are far fewer in number per day and tend to be incomplete (A. A. Saunders, 1948b). With the advent of winter, singing usually ceases altogether except in a few species, such as the Carolina Wren (*Thryothorus ludovicianus*), European Robin (*Erithacus rubecula*), and Cardinal (*Richmondena cardinalis*), which sing more or less the year round. Most migratory species sing irregularly during migration; those which do not usually arrive silently on their breeding grounds and are not likely to start singing until at least a few days have passed (A. A. Saunders, 1954).

Singing in Relation to Habitat

While they are singing, males of many species make themselves conspicuous by standing or perching on prominent objects. Very often an individual will choose one or more favorite perches for much of his singing.

Males of a number of species, especially those in open country, give **flight songs,** achieving conspicuousness by singing on the wing. Their flight songs are almost always accompanied by aerial maneuvers that increase conspicuousness. In some species, flight songs include mechanical sounds from the action of flight feathers against the air. For instance, the American Woodcock (*Philohela minor*) produces whistling sounds with its wings as it spirals skyward, then chippers vocally during its abrupt, zigzag descent. In a few species, the audible part of the flight song is entirely mechanical. The only sound made by the Common Snipe (*Capella gallinago*), as it circles high over a marsh, is a "winnowing" from the air filtering through its tail feathers.

The flight songs of open-country birds should not be confused with **irregular flight songs** ("ecstasy flights") of species which generally stand or perch when singing. Quite a large number of passerine species occasionally rise into the air and produce rapid vocal sounds, markedly different from their usual songs and of an ebullient nature, suggesting the release of over-abundant energy. The birds give such songs irregularly at infrequent intervals.

Singing in Relation to Nests

As a rule, species do not sing near their nests. When males approach their nests for one purpose or another, they are inclined to be silent or to sing much more softly. Males in a few species will, however, sing while on their nests. Weston (1947) reported that the male Black-headed Grosbeak (*Pheucticus melanocephalus*) occasionally sings while incubating and brooding. Though the volume of the song is lower than when the male is away from the nest, it carries several hundred feet. Weston located grosbeak nests by listening for the songs of the incubating males.

Amount of Singing

Males of certain species give an impressive number of songs during the daylight hours. In one day a Song Sparrow (*Melospiza melodia*) gave as many as 2,305 songs (Nice, 1943), and a Kirtland's Warbler (*Dendroica kirtlandii*) gave 2,212 songs (Mayfield, 1960). But probably a Red-eyed Vireo (*Vireo olivaceus*) holds the record with 22,197 songs in nearly 10 hours (de Kiriline, 1954).

Certain individuals in a species will sing much more than others, often because of strife over territory. For instance, if a male has established his territory in an area where his claim is challenged by one or more rival males, he will be stimulated to sing more often in order to maintain his sovereignty.

Relation of Singing to Light

A few species such as the Red-eyed Vireo sing more or less continuously all day, but a large proportion of diurnal species sing more energetically — i.e., give more songs per hour — in the early morning and in the evening when there is less light. In the morning and sometimes in the evening a small number of species produce **twilight songs** which differ in minor ways from their regular daytime songs. The Eastern Kingbird (*Tyrannus tyrannus*), Eastern Wood Pewee (*Contopus virens*), Least Flycatcher (*Empidonax minimus*), and other tyrannid flycatchers sing at a much quickened tempo as day is breaking. A few diurnal species such as the Nightingale (*Luscinia megarhynchos*) and Mockingbird (*Minus polyglottos*) commonly sing at night — the Nightingale as much at night as in the day, the Mockingbird more often on moonlit nights. The American Woodcock performs its spiral flight song in the twilight of the late evening and early morning, also at night if there is moonlight of sufficient intensity to match the brightness at twilight. When its singing field happens to be near an airport or some other installation that is greatly illuminated, the woodcock may perform through the night.

The amount of light rather than the time of day determines the beginning of singing in the morning and the end of singing in the evening. Cloudiness will delay singing in the morning and hasten its cessation in the evening. Species vary as to the amounts of light and darkness that stimulate and inhibit singing. Consequently, in any given area, different species start and stop singing at different times, forming a more or less orderly sequence. By careful study in an area one can predict the order in which the various species will start singing in the morning and stop in the evening. With some species, at least, singing begins earlier in the morning as the breeding season progresses (Nice, 1943).

Nocturnal birds may sing throughout the night. Like diurnal birds, particular amounts of light stimulate and inhibit their singing. They begin singing earlier in the evening and stop later in the morning when evenings and mornings are cloudy. Crepuscular birds sing energetically only under twilight conditions: morning and evening and during the night when there is a moon to provide twilight conditions.

The total effect of amounts of light on singing is dramatically illustrated at the time of a total solar eclipse (Ehrström, 1956; Hundley, 1964; Kellogg and Hutchinson, 1964). As totality nears, diurnal species gradually cease singing and crepuscular species begin. In the dim light during totality, nocturnal species join the crepuscular species and the few diurnal species that would normally continue singing into the late twilight.

Effects of Weather on Singing

Weather may influence the amount of singing from day to day. Excessive coolness approaching frost, as in the early morning, or intense heat, as at mid-day, may inhibit singing, while mild temperatures may encourage it. Wind is unquestionably a disturbing factor and, if strong enough, will stop singing. Although periods of high humidity, as before and after a rain, may induce vigorous singing, heavy precipitation often markedly reduces it.

Variation in Song

Certain species of birds, notably some of the parulid warblers, have at least two distinctively different songs, both given commonly by all males regardless of their age, the period in the breeding season, or the time of day. Both songs have apparently the same functions. But these birds are rather exceptional. Most species have just one territorial song and this conforms in general to a recognizable pattern by which the singer advertises himself. Yet within the pattern may occur numerous differences, all generally minor.

Individual males normally differ slightly from one another in their songs as to length, number and slurring of notes or syllables, cadence, pitch, and other details. This serves to advertise the individual just as the song itself advertises the species and sex. By closely studying the songs of several individuals in the same locality the student will find it possible to recognize individuals by their respective songs.

In a wide-ranging species the song may vary among populations, each having its own dialect. In a sedentary species, the chances of its populations developing their own dialects are greater than in migratory species. Lack and Southern (1949), Marler (1957), Marler and Boatman (1951), and others suggest that the difference in song between populations of the same species may be due to "selection pressure." Thus, in one locality where the species associates with closely related forms, the song is stereotyped—i.e., is much less variable—to assure ready recognition of the species; in another locality, where the species has few or no congeners, the song is individually more variable since the bird gains more advantage in being recognized as an individual singer than as a species.

FEMALE SONGS

The singing ability of females varies widely among species. Apparently females of many species do not produce sounds that can be called songs. In the Song Sparrow (Nice, 1943), the female occasionally sings early in the breeding season prior to nest-building; the song, always given from an elevated perch, is "short, simple, and unmusical." Undoubtedly, quite a number of species are like the Song Sparrow in this respect. In a few species—e.g., the Cardinal, Rose-breasted Grosbeak *(Pheucticus ludovicianus)*, and the Black-headed Grosbeak—the female produces songs that are about as elaborate as the male's. Weston (1947) found that female Black-headed Grosbeaks even sing while incubating eggs and brooding young. Such songs in females presumably serve to maintain the mating bond. Only in a few species—e.g., the phalaropes in which the female plays the more active role in mating—is the female's song more elaborate than the male's. There are, however, species in Central and South America, Africa, and elsewhere in which the female sings simultaneously and on a par with her mate. Called **duetting**, this behavior plays a prominent role in courtship display as well as in maintaining the mating bond. A more specialized form of duetting is **antiphonal singing** in which the female and male give alternating, usually dissimilar notes and phrases, all so precisely timed that the total effort sounds like the song of one bird. Thorpe (1967) suggested that duetting and antiphonal singing may serve to maintain the sex-partnership where the territorial function of song is reduced or absent. And Diamond and Terborgh (1969) have suggested that duetting may fill "the need" for birds living in dense vegetation, where visual contact is difficult, to evolve intricate vocal displays rather than visual displays.

MIMICRY

Certain parrots (Psittacidae) and mynahs *(Gracula* spp.*)* in captivity imitate human speech and other sounds. Presumably these birds, with human beings as substitutes for their normal social contact, learn that producing particular sounds brings attention.

This may explain why captive parrots and mynahs mimic loudly when they are ignored and soon quiet down when they receive attention.

While a large number of species in the wild sometimes include in their vocal repertoires the call notes and parts of songs of other species, only a very few in the New World are accomplished mimics. The most notorious is without any doubt the Mockingbird. The songs of some individuals are made up almost entirely of imitations of other species. Two males, singing on their nesting territories in northern Lower Michigan, included in their respective repertoires the imitations of 14 and 15 different species, several of which do not occur that far north (Adkisson, 1966). Probably both birds had come from southern areas where they had heard the species and learned to imitate them. What advantage mimicry can be to the Mockingbird or any other wild species is yet to be determined. Perhaps imitative powers may play a role in establishing and strengthening the mating bond when song no longer functions in proclaiming territory (Thorpe, 1967), but this would not seem to be the case with the Mockingbird.

SUBSONGS

Before producing their full primary songs, birds make subdued ("half-hearted") vocal efforts called **subsongs,** sometimes designated as secondary songs. From his study of the Chaffinch (*Fringilla coelebs*), Thorpe (1961) characterized subsongs as averaging quieter, longer, lower in pitch, and given by birds of any age — by adults with a low but increasing sexual drive early in the breeding season and by young birds.

Subsongs in Young Birds

The ability to sing makes its first appearance in birds at a relatively early age. According to data brought together by Nice (1943), young birds in 16 different species start to sing — i.e., to give their first subsongs — at ages ranging from 13 to 24 days, and in 15 species at ages from 4 to 8 weeks. In the Song Sparrow, Nice found that the young bird's first sounds consisted of continuous warbling. This was later followed by short songs, along with much warbling. Gradually the short songs became predominant and finally took on the pattern of the adult's full song. There is some evidence (Thorpe, 1961) that subsong in young birds is in the nature of practice for full song.

INHERITANCE OF SONG

Birds inherit only the simple components of song, characteristics of their respective species, along with the *ability* to give the full song (see the section on "Behavior" in this book, page 247). To produce the full song, they must first hear it sung by adults of their species. Failing to hear the full, species-characteristic song, they may develop a song of considerable complexity quite unlike anything heard in the wild (Thorpe, 1961). Individual variations in full song are presumably acquired rather than inherited.

THE STUDY OF BIRD SONG

Until recent years the study of bird song depended entirely on one's hearing and the ability to record what one heard by the musical scale, phonetics, and diagrams. Now thanks to modern instruments, it is possible to record bird sounds on magnetic tape (see "Recording Bird Vocalizations" in Appendix A, page 429); then, by means of the sound spectrograph, to reproduce the sounds as actual images, or spectrograms, for visual analysis. This brings to the study of bird song a far more accurate method with wide applications.

Since many recently published studies on bird song include spectrograms, the student should know how to interpret them. To assist him, two sample spectrograms, both simplified, are included here with brief explanations. Figure 1 is the spectrogram of the Chaffinch's song; Figure 2, the spectrogram of the Nightingale's.

The songs appear on the spectrograms as the dark areas. Each separate area is a note. The horizontal line shows the duration of the song in seconds, the vertical line gives the song frequency — i.e., the number of vibrations or "sound waves" — in kilocycles per second, or kilohertz.

Some of the information which the spectrograms yield: (1) The pitch of the notes, shown by their height and vertical distances on the graph. The higher the note is on the graph, the higher its pitch. (2) Amplitude or loudness of the notes, shown by the darkness or density of the notes. The thicker the note and the more time it takes to utter, the louder it is. (3) The number of notes in the song and the speed of delivery. The thinner and closer the notes are, the greater their speed of delivery.

A comparison of the two spectrograms reveals that the Chaffinch utters more than one note at a time at different pitches, in quick succession, and of great complexity, whereas the Nightingale gives one loud note at a time at a steady pitch in a somewhat deliberate manner. The Nightingale's song ends in two down-slurred notes, the first pitched much lower than the second.

For further information on the application of spectrograms to studies of song, the student is referred to the paper by Borror (1960).

Two suggested studies of song are outlined below.

1. Make a thorough study of the song of one species.

Procedure

Select a species which is fairly common. Begin while territories are being established and continue through the nesting season. If possible, make tape recordings on the spot and spectrograms later for analysis. Otherwise, adopt one of the several methods suggested for learning songs given in the section on "Field Identification" in this book.

Presentation of Results

Prepare a paper in which all information obtained is carefully analyzed and summarized. (See Appendix B for directions on preparing the manuscript.) Some of the topics that may be covered are listed below.

Types and variations of male song. Describe the typical songs in detail and point out individual variations. Note any correlations between the types of song and the season, or time of day.

Amount of singing and its relation to different stages in the nesting cycle.

Song in relation to habitat, territory, and nest.

Song in relation to light and weather conditions.

Song in the female.

Development of song in young birds.

2. Select an area where there is a wide variety of species, and determine the order in which the species start singing in the morning.

Procedure

Begin early in the nesting season when most species are establishing their territories. With a watch, a photometer such as one commonly used in outdoor photography, and

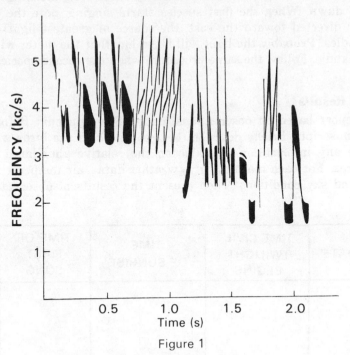

Figure 1

Song of the Chaffinch. Simplified form of sound spectrogram. (Courtesy of W. H. Thorpe, from Figure 2 in *Bird-song*, Cambridge University Press, 1961.)

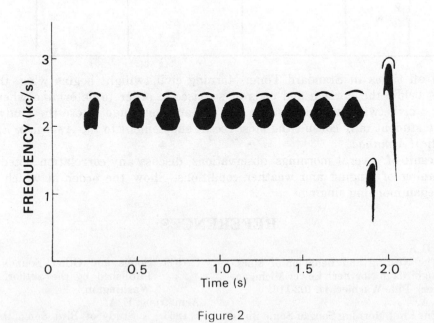

Figure 2

Song of the Nightingale. Simplified form of sound spectrogram. (Courtesy of W. H. Thorpe, from Figure 3 in *Bird-song*, Cambridge University Press, 1961.)

notebook, go to the area on at least six (preferably successive) mornings before there is any evidence of dawn. When the first species starts singing, note the exact time and (with the meter directed toward the east, the source of greatest light) the amount of light in foot candles. Probably the light will be so low that the meter will record only a fraction of one candle. Follow the same procedure when the second species starts to sing, and so on.

Presentation of Results

Draw up a report based on observations made. (See Appendix B for directions on preparing a manuscript.) Briefly describe the area in which the birds were heard singing and include any information available on the relative abundance of the various species in the area. For each morning, give weather data—air temperature, wind velocity, humidity, and sky conditions. Then present the results of observations in tabular form as follows:

SPECIES	DATE	TIME CIVIL TWILIGHT BEGINS	TIME OF SUNRISE	TIME OF FIRST SONG	MINUTES BEFORE SUNRISE
	1. 2. 3. 4. 5. 6.				
	1. 2. 3.				

Record all times in Standard Time. Morning civil twilight begins when the sun is 6 degrees below the horizon and ends at sunrise. In order to determine when sunrise occurs and civil twilight begins in the latitude and longitude where the study is being made, the student may consult the most recent supplement to the *American Ephemeris and Nautical Almanac*.

As a result of several mornings' observations, discuss any correlation noted between the beginning of singing and weather conditions. Show the order in which different species began morning singing.

REFERENCES

Adkisson, C. S.
1966 The Nesting and Behavior of Mockingbirds in Northern Lower Michigan. Jack-Pine Warbler, 44: 102-116.
Allard, H. A.
1930 The First Morning Song of Some Birds of Washington, D. C.: Its Relation to Light. Amer. Nat., 64: 436-469.
Arlton, A. V.
1949 Songs and Other Sounds of Birds. Published by the author, Parkland, Washington.
Armstrong, E. A.
1963 A Study of Bird Song. Oxford University Press, London. (A thorough and indispensable review of the literature.)

Berger, A. J.
1961   Bird Study. John Wiley and Sons, New York. (Chapter 6: "Song" — good supplementary reading.)
Borror, D. J.
1960   The Analysis of Animal Sounds. In *Animal Sounds and Communication.* Edited by W. E. Lanyon and W. N. Tavolga. American Institute of Biological Sciences, Washington, D. C.
1967   Songs of the Yellowthroat. Living Bird, 6: 141-161. (A comparative study of recorded songs of 411 Yellowthroats, representing 10 North American subspecies.)
Borror, D. J., and C. R. Reese
1954   Analytical Studies of Henslow's Sparrow Songs. Wilson Bull., 66: 243-252.
Craig, W.
1943   The Song of the Wood Pewee *Myiochanes virens* Linnaeus: A Study of Bird Music. New York State Mus. Bull. No. 334.
de Kiriline, L.
1954   The Voluble Singer of the Treetops. Audubon Mag., 56: 109-111.
Diamond, J. M., and J. W. Terborgh
1969   Dual Singing by New Guinea Birds. Auk, 85: 62-82. (Dual singing [duetting] occurs regularly in four species and seems to be represented in five others.)
Ehrströ C.
1956   Faglarnas Uppträdande under Solformorkelsen den 30 Juni 1954. Vär Fagelvarld, 15: 1-28.
Hundley, M. H.
1964   Observations on Reactions of Avifauna in Maine to Total Eclipse of the Sun, July 20, 1963. Florida Nat., 37: 8-25.
Kellogg, P. P.
1960   Considerations and Techniques in Recording Sound for Bio-acoustics Studies. In *Animal Sounds and Communications.* Edited by W. E. Lanyon and W. N. Tavolga. American Institute of Biological Sciences, Washington, D. C.
Kellogg, P. P., and C. M. Hutchinson
1964   The Solar Eclipse and Bird Song. Living Bird, 3: 185-192.
Kellogg, P. P., and R. C. Stein
1953   Audio-spectographic Analysis of the Songs of the Alder Flycatcher. Wilson Bull. 65: 75-80.
Kullenberg, B.
1946   Om Fägellätenas Biologiska Funktion. Var Fägelvärld, 5: 49-64. (A good discussion of the biological function of bird song.)
Lack, D., and H. N. Southern
1949   Birds on Tenerife. Ibis, 91: 605-607.
Lanyon, W. E.
1960   The Ontogeny of Vocalizations in Birds. In *Animal Sounds and Communication.* Edited by W. E. Lanyon and W. N. Tavolga. American Institute of Biological Sciences, Washington, D. C.
Lanyon, W. E., and F. B. Gill
1964   Spectographic Analysis of Variation in the Songs of a Population of Bluewinged Warblers *(Vermivora pinus).* Amer. Mus. Novitates No. 2176.
Lanyon, W. E., and W. N. Tavolga, Editors
1960   Animal Sounds and Communication. American Institute of Biological Sciences, Washington, D. C.
MacQueen, P. M.
1950   Territory and Song in the Least Flycatcher. Wilson Bull., 62: 194-205.
Marler, P.
1952   Variation in the Song of the Chaffinch *Fringilla coelebs.* Ibis, 94: 458-472.
1957   Specific Distinctiveness in the Communication Signals of Birds. Behaviour, 11: 13-29.
1960   Bird Songs and Mate Selection. In *Animal Sounds and Communication.* Edited by W. E. Lanyon and W. N. Tavolga. American Institute of Biological Sciences, Washington, D. C.
Marler, P., and D. J. Boatman
1951   Observations on the Birds of Pico, Azores. Ibis, 93: 90-99.
Mayfield, H.
1960   The Kirtland's Warbler. Cranbrook Inst. Sci. Bull. No. 40, Bloomfield Hills, Michigan.
McCabe, R. A.
1951   The Song and Song-Flight of the Alder Flycatcher. Wilson Bull., 63: 89-98.
Mehner, J. F.
1952   Notes on Song Cessation. Auk, 69: 466-469.
Nice, M. M.
1943   Studies in the Life History of the Song Sparrow, II. Trans. Linnaean Soc. New York, 6: i-viii; 1-328. (Chapters 9-11.)
Reynard, G. B.
1962   The Rediscovery of the Puerto Rican Whip-poor-will. Living Bird, 1: 51-60. (How a species, once thought extinct, was found again by sound recording.)

1963 The Cadence of Bird Song. Living Bird, 2: 139-148.

Saunders, A. A.
1947 The Seasons of Bird Song. The Beginning of Song in Spring. Auk, 64: 97-107.
1948a The Seasons of Bird Song—The Cessation of Song after the Nesting Season. Auk, 65: 19-30.
1948b The Seasons of Bird Song. Revival of Song after the Postnuptial Molt. Auk, 65: 373-383.
1951 A Guide to Bird Songs. Doubleday and Company, Garden City, New York.
1954 The Lives of Wild Birds. Doubleday and Company, Garden City, New York.

Saunders, D. C.
1951 Territorial Songs of the White-winged Dove. Wilson Bull., 63: 330-332.

Shaver, J. M., and R. Walker
1931 A Preliminary Report on the Influence of Light Intensity upon the Time of Ending of the Evening Song of the Robin and Mockingbird. Wilson Bull., 43: 9-18.

Smith, W. J.
1966 Communication and Relationships in the Genus *Tyrannus*. Publ. Nuttall Ornith. Club No. 6, Cambridge, Massachusetts.

Stein, R. C.
1962 A Comparative Study of Songs Recorded from Five Closely Related Warblers. Living Bird, 1: 61-70. (A study to determine interspecific relationships by song patterns on spectrograms.)

Stokes, A. W., and H. W. Williams
1969 Antiphonal Calling in Quail. Auk, 85: 83-89.

Thorpe, W. E.
1961 Bird-song: The Biology of Vocal Communication and Expression in Birds. University Press, Cambridge.
1967 Vocal Imitation and Antiphonal Song and Its Implications. Proc. XIVth Internatl. Ornith. Congr., pp. 245-263.

Tinbergen, N.
1939 The Behavior of the Snow Bunting in Spring. Trans. Linnaean Soc. New York, 5: 1-94.

Vaurie, C.
1946 Early Morning Song during Middle and Late Summer. Auk, 63: 163-171. (A quantitative study covering 47 days and based on 21 bird species.)

Ward, R.
1966 Regional Variation in the Song of the Carolina Chickadee. Living Bird, 5: 127-150.

Weston, H. G., Jr.
1947 Breeding Behavior of the Black-headed Grosbeak. Condor, 49: 54-73.

Wright, H. W.
1912 Morning Awakening and Even-Song. Auk, 29: 307-327.

MATING

The pairing of birds is commonly spoken of as **mating**. This term is distinguished from the sex relation, which is **copulation** or **coition**, and from **fertilization**, which is the union of the sperm and ovum.

PRELIMINARY STAGES

In migratory species, the spring migration to the breeding grounds usually precedes mating. The sequence of events between migration and mating is outlined below:

1. *Arrival of the Male.* Males in the majority of species are the first individuals to appear, but, in a number of species, males and females appear at the same time. Occasionally males wander for a brief period after their arrival; as a rule, they immediately establish territory. If they arrive in flocks, they usually remain in flocks for the wandering period. At this time singing is infrequent and subdued.

2. *Establishment of Territory.* Though the females may still be absent, the males establish territory. When the males are in flocks, they gradually become vociferous and antagonistic toward one another, and the flock breaks up before territory establishment. While they establish territory, they sing frequently and intensely, and they may compete vigorously with other males for specific areas.

3. *Arrival of the Female.* Like the males, females occasionally wander for a period after arrival. Ordinarily, they go directly to male territories. They may or may not arrive in flocks. Sometimes they join the still unbroken flocks of males and wander with them until solitary behavior finally becomes apparent and they enter male territories.

Stages preliminary to mating in non-migratory species needs investigation. While many such species remain mated throughout the year, events leading to initial mating are not well known or understood.

MATING

Mating usually occurs in association with established territory. When the female arrives on the territory, almost immediately the male receives her by displays and pursues her. Though the male occasionally attempts copulation at this time, he is usually unsuccessful, the female being unreceptive. If the female remains in the male's territory after the first displays and sexual pursuit, mating is accomplished. The two birds then stay more or less within the confines of the territory and in close association. Sometimes the male sings less intensely or stops singing altogether.

After a period of days, the female becomes receptive and copulation occurs. Copulation, once initiated, may be repeated several times a day through subsequent periods of nest-building and egg-laying. The singing of the male is intense; if it decreased in intensity or ceased altogether after mating, it is now resumed.

Mating may not occur in association with territory. In certain migratory species, pairing may take place on the wintering grounds. It may also take place immediately after migration, before territory is established. In these cases the two birds settle on ·the territory together.

KINDS OF MATING RELATIONS

Three kinds of mating relations occur among birds:

1. **Monogamy.** The majority of species are monogamous. Faithfulness of the pair to each other is usually constant because of the attachment of the sex-partners to each other, to their territory, and later to their young. However, remating may take place when a partner is lost by death, desertion, eviction from the territory by a rival, or other circumstances.

2. **Polygamy.** A number of species are normally polygamous, an individual of one sex (usually a male) mating at the same time with two or more individuals of the opposite sex. At times a monogamous species may show polygamy, if the sex ratio of a population is imbalanced, or, as suggested by Verner (1964), if a male succeeds in establishing a large territory in a habitat particularly attractive to more than one female. There are two conditions of polygamy:

   a. **Polygyny** in which a male mates with two or more females. This is exhibited in Ostriches *(Struthio camelus)*, Ring-necked Pheasants *(Phasianus colchicus)*, and several icterids including the Red-winged Blackbird *(Agelaius phoeniceus)*. Verner and Willson (1966) make the point that polygyny is to be expected more often in species nesting in open country—e.g., prairies and marshes.

   b. **Polyandry** in which a female mates with two or more males. The condition is believed to occur occasionally in local populations with abnormal preponderance of males.

3. **Promiscuity.** Though classified as a mating relation, promiscuity is actually copulation without relation. This condition is characteristic of several grouse (e.g., Sage Grouse, *Centrocercus urophasianus*; Greater Prairie Chicken, *Tympanuchus cupido*), the Ruff *(Philomachus pugnax)*, many tropical hummingbirds, many manakins (Pipridae), probably bowerbirds (Ptilonorhynchidae), many birds-of-paradise (Paradisaeidae), Boat-tailed Grackles *(Cassidix mexicanus)*, and Brown-headed Cowbirds *(Molothrus ater)*.

DURATION OF MATING

Duration of mating in birds shows the following variations:

1. A pair may not remain mated after the last eggs of the season are laid (as in polygamous birds with breeding territories of Type C), or after the young are reared.

2. Birds may remain mated, though less closely, through the winter between one breeding season and another. (A few species are said to "remain mated for life.")

3. Birds mated for one season may re-mate the next. This variation may sometimes be due to the birds having an attachment to the same territory; at other times it may be due to the birds having a close attachment to each other.

4. Birds which raise two or more broods yearly may remain mated between broods. Two factors may account for this variation: (a) Close attachment of the birds to each other and to their territory; and (b) overlapping of two broods (i.e., when the female

starts a new nest while the male is feeding the young of the previous nest, with the result that there is no lapse in breeding activities).

5. Birds which raise two or more broods yearly may change their mates between broods, a variation occurring only occasionally.

MATING DISPLAYS

Mating displays are behavioral adaptations or signals that facilitate the successful completion of the reproductive cycle. In a general sense they include displays that directly or indirectly promote fertilization, as well as displays that follow fertilization to the conclusion of the reproductive cycle. Mating displays may incorporate all the signals — visual, auditory, or tactile, single or in combination — that are available to the species for intraspecific communication.

Many mating displays consitute signals, frequently derived from displacement activities and maintenance activities (see the section, "Behavior", pages 246 and 249), that have evolved to the point of being species-typical and therefore ritualized, and described according to their predictable characteristics. All such ritualized displays have a dual importance; to assure the reproductive efficiency of the species and to avoid mating with other species.

Each species has a number of displays concerned with reproduction that can be classified according to the order of their occurrence and probable function in the reproductive cycle.

I. *Pre-fertilization Displays.* These include, besides the aggressive behaviors involved in the establishment of territories (see the section, "Bird Territory," page 312), the activities that serve to bring the sexes together and, in many species, to form and maintain the pair-bond.

A. *Advertising Displays.* Sometimes called "courtship" or epigamic displays, these activities attract the attention of the opposite sex. In many birds, species-specific adornments of plumage (e.g., crests, ruffs, and modified scapulars, tertiaries, and tail coverts), brilliantly colored areas of plumage, lurid gapes and exposed areas of skin, and structural modifications (sometimes highly colored) of bill and face enhance the displays by males. Often eccentric body attitudes and movements of feet, wings, and tail; special sounds both vocal and mechanical; and peculiar aerial and aquatic maneuvers accompany the displays as the performing males direct their most startling adornments and actions toward the nearest and most attentive females. The females' responses, while seemingly indifferent, may consist of feeding movements, preening, wiping the bill, etc. — actually sexually motivated activities. For more detailed descriptions of numerous advertising displays, consult Armstrong (1947), Stoner (1940), and Welty (1962).

1. *Solo Displays,* performed by solitary males in either monogamous or polygamous species which may or may not be sexually dimorphic — i.e., one sex differing in appearance from the other. The females respond in a much less showy manner.

2. *Mutual Displays,* performed by both sexes, almost invariably by monogamous species which are not sexually dimorphic. The displays of each sex are practically identical.

3. *Collective Displays,* performed by two or more males of polygamous or promiscuous species, usually sexually dimorphic, in "arenas," "tournament grounds," or "leks."

B. **Pair-bonding** and **Pair-maintaining Displays.** Activities among monogamous species that assist in forming and strengthening the pair-bond and maintaining it through the reproductive cycle. The birds may repeat them indefinitely in association with advertising displays and copulation and later with nesting.

   1. **Sexual Pursuit.** In most species the male pursues the female in the air, but in some species on water or land. Sexual pursuit may or may not accompany advertising displays. Presumably the action is either an attempt on the part of the male to copulate or a signal indicating his readiness to copulate.

   2. **"Symbolic" Nest-building.** The male is usually the chief actor even though he does not normally take part in actual nest-building. Symbolic building may consist of picking up, manipulating, and carrying material to the nest without attempting to build; passing material to the sex-partner; and bringing material when coming to the nest to relieve the sex-partner during incubation.

   3. **Courtship-feeding.** Again, the male is usually the chief actor. He gathers food and gives it to the female in the manner used when feeding young. He may feed the female prior to copulation, or while nest-building, incubating eggs, or brooding young. (For a paper on the subject of courtship-feeding, see Lack, 1940a.)

II. **Fertilization Displays.** Displays that are directly associated with copulation.

   A. **Precopulatory Displays,** essentially invitatory or solicitation performances. While often similar in both sexes, they are not necessarily given at the same time. When performing precopulatory displays on the water, a pair of Canada Geese *(Branta canadensis)* face each other and alternately curve their neck backward like a shepherd's crook and dip head and neck well below the surface. The action increases in tempo until the male swims alongside the female, ceases displaying, and proceeds to mount her (Klopman, 1962). In the Herring Gull *(Larus argentatus),* both partners give begging calls and head-tossing movements (Tinbergen, 1953). Among many passerines, the males fluff their body feathers, partially spread the wings, lower and spread the tail, and bow; the females, taking a more submissive role, sleek their body feathers, shiver the wings, elevate the tail, and crouch. Both partners may gape and give special calls heard at no other time.

   B. **Copulatory Displays,** the movements of both sexes from the time the male mounts the female until coition is effected and the male dismounts.

   C. **Postcopulatory Displays,** the actions that take place immediately after the male dismounts. They differ markedly from precopulatory displays and may or may not be mutual. In the Canada Goose (Klopman, *op. cit.*), the male raises his breast out of water, throws the neck backward, holds the head somewhat vertically, and keeps the wings partially opened and arched. The female gives the same display and both birds utter "wheezy groans." This mutual display is soon followed, first by both birds stretching their wings and flapping them, and then by vigorous preening and bathing. The postcopulatory display of the Herring Gull consists mainly of preening (Tinbergen, *op. cit.*). Among passerine birds the display is just as simple, consisting merely of fluffing and shaping the feathers and preening.

III. **Post-fertilization Displays.** The interactions of the sex-partners during egg-laying, incubation, and care of the young. They include symbolic nest-building and courtship-feeding and, in a broader sense, all the interactions between the sex-partners and their young. Only the **Nest-relief Displays**, manifest among all species in which both partners incubate the eggs, warrant consideration here. Among sea birds—e.g., penguins, albatrosses, and boobies—where one member of the pair is attentive at the nest for long periods, the arrival of the partner to take over incubation triggers elaborate displays by both birds that include, depending on the species, loud vocalizations, bizarre head and neck motions, and occasionally mutual nibbling — activities that may continue for many minutes. (For further information, consult the works by Rice and Kenyon, 1962; Richdale, 1951; Simmons, 1967; and Sladen, 1958.) In the Herring Gull (Tinbergen, *op. cit.*), the arriving sex-partner often precedes the take-over with mewing calls and a "choking" action, particularly if the occupant bird is slow or reluctant to leave. In passerine species, the nest-relief ceremony may be little more than mutual gaping before the incubating bird departs.

THE STUDY OF MATING

Select one or more species for a study of mating.

Procedure

If the species is migratory, begin the study immediately after the conclusion of spring migration; if the species is non-migratory, begin at "the first sign" of spring. Follow the individuals as closely and as often as possible. (The chief method is to watch them with a binocular from far enough away to avoid disturbance.) Later, when mating has occurred and territories are established, erect blinds in or near territories for close observation of behavior. Try to obtain photographs or make sketches of significant activities.

Presentation of Results

For each species prepare a separate report based on the observations made. Record, describe, and discuss in detail as many of the following topics as the acquired information permits. Include photographs or sketches to illustrate text material. Follow the directions for preparing a manuscript in Appendix B.

Preliminary Stages. **Arrival of Males:** Time; singly or in flocks; behavior; time when singing and other displays begin. **Establishment of Territory:** Time between arrival and establishment; change in behavior between arrival and establishment. **Arrival of Females:** Time; with or after males; singly, in flocks with males, or mated with males; behavior. (If species is non-migratory, omit sections relative to arrival; otherwise follow the outline.)

Mating. **Meeting of Male and Female:** Time; place; resulting behavior. **Unreceptive Period of Female:** Length of time involved; behavior during; length of time of song cessation. **Receptive Period of Female:** Length of time involved; number of copulations per day; method of copulation and associated behavior.

Mating Relation. Kind; variations.

Duration of Mating. Length of time birds remain mated; changes in "strength" of mating bond during the nesting cycle.

Mating Displays. Types; behavior of both sexes before, during, and after; probable origin and functions of the displays.

REFERENCES

Adams, D. A.
   1960 Communal Courtship in the Ruffed Grouse, *Bonasa umbellus* L. Auk, 77: 86-87.
Allen, A. A.
   1934 Sex Rhythm in the Ruffed Grouse *(Bonasa umbellus* Linn.) and Other Birds. Auk, 51: 180-199. (A controversial paper worthy of consideration.)
Allen, R. P., and F. P. Mangels
   1940 Studies of the Nesting of the Black-crowned Night Heron. Proc. Linnaean

Soc. New York, Nos. 50-51: 1-28. (Detailed observations on mating and display.)

Andrew, R. J.
1961 The Displays Given by Passerines in Courtship and Reproductive Fighting: A Review. Ibis, 103a: 315-348; 549-579.

Armstrong, E. A.
1947 Bird Display and Behaviour: An Introduction to the Study of Bird Psychology. Revised edition. Lindsay Drummond, London.

Brackbill, H.
1959 Remating Percentage of Some Migratory Birds. Bird-Banding, 30: 123.

Chapman, F. M.
1935 The Courtship of Gould's Manakin *(Manacus vitellinus vitellinus)* on Barro Colorado Island, Canal Zone. Bull. Amer. Mus. Nat. Hist., 68: 471-525.

Darling, F. F.
1938 Bird Flocks and the Breeding Cycle. A Contribution to the Study of Avian Sociality. University Press, Cambridge.

Emlen, J. T., Jr.
1954 Territory, Nest Building, and Pair Formation in the Cliff Swallow. Auk, 71: 16-35.

Ficken, M. S.
1963 Courtship of the American Redstart. Auk, 80: 307-317. (Solo displays.)

Ficken, M. S., and R. W. Ficken
1962 The Comparative Ethology of the Wood Warblers: *A Review.* Living Bird, 1: 103-122. (Solo displays.)

Gullion, G. W.
1952 The Displays and Calls of the American Coot. Wilson Bull., 64: 83-97.

Hardy, J. W.
1963 Epigamic and Reproductive Behavior of the Orange-fronted Parakeet. Condor, 65: 169-199.

Hochbaum, H. A.
1959 The Canvasback on a Prairie Marsh. Revised edition. Stackpole Company, Harrisburg, Pennsylvania, and Wildlife Management Institute, Washington, D. C. (Chapter 3 contains an excellent account of mating and related activities in the Canvasback and other ducks.)

Hodges, J.
1948 A Case of Polygamy in the American Redstart. Bird-Banding, 19: 74-75.

Howard, H. E.
1929 An Introduction to the Study of Bird Behaviour. University Press, Cambridge.

Iredale, T.
1950 Birds of Paradise and Bower Birds. Georgian House, Melbourne.

Johnsgard, P. A.
1965 Handbook of Waterfowl Behavior. Cornell University Press, Ithaca, New York.

Klopman, R. B.
1962 Sexual Behavior in the Canada Goose. Living Bird, 1: 123-129.

Lack, D.
1940a Courtship Feeding in Birds. Auk, 57: 169-178.
1940b Pair-formation in Birds. Condor, 42: 269-286.

MacDonald, S. D.
1968 The Courtship and Territorial Behavior of Franklin's Race of the Spruce Grouse. Living Bird, 7: 5-25. (Solo displays.)

Marshall, A. J.
1954 Bower-birds: Their Displays and Breeding Cycles. Oxford University Press, London. (Collective displays.)

McAllister, N. M.
1958 Courtship, Hostile Behavior, Nest-establishment, and Egg Laying in the Eared Grebe *(Podiceps caspicus).* Auk, 75: 290-311. (Mutual displays.)

McIlhenny, E. A.
1937 Life History of the Boat-tailed Grackle in Louisiana. Auk, 54: 274-295. (Promiscuity.)

McKinney, F.
1961 An Analysis of the Displays of the European Eider *Somateria mollissima mollissima* (Linnaeus) and the Pacific Eider *Somateria mollissima v. nigra* Bonaparte. Behaviour Supplement 7. E. J. Brill, Leiden.

Meanley, B.
1955 A Nesting Study of the Little Blue Heron in Eastern Arkansas. Wilson Bull., 67: 84-99. (Promiscuity.)

Moynihan, M.
1955 Types of Hostile Display. Auk, 72: 247-259.

Nero, R. W.
1956 A Behavior Study of the Red-winged Blackbird. I. Mating and Nesting Activities. Wilson Bull., 68: 5-37. (Solo displays.)

Nero, R. W., and J. T. Emlen, Jr.
1951 An Experimental Study of Territorial Behavior in Breeding Red-winged Blackbirds. Condor, 53: 105-116. (Polygamy.)

Nice, M. M.
1930 Do Birds Usually Change Mates for the Second Brood? Bird-Banding, 1: 70-72.
1943 Studies in the Life History of the Song Sparrow, II. Trans. Linnaean Soc. New York, 6: i-viii; 1-328. (Chapters 14-16.)

Patterson, R. L.
1952 The Sage Grouse in Wyoming. Sage Books, Denver, Colorado. (Collective displays.)

Pettingill, O. S., Jr.
1937 Behavior of Black Skimmers at Cardwell Island, Virginia. Auk, 54: 237-244. (Mating.)

Pitelka, F. A.
1943 Territoriality, Display, and Certain Ecological Relations of the American Woodcock. Wilson Bull., 55: 88-114.

Rice, D. W., and K. W. Kenyon
1962 Breeding Cycles and Behavior of Laysan and Black-footed Albatrosses. Auk, 79: 517-567. (Mutual displays; nest-relief displays.)

Richdale, L. E.
1951 Sexual Behavior in Penguins. University of Kansas Press, Lawrence. (Mutual displays; nest-relief displays.)

Sauer, E. G. F., and E. M. Sauer
1966 The Behavior and Ecology of the South African Ostrich. Living Bird, 5: 45-75. (Example of polygyny; solo displays.)

Scott, J. W.
1942 Mating Behavior in the Sage Grouse. Auk, 59: 477-498. (Collective displays.)

Sick, H.
1967 Courtship Behavior in the Manakins (Pipridae): A Review. Living Bird, 6: 5-22. (Collective displays.)

Simmons, K. E. L.
1967 Ecological Adaptations in the Life History of the Brown Booby at Ascension Island. Living Bird, 6: 187-212. (Mutual displays; nest-relief displays.)

Sladen, W. J. L.
1958 The Pygoscelid Penguins: I. Methods of Study. II. The Adelie Penguin *Pygoscelis adeliae* (Hombron and Jacquinto), Falkland Islands Dependencies Surv. Sci. Repts. No. 17.

Snow, D. W.
1961 The Displays of the Manakins *Pipra pipra* and *Tyranneutes virescens*. Ibis, 103: 110-113.
1968 The Singing Assemblies of Little Hermits. Living Bird, 7: 47-55. (Possible promiscuity in the hummingbird, *Phaethornis longuemareus*; collective displays.

Stoner, C. R.
1940 Courtship and Display among Birds. Country Life, London.

Tinbergen, N.
1939 The Behavior of the Snow Bunting in Spring. Trans. Linnaean Soc. New York, 5: 1-94. (Contains a valuable account of mating procedures, etc.)
1953 The Herring Gull's World: A Study of the Social Behaviour of Birds. Collins, London. (Part 3, on pair formation and pairing, is especially instructive.)

Verner, J.
1964 Evolution of Polygamy in the Long-billed Marsh Wren. Evolution, 18: 252-261. (A suggestion that the size of a male's territory and total amount of emergent vegetation in it are correlated with his success in acquiring mates. Females may rear more young by pairing with a male on a superior territory than with a bachelor in an inferior one. The imbalance in sex ratio is not necessarily the only reason for polygamy.)

Verner, J., and M. F. Willson
1966 The Influence of Habitats on Mating Systems of North American Passerine Birds. Ecology, 47: 143-147.

von Haartman, L.
1951 Successive Polygamy. Behaviour, 3: 256-274. (In the Pied Flycatcher, *Muscicapa hypoleuca*, one male has several territories in succession. When the female in one starts to lay eggs, the male leaves her, moves to another territory, there acquires a new female...He may eventually abandon her, return to the first, and so on.)

Welty, J. C.
1962 The Life of Birds. W. B. Saunders Company, Philadelphia.

Woolfenden, G. E.
1956 Comparative Breeding Behavior of *Ammospiza caudacuta* and *A. maritima*. Univ. Kansas Publ. Mus. Nat. Hist., 10: 45-75. (Promiscuity in sharp-tailed sparrows.)

Yocom, C. F.
1944 Evidence of Polygamy among Marsh Hawks. Wilson Bull., 56: 116-117.

Zimmerman, J. L.
1966 Polygyny in the Dickcissel. Auk, 83: 534-546.

NESTS AND NEST-BUILDING

All receptacles for eggs laid by birds are called **nests**. In each bird species, nests are remarkably similar in form and location; among different species they show wide diversity.

THE DEVELOPMENT OF NESTS

Early in their acquisition of homoiothermy ("warm-bloodedness") birds could no longer abandon their eggs, as did their reptilian forebears, leaving them to hatch in the heat of the environment. The embryos within required the steady warmth of the parental body. This imposed on birds the necessity of incubating their eggs—sitting immobile on them for long periods of time. To offset what might well have been lethal exposure to predation and other adversities of the environment, birds simultaneously developed protective measures. One was to select sites for eggs that provided adequate cover and freedom from adversities; another was to build nests that would accommodate and shelter their eggs and themselves.

From the meager information available on the ancestry and descent of birds, and from what is known about nest-building habits today, one may assume that the earliest nests of birds were on the ground in depressions which the birds scraped out with the bill and feet and molded to the shape of their bodies by repeated turning. Some of the nests may also have been on the floors of natural cavities or cavities (including burrows, holes in trees) which the birds excavated themselves. All such nests were without structure although they may have been lined with materials—e.g., plant stems, leaves, bits of shell, etc.—gathered from the immediate vicinity, and sometimes with feathers from the incubating bird's body.

Among modern birds one finds both ground and cavity nests still in abundant use. Ground nests, for instance, are characteristic of loons, pelicans, gannets, swans, geese, most ducks, grouse, quail, pheasants, shore birds, gulls, terns, murres, and goatsuckers. Probably the floating nests of grebes and the marsh nests of cranes, rails, gallinules, and coots are derived from ground nests. Most birds with ground nests either have cryptic coloration to conceal them while they are on their nests, or place their nests on islands or in other areas generally inaccessible to predators. Cavity nests are typical of petrels, a few ducks such as goldeneyes and mergansers, most vultures, puffins, auklets, parrots, trogons, kingfishers, and woodpeckers.

The earliest nests to be elevated in vegetation—shrubs, trees, and even marsh plants—were essentially platforms of loosely assembled plant materials, without evident structure, but with a shallow depression for the eggs. **Platform nests**, perhaps representing the second stage in nest-building, differed principally from the first stage, **ground and cavity nests,** in being independent of a uniformly firm surface and comprised entirely of accumulated materials for holding the eggs as well as for lining the nest.

Modern birds which build platform nests include the anhingas, herons, storks, ibises, pigeons, and cuckoos. Platform nests are also built by cormorants, hawks, eagles, and the Osprey *(Pandion haliaetus),* but not always in trees. Some cormorants place their nests consistently on cliffs. A few species of hawks, both the Golden and Bald Eagles *(Aquila chrysaetos* and *Haliaeetus leucocephalus),* and the Osprey may place their huge nests either in trees (including stubs), on cliffs, or even (in the case of the Osprey) on the ground. The Marsh Hawk *(Circus cyaneus)* nests regularly on the ground without building a platform.

The third and final stage of nest-building was the development of **cupped nests,** with true structure, consisting of materials arranged and compacted for the bottom and sides and softer materials inside for lining.

In building cupped nests, birds succeeded in directly adapting structures to shrubs and trees from the crotches to the tips of branches. Presumably the first cupped nests were statant, supported mainly from below, with the rims standing firmly upright. The majority of hummingbirds and passerine species today build **statant nests** though a few species such as magpies have modified them by extending the sides upward and arching over the top in a dome.

Eventually, a number of species suspended their cupped nests from branches by rims and sides, without supporting them from below. These nests took two forms: (1) **Pensile nests,** as built by modern vireos, suspended from stiffly woven rims and sides. (2) **Pendulous nests,** as built by modern orioles, suspended from rims and flexibly woven sides, with the deeply cupped lower parts swinging freely.

In the rapid multiplication of passerine species and the consequent increase in competition for living space, many new species moved into heretofore unoccupied niches of the environment and re-adapted cupped nests to suite the particular situations other than the crotches and branches of trees. Such birds as phoebes and several swallows, using adhesive substances, built their nests **adherent** to cliff walls and similarly vertical surfaces. Larks, pipits, several thrushes, and numerous parulid warblers, icterid blackbirds, and fringillids constructed cupped nests on the ground in forests, in open country, or in marsh vegetation. Titmice, nuthatches, creepers, bluebirds, several swallows and wrens, and the Prothonotary Warbler *(Protonotaria citrea)* put their cupped nests in preformed cavities, or in rare cases — e.g., the Bank Swallow *(Riparia riparia)* — in cavities which they excavated themselves.

The nests of several groups of modern birds defy categorizing as to stages of development. For example, owls rarely, if ever, build nests of any sort. Unless they nest on the ground as do Snowy and Short-eared Owls *(Nyctea scandiaca* and *Asio flammeus),* they appropriate tree nests of other birds or use cavities. Chimney Swifts *(Chaetura pelagica)* construct "half-cupped" nests on vertical surfaces. To make the nesting materials stick together and the nests adhere to vertical surfaces, the birds apply their own saliva. Other species of swifts show wide diversity in nests and the use of saliva in forming them — e.g., the cave swiftlets *(Collocalia* spp.) of Asia form their nests almost entirely of coagulated saliva (see Lack, 1956, for a review of nest-building by world swifts). In eastern Indonesia, Polynesia, New Guinea, and Australia, gallinaceous birds called megapodes (Megapodiidae) dig pits in the soil or build mounds of rotting vegetable matter in which they lay their eggs and then cover them, leaving them to be incubated by the surrounding heat. The mounds of one species *(Megapodius freycinet)* may measure up to 60 feet long, 15 feet wide, and 10 feet high (Frith, 1962)—probably the largest "nests" of any bird.

CLASSIFICATION OF NESTS

It is possible to set up a classification of nests indicating their probable course of development and illustrating the diversity of nest types. Such a classification for North American birds (suggested in part by Herrick, 1911) is presented below, with spaces for the names of species showing each nest type. Bear in mind that the classification is purely artificial and does not necessarily portray the phylogenetic relationships of the nest-builders.

Stage I. **Ground** and **Cavity Nests.** Nests without structure.

  A. **Ground Nests.** Simple depressions with or without lining. Birds with ground nests either have cryptic coloration or select nest sites inaccessible to predators.

    _____   _____

    _____   _____

    _____   _____

    _____   _____

  B. **Cavity Nests.** Nests in caves, crevices, burrows, holes in trees, or birdboxes, with or without lining.

    1. In preformed or natural cavities.

      _____   _____

      _____   _____

      _____   _____

    2. In cavities excavated by occupant birds.

      _____   _____

      _____   _____

      _____   _____

Stage II. **Platform Nests.** Nests elevated, without structure, consisting of loosely assembled materials with a shallow depression for the eggs.

    _____   _____

    _____   _____

    _____   _____

Stage III. **Cupped Nests.** Nests adapted to crotches and branches of trees, with definite structure, the materials arranged and compacted to form a cup.

  A. Used in crotches and branches of trees to which they are adapted.

    1. **Statant Cupped Nests.** Nests with rims standing firmly upright, supported

mainly from below. The sides may or may not be extended upward and arched over the top in a dome.

_____   _____

_____   _____

_____   _____

_____   _____

2. **Suspended Cupped Nests.** Nests not supported from below but from the rims, or sides, or both.
   a. **Pensile.** Nests suspended from the rims and sides; rather stiff.

   _____   _____

   _____   _____

   b. **Pendulous.** Nests suspended from the rims and sides; rather flexible and extremely deep-cupped with lower part swinging freely.

   _____   _____

   _____   _____

B. Re-adapted to other situations.
   1. **Adherent Nests.** Cupped nests whose sides are attached by adhesive substance to a vertical surface.

   _____   _____

   _____   _____

   _____   _____

   2. **Ground Nests.** Cupped nests on the ground. The sides may or may not be extended upward and arched over the top, making a domed structure.

   _____   _____

   _____   _____

   _____   _____

   3. **Cavity Nests.** Cupped nests in crevices, holes, birdboxes, etc.
      a. In preformed or natural cavities.

   _____   _____

   _____   _____

   _____   _____

      b. In cavities excavated by occupant birds.

   _____   _____

   _____   _____

IDENTIFICATION OF NESTS

Because nests of many closely allied species are very similar in location, materials, and structure, the student must be careful of identifying any nest, with complete certainty, without knowing the bird that constructed it. But nests of certain species, or groups of closely allied species, are sufficiently distinctive to allow identification with reasonable certainty. The student attempting to identify nests in the conterminous United States is referred to the keys to nests by Headstrom (1949, 1951). Also a good reference is the work by Campbell (1953). Although it deals with British birds, the introductory chapters suggest methods and techniques for finding and studying nests.

NEST-BUILDING

The various phases of nest-building have been investigated the least of any subject in the breeding cycle of birds. The author's own studies and the works of many others have contributed to the outline of nest-building procedures given below.

Selection of the Nesting Site

The need of suitable support and protection governs the selection of the nesting site. Birds do not deliberately select the site to conceal the nest, although frequently concealment results from the placing of the nest in a position, protected from the destructive forces of the environment (e.g., sunlight, wind, cool night temperatures).

Almost invariably, a period of appetitive searching during which the birds move from one potential site to another—usually within the confines of their established territory—precedes the final selection of a nesting site. In species that build cupped nests in trees, the birds try fitting their bodies into crotches between branches; in species building the same type of nests on the ground, the birds scratch small depressions in the surface and attempt to mold them to suit the body contours. Often they accumulate a few nesting materials, and occasionally they partly construct nests at several sites before making the final selection.

The searching for a nest site is first strongly manifest after mating when the males are singing frequently and intensively and the females are receptive to males. They defend the territory strongly during this period. Nest-searching activities frequently appear following each sexual relation. The number of copulations during the days of the searching period is unknown.

Neither has the duration of the searching period been satisfactorily determined. In some cases it lasts from three to five days; in a few cases it is more prolonged. Searching, by no means continuous during the period, occurs at intervals, especially during the early hours of the morning. These intervals of searching increase in length daily as the nest-building drive matures.

The role of each sex in searching for and selecting the nesting site varies with the species. In birds maintaining territories of Types A, B, and D, one or both sexes participate. When both sexes participate, generally the female takes the more aggressive role. In birds with territories of Type C, the female alone searches for and selects the nesting site.

The searching period represents the phase in the breeding cycle during which the nest-building drive develops. Even though birds find suitable nesting sites during this period, the selection is not final. They select the nesting site when their arrival on a suitable site coincides with the blocking of the nest-building drive by consummatory behavior.

Beginning of Nest-building

In many birds nest-building begins with the initiation of rapid growth in the ovum (Riddle, 1911) and continues more or less intensively while the ovum matures. The first egg is usually laid one or several days after the completion of the nest, but in a few species nest-building continues to some extent after laying has begun. Climatic conditions may influence the beginning of nest-building. High temperatures at the start of the nesting season stimulate nest-building; low temperatures tend to inhibit it (Nice, 1937). Delayed development of vegetation may delay the building.

Process of Nest-building

The process of nest-building may be roughly divided into three steps: (1) Preparing the site or support (e.g., scratching a depression, "cleaning out" a preformed cavity, excavating a new burrow or hole). (2) Constructing the floor and sides (i.e., the "outside"). (3) Lining the nest. One or two of the stages do not occur, or occur only in part, in certain species.

Usually, the birds gather all "outside" materials for nests of Stages II and III in the vicinity. For nests of Stage I, the materials are frequently those within reach of the bird as it stands on the nest. But materials for lining the nests of Stages II and III are often sought at great distances, presumably because the special materials are not readily available close at hand. Tree Swallows *(Iridoprocne bicolor)* may sometimes fly several miles to a chicken farm to obtain the much-preferred white feathers.

Most birds carry materials in their bill, although diurnal birds of prey — eagles, hawks, etc. — use their feet more commonly. Several species of African lovebirds *(Agapornis)* carry nesting materials in their rump feathers.

Each bird builds its nest with a number of stereotyped movements. If the nest is to have structure, the movements required are necessarily more numerous and complex. In constructing a typical cupped statant nest in a tree, the passerine bird sits in the center of the cup that is taking shape and performs the following characteristic movements: *Pulling* a long slender piece of material (e.g., a plant fiber) with its bill inward over the rim and *tucking* it into the wall under its breast or *drawing* it alongside against the wall; *looping* a similar piece of material around a supporting branch by starting it around one side and then *reaching* around the opposite side and *drawing* it back so that it encircles the branch — and sometimes repeating the performance until the material encircles the branch several times; *inserting* short pieces of material into the rim or wall with jabs of the bill; shaping the cup by *squatting* in it while alternately *fluffing* and *compressing* the body feathers; *pressing* its head, tail, and partly opened wings down against the rim of the cup, while alternately *pushing* with one foot and then the other against the floor and sides of the cup; *turning* frequently between and sometimes during any of the above movements.

Learning plays a role in nest-building to the extent that a bird must determine, for instance, by trial and error (see the section, "Behavior," page 247) which materials are the most suitable for the structure and the lining. The principal movements in nest-building, however, are innate, derived and then stereotyped from such sources as maintenance activities or even irrelevant behaviors. To cite one example: Harrison (1967) has suggested that the behavior called "sideways-building," in which the bird pulls materials into its nest and draws it alongside, is derived from "sideways-throwing," an irrelevant behavior common in a number of shore birds and gulls when the bird picks up a leaf, stick, shell, or any other small object and flicks it backward on one side with a quick motion of the head.

Many more detailed studies of nest-building in different groups of birds must be made before the origin and homologies of all the movements can be established. For some of the few detailed studies available to date, the student is referred to the works by Armstrong (1955), Herrick (1911, 1935), Nice (1937, 1943), Marler (1956), Tinbergen (1939, 1953), and others.

Participation and Behavior of the Sexes

In a great many species the female builds the nest alone. In many others the female builds the nest with the assistance of the male. The female gathers the materials, works them into place, and molds the depression. The male's assistance usually amounts to gathering materials with or without the female and, on returning with them, passing them to the female for use in the nest.

In a number of species the role of the sexes in nest-building does not conform to the above. The male may (1) build the entire nest, although this is rare; (2) build certain parts of it, such as the floor and sides, or excavate the burrow or hole; (3) share all nest-building activities with the female. Within species, the role of the sexes is subject to variation, but the variation is seldom as extreme as that between species.

There is apparently no dependable correlation among species between coloration of the male and his nest-building proclivities, nor a correlation between his nest-building proclivities and his part in subsequent incubation. There is, on the other hand, a rather positive correlation between a male's nest-building proclivities and his participation in subsequent parental care, i.e., if a male assists in building, he will also assist in feeding the young.

During nest-building the male customarily sings as vigorously as at any other time during the breeding cycle. Copulation occurs throughout nest-building, though the frequency is not known.

Length of Time Involved

The total length of time involved in nest-building is difficult to determine because seldom is the beginning observed. Nests, because they are constructed in protected sites, are usually discovered *as a result* of building activities already under way. Furthermore, the cessation is sometimes indeterminable. Nest-building may stop suddenly a day or more before egg-laying, or it may continue for some time after egg-laying has begun.

The *number of days* is used as a measure of the time involved in the construction of nests. Due to the irregularities of nest-building in most species, no more accurate measure can be applied unless the observer has sufficient persistence and good fortune to follow daily activities from start to finish, in which case he can count the total *number of hours* involved.

Remarkably few records are available to show even the number of days involved in nest-building. Most passerine birds may require about six (Allen, 1961), three for constructing the outside of the nest and three more for finishing the interior and lining it. But judging by the few precise records at hand, the length of time is subject to wide interspecific and intraspecific variation.

Interspecific variation is accounted for in part by two factors: (1) **Type of Nest.** Obviously, certain nests that are more elaborate than others require more time for construction. To build their long, pendulous nests, the Alta Mira or Lichtenstein's Oriole *(Icterus gularis)* in Mexico took "at least 18 and perhaps as many as 26 days" (Sutton and Pettingill, 1943) and the Wagler's Oropendola *(Zarhynchus wagleri)* in the Canal Zone, Panama, "about one month" (Chapman, 1928). (2) **Climate.** Species in the tropics take more time than closely allied species in the temperate regions. Derby of Kiskadee Fly-catchers *(Pitangus sulphuratus)* in Mexico took 24 days (Pettingill, 1942) whereas most tyrannid flycatchers farther north need only three to rarely more than 13 days for finishing their nests.

Two factors account in part for the intraspecific variation in time required for nest-building: (1) **Weather Conditions.** Cool or inclement weather may retard nest-building. On the other hand, rainfall may stimulate it, at least in the case of the European Wren *(Troglodytes troglodytes)* because it makes the nesting material more flexible and easier to manipulate (Armstrong, 1955). (2) **Renesting.** An individual, or pair, building a second or third nest in the same season takes less time.

Though the length of time taken in nest-building is measured in number of days, nest-building does not proceed steadily during those days. Usually only the early parts of the day are involved, at which time there are periods of building (i.e., attentive periods) and periods of no building (i.e., inattentive periods). As nest-building progresses, the attentive periods lengthen, and inattentive periods shorten. At the height of nest-building the average length of attentive and inattentive periods often quite closely corresponds to the average length of such periods during incubation.

A few birds modify or repair their nests after they have laid eggs and begun incubation, or even later. Bald Eagles *(Haliaeetus leucocephalus)* frequently add materials to their large aeries throughout the nesting season. The marsh-dwelling rails, gallinules, and coots are noted for their ability to build up their nests, whenever the water rises, to keep the flood from their eggs. The King Rail *(Rallus elegans)* has been reported to elevate its eggs in this way by as much as a foot (Meanley, 1953).

Nest-building in Young Birds

The nest-building abilities of young birds need investigation. From meager evidence it appears that younger birds may build their nests as quickly and expertly as older birds do (Nice, 1943). This is perhaps not true among species that build elaborate nests.

In rare instances, young from a nest built early in a season assist adults in building a second nest later in the season. Independent attempts at nest-building have been seen among young birds-of-the-year.

FALSE NESTS

Certain species of birds build structures called "nests" which are not true nests, because they are not constructed for the purpose of containing eggs.

Cock "Nests"

Sometimes called **"dummy nests."** These are nests constructed by males, notably wrens, in the vicinity of the regular nest. They resemble the regular nest, but they are not completely constructed and are without lining. Herrick (1935) suggests that such nests "are simply the work of male birds in response to an inherited predisposition to this sort of activity while singing and waiting upon their chosen territory for a suitable mate, or even after, when such a mate is engaged in the prosaic task of incubation."

Refuge "Nests"

A few birds, especially burrow- and hole-nesters, often create "nests," similar to their regular nests, for roosting and for shelter during unfavorable weather. They may be constructed at any time of year but particularly in the fall. Some birds merely use cavities already available.

RE-USE OF NESTS

A few species of birds, notably some of the large hawks, the Golden and Bald Eagles, and the Osprey, use the same nest year after year. When the same nest is used, it is generally either "repaired" and enlarged each year, or a completely new nest is built on top of the old.

Certain species appropriate the nests used by other species in the preceding season. Sometimes the new tenants alter the nests to suit their own requirements. A Mourning Dove *(Zenaidura macroura)*, for instance, may build its platform nest on top of the cupped nest of a Robin *(Turdus migratorius)*. Or the new tenants may accept the nest as it is. This is true of a Great Horned Owl *(Bubo virginanus)* taking over a hawk or eagle nest, and of the Solitary Sandpiper *(Tringa solitaria)* using the cupped tree nest of a Robin, Rusty Blackbird *(Euphagus carolinus)*, or Eastern Kingbird *(Tyrannus tyrannus)*.

Among passerine birds, as well as among many other groups, it is exceptional for pairs to renest in the same nest or at the same nesting site.

PROTECTION OF NESTS

Nests are often concealed through the choice of nesting sites. If they are not concealed, then they may be inaccessible as when nests are placed on small islands, on the shelves of cliffs, and near the ends of slender branches. The theory that snake skins, sometimes found in nests, serve as a means of frightening enemies has been discredited (see Rahd, 1953).

Nests may gain protection against predators when they are placed close together in a colony, or singly in a colony of other species, because there are many more occupant birds on the alert to attack and take deterrent action. Nests in tropical regions also gain protection when they are placed—deliberately by some species—in the immediate vicinity of aggressive social ants and wasps, or near the nests of larger, more aggressive bird species. In Mexico, Pettingill (1942) found one nest each of the Social or Vermilion-crowned Flycatcher *(Myiozetetes similis)* and the larger Kiskadee Flycatcher in a bull's horn acacia, a shrub about 12 feet high, that was tenanted by countless thousands of small ants ready to bite and sting. Although the ants did not annoy the nest occupants, they viciously swarmed over anyone touching the shrub. Both nests obviously benefited from the protection afforded by the bellicose ants while the nest of the Social Flycatcher benefited further by being close to the nest of the larger and more belligerent Kiskadee Flycatcher.

NEST FAUNA

Nests of birds, particularly those in burrows, cavities, or birdboxes, or under the eaves of buildings, are snug havens for many small invertebrates, especially arthropods—mites, spiders, insects, etc. Most of them are visitors stopping only for shelter. (For an idea of the variety of insects that may be found in birds' nests, see the check-list by Hicks, 1959.) But along with the visitors are a number of flies and fleas that pass their life cycles specifically in birds' nests and at one stage are parasitic on the birds. They are to be distinguished from the obligate ectoparasites on the feathers and/or other parts of birds.

Chief among the nest parasites, from the viewpoint of their serious effect on the occupant birds, are blowflies *(Protocalliphora* spp.*)* whose larvae live in the nest cup and, beginning at twilight, attach themselves on the nestlings for a blood meal during the night. If the nest infestation by larvae is heavy, the nestlings can be greatly weakened, even killed, from loss of blood.

See Appendix I for a review of the ectoparasites of birds. For additional information on the subject of nest fauna, refer to Boyd (1951), Coombs (1960), George and Mitchell (1948), Guberlet and Hotson (1940), Herman (1937), Hill and Work (1947), Kenaga (1961), Mason (1936, 1944), Nolan (1955), Owen (1954), and Rothschild and Clay (1957).

THE STUDY OF NESTS AND NEST-BUILDING

Select a species for the study of nests and nest-building and concentrate on the activities of one pair.

Procedure

When possible, begin the study soon after the territory is established, thus observing the selection of the nesting site and the start of nest-building. Follow all activities of the pair through the searching and nest-building periods. Practice extreme caution in watching the birds during the nest-building, since they are not strongly attached to the nesting site and will abandon it on the slightest provocation. Therefore, use a binocular or telescope and remain as far away as possible.

Take full notes on the activities of both sexes; time their respective attentive and inattentive periods, and the periods when the male is singing. When the birds are not present at the nest, take photographs of stages in construction, and make careful notes on the appearance of the nest at each stage.

Obtain full data on the completed nest, particularly the following:

Measurements: **If ground nest without structure:** inside diameter (i.e., diameter of the depression); inside depth (i.e., depth of the depression). **If cavity nest without structure:** length, or depth, of cavity from lower edge of entrance to floor; diameter (or diameters) of the entrance and diameter (or diameters) of the part of cavity containing the eggs. **If elevated nest** or **re-adapted ground nest:** outside diameter, outside depth; inside diameter, inside depth. (If nests are domed, consider outside depth to be from top of arch, and inside depth to be from the under surface of arch.) **If re-adapted cavity nest:** use the measurements indicated above for directly adaptive cavity nests, plus the measurements of the inside depth and inside diameter of the nest structure built in the cavity.

Location: Take notes on the location and support of the nest, naming all dominant vegetation in the vicinity and observing the condition and protective value. If the nest is in a tree, measure its height from the ground. Use an altimeter or clinometer (see Appendix A, page 431) in case the height is too great for accurate determination by measuring stick or tape. Also measure or estimate the distance of the nest from the main trunk of the tree. If the nest is in a cavity, measure the distance from the lower edge of the entrance to the ground and note the direction (using a compass) that the entrance faces. Photograph the environment, with the nest occupying a central position in the background.

Description of the Nest: Take two or three photographs that show the support of the nest and the appearance of the nest close up. Disturb the surrounding cover as little as possible. When taking the close-up pictures of an open nest, place the camera partly above and partly to the side so that the far inside wall and near outside wall will show.

After the young have left, the nest should be taken to the laboratory and studied in detail. Its structure should be analyzed, all materials identified, and the relative quantity of each material determined.

If, after a nest is completed, it is for some reason destroyed or deserted, attempt to follow the same pair and gather evidence of renesting. The following data are particularly desirable: (1) Time involved between the destruction or desertion of the first nest and the completion of the second. (2) Variation among individuals in the selection of the second nesting site and manner of building. (3) Proximity of the first and second nests. Compare the data with the findings of other investigators.

Presentation of Results

Draw up a report on the nests and nest-building of the pair observed. Prepare the manuscript as directed in Appendix B. An outline of suggested topics is given below.

Territory. Type; brief description; date discovered.

Selection of Nesting Site. Time and date of searching period; relation to weather conditions; role and activities of sexes, including number and duration of singing periods of male and number of copulatory acts per day; number and description of nest-building attempts; time of day and total length of time (in hours) devoted to searching; searching as related to territory and territorial behavior.

Beginning and Duration of Nest-building. Brief statements of date of beginning and the duration. Discussion of weather, vegetation, and other ecological conditions as related to the time of beginning.

Nest-building. Detailed chronological account of the stages and mechanics of nest-building. This should include the role of the sexes, number of copulations per day, times when the male sings, manner of gathering materials, sources of materials, and number of material-gathering trips. Discuss the length of time involved separately. Include the hours of the day when nest-building takes place, a table summarizing attentive and inattentive periods, and statements indicating the total amount of time spent in building the nest. (In deriving the table, follow the directions for the table on incubation periods given in the next section of this book.) Use photographs to illustrate stages in construction.

Description of Nest. **Location:** Describe in full, and give height from ground, if elevated. Use photographs to illustrate the nesting site. Discuss protective factors, etc. **Measurements:** Include the measurements as directed above. **Structure:** Note any details of structure not determined when watching the nest under construction. Analyze the materials, using a wheel diagram to show relative quantities of different materials. Include photographs of the nest.

Miscellaneous Observations. Present any observations made on renesting, building ability of younger birds, presence of false nests, protection afforded the nest, and nest fauna.

REFERENCES

Alexander, W. B.
1931   Association of Birds' Nests with Nests of Insects in Australia. Proc. Ent. Soc. London, 5: 111-114.

Allen, A. A.
1961   The Book of Bird Life. Second edition. D. Van Nostrand Company, New York. (Chapter 16 for information on finding nests and identifying them. A key to the nests of birds of northeastern North America is given on pages 317-324.)

Amadon, D.
1944   A Preliminary Life History Study of the Florida Jay, *Cyanocitta c. coerulescens*. Amer. Mus. Novitates No. 1252. (Nest-building, etc.)

Armstrong, E. A.
1955   The Wren. Collins, London. (Chapter 9: Nest-building.)

Astley, A.
1935   Young Swallows Assisting in Building the Second Nest. Brit. Birds, 28: 204.

Baker, E. C. S.
1931 Nesting Association between Birds and Wasps, Ants, or Termites, in the Oriental Region. Proc. Ent. Soc. London, 6: 34-37.

Blanchard, B. D.
1941 The White-crowned Sparrows *(Zonotrichia leucophrys)* of the Pacific Seaboard: Environment and Annual Cycle. Univ. California Publ. in Zool., 46: 1-178. (Compression of nesting cycle in northern latitudes as compared with southern.)

Boyd, E. M.
1951 The External Parasites of Birds: A Review. Wilson Bull., 63: 363-369.

Brackbill, H.
1950 Successive Nest Sites of Individual Birds of Eight Species. Bird-Banding, 21: 6-8. (Data on 40 nests indicate fairly fixed nesting heights for some individuals and highly variable ones for others.)
1952 A Joint Nesting of Cardinals and Song Sparrows. Auk, 69: 302-307.

Campbell, B.
1953 Finding Nests. Collins, London.

Chapman, F. M.
1928 The Nesting Habits of Wagler's Oropendola *(Zarhynchus wagleri)* on Barro Colorado Island. Bull. Amer. Mus. Nat. Hist., 58: 123-166.

Chisholm, A. H.
1952 Bird-Insect Nesting Associations in Australia. Ibis, 94: 395-405.

Collias, N. E., and E. C. Collias
1964 Evolution of Nest-building in the Weaverbirds (Ploceidae). Univ. California Publ. in Zool., 73: i-viii; 1-239. (Concerned with some two dozen selected and representative species of weaverbirds which show great range in variation in their nests. An illuminating paper on the manner in which modes of nest-building have evolved and the functions they serve.)

Coombs, C. J. F.
1960 Ectoparasites and Nest Fauna of Rooks and Jackdaws in Cornwall. Ibis, 102: 326-328.

Davis, D. E.
1955 Breeding Biology of Birds. In *Recent Studies in Avian Biology.* Edited by A. Wolfson. University of Illinois Press, Urbana.

Emlen, J. T., Jr.
1954 Territory, Nest Building, and Pair Formation in the Cliff Swallow. Auk, 71: 16-35.

Erskine, A. J.
1959 A Method for Opening Nesting Holes. Bird-Banding, 30: 181-182. (Drilling a hole from the back or side of a tree. Details given.)

Favaloro, N.
1942 The Usurpation of Nests, Nesting Sites and Materials. Emu, 41: 268-276. (Many instances of Australian birds appropriating other nests while in use, using old nests, etc.)

Frith, H. J.
1962 The Mallee-fowl: The Bird That Builds an Incubator. Angus and Robertson, Sydney.

George, J. L., and R. T. Mitchell
1948 Notes on Two Species of Calliphoridae (Diptera) Parasitizing Nestling Birds. Auk, 65: 549-552.

Guberlet, J. E., and H. H. Hotson
1940 A Fly Maggott Attacking Young Birds, with Observations on Its Life History. Murrelet, 21: 65-68.

Harrison, C. J. O.
1967 Sideways-throwing and Sideways-building in Birds. Ibis, 109: 539-551.

Headstrom, R.
1949 Birds' Nests: A Field Guide. Ives Washburn, New York. (For use in the United States east of the 100th meridian.)
1951 Birds' Nests of the West: A Field Guide. Ives Washburn, New York. (For use in the contiguous United States west of the 100th meridian.)

Herman, C. M.
1937 Notes on Hippoboscid Flies. Bird-Banding, 8: 161-166.

Herrick, F. H.
1911 Nests and Nest-building in Birds. Jour. Animal Behavior, 1: 159-192; 244-277; 336-373.
1934 The American Eagle. D. Appleton-Century Company, New York. (Nest-building and re-use of nests year after year.)
1935 Wild Birds at Home. D. Appleton-Century Company, New York.

Hicks, E. A.
[1959] Check-list and Bibliography on the Occurrence of Insects in Birds' Nests. Iowa State College Press, Ames. (Contains two check-lists: one of insects found in birds' nests and one of birds in whose nests insects have been found. Both lists contain many hundreds of references to a 68-page bibliography.)

Hill, H. M., and T. H. Work
1947 Protocalliphora Larvae Infesting Nestling Birds of Prey. Condor, 49: 74-75.

Howard, O. W.
1904 The Coues Flycatcher as a Guardian of the Peace. Condor, 6: 79-80. (An example of an aggressive species with smaller species nesting nearby.)

Ingram, C.
1943 Swallows Adding to Nest after Beginning of Incubation. Brit. Birds., 37: 116.

Jensen, J. K.
1918 Subsequent Nestings. Auk, 35: 83-84.

Kenaga, E. E.
1961 Some Insect Parasites Associated with the Eastern Bluebird in Michigan. Bird-Banding, 32: 91-94. (One nest contained 2,300 insects, dependent directly or indirectly on the nestlings.)

Kendeigh, S. C.
1952 Parental Care and Its Evolution in Birds. Second corrected printing, 1955. Illinois Biol. Monogr., 22: i-x; 1-356. (Contains a considerable amount of information on nest-building in many species.)

Kramer, G.
1950 Der Nestbau biem Neuntöter *(Lanius collurio L.)*. Ornith. Berichte, 3: 1-14. (An excellent account of nest-building in captive Red-backed Shrikes.)

Kuerzi, R. G.
1941 Life History Studies of the Tree Swallow. Proc. Linnaean Soc. New York, Nos. 52-53: 1-52. (Nest-building.)

Lack, D.
1956 Swifts in a Tower. Methuen and Company, London.

Laskey, A. R.
1950 A Courting Carolina Wren Building over Nestlings. Bird-Banding, 21: 1-6.

Lawrence, L. de K.
1953 Nesting Life and Behaviour of the Red-eyed Vireo. Canadian Field-Nat., 67: 47-77. (Nest-building.)
1967 A Comparative Life-history Study of Four Species of Woodpeckers. Amer. Ornith. Union, Ornith. Monogr. No. 5. (Excavating nests and related behavior.)

Laven, H.
1940 Nestbaustudien. Ornith. Monatsber., 48: 128-131.

Lea, R. B.
1942 A Study of the Nesting Habits of the Cedar Waxwing. Wilson Bull., 54: 225-237. (Nest-building.)

Lewis, J. B.
1926 The Re-use of Old Nest Material by the Blue-gray Gnatcatcher and Ruby-throated Hummingbird. Wilson Bull., 38: 37-38.

Linsley, E. G.
1943 Insect Inhabitants of Bird and Mammal Nests. Jour. Wildlife Management, 7: 423.

Low, S. H.
1934 Nest Distribution and Survival Ratio of Tree Swallows. Bird-Banding, 5: 24-30. (Information concerning re-use of nests.)

Maclaren, P. I. R.
1950 Bird-Ant Nesting Associations. Ibis, 92: 564-566.

Marler, P.
1956 Behaviour of the Chaffinch *Fringilla coelebs*. Behaviour Supplement 5. E. J. Brill, Leiden.

Mason, E. A.
1936 Parasitism of Bird's Nests by Protocalliphora at Groton, Massachusetts. Bird-Banding, 7: 112-121.
1944 Parasitism by Protocalliphora and Management of Cavity-nesting Birds. Jour. Wildlife Management, 8: 232-247.

Maxse, V.
1951 The British Long Tailed Tit *(Aegithalos caudatus roseus)*: How Its Nest Is Built. Proc. Xth Internatl. Ornith. Congr., pp. 563-566.

Meanley, B.
1953 Nesting of the King Rail in the Arkansas Rice Fields. Auk, 70: 261-269.

Meuli, L. J.
1935 Observations on Nest Site Trials by the Eastern Robin. Wilson Bull., 47: 296-297.

Moreau, R. E.
1936 Bird-Insect Nesting Associations. Ibis, 1936: 460-471.
1942 The Nesting of African Birds in Association with Other Living Things. Ibis, 1942: 240-263.

Mousley, H.
1917 A Study of Subsequent Nestings after the Loss of the First. Auk, 34: 381-393.
1918 Subsequent Nestings. Auk, 35: 237-238.

Myers, J. G.
1929 The Nesting-together of Birds, Wasps, and Ants. Proc. Ent. Soc. London, 4: 80-88.

Nauman, E. D.
1930 The Nesting Habits of the Baltimore Oriole. Wilson Bull., 42: 295-296. (In-

formation on length of time in building, etc.)

Nethersole-Thompson, C., and
D. Nethersole-Thompson
1943   Nest-site Selection by Birds. Brit. Birds, 37: 70-74; 88-94; 108-113.

Nice, M. M.
1937   Studies in the Life History of the Song Sparrow, I. Trans. Linnaean Soc. New York, 4: i-vi; 1-247. (Chapter 10.)
1943   Studies in the Life History of the Song Sparrow, II. Trans. Linnaean Soc. New York, 6: i-viii; 1-328. (Chapter 17.)

Nickell, W. P.
1943   Secondary Uses of Birds' Nests. Jack-Pine Warbler, 21: 48-54.

Nolan, V., Jr.
1955   Invertebrate Nest Associates of the Prairie Warbler. Auk, 72: 55-61. (In 9 nests were 19 species of arthropods, of which one or more of three species were known to be parasitic in 7 nests.)

Odum, E. P.
1941   Annual Cycle of the Black-capped Chickadee-2. Auk, 58: 518-535. (Nest-building, role of sexes, etc.)

Owen, D. F.
1954   Protocalliphora in Birds' Nests. Brit. Birds, 47: 236-243.

Pettingill, O. S., Jr.
1942   The Birds of a Bull's Horn Acacia. Wilson Bull., 54: 89-96. (Nest-insect relationships; comparison of length of building periods between temperate and tropical species; small species nesting in vicinity of larger species.)

Poulton, E. C.
1932   Further Evidence of Nesting Associations between Neo-tropical Birds and Wasps. Proc. Ent. Soc. London, 7: 55-56.

Rand, A. L.
1953   Use of Snake Skins in Birds' Nests. Chicago Acad. Sci. Nat. Hist. Miscellanea, No. 125.

Riddle, O.
1911   On the Formation, Significance and Chemistry of the White and Yellow Yolk of Ova. Jour. Morph., 22: 455-490.

Roberts, N. L.
1941   Choosing the Nest Site. Emu, 41: 162-163.

Ross, W. M.
1942   How a Tree-creeper Built Its Nest. Brit. Birds, 36: 110-111.

Rothschild, M., and T. Clay
1957   Fleas, Flukes, and Cuckoos: A Study of Bird Parasites. Third edition. Collins, London. (Chapter 14: The Fauna of Birds' Nests.)

Ryves, B. H.
1944   Nest-construction by Birds. Brit. Birds, 37: 182-188; 207-209. (Participation of sexes in nest-building among birds generally.)

Skutch, A. F.
1940   Social and Sleeping Habits of Central American Wrens. Auk, 57: 293-312. (Dummy nests used as places for sleeping.)

Stevenson, E.
1942   Key to the Nests of the Pacific Coast Birds. Oregon State Monogr. Stud. Zool., No. 4.

Stoddard, H. L.
1931   The Bobwhite Quail: Its Habits, Preservation and Increase. Charles Scribner's Sons, New York. (Nests, nesting sites, and nest-building in Chapter 2.)

Strecker, J. K.
1926   On the Use, by Birds, of Snakes' Sloughs as Nesting Material. Auk, 43: 501-507.

Summers-Smith, J. D.
1963   The House Sparrows. Collins, London. (Chapter 6: Nests and Nest-building.)

Suthard, J.
1927   On the Usage of Snake Exuviae as Nesting Material. Auk, 44: 264-265.

Sutton, G. M., and O. S. Pettingill, Jr.
1943   The Alta Mira Oriole and Its Nest. Condor, 45: 125-132. (Nest-building; comparison of nest-building periods in several species.)

Teachenor, D.
1927   Snakes' Sloughs as Nesting Material. Auk, 44: 263-264.

Tinbergen, N.
1935   Field Observations of East Greenland Birds, I. The Behaviour of the Red-necked Phalarope (*Phalaropus lobatus* L.) in Spring. Ardea, 24: 1-42. (Searching; nest-building; role of sexes.)
1939   The Behavior of the Snow Bunting in Spring. Trans. Linnaean Soc. New York, 5: 1-94.
1953   The Herring Gull's World: A Study of the Social Behaviour of Birds. Collins, London. (Chapter 15 concerns nest-building.)

Tutt, H. R.
1951 Data on the Excavation of the Nest Hole and Feeding of the Young of Green Woodpecker *(Picus viridis)* Proc. Xth Internatl. Ornith. Congr., pp. 555-562.

Welter, W. A.
1935 The Natural History of the Long-billed Marsh Wren. Wilson Bull., 47: 3-34. (Nest-building; cock "nests.")

Whittle, C. L.
1927 The Role of the Snake Skin. Auk, 44: 262-263.

Williams, L.
1942 Interrelations in a Nesting Group of Four Species of Birds. Wilson Bull., 54: 238-249. (Use of man-made structures as nest sites, re-use of nests, roosting in nests, etc.)

EGGS, EGG-LAYING, AND INCUBATION

All species of birds lay eggs and all with few exceptions incubate them with the heat of their own bodies.

SIZE, SHAPE, AND COLORATION

The largest eggs are laid by the largest birds and the smallest by the smallest, but this proportion of egg size to body weight does not apply to all birds. The largest egg known, that of the extinct elephant bird *(Aepyornis)*, measures 14.5 by 9.5 inches and is assumed to have held two gallons of fluid that may have weighed as much as 27 pounds—perhaps less than 3 percent of the adult bird's weight; and the egg of an Ostrich *(Struthio camelus)* measures roughly 7 by 5.5 inches and weighs nearly 3 pounds—about 1.7 percent of the body weight. At the opposite extreme, the egg of one of the smallest hummingbirds measures 13 by 8 millimeters and weighs 0.5 gram—around 10 percent of the parent's weight. From these figures, one may deduce correctly that the larger birds lay proportionately smaller eggs. But there are exceptions. The chicken-size kiwi *(Apteryx* spp.*)* produces an egg which, measuring nearly 5.5 by 3.5 inches and weighing approximately a pound, constitutes as much as 25 percent of the body weight.

Sometimes bird species that are the same size, such as the Common Snipe *(Capella gallinago)* and the Robin *(Turdus migratorius)*, lay eggs differing greatly in size. The egg of the Common Snipe, a shore bird, is larger because it yields a well-developed downy young bird or chick whereas the egg of the Robin, a passerine bird, produces only a helpless nestling. Within species, the size of eggs normally varies in accordance with age, the younger birds in their first year of nesting laying smaller eggs.

Like the eggs of the Domestic Fowl *(Gallus gallus)*, the eggs of most birds are somewhat rounded at one end and bluntly pointed at the other. The shape may vary from this in two directions: toward the two ends matching each other in shape, or toward the two ends becoming respectively more pointed and more rounded. Thus the eggs of Ostriches, owls, and kingfishers are spherical, those of grebes equally pointed at both ends, and those of hummingbirds elongate and equally blunt at both ends. In the other direction, the eggs of shore birds, quail, and murres *(Uria* spp.*)* are remarkably pointed at one end, large and rounded at the other. Such exaggerated shapes are probably adaptations: in the shore birds, so that the normal clutch of four large eggs may be incubated with the pointed ends toward the nest's center, thereby occupying minimum space; in quail, so that the normally large clutch of 10 or more eggs will take up less space; and in the murres, so that the single egg of each pair of birds on a bare nesting ledge will turn in a circle when disturbed and not roll off to its destruction.

The shells of eggs, even though always penetrated by countless microscopic pores, vary widely in surface texture from being smooth, as in most birds, to being on the one hand quite glossy as in woodpeckers, and even spectacularly shiny as in the tinamous (Tinamidae), a New World group of birds, or being on the other hand noticeably rough or chalky as in cormorants.

Many eggs of widely different species are white, but the majority of eggs show a great range in colors from faint to intense. In some species the eggs have simply a uniform **ground color** from buff to reddish brown or from pale blue to deep blue-green. This color is deeply suffused in the calcareous material forming the shell. In other species the eggs have in the outermost stratum of the shell an array of pigments, with or without the underlying ground color, that provide the **markings**—blotches, scrawls, streaks, speckles, etc. Quite often the markings tend to be concentrated in a wreath around the large end, since, in the egg's final descent through the oviduct, the large end comes first picking up the major supply of pigments from the cellular walls.

Because reptiles, the forebears of birds, probably produced white eggs, one may assume that the eggs of the earliest birds were similarly white. Then, in the course of avian evolution, the eggs acquire first the ground color and second the markings for adaptive purposes.

Heavy pigmentation is characteristic of eggs in open nests, particularly in ground nests, where it serves a dual function: to shield the embryos within from intensive solar radiation and to provide cryptic patterns for concealing the eggs from the view of predators. Light pigmentation or none at all is characteristic of eggs in cavity nests. Bluebirds (*Sialia* spp.) which nest in cavities have pale blue eggs, while their familial relatives, Robins and hylocichlid thrushes, which construct open nests, have vividly blue eggs. Not all cavity-nesting birds lay pale or colorless eggs. Those of chickadees and nuthatches are spotted and, inexplicably the Long-billed Marsh Wren (*Telmatodytes palustris*) lays chocolate brown eggs and the Short-billed Marsh Wren (*Cistothorus platensis*) nearly white eggs in the dark interiors of their similarly ball-shaped nests.

IDENTIFICATION OF EGGS

An examination of a collection of birds' eggs shows the wide range in size and shape and the almost infinite variety of colors and markings. At the same time it reveals the hopelessness of attempting either a classification or a key based on their differences. Relatively few eggs are as distinct as the species producing them. Within a species there can be remarkable diversity in color. A collection of murre eggs from one colony will show every possible combination of ground colors and markings. Between species that are closely related the intergradation is sometimes so perfect as to exclude distinction. The student is advised, therefore, never to attempt positive identification of eggs without direct knowledge of the birds that laid them. For descriptive information, refer to the work of Reed (1904), as well as other works listed in Appendix D.

NUMBER OF EGGS IN A CLUTCH

A **clutch,** or **set,** of eggs is the total number of eggs laid by one bird in one nesting. A few groups of species characteristically lay one egg (e.g., albatrosses, shearwaters, storm petrels, diving petrels, tropicbirds, frigatebirds, most alcids), a few other groups lay two (loons, goatsuckers, most pigeons, hummingbirds), and most shore birds lay four. With the exception of shore birds, practically all other species normally laying more than two eggs show marked variation in clutch size.

Appendix H gives the range in clutch sizes among families of North American birds. All the figures, obtained from various sources, apply only to the representatives of the families occurring regularly north of Mexico.

As a rule, the greater the characteristic size of a clutch in a species, the greater the variation. Nice (1943) showed that clutch size in the Song Sparrow *(Melospiza melodia)* is influenced by the following factors: (1) **Age.** Young females may lay smaller clutches the first season. (2) **Weather Conditions.** Cold weather may reduce the size of a clutch. (3) **Time of Season.** Smaller clutches may be laid at the end of the season by birds that have laid larger clutches earlier. This has been confirmed by von Haartman (1967) in his studies of the Pied Flycatcher *(Ficedula hypoleuca)* nesting in Finland. See also Howard (1967) for seasonal variation in the Robin *(Turdus migratorius).* (4) **Individual Variation.** Some adults lay typically smaller clutches than others.

The same or closely allied species, nesting in areas widely separated geographically and/or ecologically, may show significant differences in clutch sizes. Thus birds nesting on sea islands may have smaller clutches than those on continents (e.g., see Pettingill, 1960), and birds nesting on the sea coasts or in the tropics generally produce fewer eggs per clutch than those in the interior of continents or in temperate regions. There are at least two explanations for this phenomenon. Clutch size, in the view of Lack (1954, 1966, 1968), is adapted to the amount of available food which the parents can provide for the resulting brood so that all of its members may survive. In other words, the more food for the young, as in temperate regions, the larger the number of eggs that can be laid. Cody (1966) theorizes that the stability of environment determines the clutch size. Predator-free conditions as on sea islands and a more uniform climate as on sea islands, along sea coasts, and in the tropics favor a smaller clutch because the resulting young are subjected to fewer risks; therefore, they do not "need" as many young to perpetuate the species. Instability, conversely, means larger clutches. Cody cites as an example, the Bay-breasted, Cape May, and Tennessee Warblers *(Dendroica castanea,* D. *tigrina,* and *Vermivora peregrina),* all of which depend on an unstable food supply, such as the irregular outbreaks of spruce budworm, and consequently lay clutches of 5-6, 5-7, and 5-6 eggs, respectively— more eggs per clutch than most other Parulidae whose normal clutch size is 4, rarely 5.

Among passerine species generally there tends to be a positive correlation between the size of the clutch and the type of nest, the clutches averaging larger in cavity nests than in all other types.

Occasionally nests of certain ducks and a few other birds in one area contain many more eggs than occur in a normal clutch. This is the result of two or more females laying in the same nest.

A few wild birds continue laying beyond the usual number of their clutch when eggs are removed from the nests (e.g., see Phillips, 1887). Such birds are called **indeterminate egg-layers** (D. E. Davis, 1942b). Most birds are **determinate egg-layers**—they lay a definite number of eggs per clutch. If the eggs are removed, the birds desert the nests after they have laid the normal number.

NUMBER OF CLUTCHES PER BREEDING SEASON

Most species characteristically lay only one clutch a year since the time remaining in the breeding season, after the young are reared, is not long enough for repeating the nesting cycle. For a few species of large birds whose young develop at an exceptionally slow rate, even one year is not long enough. This is the case with the Wandering Albatross *(Diomedea exulans)* and California Condor *(Gymnogyps californianus)* which nest every other year (Tickell, 1968; Koford, 1953) and the King Penguin *(Aptenodytes pata-gonica)* which nests twice in three years (Stonehouse, 1960).

Among many passerine birds and pigeons and in a few other species, two or more clutches a year are common, particularly in regions where ecological conditions favor a long or unending season for breeding. Species, for example, nesting in southern United States produce more clutches per year than other species in northern United States and Canada; and, similarly, individuals of some species, whose breeding range extends from southern United States into Canada, produce more clutches per year in the south than in the north. This same situation applies to a few sea birds nesting on islands with an equable year-round climate and a continuously available food supply. On Ascension Island in the tropical Atlantic where there is little seasonal change, the Sooty Tern *(Sterna fuscata)* breeds at intervals of 9.6 months or roughly five times in every four-year period (Chapin, 1954; Ashmole, 1963), but on other islands with more seasonal conditions it breeds annually.

Probably all but a few species will replace the first clutch, if it is destroyed, although no thorough investigation has ever been made to determine the "limits" of clutch replacement following successive destructions. Presumably the more clutches per season the species characteristically lays, the more clutches it will replace if they are destroyed. The time required to replace the clutch is worthy of study. Nice (1943) found that, in the case of the Song Sparrow, the first egg of the new clutch is laid five days after destruction of a clutch.

EGG-LAYING

Egg-laying, like nest-building, is a phase of the breeding cycle in need of careful investigation. Several topics particularly worthy of attention are outlined below. Many of the statements are based on findings in a few passerine species and others, and must not be construed as being applicable to all birds.

Start of Egg-laying

Generally egg-laying begins when the nest is completed. Actually egg-laying may start sometimes before and frequently from one to several days after the nest is completed.

Time of Egg-laying

Birds usually lay very early in the morning. Sometimes the female enters the nest the evening before; ordinarily she does so between the break of day and sunrise. She lays one egg each morning (i.e., every twenty-four hours) until the clutch is complete. The female sits on the nest for a brief time, before and after laying, for an average of perhaps fifty minutes. The time on the nest tends to lengthen by a few minutes with each egg laid. Full incubation behavior does not usually begin until she has completed the clutch.

Behavior of the Sexes

During the egg-laying period—i.e., the period of days during which the female completes the clutch—her behavior is in marked contrast to that during nest-building. She spends a considerable amount of time in leisurely preening, especially following the interval spent on the nest for egg-laying; she moves casually over the territory and searches for food; she rests for relatively long intervals. She seemingly ignores the nest and nesting site between layings.

Except in birds with breeding territories of Type C, the male is usually in close attendance upon the female when she is not on the nest. He accompanies her to the nest and rejoins her when she leaves. He customarily sings vigorously during the egg-laying period. Copulations are frequent at any time of day except immediately after egg-laying.

In birds with territories of Type C, the female may join the male in the mating area after egg-laying and remain for an undetermined period. Elsewhere than on the mating area the sexes are not in contact either during egg-laying or during subsequent phases of the breeding cycle.

INCUBATION

Incubation—the process by which the bird applies its body heat to the eggs—and the accompanying behavior have been given more attention than many other phases of the breeding cycle, but the subject is far from exhausted; information is still lacking for a wide variety of species.

Development of the Incubation Patch

In most species of birds, prior to incubation, one or more **incubation patches** develop on the ventral surface of the body. Each patch consists of a feather-free area with thickened skin and a rich supply of blood vessels to facilitate the transfer of heat from the body of the incubating bird to the eggs. In the Song Sparrow, the feathers where the patch is to develop are lost four to six days before the bird lays the first egg (Nice, 1937). The patches persist through the incubation period and into the early part of the brooding period; the feathers grow back again with the acquisition of the winter plumage. The majority of species, including passerines, have only one patch, situated in the median apterium; but some species have two patches, one in each lateral apterium; and other species have three patches, one in each lateral apterium and one in the median. A few species lack patches entirely. The student will find instructive the paper by Bailey (1952) which shows how the incubation patch develops in passerine birds.

Start of Incubation

As a rule, the bird begins sitting on the nest before the last egg is laid. Sometimes, however, it may begin from one to three days after the last egg is laid, or sometimes, as in several lower orders, it may begin after the laying of the first egg. Sitting on its nest, before the clutch is complete, may or may not involve fully warming the eggs, particularly if the incubation patch is not developed; thus its presence on the nest may not always indicate that incubation has actually begun. Incubation is actually under way only when the bird applies *maximum* body heat to the eggs.

Participation of the Sexes and Behavior

Among all bird species the role of the sexes in incubation is subject to every conceivable variation. Skutch (1957), after analyzing this diversity from his own observations and published information, prepared a synopsis or "key" to incubation patterns which the student will find instructive. He concluded, as Kendeigh (1952) had concluded earlier, that because the prevailing mode of incubation is by both sexes, it is therefore probably the primitive method. From this initial pattern all the other modes have diverged, from greater male to greater female participation to sole participation by one sex or the other.

In most North American species, except for a few in which only the male incubates, either the female alone incubates or the female and male share incubation. Incubation by the female alone is far commoner than incubation shared by both sexes.

There is some correlation between the coloration of the male and his role in incubation. If he is more colorful than the female, he usually never sits on the eggs; if his coloration is similar to that of the female, he sometimes participates; if he is less colorful than the female, he alone sits on the eggs.

There is also some correlation between the presence or absence of incubation patches and the participation of one or both sexes in incubation. If only one sex has patches, it usually seems to be true that only the sex with the patches incubates; if both sexes have patches, both incubate. In certain passerine species the males are apparently without patches and yet sit on the nest quite regularly. Bailey (1952) has suggested that male passerines with this habit probably do not sit on the eggs at night when the greater heat-giving efficiency of the incubation patch is necessary.

In most passerine species, the behavior of the male during incubation follows the same general pattern. He moves leisurely over the territory, feeds, and rests. He sings as vigorously as during the previous periods of territory establishment, nest-building, and egg-laying, or even more vigorously. He appears to be fully aware of the location of the nest and the incubating female. Sometimes he accompanies the female to the nest. If he does not incubate, he frequently visits the nest carrying food (courtship-feeding), or nesting material (symbolic nest-building), or simply approaches the immediate vicinity of the nest to "warn" (i.e., "signal") or to "call off" the female. If he incubates, he comes to the nest with more or less regularity to relieve the female. The arrival of the male at the nest, whether for courtship feeding, symbolic nest-building, or nest-relief, initiates some form of mutual display by both sexes that is peculiar to the species.

Among all birds, the behavior of the incubating bird, regardless of sex, is quite similar. While the bird is on the nest there are moments of restlessness, during which it changes sitting position, rises and settles, moves and turns the eggs with the bill, tampers with the nesting material and overhanging cover, and pokes at the bottom of the nest, sometimes with a trembling motion (see Hartshorne, 1962). Generally the incubating bird is vocally quiet, although it sometimes responds in subdued tones to the calls of the sex-partner. Departure and return are deliberate and secretive. If the bird is a ground nester, it walks away for a short distance and then flies up; if an elevated nester, it flies off quickly in a downward direction and continues near the ground for a short distance, occasionally taking advantage of concealment provided by underbrush and lower branches of trees. The method of return resembles the method of departure.

Only a few species deliberately cover their eggs when they leave the nest. Waterfowl cover their eggs with down from the nest's lining. Plucked from their lower breast earlier, the down serves to insulate as well as conceal the eggs. The Kittlitz's Sandplover (*Charadrius pecuarius*) of Africa, when taking leave of its nest, hastily kicks sand over its eggs until they are completely out of sight (Conway and Bell, 1968).

While off the nest the incubating bird commonly seeks and accompanies the sex-partner if the sex-partner does not replace the incubating bird on the nest. Off-the-nest periods are devoted mainly to feeding, preening, and bathing, or dusting in areas seldom near the nest. Copulation during the incubation period is infrequent.

The reaction of incubating birds to intrusions, human or other, vary markedly among different species. Reactions also vary within a species and show changes during the progress of incubation. Some species, particularly those with cryptic coloration or well-concealed nests, permit the close approach of an intruder; others leave the nest in haste far ahead of the intruder. Some individuals allow themselves to be touched with the hand, while others of the same species flush before the hand touches them. Birds which, early in incubation, flush well in advance of the intruder may, toward hatching, sit more closely, remaining on the nest until almost touched.

A few non-passerine species are able to move their eggs. The Herring Gull *(Larus argentatus)*, and probably many other gulls as well as terns, can retrieve an egg inadvertently kicked out of the nest by putting its bill over the egg and rolling it back into the nest (Tinbergen, 1953). The Clapper Rail *(Rallus longirostris)* can retrieve an egg similarly displaced by picking it up between its mandibles (Pettingill, 1938) and so can the Virginia Rail *(Rallus limicola)* as shown in a film taken by O. S. Pettingill, Jr. In instances of this sort, the eggs are within reach of birds sitting on their nests. Undoubtedly the birds are incapable of retrieving eggs that are beyond their reach. Common Nighthawks *(Chordeiles minor)*, and probably other goatsuckers, can move their eggs several feet to a new nest site if for some reason the original site becomes unfavorable (Gross, 1940; Sutton and Spencer, 1949). They accomplish this by rolling the eggs in front of their feet (Weller, 1958). For no explicable reason, a female Pileated Woodpecker *(Dryocopus pileatus)* in Florida, whose cavity nest in a tree became exposed when the tree broke off at the nest hole, made three trips from the nest hole to an unknown destination, each time carrying one of her three eggs in her bill (see Truslow, 1967, for details on the action and photographs).

Birds do not recognize their eggs as their own, even toward the end of the incubation period. If one substitutes eggs of different color but approximately the same size, the incubating bird will readily accept them. Some birds such as gulls and albatrosses will "incubate" electric light bulbs or other odd objects in place of their eggs for an indefinite period. The persisting adage, that a bird will desert its nest if its eggs are touched or handled, obviously has no foundation in fact. It is the disturbance by an intruder at the nest that causes the owner's desertion.

Why birds do not recognize their eggs as their own is understandable. Eggs are stationary; therefore, there has been no "need" for birds to evolve a means of keeping track of them. As Tinbergen (1953) has suggested, all a bird needs to recognize or know about its eggs is their location, the nest, not their properties.

Length of Time Involved

Incubation, once begun, usually continues until the last egg is hatched. The number of days involved in incubation varies with different species, more time being required for those species producing well-developed young or chicks. Within species, incubation time varies slightly. For example, eggs subjected to an excessive amount of cooling, while the incubating bird is off the nest for extended periods, are slower to hatch. The student desiring a discussion of several factors affecting the length of incubation is referred to the work of Kendeigh (1940).

The **incubation period** of an egg is the time between the start of a regular, uninterrupted incubation and the emergence of the young. Some of the longest incubation periods on record are 75-82 days in the Wandering Albatross *(Diomedea exulans)*, 75-80 in the Brown Kiwi *(Apteryx australis)*, and 54-64 in the King and Emperor Penguins *(Aptenodytes patagonica* and *A. fosteri)*. The eggs of the Mallee Fowl *(Leipoa ocellata)*, an Australian megapode, after being laid and buried in a mound of decaying vegetable matter, need to incubate from 49 to 90 days, depending on the temperature of the mound (Frith, 1962). In no species of birds are incubation periods normally shorter than 11 days; authenticated periods of less than 11 days are very rare (Nice, 1953).

Appendix H gives the ranges of incubation periods in families of North American birds. The figures, obtained from Kendeigh (1952), Reilly (1968), and other sources, apply only to representatives of the families occurring north of Mexico. In the families that include many species of greatly different sizes—e.g., in the Ardeidae, Anatidae, Accipitridae, Scolopacidae, Laridae, Strigidae, Picidae, and Corvidae—the incubation periods are correspondingly more wide-ranging because the embryos of the larger species require more time for development than the smaller.

But the incubation periods are not always correlated with the size of the birds. In families of passerine birds, species nesting in cavities have longer incubation periods than species of similar size nesting in open cupped nests. The Eastern Bluebird *(Sialia sialis)*, for example, has an incubation period of about 14 days whereas the incubation period of *Hylocichla* thrushes averages 12 days. Between families of birds, the incubation periods may vary widely even though the birds may be similar in size. The Leach's Petrel *(Oceanodroma leucorhoa)*, a burrow-nesting sea bird, has an incubation period of 41-42 days, about twice as long as that of the similarly sized Least Tern *(Sterna albifrons)*. Lack (1968) attributes these differences to the greater protection from predators offered by nest sites and hence the less "need" for a quickened pace in the nesting cycle.

The accepted method for determining the incubation period is to measure the time in days—in hours when possible—from the laying of the last egg to the time when all eggs in the clutch have hatched. But this method, while practical, is not entirely accurate, as Kendeigh (1963) points out, because "the first eggs of a clutch may receive some heat and undergo a certain amount of development before the clutch is completed or full incubation begins. The eggs tend, as a consequence, to hatch in the order laid, but the intervals between hatching of the eggs are shorter than the intervals between their laying so that the time that elapses between laying of an egg and its hatching progessively decreases with each additional egg in the clutch." Kendeigh *(op. cit.)* describes a more precise method of measurement by recording mechanically the total heat applied to the eggs from the time they are laid until hatching. For details, see his paper.

Studies of some species—for example, the European Wren *(Troglodytes troglodytes)*, see Armstrong, 1955; the Long-billed Marsh Wren *(Telmatodytes palustris)*, see Verner, 1965—have revealed that incubation periods shorten as the season advances, possibly because the air temperature averages warmer, preventing the eggs from losing heat when the incubating bird is off the nest. There may be other reasons such as an increased and more available food supply, enabling the incubating bird to spend less time in foraging and more time on the nest.

The student should bear in mind that, in most North American species, the incubation period is "known" in the sense that there are a few (sometimes only one or two) records obtained by the method defined above. Many "records" for some species are unreliable estimates or guesses (Nice, 1954). For the most recent listing of incubation periods in different species, refer to Reilly (1968). On consulting this, the student will be impressed by the great number of species for which the incubation period is either unknown or simply estimated.

Incubation Rhythm

At intervals during incubation the incubating bird leaves the nest. The eggs are either left uncovered temporarily or are covered in the meantime by the sex-partner. Thus the incubating bird has periods of **attentiveness** (i.e., periods *on* the nest) alternating with periods of **inattentiveness** (i.e., periods *off* the nest). This alternation of periods is spoken of as the **incubation rhythm**.

The frequency and length of these periods presents a fascinating study because of the variations involved. When both sexes incubate, (1) the male and female may take turns at attentiveness for almost equal periods, or (2) one sex may be attentive more often and for longer periods than the other. The length of attentive periods ranges from a few minutes in some species to several days in others and is probably inherent. In the Wandering Albatross *(Diomedea exulans)*, attentive periods may be as short as two days to as long as 38 days (Tickell, 1968). When one sex incubates, as does the female in many passerine species, the periods of attentiveness show great extremes, ranging from numerous brief periods per day to one period many hours in length. If the female passerine's attentive periods are long, usually the male is feeding her. The sex attentive for the night period (in the case of diurnal birds) is not known in many species. In at least two passerine species (see Erickson, 1938; and Weston, 1947) only the female is known to incubate at night, even though both sexes share in covering the eggs during the day. In woodpeckers, the male commonly incubates at night.

Much more time during the day is spent by the incubating bird in attentive than in inattentive periods. From observational data on the incubation rhythm of 137 individuals representing 82 species of tropical birds—mostly passerine, in which the female alone incubates without receiving much if any food from her mate—Skutch (1962) found that 101 kept their eggs covered from 60 to 80 percent of the time. This he regarded as "average or normal constancy." Constancy above 80 percent was shown chiefly by birds that were well fed by their mates or could obtain food in ready abundance when they left their nests. In temperate regions, most passerine species, in which the female alone incubates without being fed appreciably by the male, normally cover their eggs within the same range of constancy as in the tropics. For example, in Michigan, female Scarlet Tanagers *(Piranga olivacea)* revealed a constancy of 77 percent (Prescott, 1964). Also in Michigan, one female Robin *(Turdus migratorius)*, watched continuously at the nest on two days from daybreak to darkness, was attentive 77 percent of the time one day and 80 percent the other (Pettingill, 1963). In north-central Alaska where, during the nesting season, there is continuous daylight save for a few hours of civil twilight between sunset and sunrise, female Tree Sparrows *(Spizella arborea)* were attentive 76 to 77 percent of the day (Weeden, 1966). When the male feeds the female frequently while she incubates, she covers the eggs for a greater percent of the time. This is the case with the female Cedar Waxwing *(Bombycilla cedrorum)* whose attentiveness rarely falls below 90 percent (Putnam, 1949).

Attentive periods tend to be longer toward evening and in the early morning when it is cooler and, for insectivorous birds, there is less food readily available. With the increase in daytime air temperature, attentive periods shorten and inattentive periods increase in number, resulting in much less total time spent on the nest. Climatic conditions may influence the length of attentive and inattentive periods. Warm weather usually shortens attentive periods; excessively hot weather prolongs attentive periods if nests are exposed and require the shade of the bird's body; cool weather almost invariably lengthens attentive periods. Attentive periods may be consistently long throughout the day if the incubating bird is fed by its mate. The female Cedar Waxwing has attentive periods often extending more than two hours (Putnam, 1949).

From his studies of incubation rhythm in the House Wren *(Trolodytes aedon)*, Kendeigh (1952) concluded that the length of the attentive periods is much more variable than the length of the inattentive and that the relationship between the two periods is probably a psychological one, "since with shorter attentive periods, less rest and less food are required, while with longer periods there needs to be more rest and more food."

Restlessness on the nest, described earlier, is much reduced during long attentive periods, as toward evening and during the early morning. Individual birds vary greatly in the amount of restlessness, some being much more active than others. Beer (1965), in paying special attention to this behavior in the Black-billed Gulls *(Larus bulleri)* of

New Zealand, noted that, when the clutch of eggs contained the "optimum" number, the birds sat on the nest more quietly (with "more uninterrupted sitting") than when the clutch was smaller.

ABNORMALITIES OF INCUBATION

When eggs fail to hatch, birds will continue to incubate for a variable length of time. Most birds, according to Skutch (1962), continue for at least 50 percent longer than the normal period and they may even continue for twice or even three times longer (for example, see Kirkman, 1937). An exception, he points out, are pigeons, some of which will not continue to incubate "even a day beyond the usual time of hatching."

For further information on the subject of incubation, read Kendeigh (1952), Nice (1943), and Skutch (1962). Weeden (1966) gives a good review of recent literature on attentiveness.

BROOD PARASITISM

At least five groups of birds in the world contain **brood parasites**—bird species adapted to laying their eggs in the nest of another (host) species which incubates the eggs and rears the young.

1. Among ducks, several species, notably the Redhead *(Aythya americana)* and Ruddy Duck *(Oxyura jamaicensis)*, often lay their eggs in nests of the same species or different species, and some females may never nest themselves (Weller, 1959). The only species, however, known to be completely parasitic is the Black-headed Duck *(Heteronetta atricapilla)* of southern South America which parasitizes, besides other ducks, such birds as coots and ibises (see Weller, 1968).

2. Among the honey-guides (Indicatoridae) of Africa and southeastern Asia, at least two African species, probably others in the family, lay a white egg in the nest of a host, usually a hole-nesting species such as a barbet (Capitonidae) or woodpecker, which also has white eggs (see Friedmann, 1955, 1968a).

3. Among the weaverbirds (Ploceidae), about 20 African species parasitize other ploceids or even warblers (Sylviidae) with unmarked white eggs similar to their hosts' (see Friedmann, 1960).

4. Among the cuckoos, some 50 species in the Old World and three in the New World tropics lay their eggs, quite small in proportion to body size, in the nests of usually smaller species, mainly passerine. The Common Cuckoo *(Cuculus canorus)* of Eurasia is an extraordinarily specialized brood parasite (Chance, 1940; Baker, 1942). The egg it produces often matches the small size and sometimes mimics the coloration of its host's eggs. In a given area, the population of Common Cuckoos is normally comprised of several groups, each distinguished by producing eggs that resemble those of its particular host. Each group is thus restricted to parasitizing the species it mimics. How the females of each group determine their host is commonly explained by the presumption that they were reared by the host and were consequently imprinted on the host when they were young birds. (For brood parasitism in other cuckoos, see Friedmann, 1948, 1956, 1964, 1968b).

5. Among the cowbirds are five parasitic species, none of which are as specialized as cuckoos. Their speckled eggs are neither unusually small in proportion to body size nor always similar in appearance to those of their hosts. Cowbirds are the only New World *passerine* parasites and their hosts are, with rare exceptions, passerine (Friedmann, 1963).

Of the five parasitic cowbirds, the Brown-headed *(Molothrus ater)* is the only one widely distributed in parts of North America north of Mexico. Its hosts number at least 206, although more than half are uncommon, rare, or accidental (Friedmann, 1963). The species most often victimized include the vireos, tanagers, *Hylocichla* thrushes, several of the smaller tyrannid flycatchers, a few icterids, and many warblers and fringillids.

Three persons (Hann, 1941; Mayfield, 1960; Josselyn Van Tyne, *in* Mayfield, 1960) made extensive observations and reviewed the literature on the cowbird's procedure in parasitizing a nest. The female finds the nest by first noticing building activities and subsequently makes frequent "trips of inspection" while her prospective host is absent. Usually after her host has laid two or more eggs and before her host begins incubation, she enters the nest, about a half hour before sunrise while the nest is still unoccupied, and deposits her egg in a few seconds. If she removes one or more eggs, as is normally the case, she does so later in the day, the day before, or the day after, but not at the time of laying. She carries away the egg by piercing it with her bill and later eats it.

The female cowbird's interest in the nest may continue after the host begins incubation, even to the extent of her removing additional eggs, although her approach to the nest is likely to be discouraged by the host's behavior which is more hostile than at the time of egg-laying. There are rare instances reported of a cowbird later removing the nestling of the host. Tate (1967) saw a female cowbird carrying away and dropping one of three nestling Black-throated Green Warblers *(Dendroica virens)* from a nest that also contained a nestling cowbird. This episode, Tate suggested, may account sometimes for the unexplained loss of nestlings from a parasitized nest.

The female cowbird ordinarily lays one egg in each of several nests. Her hosts may be one or more species. In a given season one cowbird may lay between 10 to 12 eggs in several "clutches" of one to six eggs each (Payne, 1965), with intervals of "a few days to a few weeks" (Payne, *op cit.),* or from six to 12 days (Nice, 1937). However, one individual was reported (Walkinshaw, 1949) to lay 25 eggs in one season, probably in seven clutches (Nice, 1949).

Nests may often have more than one egg. If a nest has two or three, they may have been laid by one female which had difficulty finding enough nests to parasitize; but if a nest has more than three, in all likelihood two females have been laying in it.

Most host species during nest-building and egg-laying do not react to the cowbird as a harmful intruder; indeed, they show no more hostile behavior toward it than to any other species of like size approaching the nest. But their reaction to the cowbird's egg or eggs in the nest may be another matter. While some species accept the parasite's contribution as though it were their own, other species may react in one of three ways: (1) Commonly, by deserting the nest. This action may or may not be taken by all individuals in a species, depending on the circumstances. An egg laid in a newly constructed nest before the host's own, or a number of eggs in excess of the host's, or even the mere sight of a cowbird at the nest, may cause desertion. (2) Less commonly, by removing the cowbird's egg. Robins *(Turdus migratorius)* and Catbirds *(Dumetella carolinensis)* are among a few species that quite regularly do so. (3) Rarely, by building over the cowbird's egg, sometimes along with its own. With each additional cowbird's egg, the host repeats the act, forming a nest of two or three stories. The Yellow Warbler *(Dendroica petechia)* is noted particularly for this behavior, and in one record instance built a six-storied nest in which 11 cowbird's eggs were covered over by floors (Berger, 1955). Possibly the behavior in most instances may be due to the nest being incomplete at the time the cowbird laid the egg and the consequent "building over" is simply the result of the host proceeding with building.

THE STUDY OF EGGS, EGG-LAYING, AND INCUBATION

Select, if possible, a nest under construction, and follow the succession of events through egg-laying and incubation.

Procedure

Much valuable information can be obtained during the egg-laying period, Note the start of egg-laying (exact hour of the day when the first egg is laid), and note its relation to stage in nest-building and to environmental factors. Determine the exact time when the other eggs of the clutch are laid, the time spent on the nest by the female at each laying, and the first evidence of actual incubation. (After each egg is laid, mark it indelibly so that it may be recognized at the time of hatching.) Follow as closely as possible the activities and behavior of the pair from day to day.

Throughout incubation, make detailed direct observations on the participation and behavior of the sexes. Attempt to determine the incubation period by measuring the time in hours from the laying of the last egg in the clutch to the hatching of the last egg. Gather specific data on the incubation rhythm (1) at different times of the day, (2) at different stages of incubation, and (3) under different weather conditions. In order to yield significant results, each observation period should be at least *two hours* in length. Use the following methods for organizing data in tabular form, first applied by Pitelka (1940) and later modified by him (1941). Adjust this method to suit the particular relationships of the sexes in the species studied.

For each observation period, record:

> Date
> Stage of incubation (i.e., day of incubation)
> Time of day; also total hours and minutes
> Air temperature at beginning and ending
> Wind velocity
> Weather conditions
>
> Attentive periods of both sexes:
> Total number
> Average time in minutes
> Extremes in minutes
> Percentage of total time
>
> Inattentive periods of both sexes:
> Total number
> Average time in minutes
> Extremes in minutes
> Percentage of total time
>
> Attentive and inattentive periods of male:
> (Repeat as above)
>
> Attentive and inattentive periods of female:
> (Repeat as above)

In addition to tabulating the data as directed above, take notes on the activities and behavior of the sexes during each of the attentive and inattentive periods. Qualitative as well as quantitative data are desirable. If possible, use a portable tape recorder in dictating the notes and transcribe them later. This method allows the observer to keep his eye on activities and not miss an action while writing his notes.

For a complete all-day record of nest attentiveness, the use of an automatic recorder is desirable, provided it can be adapted to the type of nest. For most nests a thermocouple and potentiometer can be used (see Kendeigh, 1952, for details). By means of a very thin, flexible wire threaded through the nest so that it lies just above or on the eggs, the device records the heat generated by the bird when it settles on the eggs to incubate them and instantly records the drop in temperature when it leaves the eggs. For ground nests and hole nests, use the itograph (also see Kendeigh, 1952). This consists of a double set of perch contacts at the nest's edge or entrance attached to a set of dry-cell batteries and an electromagnet with a pen that registers all the trips to the nest. For nests that do not show an appreciable amount of light through the bottoms and sides, use a small (three-quarters of an inch in diameter) photo-resistor positioned in the floor of a nest and connected by an extension cord through the floor to an adjustable resistor and recording element of an Esterline-Angus recorder (see Weeden, 1966). This device registers rapidly the decrease of light when the bird sits on the nest and increase when it gets off. It also registers the bird's restlessness as it rises up and re-settles.

Should the nest and eggs be destroyed, strive to follow the pair in order to learn the length of time required to start the laying of a new clutch. All too often students cease studying a pair if disaster comes to the nest. The result is that surprisingly little is known of what the occupant birds do thereafter.

Pay attention to the size of the clutch and search the literature to determine whether or not the size is average. If it is not average, attempt to determine why.

Full description of the eggs is sometimes advisable, particularly if the student contemplates an extensive study of the species or has some special problem in mind that concerns the eggs. (See Nice, 1937, for a discussion of problems connected with the eggs of the Song Sparrow.) Use the following directions for gathering the descriptive information: **Coloration.** Determine the colors of each egg by comparing it directly with a color chart (see Appendix A for available color charts). Note the distribution of color and markings. **Measurements.** Measure the eggs, using dividers, or calipers, and a millimeter ruler. Length and greatest width are the two measurements taken. Extreme accuracy is essential, since variations, especially in small eggs, are slight. **Weights.** Weigh the eggs in grams to the second decimal place. It is important to note the stage in incubation, since eggs become gradually lighter as incubation proceeds. (Sometimes eggs will lose as much as one-third of their original weight.) Weighing eggs in the field is generally unsatisfactory because it is difficult to be accurate. Proper balance is affected by air currents and off-level position of the scales. If eggs must be weighed during incubation, it should be done in the laboratory while substitute eggs are placed in the nest to prevent desertion by the adults. Obviously, eggs must be weighed as quickly as possible to avoid extreme cooling. Caution: Handle with great care. Eggs are generally thin-shelled. The slightest cracking of the shell will very likely inhibit the growth of the embryo.

If the species selected for study is known to be victimized by the Brown-headed Cowbird, watch for first evidences of cowbird relationships: how and when the cowbird discovers the nests; time when the cowbird lays its eggs. Should parasitism actually result, observe the reactions of the hosts to the eggs and any evidences of harm that the eggs may have caused. Consult the papers by Hann (1941), Norris (1947), Berger

(1951), Mayfield (1960), and Friedmann (1963) for useful information on cowbird activities.

Photographs showing activities of adults during incubation (e.g., courtship-feeding, turning eggs, incubating positions, reactions to intrusions) are useful as illustrations in the presentation of results.

Presentation of Results

Draw up a report on the results obtained from a study of egg-laying and incubation. Prepare the manuscript as directed in Appendix B. An outline of suggested topics is given below.

Egg-laying. Date of start. Relation of start to construction of nest and to environmental factors. Time of laying of each egg and interval between layings. Amount of time spent on nest by female at each laying, and accompanying activities. Time of first evidence of incubation. Behavior of the sexes: activities of the female off the nest and reaction to male; number and times of copulations; activities of the male, including number of singing periods and vigor of singing; territorial activities of the pair; reactions of the pair to intrusions.

The Clutch. **Size.** Number of eggs finally laid. Reason for size of clutch, if abnormal. **Number of Clutches per Season.** Present whatever data have been obtained. **Description of Clutch.** Present descriptive material, if significant. Compare with information in the literature.

Incubation. **Start of Incubation.** Relation of start to size of clutch at the time. Brief statement of duration. If the eggs are of a large species and incubation is already in progress, it is sometimes possible to determine the stage of incubation. For a method see Weller (1956). **Participation and Behavior of the Sexes.** Brief statement of sexes participating, followed by detailed accounts, *viz:* (1) Activities of male: territorial behavior; singing; relation to nest, incubation, and incubating female (visits, courtship-feeding, symbolic building, calls to female and resulting responses, nest-relief displays, reactions to intrusion). (2) Activities of incubating bird: general behavior and activities; manner of leaving nest and returning; behavior and activities while off the nest. (3) Length of time involved: total length of incubation (in hours) for each egg. Present tables for each observation on incubation rhythm, followed by tables of similar design summarizing all the data from all observations. (See Pettingill, 1963, for a chart showing the actions and attentiveness of an incubating bird during a whole day.) After the tables, discuss the information, noting variations in rhythm at different times of the day, at different stages of incubation, and under different weather conditions. Graphs may be used to illustrate certain points in the discussion. Indicate which sex incubates at night. **Abnormalities of Incubation.**

Cowbird relationships. Present whatever information has been obtained.

REFERENCES

Amadon, D.
  1943 Bird Weights and Egg Weights. Auk, 60: 221-234.
  1944 A Preliminary Life History of the Florida Jay, *Cyanocitta c. coerulescens.* Amer. Mus. Novitates No. 1252. (Incubation before laying first egg; other information on egg-laying and incubation.)

Armstrong, E. A.
  1955 The Wren. Collins, London.

Ashmole, N. P.
  1963 The Biology of the Wideawake or Sooty Tern *Sterna fuscata* on Ascension Island. Ibis, 103b: 297-364.

Averill, C. K.
  1933 Geographical Distribution in Rela-

tion to Number of Eggs. Condor, 35: 93-97.

Bailey, R. E.
1952 The Incubation Patch of Passerine Birds. Condor, 54: 121-136.

Baker, E. C. S.
1942 Cuckoo Problems. H. F. and G. Witherby, London.

Baldwin, S. P., and S. C. Kendeigh
1927 Attentiveness and Inattentiveness in the Nesting Behavior of the House Wren. Auk, 44: 206-216.

Beebe, C. W.
1906 The Bird: Its Form and Function. Henry Holt and Company, New York. (Dover reprint available. Chapter 16 deals with the varieties of eggs in birds; Chapter 17 gives a popularized description of bird embryology.)

Beer, C. G.
1965 Clutch Size and Incubation Behavior in Black-billed Gulls *(Larus bulleri).* Auk, 82: 1-18.

Berger, A. J.
1951 The Cowbird and Certain Host Species in Michigan. Wilson Bull., 63: 26-34.
1955 Six-storied Yellow Warbler Nest with Eleven Cowbird Eggs. Jack-Pine Warbler, 33: 84.
1960 Some Uncommon Cowbird Hosts. Jack-Pine Warbler, 38: 118.

Bergtold, W. H.
1917 A Study of the Incubation Periods of Birds: What Determines Their Lengths? Kendrick-Bellamy Company, Denver, Colorado.

Brackbill, H.
1952 Three-brooded American Robins Bird-Banding, 23: 29.

Burns, F. L.
1915 Comparative Periods of Deposition and Incubation of Some North American Birds. Wilson Bull., 27: 275-286.

Bussmann, J.
1933 Experiments with the Terragraph on the Activities of Nesting Birds. Bird-Banding, 4: 33-40.

Chance, E. P.
1940 The Truth About the Cuckoo. Country Life, London.

Chapin, J. P.
1954 The Calendar of Wideawake Fair. Auk, 71: 1-15.

Cody, M. L.
1966 A General Theory of Clutch Size. Evolution, 20: 174-184.

Conway, W. G., and J. Bell

1968 Observations on the Behavior of Kittlitz's Sandplovers at the New York Zoological Park. Living Bird, 7: 57-70.

Cox, G. W.
1960 A Life History of the Mourning Warbler. Wilson Bull., 72: 5-28. (Incubation period, attentiveness, etc.)

Davis, D. E.
1942a The Number of Eggs Laid by Cowbirds. Condor, 44: 10-12.
1942b Number of Eggs Laid by Herring Gulls. Auk, 59: 549-554. (Attempts to restrain gulls from laying the full normal clutch and to induce them to lay more.)
1955a Determinate Laying in Barn Swallows and Black-billed Magpies. Condor, 57: 81-87.
1955b Breeding Biology of Birds. In *Recent Studies in Avian Biology* Edited by A. Wolfson. University of Illinois Press, Urbana.

Davis, J.
1960 Nesting Behavior of the Rufous-sided Towhee in Coastal California. Condor, 62: 434-456. (Important information on egg-laying, incubation, and related activities.)

Davis, J., G. F. Fisler, and B. S. Davis
1963 The Breeding Biology of the Western Flycatcher. Condor 65: 337-382. (Extensive data on nesting, including incubation periods, role of sexes, attentiveness, etc.)

Emlen, J. T., Jr.
1941 An Experimental Analysis of the Breeding Cycle of the Tricolored Red-wing. Condor, 43: 209-219. (Experiments in reducing and extending the incubation cycle.)

Erickson, M. M.
1938 Territory, Annual Cycle, and Numbers in a Population of Wren-tits *(Chamaea fasciata).* Univ. California Publ. in Zool., 42: 247-334.

Fautin, R. W.
1941 Incubation Studies of the Yellow-headed Blackbird. Wilson Bull., 53: 107-122.

Friedmann, H.
1929 The Cowbirds: A Study in the Biology of Social Parasitism. Charles C. Thomas, Springfield, Illinois.
1948 The Parasitic Cuckoos of Africa. Washington Academy of Science, Washington D. C.
1955 The Honey-guides. U. S. Natl. Mus. Bull. 208.

1956 Further Data on African Parasitic Cuckoos. Proc. U. S. Natl. Mus., 106: 377-408.

1960 The Parasitic Weaverbirds. U. S. Natl. Mus. Bull. 223.

1963 Host Relations of the Parasitic Cowbirds. U. S. Natl. Mus. Bull. 233.

1964 Evolutionary Trends in the Avian Genus Clamator. Smithsonian Misc. Coll., 146 (4): 1-127.

1968a Additional Data on Brood Parasitism in the Honey-guides. Proc. U. S. Natl. Mus., 124 (3648): 1-8.

1968b The Evolutionary History of the Avian Genus Chrysococcyx. U. S. Natl. Mus. Bull. 265.

Frith, H. J.
1962 The Mallee-fowl: The Bird That Builds an Incubator. Angus and Robertson, Sydney.

Grinnell, J.
1927 A Critical Factor in the Existence of Southwestern Game-birds. Science, 65: 528-529. (Remarks on the effect of drought on reproduction.)

Gross, A. O.
1940 Eastern Nighthawk. In *Life Histories of North American Cuckoos, Goatsuckers, Hummingbirds and Their Allies*. By A. C. Bent. U. S. Natl. Mus. Bull. 176: 206-234.

Hann, H. W.
1937 Life History of the Oven-bird in Southern Michigan. Wilson Bull., 49: 145-237. (Egg-laying; incubation; parasitism by cowbird; etc.)

1941 The Cowbird at the Nest. Wilson Bull., 53: 211-221. (Fifteen observations of the female cowbird at the Ovenbird's nest.)

Hardy, E.
1941 The Gestation Period of Wild Birds. Ibis, 1941: 462-463.

Hartshorne, J. M.
1962 Behavior of the Eastern Bluebird at the Nest. Living Bird, 1: 131-149.

Herrick, F. H.
1935 Wild Birds at Home. D. Appleton-Century Company, New York. (Important information, particularly in Chapter 10.)

Hochbaum, H. A.
1959 The Canvasback on a Prairie Marsh. Second edition. Stackpole Company, Harrisburg, Pennsylvania, and Wildlife Management Institute, Washington, D. C. (Information on egg-laying; cases of different species of ducks laying in each other's nests.)

Howard, D. V.
1967 Variation in the Breeding Season and Clutch-size of the Robin in the Northeastern United States and the Maritime Provinces of Canada. Wilson Bull., 79: 432-440.

Huggins, R. A., and S. E. Huggins
1941 Possible Factors Controlling Length of Incubation in Birds. Amer. Nat., 75: 282-285.

Huxley, J. S.
1927 On the Relation between Egg-weight and Body-weight in Birds. Jour. Linnaean Soc. Zool. London, 36: 457-466.

Johnston, R. F.
1954 Variation in Breeding Season and Clutch Size in Song Sparrows of the Pacific Coast. Condor, 56: 268-273.

Kendeigh, S. C.
1940 Factors Affecting Length of Incubation. Auk, 57: 499-513.

1952 Parental Care and Its Evolution in Birds. Second corrected printing, 1955. Illinois Biol. Monogr., 22: i-x; 1-356. (Contains a wealth of information on incubation activities in many species.)

1963 New Ways of Measuring the Incubation Period of Birds. Auk, 80: 453-461.

Kessler, F.
1962 Measurement of Nest Attentiveness in the Ring-necked Pheasant. Auk, 79: 702-705.

King, J. R.
1954 Victims of the Brown-headed Cowbird in Whitman County, Washington. Condor, 56: 150-154.

Kirkman, F. B.
1937 Bird Behaviour. A Contribution Based Chiefly on a Study of the Black-headed Gull. T. Nelson and Sons, London, (Chapter 6 has information on incubation behavior and reactions to numerous experiments.)

1937 Black-headed Gulls Incubating for 75 days. Brit. Birds, 34: 22.

Kluijver, H. N.
1951 The Population Ecology of the Great Tit, *Parus m. major* L. Ardea, 38: 1-135. (Contains much detailed information on several aspects of egg-laying and incubation that have been only slightly investigated.)

Koford, C. B.
1953 The California Condor. Natl. Audubon Soc. Res. Rept. No. 4, New York.

Labisky, R. F., and G. L. Jackson
1966 Characteristics of Egg-laying and Eggs of Yearling Pheasants. Wilson Bull., 78: 379-399. (An analysis of eggs

laid by nine yearling Ring-necked Pheasants in captivity, showing variation in rate of laying, shell color, size, weight, shape, and seasonal pattern.)

Lack, D.
1947- The Significance of Clutch-size. Ibis,
48   89: 302-352; 90: 25-45.
1954 The Natural Regulation of Animal Numbers. Oxford University Press, London.
1966 Population Studies of Birds. Oxford University Press, London.
1968 Ecological Adaptations for Breeding Birds. Methuen and Company, London.

Laskey, A. R.
1946 Some Bewick Wren Nesting Data. Migrant, 17: 39-43. (Information on 52 nests.)

Lawrence, L. de K.
1953 Nesting Life and Behaviour of the Red-eyed Vireo. Canadian Field-Nat., 67: 47-77. (Egg-laying, incubation, etc.)

Marshall, N.
1943 Factors in the Incubation Behavior of the Common Tern. Auk, 60: 574-588.

Mayfield, H.
1960 The Kirtland's Warbler. Cranbrook Inst. Sci. Bull. No. 40, Bloomfield Hills, Michigan.

Moore, T.
1947 The Victims of the Cowbird. Flicker, 19: 39-42.

Moreau, R. E.
1939 Numerical Data on African Birds' Behaviour at the Nest. *Hirundo s. smithii* Leach, the Wire-tailed Swallow. Proc. Zool. Soc. London, 109A: 109-125.
1940 Numerical Data on African Birds' Behaviour at the Nest. II. *Psalidoprocne holomelaena massaica* Neum., the Rough-wing Bank Martin. Ibis, 1940: 234-248.
1944 Clutch-size: A Comparative Study, with Special Reference to African Birds. Ibis, 1944: 286-347. (An important paper comparing clutch-sizes in northern and southern latitudes.)

Nelson, J. B.
1964 Factors Influencing Clutch-size and Chick Growth in the North Atlantic Gannet *Sula bassana*. Ibis, 106: 63-77. (Experiments showing that this species is capable of hatching two eggs and rearing two young even though it normally lays one egg.)

Nice, M. M.
1932 Observations on the Nesting of the Blue-gray Gnatcatcher. Condor, 34: 18-22. (Suggestions for study of incubation rhythm.)
1937 Studies in the Life History of the Song Sparrow, I. Trans. Linnaean Soc. New York, 4: i-vi; 1-247. (Chapters 11-13.)
1943 Studies in the Life History of the Song Sparrow, II. Trans. Linnaean Soc. New York, 6: i-viii; 1-328. (Chapter 18.)
1949 The Laying Rhythm of Cowbirds. Wilson Bull., 61: 231-234.
1953 The Question of Ten-day Incubation Periods. Wilson Bull., 65: 81-93.
1954 Problems of Incubation Periods in North American Birds. Condor, 56: 173-197.

Norris, R. T.
1947 The Cowbirds of Preston Frith. Wilson Bull., 59: 83-103.

Odum, E. P.
1941 Annual Cycle of the Black-capped Chickadee-2. Auk, 58: 518-535. (Egg-laying; incubation.)
1944 Circulatory Congestion as a Possible Factor Regulating Incubation Behavior. Wilson Bull., 56: 48-49. (A suggestion that circulatory congestion may be a more important factor than hunger in regulating attentive and inattentive periods.)

Olsen, M. W.
1942 The Effect of Age and Weight of Turkey Eggs on the Length of the Incubation Period. Poultry Sci., 21: 532-535.

Parmelee, D. F., H. A. Stephens, and R. H. Schmidt
1967 The Birds of Southeastern Victoria Island and Adjacent Small Islands. Natl. Mus. Canada Bull. 222. (Data on nesting, incubation periods, and role of sexes in species hitherto studied very little.)

Payne, R. B.
1965 Clutch Size and Numbers of Eggs Laid by Brown-headed Cowbirds. Condor, 67: 44-60.

Paynter, R. A., Jr.
1949 Clutch-size and the Egg and Chick Mortality of Kent Island Herring Gulls. Ecology, 30: 146-166.
1951 Clutch-size and Egg Mortality of Kent Island Eiders. Ecology, 32: 497-507.

1954   Interrelation between Clutch-size, Brood-size, Prefledging Survival, and Weight in Kent Island Tree Swallows. Bird-Banding, 25: 35-58; 102-110; 136-148.

Petrides, G. A.
1944   Unusual Incubation of the Red-eyed Vireo. Auk, 61: 298. (Incubation beginning after laying of first egg.)

Pettingill, O. S., Jr.
1938   Intelligent Behavior in the Clapper Rail. Auk, 55: 411-415.
1960   The Effects of Climate and Weather on the Birds of the Falkland Islands. Proc. XIIth Internatl. Ornith. Congr., pp. 604-614.
1963   All-day Observations at a Robin's Nest. Living Bird, 2: 47-55.

Phillips, C. L.
1887   Egg-laying Extraordinary in Colaptes auratus. Auk, 4: 346. (Flicker induced to lay 71 eggs in 73 days. After the bird had laid two eggs, one egg was removed, the other was left as a nest-egg. This process was continued until the remarkable number was obtained.)

Pitelka, F. A.
1940   Breeding Behavior of the Black-throated Green Warbler. Wilson Bull., 52: 3-18.
1941   Presentation of Nesting Data. Auk, 58: 608-612.

Poulsen, H.
1953   A Study of Incubation Responses and Some Other Behaviour Patterns in Birds. Vidensk. Medd. Dansk Naturh. Foren., 115: 1-131.

Prescott, K. W.
1964   Constancy of Incubation for the Scarlet Tanager. Wilson Bull., 76: 37-42.

Putnam, L. S.
1949   The Life History of the Cedar Waxwing. Wilson Bull., 61: 141-182. (Contains considerable information on various aspects of incubation.)

Reed, C. A.
1904   North American Birds Eggs. Doubleday, Page and Company, New York. (A "revised and unabridged republication" of this work, called "the Dover edition," was issued in 1965 by Dover Publications, New York.)

Reilly, E. M., Jr.
1968   The Audubon Illustrated Handbook of American Birds. McGraw-Hill Book Company, New York.

Romanoff, A. L., and A. J. Romanoff

1949   The Avian Egg. John Wiley and Sons, New York.

Ryves, B. H.
1943   An Investigation into the Rôles of Males in Relation to Incubation. Brit. Birds, 37: 10-16.
1943   An Examination of Incubation in Its Wider Aspects Based on Observation in North Cornwall. Brit. Birds, 37: 42-49. (Important paper dealing mainly with start of egg-laying and related phenomena.)
1946   Some Criticisms on the Recording of Incubation-periods of Birds. Brit. Birds, 39: 49-51.

Schantz, W. E.
1939   A Detailed Study of a Family of Robins. Wilson Bull., 51: 157-169. (Egg-laying, incubation, etc.)

Skutch, A. F.
1933a A Male Kingfisher Incubating at Night. Auk, 50: 437.
1933b Male Woodpeckers Incubating at Night. Auk, 50: 437.
1945   Incubation and Nestling Periods of Central American Birds. Auk, 62: 8-37.
1952   On the Hour of Laying and Hatching of Birds' Eggs. Ibis, 94: 49-61.
1957   The Incubation Patterns of Birds. Ibis, 99: 69-93.
1962   The Constancy of Incubation. Wilson Bull., 74: 115-152.
1967   Adaptive Limitation of the Reproductive Rate of Birds. Ibis, 109: 579-599.

Spencer, O. R.
1943   Nesting Habits of the Black-billed Cuckoo. Wilson Bull., 55: 11-22. (Egg-laying and incubation.)

Stonehouse, B.
1953   The Emperor Penguin, *Aptenodytes forsteri* Gray. 1. Breeding Behaviour and Development. Falkland Islands Dependencies Surv. Sci. Repts. No. 6.
1960   The King Penguin, *Aptenodytes patagonica*, of South Georgia. 1. Breeding Behaviour and Development. Falkland Islands Dependencies Surv. Sci. Repts. No. 23.

Sutton, G. M., and H. H. Spencer
1949   Observations at a Nighthawk's Nest. Bird-Banding, 20: 141-149.

Swanberg, P. O.
1950   The Concept of "Incubation Period." Vår Fågelvärld, 9: 63-80.

Tate, J., Jr.
1967   Cowbird Removes Warbler Nestling from Nest. Auk, 84: 422.

Tickell, W. L. N.
  1968 The Biology of the Great Albatrosses, *Diomedea exulans* and *Diomedea epomophora*. American Geophysical Union, Antarctic Res. Ser., 12: 1-55.

Tinbergen, N.
  1939 The Behavior of the Snow Bunting in Spring. Trans. Linnaean Soc. New York, 5: 1-94.
  1953 The Herring Gull's World: A Study of the Social Behaviour of Birds. Collins, London.

Truslow, F. K.
  1967 Egg-carrying by the Pileated Woodpecker. Living Bird, 6: 227-236.

Tucker, B. W.
  1943 Brood-Patches and the Physiology of Incubation. Brit. Birds, 37: 22-28.

Van Someren, V. G. L.
  1944 Incubation and Nestling Periods. Ibis, 1944: 223-224.

Verner, J.
  1965 Breeding Biology of the Long-billed Marsh Wren. Condor, 67: 6-30.

von Haartman, L.
  1967 Clutch-size in the Pied Flycatcher. Proc. XIVth Internatl. Ornith. Congr., pp. 155-164.

Walkinshaw, L. H.
  1949 Twenty-five Eggs Apparently Laid by a Cowbird. Wilson Bull., 61: 82-85.
  1965 Attentiveness of Cranes at Their Nests. Auk, 82: 465-476.

Weeden, J. S.
  1966 Diurnal Rhythm of Attentiveness of Incubating Female Tree Sparrows *(Spizella arborea)* at a Northern Latitude. Auk, 83: 368-388.

Weller, M. W.
  1956 A Simple Field Candler for Waterfowl Eggs. Jour. Wildlife Management, 20: 111-113. (Use a mailing tube. Hold the eggs at one end against the sun and look through from the other. Apply the criteria in the paper for determining the age of the embryo. If the shell of the egg is too dense, the size of the air cell can be used as a criterion.)
  1958 Observations on the Incubation Behavior of a Common Nighthawk. Auk, 75: 48-59.
  1959 Parasitic Egg Laying in the Redhead *(Aythya americana)* and Other North American Anatidae. Ecol. Monogr., 29: 333-365.
  1968 The Breeding Biology of the Parasitic Black-headed Duck. Living Bird, 7: 169-207.

Weston, H. G., Jr.
  1947 Breeding Behavior of the Black-headed Grosbeak. Condor, 49: 54-73.

Williamson, K.
  1945 The Relation between Duration of Hatching and the Incubation Period. Ibis, 87: 280-282.

YOUNG AND
THEIR DEVELOPMENT

The physical and behavioral development of young birds continues to be neglected. Ornithological treatises, except a few dealing extensively with certain species, describe development only in broad generalities. Even such elementary data as the exact ages at which the birds acquire various attributes and the exact length of nestling life are lacking for most species.

HATCHING

The shell of the egg gradually weakens during the course of incubation and thus becomes susceptible to the increasing pressure brought to bear by the maturing embryo. The first external evidence of hatching is a star-shaped crack that appears on the side of the egg, toward the larger end. The crack is caused by the embryo's **egg-tooth**—a calcareous, scale-like structure on the tip of the upper mandible. Muscular action, particularly by a special "hatching muscle" (Fisher, 1966), scrapes and presses the egg-tooth against the inside of the already weakened shell until a crack results. An egg is then said to be **pipped**.

An egg may remain pipped for a variable length of time, ranging from a few hours to forty-eight hours or more. Generally the pipped condition persists longest among species of lower orders.

Final hatching takes place in a comparatively short time. From the star-shaped crack, a fissure develops, usually around the larger end. Muscular action of the embryo, chiefly in the legs and neck, forces the shell apart at the circular fissure. The embryo on emerging becomes a **young bird**. Its egg-tooth is sloughed during the first week after hatching in some species, after the first week in others (Clark, 1961).

Eggs in a clutch do not usually hatch simultaneously, especially if incubation began before the clutch was complete: Nests may, therefore, contain young of different ages. Hatching may occur at any time during the day or night.

CONDITION OF DEVELOPMENT AT HATCHING

Young birds of different species show different degrees of development at hatching. Some are poorly developed and helpless while others are decidedly advanced in development and somewhat self-reliant. These contrasting conditions are expressed in three sets of terms:

1. **Altricial and Precocial.** Species of birds are altricial whose young at hatching have their eyes closed, are incapable of locomotion, and depend on their parents for food. Such

young are commonly called **nestlings**. The young of precocial birds have their eyes open at hatching, have down (neossoptiles), and are capable of locomotion soon after hatching. They may be dependent on their parents for food or may immediately join the parents in searching for food. These young are called **downy young** or **chicks**.

2. **Nidicolous and Nidifugous.** A nidicolous bird remains in the nest for an extended period after hatching and is fed by its parents. A nidifugous bird leaves the nest soon after hatching and joins the parents in searching for food. It usually has some type of cryptic coloration.

3. **Psilopaedic and Ptilopaedic.** A psilopaedic bird is either naked at hatching or has only very sparse down on the dorsal region. A ptilopaedic bird at hatching is covered with down (usually dense) dorsally and ventrally.

A bird that is *typically* altricial is also nidicolous and psilopaedic while a bird that is *typically* precocial is also nidifugous and ptilopaedic. A number of birds are neither typically altricial nor typically precocial. For conditions of this sort, Nice (1962) has proposed two terms: **semi-altricial**, when the young have down yet are unable to leave the nest; **semi-precocial**, when the young have down and stay in the nest even though able to walk.

A table is presented on the following page showing the condition of development of young among the various North American orders. Orders marked with an asterisk (*) are neither typically altricial nor typically precocial.

Several terms, commonly used to designate youth in birds, require explanation. **Fledgling** is applied broadly to an altricial bird from the time it leaves the nest until it becomes independent of its parents. A young bird actually **fledges** (or has **fledged**) when it takes flight for the first time. A **juvenile bird** (or simply **juvenile**) refers to any young bird with teleoptiles, that has not attained sexual maturity and is therefore incapable of breeding. The term **juvenal** refers only to a particular plumage (see page 188) and should not be used, as is sometimes the case, to designate youthfulness. An **immature bird** (or simply **immature**) indicates any young bird with teleoptiles, which has not acquired its fully adult plumage (definitive feathering) but may nevertheless be capable of breeding.

DEVELOPMENT OF ALTRICIAL YOUNG

The Newly Hatched Young

The altricial nestling of a passerine species immediately after hatching is ungainly, usually having an enormous head and abdomen, in marked contrast to very short and undeveloped limbs. The mouth is large and is conspicuous because of the **rictal flanges** which protrude laterally as if the rictal region were swollen.

The nestling is usually naked (psilopaedic), but in certain species, long, sparse down feathers (neossoptiles) are present on one or more of the dorsal pterylae. The down is usually more prominent on the spinal and capital pterylae. The down may be of different colors in different species and may increase slightly in length after hatching. There are no macroscopic indications of developing teleoptiles, although their papillae may in some species be sufficiently large to indicate faintly the outlines of all pterylae.

The coloration of the skin is generally pinkish, darker above than below. The viscera show through the skin of the abdomen. The soft parts (i.e., parts of the bill and feet) are variously colored, sometimes intensively. The lining of the mouth is often strikingly colored (see Ficken, 1965). The tip of the bill, egg-tooth, and rictal flanges, however, are predominantly white, or yellowish white, in all species.

CONDITION OF DEVELOPMENT AT HATCHING

	ALTRICIAL	PRECOCIAL	PSILOPAEDIC	PTILOPAEDIC	EYES CLOSED	EYES OPEN	NIDICOLOUS	NIDIFUGOUS
Gaviiformes		X		X		X		X
Podicipediformes		X		X		X		X
*Procellariiformes	X			X		X[1]	X	
Pelecaniformes	X		X[2]		X		X	
*Ciconiiformes	X			X		X	X	
Anseriformes		X		X		X		X
*Falconiformes	X			X		X	X	
Galliformes		X		X		X		X
Gruiformes		X		X		X		X
Charadriiformes: Charadrii		X		X		X		X
*Charadriiformes: Lari		X		X		X	X	
*Charadriiformes: Alcae		X		X		X	X	
Columbiformes	X		X		X		X	
Psittaciformes	X		X		X		X	
Cuculiformes	X		X		X		X	
*Strigiformes	X			X	X		X	
*Caprimulgiformes		X		X		X	X	
Apodiformes	X		X		X		X	
Trogoniformes	X		X		X		X	
Coraciiformes	X		X		X		X	
Piciformes	X		X		X		X	
Passeriformes	X		X		X		X	

[1] Eyes closed in most petrels
[2] Tropicbirds (Phaëthontidae) are ptilopaedic

The eyes are closed, but the eyeballs show through the lids. The nestling rests on its abdomen with head slightly under the breast and the legs kept forward. Activities consist chiefly of simple grasping with the feet, for feeding, for defecating, and for keeping the body upright. The bird is responsive mainly to vibrations of low intensity.

The nestling may or may not be vocally silent. In some species it may give faint sounds even before it emerges from the egg.

The nestling is "cold-blooded" (poikilothermic), which means that its body temperature corresponds to the surrounding air temperature. It shows no ability to regulate its body temperature. The brooding parent must therefore supply the heat. Should the parent bird leave the nest for an extended period after brooding, the nestling's body temperature will drop soon to the temperature level of the surrounding air. Unless the air temperature is excessively low, the effect on the nestling will not be lethal as it can endure considerable cooling.

Length of Time in Development

The development of an altricial bird is regarded as divisible into two phases: (1) development in the nest and (2) development after leaving the nest. Both phases vary greatly in length among different species.

The length of time in development is commonly measured *in days* from the day of hatching (called 0 day) to the day when the young bird is independent of parental care, and may range from approximately 25 days to several months. In passerine species, Nice (1943) found that the majority attain flight proficiency at about 17 days and become independent of parental care at about 28 days. The end of the phase spent in the nest may not coincide with the attainment of flight, some species leaving before they are able to fly.

In most altricial species there is some correlation between the length of the incubation period and the length of the period spent in the nest. If the incubation period is a long one, so is the period of nest life. There are also two factors with which time of leaving the nest seems to be correlated: (1) the safety of the nest and (2) the size of the bird. Thus a bird nesting in a cavity undergoes a longer period of nest life than one nesting on the ground; a large bird takes a longer time to develop than a smaller one. The young Eastern Bluebird *(Sialia sialis)* fledges from its cavity nest in about 15 days, whereas the young Hermit Thrush *(Hylocichla guttata)* leaves its ground nest in nine to 10 days though it may not take its first flight until two or three days later. The young Bald Eagle *(Haliaeetus leucocephalus)* fledges in 10 to 12 weeks while the young of the California Condor *(Gymnogyps californianus)*, one of the largest flying land birds, may take as long as five months before first venturing on the wing (Koford, 1953). Perhaps the longest period of nest life of any bird is that of the Wandering Albatross *(Diomedea exulans)*, one of the largest flying sea birds, whose young takes nine months (278 days) to fledge (Tickell, 1968). For a summary of the ages at which North American species take first flight, see Appendix H.

Weight

The weight of the nestling of most passerine birds at hatching averages approximately two-thirds of the weight of the fresh egg (Heinroth, *in* Nice, 1943) and, except in the Corvidae, is about 6 to 8 percent of the weight of the adult female (Nice, 1943). The weight of the nestling rapidly increases during nest life until, at the time of nest-leaving, its weight is from 20 to 30 percent less than that of the adult female, or sometimes as much as 50 percent less. At nest-leaving, a slight drop in weight occasionally occurs; thereafter the weight increases, though more slowly than before nest-leaving, until by the time the young bird has attained independence its weight nearly equals that of the adult female.

In some birds, particularly non-passerine species with a slow rate of growth, the nestling stores a vast amount of fat, thereby attaining a weight after the halfway mark in its nestling life that may greatly exceed the weight of the adult. Sea birds such as procellariiform species are notable in this respect. The young Leach's Petrel *(Oceanodroma leucorhoa)* may weigh half again as much as the adult at 40 days of age and 23 to 30 days before fledging (estimated from W. A. O. Gross, 1935). The young Wandering Albatross may reach maximum weight at least one-third in excess of an adult at 221 to 230 days of age and 50 to 60 days before departure (estimated from Tickell, 1968). Following this peak in weight the young of both species are fed less frequently, finally not at all, since the parents begin voluntarily to desert them. Consequently their weight steadily declines as they use their fat stores, until at fledging their weight approximates that of the adults. Possibly fat storage by these young may be an adaptation for survival during long, stormy periods when the parents fail to obtain food from the tempestuous sea. Roberts (1940) found that young of the Wilson's Petrel *(Oceanites oceanicus)* in the Antarctic may weigh twice as much as an adult early in their long period of confinement to burrows and that their weight fluctuates markedly thereafter due, in this case, to heavy snowfalls successively covering their burrows and preventing the parents from entering to feed them. One young bird was able to survive as long as 20 days.

Plumage

Teleoptiles of the juvenal plumage first show evidence of development when the feather papillae darken and enlarge. Soon sheaths ("pinfeathers") emerge from the papillae, first on certain dorsal pterylae, particularly on the alar, humeral, and spinal tracts, then on the remaining dorsal pterylae, and finally on the ventral pterylae. Later, and in this same order, the sheaths open, allowing the teleoptiles to continue growth. By the time of nest-leaving, the teleoptiles are sufficiently lengthened, unfolded, and expanded to cover all the apteria. Down feathers, when present, grow from the same papillae as the teleoptiles and are pushed out by them. They remain for a time attached to the tips of the teleoptiles. Frequently, young birds with well-developed juvenal plumage show the down feathers still adhering. Eventually the down feathers drop off.

For an excellent day by day account and description of the development of plumage in a passerine bird, see the paper by Banks (1959).

Body-temperature Control

In passerine species, the nestling develops "warm-bloodedness" (homoiothermy) early, while undergoing its most rapid growth, and attains temperature control soon after the mid-point in the period of nest life. Young House Wrens *(Troglodytes aedon)* develop temperature control by nine days of age (Baldwin and Kendeigh, 1932). Three fringillids, the Chipping, Field, and Vesper Sparrows *(Spizella passerina, S. pusilla,* and *Pooecetes gramineus)*, which leave the nest in 10 days or less, show effective temperature control by six to seven days after hatching (Dawson and Evans, 1957, 1960). The acquisition of a feather covering is not necessary for the establishment of temperature control as the nestling already has some control at moderate air temperature when the teleoptiles start emerging from their sheaths. The insulating plumage which the nestling later acquires serves mainly to prevent excessive loss of body heat when the air temperature is much lower than the body temperature. Excessively warm air tends to kill nestlings more quickly than cold.

Coloration of Soft Parts

Colors of the soft parts change gradually during early development. The whitish parts of the bill and the brilliantly colored mouth lining tend to darken as nest life advances. The iris and the feet slowly assume a coloration resembling that of the adult.

Behavior

As the young altricial bird develops in the nest, its many and complex behavioral traits of later life begin to emerge. The following descriptions apply primarily to the passerine bird.

Within minutes after hatching the nestling spontaneously lifts its head straight up and "gapes"—opens its mouth wide. This is the primordial "begging" action for food. It may be repeated spontaneously many times and will re-occur on receiving any vibratory or tactile stimulus from its parent, or from a like stimulus simulated by a person at the nest. If not on the first day, then two or more days later, depending on the species, faint vocal sounds accompany the action. As the nestling develops and its eyes begin to open, it responds to visual stimuli and directs its gaping toward them. Meanwhile, increasingly sensitive to sounds of varying intensity, it begs readily on hearing its parent's "food calls" at the nest. The begging action itself gradually modifies: the nestling stands in the nest, vocalizes strongly, fans its wings, and stretches accurately toward the food source. If fed, the nestling, during its first few days, simply raises its posterior abdominal region slightly and voids the fecal matter; during its later life in the nest, it raises its abdominal region higher and more conspicuously, sometimes with a wavering action as if to draw attention to the process, and may turn so as to void the fecal material toward the edge of the nest.

When the student takes a young bird no older than two days from its nest and places it on its back on a flat surface, he will note that it reacts by clutching at the air with a grasping action of its toes and attempts, albeit futilely, to turn itself over. At this early stage the nestling cannot grasp the bottom of the nest effectively and is just barely able to keep itself upright; if it rolls over on its back inadvertently, it can right itself only by persistently struggling until it manages to push against the wall of the nest or one of its nest-mates. Two or three days later the same nestling, when put on a flat surface, can quickly right itself. In the nest it retains such a firm grasp of the bottom that one can disengage its feet only with difficulty when he attempts to lift it out. For exact ages at which the grasping appears in several species, see Holcomb (1966a, 1966b).

As it approaches the mid-point in its nest life, the young bird begins several maintenance activities, particularly preening and head scratching; at the same time it displays such comfort movements as stretching, yawning, and shaking. It spends a considerable amount of time sleeping, the head forward on the nest itself or on its nest-mates. Toward the end of nest life, it stays awake for long periods during which it may indulge in playful activities (see the section, "Behavior," page 254), in numerous sessions of wing-fanning, and in fighting. If it happens to be larger than its nest-mates, it may be quite aggressive, pecking at them, crowding them, and taking more than its share of food from the parents. In eagles and some of the other large predators, that have only two young unequal in age and size, the larger one may early in nest life so severely harass the smaller one as to kill it.

At the start of nest life, the young bird responds indiscriminately to stimuli, whether from the parent or any other source, and begs. Later, as it acquires vision, the nestling becomes increasingly more selective in its response to stimuli, finally responding only to stimuli coming from the parents. It may gape at the human intruder, but not stretch in his direction or vibrate its wings, and give vocal sounds. Eventually, near the time of nest-leaving, it will not gape at all; instead it cowers, sinking lower in the nest, sleeking the plumage, and sometimes even closing the eyes. If removed from the nest, it ordinarily "comes alive," giving "fear calls" and attempting escape. The commotion may induce its nest-mates to come alive, burst from the nest, and disperse. Thus alarmed, neither the nestling, taken from the nest, nor its fellows which departed of their own accord can be put back in the nest with the expectation of their staying. The student should therefore be cautioned not to handle nestlings that show cowering behavior least he cause them to leave the nest prematurely—sometimes before they are capable of faring for themselves.

The young bird leaves the nest of its own accord on the maturing of an inherent impulse or instinct to do so. The action may be triggered by any one of a number of factors such as discomfort from high air temperature, hunger, exuberance during a session of wing-fanning, or a period of restlessness among the nest-mates. Its departure may in turn trigger the departure of its nest-mates, if they are the same age. Although observers have sometimes credited parents with prompting their offspring to fledge by intentionally withholding food and thus enticing them to "come and get it," in all probability there is no such conscious purpose involved. What *seems* to be an intent on the part of the parents is simply their hesitation, occasioned by some undetermined cause, in bringing the food, or their lowered feeding rate (see Lawrence, 1967). Any such break in routine increases the young birds' hunger and their consequent aggressiveness. In most species, once the young bird fledges from the nest of its own accord, it rarely returns. Young eagles, however, and a few other large predators return to the nest repeatedly, using it as a "home base" for several days. Young White Storks *(Ciconia ciconia)* return to the nest to be fed and continue to use the nest as a sleeping place (Haverschmidt, 1949) until they migrate from the area.

Course of Development

For purposes of comparison with other species, it is convenient to consider the course of development of a species as divisible into **stages**. In the Song Sparrow *(Melospiza melodia)*, a typical altricial species, Nice (1943) has described five stages which may well serve as a guide in investigating development in other species. Her summary of these stages, in which she points out their duration and characteristics and indicates the time that certain activities were first manifest, is given below verbatim.

The **first** stage embraces the first 4 days of nest life; it is characterized by rapid growth and the start of feather development; the chief motor coordinations are the food response and defecation; the first food call has been heard at 2 days.

The **second** stage—5 and 6 days—is one of rapid growth in weight and feathering and the establishment of temperature control; the eyes open, and the beginnings of several motor coordinations are seen—standing, stretching up of the legs, and the first intimations of preening.

The **third** stage—7, 8 and 9 days—shows rapid development of motor coordinations; cowering typically appears; the birds are well covered with feathers and are able to leave the nest at any time. They stretch their wings up and sidewise, they scratch their heads and shake themselves, they fan their wings and flutter them when begging. New notes appear—the scream, the location call, and several new feeding notes. In the 9 to 10 days of its nest life the Song Sparrow has changed from a nearly naked, blind, and practically cold-blooded creature of about 1.5 grams, absolutely dependent on parents and nest for food, warmth and shelter, to a warm-blooded, well-feathered individual of 15 to 18 grams, independent of nest and nest-mates, able to care for its feathers, to move about, to escape enemies to some extent, to inform its parents of its whereabouts so they can feed it, and to respond to its parents' notes of alarm and fear.

The **fourth** stage starts at the age of 9 to 10 days when the fledglings leave the nest. It is spent in retirement by the young bird away from its nest-mates, is characterized by silence (except when calling for food) and by general immobility. The chief advance is the acquisition of flight. A new method of begging for food appears: the birds no longer make themselves conspicuous, but call their parents and beg from a horizontal rather than vertical position. One brood was "psychologically" still in the nest for a day after leaving it. Sleeping in the adult position typically appears at 10 days. A beginning is made in independent feeding activities: wiping the bill, pecking at objects, picking up food, catching insects, working at grass heads, and scratching the ground. Bathing reactions appeared in several of the hand-raised birds. Singing appeared with 3 birds at 13, 14 and 15 days.

The **fifth** stage, starting with the attainment of flight and ending with independence, lasts about 10 days. At 17 days Song Sparrows in Ohio come out from retirement; they come into contact again with nest-mates and soon pursue their parents for food. They may beg from their parents until they are 5 weeks old, but the latest that I saw one fed was at 30 days. The Massachusetts Song Sparrows attempted to shell seeds at 17 days, but were not able to do so until the age of 25 days. The complete bathing technique, consisting of 3 chief motions, is attained during this stage. Sunning was first recorded at 29 days. Frolicking [play-fleeing] was first seen at 17 and 18 days. The fear note appears at 19 to 21 days. This fifth period may be the chief time when the young are conditioned by parental behavior with regard to what to fear and what not to fear. A note of antagonism was first noted at 17 days; pecking others at 18 days; threatening postures and fighting at 19 days. *Tsip*, denoting a social bond, first appeared at 19-20 days. The characteristic adult note *tchunk* was heard at 28 and 29 days.

Effects of Brood Parasitism

Young brood parasites usually have inherent means which give them an advantage over the young of their hosts or at least put them on an equal footing with their nest-mates. The nestling honey-guide (*Indicator* sp.) has sharp hooks on the tips of its mandibles which it retains long enough for biting and successfully killing the young of its host (Friedmann, 1955). The nestling Common Cuckoo *(Cuculus canorus)* is endowed with special reflexes which enable it to move under, push up on its back, and roll or thrust over the nest's edge, any egg or other nestling. In both cases the parasite gains thereafter the full attention of its foster parents.

The Brown-headed Cowbird *(Molothrus ater)* has no such specializations. But it does have an incubation period of about 12 days which is often shorter than its host's and thus give the nestling a head start in development. Furthermore, when the cowbird parasitizes warblers and other species much smaller than itself, the nestling at the outset has a distinct advantage in size with the result that, more often than not, it "swamps out" the young of its host, causing their death within two or three days after hatching by constantly trampling on them and outreaching them for food brought by their parents. This is the usual fate of nestling Kirtland's Warblers *(Dendroica kirtlandii)* which normally hatch two days after the cowbird (Mayfield, 1960). Larger host species fare better and are probably better hosts because they provide more food for the young parasites. Nestlings of the Scarlet Tanager *(Piranga olivacea)*—species nearly as large as the Brown-headed Cowbird—compete successfully with young cowbirds and do not seem to be in any way affected by their presence (Prescott, 1965). Nestlings of the Red-eyed Vireo *(Vireo olivaceus)*—a species intermediate in size between the Kirtland's Warbler and Scarlet Tanager—compete poorly to only moderately well with the rapidly growing parasites whose increasing tendency is to push and crowd them out of the nest. Ordinarily one, and never more than two, nesting vireos ever manage to stay in the nest and survive to fledging (Southern, 1958).

The common tendency in observing brood parasitism is to abhor the parasite and pity the host as an unfortunate victim of a degenerate habit. The truth is that brood parasitism is a highly developed habit and a wholly successful mode of survival at no appreciable cost to the host species. If the parasitic habit were consistently harmful, it could not be successful for long because it would eventually eliminate the host species.

The Brown-headed Cowbird is a remarkably successful parasite as evidenced by the long list of its host species, vast population, and wide geographical range. And it is becoming increasingly successful, as Mayfield (1965) has shown. Formerly an inhabitant only of the grassy plains in mid-North America, it has recently penetrated the eastern forested regions where it has found many new host species which lack an inherent defense against brood parasitism. In no case does the cowbird seem to have caused a marked decline in its new hosts, even in one of its most molested, the Red-eyed Vireo. Two studies of this species, one in northern Michigan (Southern, 1958), where its nests were parasitized, and one in southern Ontario (Lawrence, 1953), where its nests were not, showed a difference in success of only three percent. If there is any species seriously affected by the cowbird, it is the heavily parasitized Kirtland's Warbler. Mayfield (1960) has estimated that its whole population, numbering about 1,000 adults, would produce about 60 percent more fledglings if there were no cowbird parasitism.

DEVELOPMENT OF PRECOCIAL YOUNG

The development of the typically precocial chick of wild birds has been given much less attention than the development of the typically altricial nestling. One obvious reason for this is the great difficulty in making extended observations on a creature that, in a few hours after hatching, becomes extremely ambulatory and elusive. A résumé of the development of the typically precocial chick of North American orders is attempted below, the emphasis being placed on characteristics which are in contrast to the development of altricial young.

The Newly Hatched Young

The newly hatched chick is more advanced in development than the altricial nestling. Its proportions more closely match those of its parents, except for the wings, which are relatively small and undeveloped. The chick's more advanced development, compared to the nestling's, is correlated with a larger supply of yolk that it had for nourishment as an embryo.

The chick is thickly covered with neossoptiles on all pterylae (ptilopaedic), the down feathers being extremely long on the spinal pterylae in land chicks (i.e., chicks of Galliformes and Charadrii), and very thick on the ventral pterylae of aquatic chicks (i.e., chicks of Gaviiformes, Podicipediformes, Anseriformes, and Gruiformes). The down becomes dry and fluffy within two or three hours after hatching.

The whitish egg-tooth and white tip of the lower mandible are conspicuous. The rictal region is without flanges. The lining of the mouth is no more brilliant in color than the adult's and is often less brilliant.

The eyes are open at hatching. Both vision and hearing are already developed, and the chick shows a quick response to visual and auditory stimuli. The chick rests on its tarsi; the neck is too weak to hold up the head, and consequently the head rests forward on the floor of the nest. Within minutes after hatching, the chick of a shore bird will "teeter" or "bob" if the trait is characteristic of the species.

At hatching the chick gives brood calls—i.e., "peeping" sounds—that can be heard earlier before it emerges from the egg.

Although the chick has body-temperature control partly established, it is essentially cold-blooded. When left unbrooded in cool weather, its body temperature soon begins to drop to the level of the surrounding air temperature with the result that it becomes increasingly immobile.

Length of Time in Development

The development of the precocial bird is divisible into two phases: (1) the development between hatching and the attainment of flight and (2) development after the attainment of flight. The young bird becomes partially independent of parental care (i.e., brooding, feeding, etc.) during the first phase, but it may remain dependent socially on the family group during the second phase. Thus the two phases in the development of the precocial bird differ from the two phases in the development of the altricial bird.

The length of time before the chick attains flight ranges widely: 7 to 27 days in gallinaceous birds; 14 to 34 days in shore birds; 4 to 12 weeks in most aquatic birds. See Appendix H for length of time in many groups of birds.

Weight

The weight of the precocial chick at hatching has the same proportion to the weight of the fresh egg as the weight of the altricial nestling. In other respects the weight of the precocial chick differs from the weight of the altricial nestling, *viz:* (1) The weight of the precocial chick is about one to six percent the weight of the adult female, that is, the chick is somewhat smaller than altricial young in proportion to the size of the adult female. (2) The weight of the precocial chick decreases between hatching and the first feeding; thereafter it increases *steadily* throughout development. There is no retarding of weight increase during and after the attainment of flight. (3) At the time flight is attained, the chick of land birds weighs from 60 to 70 percent less than the adult female. In other words, the chick is much smaller than altricial young in proportion to the size of the adult female at the time of flight attainment.

Plumage

In land chicks, the teleoptiles are first evident on the alar, scapular, and caudal pterylae. They appear within the first five or six days after hatching and continue well ahead of the other teleoptiles in growth. In certain species of Galliformes the teleoptiles of the alar pterylae (particularly the remiges and their respective coverts) are evident at hatching and may even be opened out of their sheaths. In aquatic chicks, the teleoptiles are first evident on the ventral pteryla, then on the humeral and the caudal pterylae. They appear between the first and third weeks and continue well ahead of the other teleoptiles in growth. In general, the teleoptiles are first evident much later in Anseriformes than in other aquatic birds. In land chicks, the teleoptiles last to appear are those of the capital pteryla, cervical region, and pelvic region. In aquatic chicks, the teleoptiles last to appear are those of the capital pteryla, interscapular region (in Anseriformes), pelvic region, and alar pteryla. The remiges and their coverts are farthest behind in point of development.

Body-temperature Control

The newly hatched chick already has body-temperature control partly established. From hatching on, temperature regulation develops slowly, much more so than in the altricial bird. In most species, the chick does not attain full temperature control until about four weeks of age.

Behavior

The activities directly adjusted to prolonged nest life do not occur in precocial young. Most precocial young birds remain on the nest from three to twenty-four hours or more. During this period the down dries and the birds become increasingly active, rapidly acquiring perfection of muscular coordinations associated with standing, walking, and running or swimming. While the parent continues to brood, they peck at objects, especially those that are brightly colored or shining, fan their wings, stretch, yawn, and preen. When resting, they customarily sit on their tarsi with heels touching the ground. They give **brood calls** frequently and are soon responsive to the **food call** or **assembly call** of the parent.

After the initial period in the nest, the chick will show one of the following four patterns of behavior that have been described by Nice (1962) in the precocial species studied by her.

1. Stay in or near the nest, even though able to walk, and be fed by its parents. Characteristic of Lari (skuas, jaegers, gulls, terns, and skimmers) and Alcae (auks, murres, and puffins). For the first two or three days the chick remains in the nest where the parents bring it food. Thereafter, unless the nesting site or burrow prohibits departure, it may wander as far as the edge of the nesting territory, sometimes seeking shade under plants or other objects; but it returns to the nest for brooding and food.

2. Follow the parents from the nest and receive food from the parents. Gaviiformes (loons), Podicipediformes (grebes), and Gruiformes (cranes, rails, gallinules, and coots). The chick, if a loon, grebe, coot, or gallinule, leaves the nest by swimming; if a rail or crane, by walking. Initially, the chick receives food passed to it by the parent; in a short time it is able to obtain its own. The young grebe or loon, for example, waits on the water's surface for one of the parents, searching for food below, to come up with food in the bill and deliver it; within a few days, the same young bird dives for its own food.

3. Follow the parent and be shown food. Probably most Galliformes (grouse, pheasants, quails, turkeys). The chick leaves the nest, usually with the hen only. She directs the chick's attention to the food by picking it up and dropping it repeatedly, meanwhile giving a food call. The chick responds by taking the food dropped and swallowing it. Within a short time the chick learns to find its own food, often by scratching in the ground litter with its feet.

4. Follow the parents and find its own food. Anseriformes (swans, geese, and ducks) and Charadrii (oystercatchers, phalaropes, sandpipers, plovers, and other shore birds). The cygnet and gosling leave the nest with both parents, the duckling usually with the hen only; the shore bird chick leaves with one or both parents, depending on the species. At the outset the chick starts feeding without parental inducement or assistance by first nibbling indiscriminately at objects and soon learning by trial and error what is edible. Among the few exceptions is the chick of the American Woodcock *(Philohela minor)* whose parent passes it earthworms obtained by probing in the soil (Pettingill, 1936).

At sudden noises or abrupt, unfamiliar actions, most chicks scatter, hide under or beside large objects when available, and "freeze"—i.e., crouch flat and remain motionless with eyes closed. Usually, but not always, they remain frozen until the parent utters the assembly call. Aquatic chicks frequently give evidence of an ability to dive under water in the face of danger even though normally they do not dive at all or not until later in life.

A definite social bond is evident in a brood. When one chick is "lost"—i.e., out of contact with the parent and the brood—it utters the brood call loudly and shows evident distress until contact is re-established. The family bond holds strongly until the chicks are well able to fly and require no further brooding. It is not unusual for the bond to persist into the fall and winter. Chasing, play-fighting, and pecking sometimes occur between members of a brood.

Calls are various in number among different precocial chicks. Nearly all species give the brood call mentioned above. It is uttered continually by each chick during the day while the brood is active. Presumably it functions as a means of keeping the family together. Nearly all chicks except those which stay in or near the nest give **alarm** or **warning calls**, or both, when danger approaches or when something is disturbing in appearance. Such calls serve to protect the entire brood by warning all members at once. All precocial young give **distress calls** when being harmed or handled. Some precocial young give **food calls** if they are accustomed to being fed. Some precocial young also give **contentment calls** or **conversational calls** when being brooded. These are faintly uttered and audible to the observer only when he is close at hand. Probably all chicks give **threat calls** when challenging a fellow member to a playful duel or when assuming dominance over another member. Very frequently, all sounds given by young, except the brood call, resemble sounds of similar purpose given by the parent. The chief difference is in the tone.

Course of Development

The course of development of the precocial young bird differs from that of the altricial in that the first three stages of development are passed in the egg; the precocial bird hatches at the beginning of the fourth stage. At this stage, both precocial and altricial young are ready to leave the nest. Both have well-developed muscular coordinations, vision, and hearing. Both are fully feathered, the precocial with neossoptiles, the altricial with teleoptiles. Both are still dependent on parental care, the precocial chiefly for brooding, the altricial for both brooding and feeding. The main difference between the two groups at this stage is that the precocial chick is still incapable of flight whereas the altricial nestling is just beginning to fly.

Development of Megapodes

Always of special interest in the study of development of precocial birds are the megapodes, gallinaceous birds whose eggs, after an exceptionally long period of incubation in mound nests (see pages 337 and 357 in this book), produce chicks no less exceptional in their precocity. Each chick, its egg-tooth already having disappeared, digs out of the mound on its own and fares alone, receiving no parental attention whatsoever and existing without the companionship of brood-mates. It is mute since brood calls can have no function. The chick of one species, the Mallee Fowl *(Leipoa ocellata)*, is able to run swiftly within two hours after emerging from the mound and can fly strongly in 24 hours (Frith, 1962). These highly precocial features are reptile-like and suggest primitiveness, yet there is no convincing evidence that this is the case. More likely, megapodes evolved from conventional galliform stock through varying selective pressures on their reproductive physiology and behavior (Clark, 1964).

AGE OF MATURITY FOR BREEDING

The period between the bird's attaining its independence from its parents and reaching maturity for breeding varies widely among different species. With very few exceptions, no wild North American bird breeds until about a year (9 to 13 months) of age. As for the exact ages at which birds of different species breed, reliable records are so few that one can only make generalizations.

The great majority of passerine species and other small land birds normally breed for the first time when just under a year old and this is also the general rule for ducks, many gallinaceous birds, pigeons, some of the smaller owls, and presumably for grebes and the smaller herons and shore birds. The larger herons and shore birds probably breed when they are two years old. Birds which are almost certainly two years old, if not older, include the loons, pelecaniform birds, and geese. Most gulls, terns, and alcids breed on reaching the three-year mark, as do some swans, although most not until the four- or even the five-year mark. Nearly all the diurnal birds of prey breed beginning at two years of age, except condors and eagles which breed at four or five years of age. The initial age for breeding of the procellariiform birds is unknown for most species—it is probably three years for the smaller petrels. For the two largest albatrosses, the Royal and Wandering *(Diomeda epomophora* and *D. exulans)*, breeding is known to start at eight or nine years (see Lack, 1968, for source data) making the span of time between hatching and sexual maturity the longest of all birds.

While species do not breed generally until they have acquired the fully adult plumage (definitive feathering), this is not always the rule. For instance, the American Redstart *(Setophaga ruticilla)* breeds the first year even though its first nuptial plumage lacks the contrastingly black and orange coloration of the successive nuptial plumage. The Wandering Albatross may take 20 to 30 years or more before acquiring its definitive feathering (Tickel, 1968).

THE STUDY OF YOUNG AND THEIR DEVELOPMENT

Select a nest containing eggs under incubation and be prepared to observe in detail the hatching, the appearance and behavior of the young at hatching, and the physical and psychological development of the young.

Procedure

Take full notes on the entire process of hatching from the time the first egg is pipped. Determine how long it takes for the pipped egg to hatch. Note time of hatching of each egg.

As soon as the young are dry, mark them so that they may be recognized as individuals. Since newly hatched young, notably those of altricial species, have small tarsi that will not hold bands, each should be temporarily marked with red nail polish. Mark one bird on the right tarsus, another on the left tarsus, another on the right outer toe, another on the left outer toe, etc. When, a few days later, the tarsi are large enough to hold bands, substitute colored bands (see Appendix A).

If hatching was not simultaneous, observe the effect of age differences on growth and development.

Describe in full the newly hatched young bird: plumage; coloration (use color charts; see Appendix A); condition of eyes; responses; body attitudes and behavior. Weigh each bird as soon after hatching as possible.

Determine the length of time *(in days)* involved in the two phases of development. Consider the day of hatching as 0 day.

Weigh each bird each day, preferably early in the morning before its first feeding. Always weigh at the same hour each day in order to get an accurate spacing of twenty-four hours. Follow the methods for weighing described in Appendix A. (When altricial young are weighed during the third stage of development, the resulting disturbance to the birds often causes them to leave the nest after being put back, thus preventing the determination of the total time involved in normal nestling life.) Strive to obtain weights of young after nest-leaving. Such weights of both altricial and precocial young are rare and most desirable.

Studies of the increase in length of wings, tail, tarsi, bill, body, and extent of wings are sometimes worthwhile in comparative investigations of species or in investigations of relative growth of parts of the body. Measurements should be taken in millimeters and according to the methods described in Appendix A.

Observe the development of plumage, noting the changes occurring on each pteryla. Sometimes it is desirable to measure the day-to-day growth of typical feathers in each pteryla. When the juvenal plumage is finally complete, describe it in full.

Note the day-to-day changes in the coloration of the fleshy parts and the time when the egg-tooth is lost. Either compare the colors directly with a color chart or record them in water-color and compare them with the chart later in the laboratory.

Make detailed observations, using a blind if necessary, of the development in activities and behavior. Extensive note-taking is necessary each day. In tracing the course of development, determine the stages as described by Nice (1943), and compare the two sets of results as to time involved and characteristics.

If the nest is parasitized by cowbirds, follow the development of the young cowbirds along with the development of the hosts' young. Attempt to determine whether or not the hosts' young suffer from the presence of the young cowbirds.

The periods of development during and after nest-leaving warrant particular attention because of the dearth of existing information. Therefore, attempt to determine what triggers the young to leave; where and how far they go during the days following nest-leaving; whether they ever return to the nest or its immediate vicinity; where and how they roost; whether they remain on their parents' territory; and how long before they scatter.

Occasionally, a considerable amount of information on growth and development can be acquired by raising young birds in captivity. The beginning student is nevertheless advised to study young birds *first* under wild conditions, thus gaining the proper conception of "natural" behavior.

If possible, take a series of photographs of young on successive days that shows their growth and development. These will be particularly valuable to illustrate the description of plumage and the general course of development. Use the same bird for successive photographs, placing it on a dark, plain background (preferably black cloth) and always in the same position and at the same distance from the camera. Also, take photographs showing various activities, and use them as illustrations when presenting the results.

Presentation of Results

Draw up a report on the results obtained from a study of young birds and their development. Prepare the manuscript as directed in Appendix B. An outline of suggested topics is given below.

Hatching. Number of eggs; date and time of day when each is pipped; date and time of day when each is hatched; process of hatching; comments on time involved.

Newly Hatched Young. Describe: Plumage; coloration; condition of eyes; responses; body attitudes; behavior. Give the weight. Comment on the condition of development (whether precocial or altricial, etc.); on the significance of coloration (whether cryptic or not, etc.); on the weight as compared with the weight of adult female.

Development of Young. Present a day-by-day account of the development of the young birds. Include in each daily account: Weight; changes in plumage and coloration; description of activities and behavior.

Discussion. After having given the above chronological account, discuss in summary form the following: **Weight.** Changes in weight during the first two phases of development. Represent the weights of each bird in a graph. See Fautin (1941) for a method used. Compare the weight at the end of each phase with the weight of the adult female. **Plumage.** General course of development of the different pterylae. Describe the complete juvenal plumage. Relation of plumage growth to activities (i.e., preening, bathing or dusting, swimming, etc.). **Coloration.** General changes in color and their relation to activities. **Activities and Behavior.** Classify and discuss briefly the various activities observed, stating the time when each was first observed and its relation to phases of development. Note indications of learned behavior. Describe in detail the activities and behavior related to nest-leaving and the subsequent period of development. **Length of Time.** Give the total length of time in the last two phases of development, and compare with that given in the literature on the same species and closely related species. **Course of Development.** Trace briefly the course of development, and compare with Nice (1943, 1962) and other available sources.

Development of Young Cowbirds. Present an account of the development of young cowbirds, stressing any effects that their presence may have had on the development of the hosts' young.

REFERENCES

Baker, M. F.
  1948 Notes on Care and Development of Young Chimney Swifts. Wilson Bull., 60: 241-242.

Baldwin, S. P., and S. C. Kendeigh
  1932 Physiology of the Temperature of Birds. Sci. Publ. Cleveland Mus. Nat. Hist., 3: i-x; 1-196. (Body temperature of nestling birds and mention of body temperature in precocial young.)

Banks, R. C.
  1959 Development of Nestling White-crowned Sparrows in Central Coastal California. Condor, 61: 96-109.

Barbour, R. W.
  1950 Growth and Feather Development of Towhee Nestlings. Amer. Midland Nat., 44: 742-749.

Bond, R. M.
  1942 Development of Young Goshawks. Wilson Bull., 54: 81-88.

Brackbill, H.
  1947 Period of Dependency in the American Robin. Wilson Bull., 59: 114-115.

Breed, F. S.
  1911 The Development of Certain Instincts and Habits in Chicks. Behavior Monogr., 1: 1-78.

Burns, F. L.
  1921 Comparative Periods of Nestling Life of Some North American Nidicolae. Wilson Bull., 33: 4-15; 90-99; 177-182. (Contains a long table of periods in different species compiled from various sources.)

Clark, G. A., Jr.
  1961 Occurrence and Timing of Egg Teeth in Birds. Wilson Bull., 73: 268-278.
  1964 Life Histories and the Evolution of Megapodes. Living Bird, 3: 149-167.
  1969 Oral Flanges in Juvenile Birds. Wilson Bull., 81: 270-279. (Oral flanges [=rictal flanges] tend to be larger in hole-nesting passerines.)

Davis, D. E.
  1955 Breeding Biology of Birds. In *Recent Studies in Avian Biology*. Edited by A. Wolfson. University of Illinois Press, Urbana.

Davis, E.
  1943 A Study of Wild and Hand Reared Killdeers. Wilson Bull., 55: 223-233. (Important observations on the development and behavior of precocial young.)

Dawson, W. R., and F. C. Evans
  1957 Relation of Growth and Development to Temperature Regulation in Nestling Field and Chipping Sparrows. Physiol. Zool., 30: 315-327.
  1960 Relation of Growth and Development to Temperature Regulation in Nestling Vesper Sparrows. Condor, 62: 329-340.

Fautin, R. W.
  1941 Development of Nestling Yellow-headed Blackbirds. Auk, 58: 215-232.

Ficken, M. S.
  1965 Mouth Color of Nestling Passerines and Its Use in Taxonomy. Wilson

Bull., 77: 71-75. (With few exceptions, it is distinctive of each passerine family.)

Fisher, H. I.
1966   Hatching and the Hatching Muscle in Some North American Ducks. Trans. Illinois State Acad. Sci., 59: 305-325. (The muscle originates in the back and sides of the upper neck and inserts on the back of the skull. Reaches maximum development at hatching and serves to raise the egg-tooth against the shell. Earlier papers by the author describe the muscle in other groups of birds.)

Friedmann, H.
1955   The Honey-guides. U. S. Natl. Mus. Bull. 208.

1963   Host Relations of the Parasitic Cowbirds. U. S. Natl. Mus. Bull. 233. (Contains references to the author's many earlier papers on host relations of the Brown-headed Cowbird, *Molothrus ater.*)

Frith, H. J.
1962   The Mallee-fowl: The Bird That Builds an Incubator. Angus and Robertson, Sydney.

Gross, A. O.
1938   Eider Ducks of Kent's Island. Auk, 55: 387-400. (Information, including weights, on newly hatched precocial young.)

Gross, W. A. O.
1935   The Life History Cycle of Leach's Petrel *(Oceanodroma leucorhoa leucorhoa)* on the Outer Sea Islands of the Bay of Fundy. Auk, 52: 382-399.

Hann, H. W.
1937   Life History of the Oven-bird in Southern Michigan. Wilson Bull., 49: 145-237. (Hatching and development of altricial young; method of leaving nest.)

Haverschmidt, F.
1949   The Life of the White Stork. E. J. Brill, Leiden.

Herrick, F. H.
1935   Wild Birds at Home. D. Appleton-Century Company, New York. (Important information, particularly in Chapter 10.)

Hochbaum, H. A.
1959   The Canvasback on a Prairie Marsh. Second edition. Stackpole Company, Harrisburg, Pennsylvania, and Wildlife Management Institute, Washington, D. C. (Development of precocial young, particularly plumage.)

Hoffmeister, D. F., and H. W. Setzer
1947   The Postnatal Development of Two Broods of Great Horned Owls *(Bubo virginianus)*. Univ. Kansas Publ. Mus. Nat. Hist., 1: 157-173.

Holcomb, L. C.
1966a The Development of Grasping and Balancing Coordination in Nestlings of Seven Species of Altricial Birds. Wilson Bull., 78: 57-63.

1966b Red-winged Blackbird Nestling Development. Wilson Bull., 78: 283-288. (Concerned mainly with development of ability to grasp and balance.)

Howell, T. R.
1964   Notes on Incubation and Nestling Temperatures and Behavior of Captive Owls. Wilson Bull., 76: 28-36.

Hoyt, J. S. Y.
1944   Preliminary Notes on the Development of Nestling Pileated Woodpeckers. Auk, 61: 376-384.

Huggins, S. E.
1940   Relative Growth in the House Wren. Growth, 4: 225-236.

Kendeigh, S. C.
1939   The Relation of Metabolism to the Development of Temperature Regulation in Birds. Jour. Exp. Zool., 82: 419-438.

1952   Parental Care and Its Evolution in Birds. Second corrected printing, 1955. Illinois Biol. Monogr., 22: i-x; 1-356.

Koford, C. B.
1953   The California Condor. Natl. Audubon Soc. Res. Rept. No. 4, New York.

Kuhlmann, F.
1909   Some Preliminary Observations on the Development of Instincts and Habits in Young Birds. Psychol. Rev., 11: 49-85.

Lack, D.
1968   Ecological Adaptations for Breeding Birds. Methuen and Company, London.

Lack, D., and E. Lack
1951   The Breeding Biology of the Swift *Apus apus.* Ibis, 93: 501-546. (Important correlations between weather and nestling survival.)

Lack, D., and E. T. Silva
1949   The Weight of Nestling Robins. Ibis, 91: 64-78. (Analysis of various data on *Erithacus rubecula.*)

Lawrence, L. de K.
1953   Nesting Life and Behaviour of the

Red-eyed Vireo. Canadian Field-Nat., 67: 47-77.

1967   A Comparative Life-history Study of Four Species of Woodpeckers. Amer. Ornith. Union, Ornith. Monogr. No. 5.

Lees, J.
1949   Weights of Robins-Part 1. Nestlings. Ibis, 91: 79-88. (Analysis of various data on *Erithacus rubecula*.)

Mayfield, H.
1960   The Kirtland's Warbler. Cranbrook Inst. Sci. Bull. No. 40, Bloomfield Hills, Michigan.

1965   The Brown-headed Cowbird, with Old and New Hosts. Living Bird, 4: 13-28.

Miller, L.
1921   The Biography of Nip and Tuck. A Study of Instincts in Birds. Condor, 23: 41-47. (A very readable and informative account of the development of behavior in two Linnets, *Carpodacus mexicanus*.)

Mosby, H. S., and C. O. Handley
1943   The Wild Turkey in Virginia: Its Status, Life History and Management. Commission of Game and Inland Fisheries, Richmond, Virginia. (Information on weights, development, and behavior of precocial young.)

Nice, M. M.
1937   Studies in the Life History of the Song Sparrow, I. Trans. Linnaean Soc., New York, 4: i-vi; 1-247.

1939   Observations on the Behavior of a Young Cowbird. Wilson Bull., 51: 233-239.

1941   Observations on the Behavior of a Young Cedar Waxwing. Condor, 43: 58-64.

1943   Studies in the Life History of the Song Sparrow, II. Trans. Linnaean Soc. New York, 6: i-viii; 1-328. (Chapters 2-5.)

1950   Development of a Redwing *(Agelaius phoeniceus)*. Wilson Bull., 62: 87-93.

1953   Some Experiences in Imprinting Ducklings. Condor, 55: 33-37.

1962   Development of Behavior in Precocial Birds. Trans. Linnaean Soc. New York, 8: i-xii; 1-211.

Odum, E. P.
1941   Annual Cycle of the Black-capped Chickadee-2. Auk, 58: 518-535. (Development of altricial young, and dispersal.)

Parmalee, P. W.
1952   Growth and Development of the Nest-

ling Crow. Amer. Midland Nat., 47: 183-201.

Pettingill, O. S., Jr.
1936   The American Woodcock *Philohela minor* (Gmelin). Mem. Boston Soc. Nat. Hist., 9: 167-391. (Notes on development and behavior of precocial young.)

1937   Behavior of Black Skimmers at Cardwell Island, Virginia. Auk, 54: 237-244. (Activities of young.)

1939   History of One Hundred Nests of Arctic Tern. Auk, 56: 420-428. (An account of destructive forces operating in a breeding colony.)

Prescott, K. W.
1965   The Scarlet Tanager. New Jersey State Museum, Investigations No. 2.

Rand, A. L.
1937   Notes on the Development of Two Young Blue Jays *(Cyanocitta cristata)*. Proc. Linnaean Soc. New York, No. 48: 27-58.

1941   Development and Enemy Recognition of the Curve-billed Thrasher *Toxostoma curvirostre*. Bull. Amer. Mus. Nat. Hist., 78: 213-242.

Ricklefs, R. E.
1967   Relative Growth, Body Constituents, and Energy Content of Nestling Barn Swallows and Red-winged Blackbirds. Auk, 84: 560-570.

Roberts, B.
1940   The Life Cycle of Wilson's Petrel *Oceanites oceanicus* (Kuhl). Brit. Graham Land 1934-37 Sci. Repts., 1 (2): 141-194.

Skutch, A. F.
1945   Incubation and Nestling Periods of Central American Birds. Auk, 62: 8-37.

Southern, W. E.
1958   Nesting of the Red-eyed Vireo in the Douglas Lake Region, Michigan. Jack-Pine Warbler, 36: 105-130; 185-207.

Stoddard, H. L.
1931   The Bobwhite Quail: Its Habits, Preservation and Increase. Charles Scribner's Sons, New York. (Important notes on development and behavior of precocial young in Chapters 2 and 4. Chapter 5 is devoted to the Bobwhite's vocabulary, including that of the young.)

Stonehouse, B.
1960   The King Penguin, *Aptenodytes patagonica*, of South Georgia. 1. Breed-

ing Behaviour and Development. Falkland Islands Dependencies Surv. Sci. Repts. No. 23.

Stoner, D.
1942 Behavior of Young Bank Swallows after First Leaving the Nest. Bird-Banding, 13: 107-110.

Sumner, E. L., Jr.
1933 The Growth of Some Young Raptorial Birds. Univ. California Publ. In Zool., 40: 277-307. (Concerns the temperature, weight, food consumption, bone measurements, and feather development of young Horned and Barn Owls and Golden Eagle.)
1934 The Behavior of Some Raptorial Birds. Univ. California Publ. in Zool., 40: 331-361. (Detailed accounts of young Horned and Barn Owls and Golden Eagle.)

Sutton, G. M.
1943 Notes on the Behavior of Certain Captive Young Fringillids. Univ. Michigan Mus. Zool. Occas. Papers No. 474.

Tickell, W. L. N.
1968 The Biology of the Great Albatrosses, *Diomedea exulans* and *Diomedea epomophora*. American Geophysical Union, Antarctic Res. Ser., 12: 1-55.

Tinbergen, N.
1939 The Behavior of the Snow Bunting in Spring. Trans. Linnaean Soc. New York, 5: 1-94. (Important data on young leaving the nest.)

1953 The Herring Gull's World: A Study of the Social Behaviour of Birds. Collins, London. (Chapters 18, 21, and 22 deal with hatching and behavior of precocial young. Highly recommended reading.)

Walkinshaw, L. H.
1937 The Virginia Rail in Michigan. Auk, 54: 464-475. (Information on behavior, weights, and plumage of precocial young.)

Weaver, R. L.
1942 Growth and Development of English Sparrows. Wilson Bull., 54: 183-191.

Webster, J. D.
1942 Notes on the Growth and Plumages of the Black Oyster-catcher. Condor, 44: 205-211. (An excellent account of growth and development of plumages in precocial young.)

Weller, M. W.
1957 Growth, Weights, and Plumages of the Redhead, *Aythya americana*. Wilson Bull., 69: 5-38.

Williamson, K.
1945 The Relation between Duration of Hatching and the Incubation Period. Ibis, 87: 280-282.

Yocom, C. F.
1950 Weather and Its Effect on Hatching of Waterfowl in Eastern Washington. Trans. N. Amer. Wildlife Conf., 15: 309-319.

PARENTAL CARE

Numerous important problems relative to the care of young by parent birds are reviewed below. Unless otherwise indicated, statements concern the parents of both altricial and precocial young.

REMOVAL OF EGGSHELLS

Parent birds may or may not remove the eggshells from the nest after the young hatch. Generally, in nidicolous species, the parents either eat the shells or carry them away and drop them. Some individuals may eat the shells of one egg and carry away the shells of another. Species which normally carry away shells may remove strange objects that inadvertently get into their nests. A Common Tern *(Sterna hirundo)* will carry off a film carton blown into its nest (Pettingill, 1931). In nidifugous species, the birds depart, leaving the shells in the nest. Whether birds are nidicolous or nidifugous, they usually leave eggs that fail to hatch in the nest.

BROODING

Brooding is fundamentally a continuation of incubation behavior through the early stages in the development of the young. Although the term, **brooding**, is sometimes applied to the process of incubating the eggs, as well as to the process of covering the young, it actually means the covering of the young only.

Start and Duration

The brooding period begins with the hatching of the first egg and continues until the parents no longer cover the young at any time, day or night.

Participation and Behavior of the Sexes

Generally the participation of the sexes in brooding is the same as in incubation. Thus, if the female alone incubates, she alone broods; if the sexes share in incubation, they also share in brooding.

The behavior of the male in passerine species often changes after the young hatch. He sings less and less each day, finally ceasing altogether; and he shows a gradual waning of aggressiveness in defense of territory.

Even though the male takes no part in incubation, he is quite likely to assist in feeding the young. In many passerine species he appears at the nest, long before the eggs hatch, carrying food with the "anticipation" of feeding the young—an act called **antic-**

ipatory food-bringing. Nolan (1958) found that the male Prairie Warbler *(Dendroica discolor)* carried food from one to three times a day throughout the incubation period, taking it to the nest whether or not the female was present. In her absence he proffered the food to the eggs, then either swallowed it or took it away with him. Krause (1965) observed similar behavior in the male Canada Warbler *(Wilsonia canadensis)*. While this activity, a vacuum activity (see the section, "Behavior," page 246), serves no on-the-scene purpose, Skutch (1953) suggested that, when only the female incubates, it may be useful in bringing the male's attention to the nest early so that later, when the young hatch, he will begin feeding them right away.

The behavior of the brooding bird in nidicolous species tends to follow a common pattern. The bird sits "closely," with relatively few and brief inattentive periods during the first one or two days, and shows evident concern for the young, standing high in the nest every now and then when they wriggle below. If the rays of the hot sun reach the nest, the bird commonly remains standing and frequently spreads the wings, thus exposing more of its body to heat loss and at the same time providing additional shade and ventilation for the young. For the first few days the passerine bird is secretive in its going and coming and departs and returns as deliberately as during incubation, but, as the young develop, it may fly off the nest and return openly. For this reason, nests with well-developed young are easier to find; the observer has merely to watch the bold activities of the bird. Activities of the brooding bird during inattentive periods are devoted more extensively to gathering food than during incubation because the bird must find food for the young, as well as for itself. Copulation stops altogether.

In nidicolous species during the first few days, the brooding bird reacts to intrusions in much the same manner as during the last part of incubation. Thereafter, it becomes more vociferous and hostile. It flushes from the nest and remains to undertake defensive action (see below).

Length of Time

The number of days involved in brooding depends on several factors: (1) **The rapidity with which young develop body-temperature control and protective plumage**. The more rapidly they develop, the shorter the time they need brooding. (2) **Climatic conditions**. Excessively cool weather, heavy rains, or exposure of the nest to intense sunlight may prolong, or cause a return to, brooding. (3) **Protection of the nest**. Birds nesting in cavities, for example, where air temperatures are more uniform, require less brooding.

Brooding is not continuous during any day (except in nidifugous birds during the period of hatching). It is interrupted by periods of inattentiveness. The periods of attentiveness alternating with periods of inattentiveness compose a **brooding rhythm**.

The frequency and length of attentive periods decrease as the growth of the young advances although the climatic conditions mentioned above may cause irregularities. During the first day or two, the brooding rhythm may correspond to the incubation rhythm and one or both may participate, as in incubation. In nidicolous species, brooding may cease altogether before the young leave the nest; in nidifugous species, brooding may cease before the young fly. Probably birds continue night brooding longer than day brooding, but actual evidence in the majority of species is lacking.

FEEDING

Feeding of the young is one of the commonly studied phases in the breeding cycle of birds because it is easily observed; at the same time, it is extremely challenging because of the numerous complexities.

Start and Duration

Feeding starts after a variable lapse of time following hatching. In typically altricial species, it usually begins within two or three hours after hatching, but in other species that require parental feeding, the interval between hatching and the first feeding is much longer. Feeding continues until the young reach independence, long after the departure from the nest. The cessation of feeding commonly indicates the attainment of independence.

Participation of the Sexes

In the majority of nidicolous species both sexes share in the feeding of the young. When the female does all of the brooding, the male takes a greater part in feeding, making many trips to the nest with food. If the female is brooding when he arrives, she may either fly from the nest for an inattentive period, or remain on the nest, standing aside to permit the male to feed the young. In many passerine species, the female may take the food from the male when he arrives at the nest and either consume it or pass it to the young. The part taken by the sexes may change as the young develop. One sex, taking an active part in feeding immediately after the hatching, may gradually or suddenly become less active, while the other sex takes over the main responsibility. Individuality of the birds may determine the role played by the sexes in feeding, especially after the brooding period has ceased.

Methods of Feeding

There are several methods by which adults feed their young. A classification of these methods follows:

I. *Indirect Feeding.* Feeding that consists merely of *attracting young to the food*, sometimes by calling, sometimes by picking up the food and holding it, and sometimes by repeatedly picking up the food, manipulating it, and dropping it, while giving food calls. Characteristic only of those nidifugous species that feed their young.

II. *Direct Feeding.* Feeding that consists of *carrying the food to the young*. Characteristic of all nidicolous species.

   A. Food is carried in the bill and placed in the opened mouths of the young. Characteristic of a wide variety of species, mostly passerine.

   B. Food is swallowed and later made available to the young in the following ways:

   1. By regurgitating (i.e., disgorging) the food:

   a. Into the opened mouths of the young. Examples: finches and waxwings.

   b. Into the throats of the young. Example: hummingbirds.

   c. Into the opened mouths of the young after their mandibles have straddled the parent's bill. Example: herons.

   d. Into the partially opened mouths of the young after the parents' mandibles have straddled the bills of the young. Example: pigeons.

   e. Onto the floor of the nest or a neighboring surface, where it is picked up by the young. Example: gulls.

   2. By opening the mouth and allowing the young to penetrate the throat region, where they pick up the food. Examples: pelicans and cormorants.

   C. Food is carried in the talons to the nest and there shredded with the bill and passed to the young. Example: birds of prey, except vultures.

Many nidicolous young receive food at first by regurgitation and later receive it directly.

Feeding trips to the nest are quick and direct. By habit the bird sometimes comes to the nest by the same route each time, stays at the nest edge long enough to complete feeding and, in some species, to pick up fecal material, and then leaves hastily, or perhaps remains to brood. The feeding trips become less and less secretive as the growth of the young advances and feeding becomes more frequent.

In typical passerine species the young become aware of the parent's approach to the nest (1) by vibrations when it alights on or near the nest, (2) by its call notes, and (3) by its visible actions. The young respond immediately with begging behavior (previously outlined in this book). Their brightly colored mouth linings, seen as they gape for food, provide "**food targets**" which "direct" the parent to place the food properly.

When a parent arrives at the nest, usually *all* the nestlings respond, although with varying degrees of intensity, depending on the state of hunger. But the food is not deliberately distributed among them. The young bird with the quickest and most effective response (presumably the hungriest) receives food until the swallowing reflex is inhibited by a full esophagus. Then the young bird with the next quickest and most effective response receives food, and so on. Thus food is distributed by a process called **automatic apportionment**. Sometimes the parent indulges in "**trial feedings**" when the responses of the young are of equal intensity. It places the food in one mouth. If the food is not swallowed, it is removed and placed in another mouth, and so on. This behavior is also a mode of automatic apportionment.

Rate of Feeding

In passerine species the rate of feeding increases daily because of the increasing needs of fast-growing young. In species which stay in the nest until attaining flight (as contrasted with those species which leave before attaining flight) there is an increase for the first week or ten days, after which there is no increase (Nice, 1943). The rate of feeding is increased by either (1) making more trips per day, or (2) supplying greater amounts of food per trip. Generally those species that customarily carry only one item of food per trip increase the rate of feeding *by making more trips*; whereas, those species that customarily either carry several items of food or feed by regurgitation increase the rate of feeding *by supplying greater amounts of food per trip*.

The feeding rate in nidicolous species, other than passerine, presumably increases as growth advances, but the increase is less evident because growth is comparatively slow.

In passerine species the number of young fed per trip often depends on the amount of food supplied and the number of young in the nest. (1) When a bird brings one item, usually it feeds only one young. (2) When a bird brings several items and there are three or more young in the nest, it does not usually feed them all. (3) When the bird supplies food by regurgitation, it usually feeds all young, regardless of number, each trip. No matter how food is supplied, the individuals in a nest generally receive an equal amount of food during the course of the day, due to automatic apportionment. Exceptions to the rule of automatic apportionment occur when young are not of equal size or vigor, and the larger, or more vigorous, young obtain more food at the expense of the smaller.

NEST SANITATION

Defecation in nidicolous young occurs often immediately after feeding. Young of certain species either void fecal material over the edge of the nest (e.g., hawks and hummingbirds), or into the nest and over the immediate vicinity (e.g., herons). Supposedly all passerine species void fecal material in the form of **fecal sacs**—masses of whitish semi-solid urine and darkish intestinal excreta enveloped in thick mucous. In some species the parents regularly dispose of the sacs throughout nest life; in other species through approximately the first three-fourths of nest life, after which the sacs accumulate on the nest. The adults usually collect the sacs following a feeding and either eat them or carry them away and drop them. Birds seem to vary greatly in respect to the methods they use in disposing of the sacs. Certain individuals may consistently eat the sacs, others may consistently carry them away, and still others may use both methods. Although the birds usually collect fecal sacs following a feeding, they do not necessarily collect them after every feeding. In 35 studies of 28 species Nice (1943) calculated that the percentage of times the excreta were removed (in terms of number of feeding trips) ranged from 9 to 67 percent, the median being 25.

NEST HELPERS

Among "irregularities" known to occur in nesting activities are the instances when an outside bird assists in one way or another in the affairs of a mated pair. Some of the reported instances are those in which (1) a young bird-of-the-year aids a late-nesting pair in building a nest or feeding the young, (2) a mated adult whose young have been recently lost assists in feeding the young of another pair of the same or of a different species, and (3) an unmated adult assists in the nest-building or in feeding the young of a pair. For a thorough discussion and review of the subject, consult Skutch (1961).

PARENTAL CARRYING OF YOUNG

A few birds carry their young. Aquatic birds such as loons, grebes, and coots let their chicks ride on their backs while they are swimming, as do some waterfowl—swans, sheldgeese, and several ducks (Johnsgard and Kear, 1968). But species that intentionally pick up their young and bear them for any distance are exceptional. Rails, particularly long-billed species, may do so. Pettingill (1938) reported an instance of a parent Clapper Rail *(Rallus longirostris)* removing one chick from its brood held captive in an open-top box. On hearing the brood calling inside, the parent bird found its way to the top edge, jumped in, grasped the chick in its bill, then jumped out with it and put it on the ground. Pettingill later tested (and filmed) the same ability in a parent Virginia Rail *(Rallus limicola)* by putting its newly hatched brood in a 12-inch high, open-top carton about eight feet from the nest. The parent removed the chicks in the same manner, one after the other, and carried them back to the nest. The African Jacana *(Actophilornis africanus)* and another jacana, the Lotus-bird *(Irediparra gallinacea)* of Australia, carry their chicks under their wings. Hopcraft (1968) describes this procedure in the African Jacana. The parent bird first crouches low and makes a churring noise, whereupon the chicks run under its wings as if to be brooded. The parent then presses its wings firmly against the sides, stands up with the feet of the chicks dangling below the wings, and walks away with them over the broad-leaved water plants.

Ornithological literature contains numerous reports of various other species intentionally carrying young in one way or another, but, as in the case of the American Woodcock *(Philohela minor)* that is said to secure a chick between its legs and carry it in flight, they are usually without adequate proof (Pettingill, 1936).

DEFENSE

Defense of young against enemies begins to appear late in incubation and reaches peak of vigor when the young leave the nest. In nidifugous species, defense is most active soon after hatching; in nidicolous species, many days later, following the prolonged period of nest life.

As a rule, when both sexes participate in brooding or feeding, both sexes also participate in defense, but when one sex alone broods and feeds, the same sex may or may not defend the young. In many species the male shows a tendency to defend the young, even though he takes little or no part in feeding them. However, studies of pairs of birds show much individual variation in intensity of defensive action. While one pair may defend the young vigorously, such defense may be scarcely evident in another; while in one pair the male may take a more active part than the female, in another pair the reverse may be true.

Some of the more common methods by which parents defend their young are classified below. For a discussion of defense methods in birds of all ages, see the section, "Behavior," pages 255-257.

I. *Direct Defense.* Defense by dealing directly with the enemy.

    A. *Threatening.* The bird gives loud, sometimes piercing calls, and displays the plumage, remaining defiantly in the presence of the enemy and sometimes feigning bodily attack.

    B. *Attacking.* The bird rushes at the intruding enemy with intent to strike with bill, or feet (sometimes with talons or tarsal spurs), or wings, or combinations of these. Frequently threat displays precede or follow attack.

    C. *Mobbing.* Occasionally, the parents and several birds of the same or closely allied species attack an enemy together in a concerted effort to distract its attention.

II. *Indirect Defense.* Defense by dealing indirectly with the enemy.

    A. *Communicating Danger.* The bird gives warning calls that silence and immobilize the young (sometimes also the sex-partner) and, in some species, make them hide and "freeze."

    B. *Giving Distraction Display* **("Feigning Injury"** or **"Rodent-running").** The bird behaves in such a way as to detract the attention of the enemy from the nest or young.

Apparently birds show different methods of defense before different enemies. They may, for example, give threatening displays before human beings and distraction displays before small mammals. The subject of defense methods by one species against different enemies needs investigation.

YOUNG-PARENT RECOGNITION

The young of a precocial species, once imprinted on its own parent, presumably "knows" the parent by appearance and repertoire of vocalizations—**contact calls** such as the clucking sounds in gallinaceous birds, **rallying calls** for food, **alarm calls**, and others. The parent, in turn, learns to know its young by similar means. A strong family bond is thus established and the family integrity is maintained for the duration of the young bird's dependency, or longer. Exceptions are broods of ducks, such as mergansers and eiders, which often convene, forming mixed families under the surveillance of one or more hens.

The best example of mutual recognition can be observed in colonies of precocial aquatic birds where many families of one species exist in close association. If a student visits a colony of Herring Gulls *(Larus argentatus)* with chicks, he will notice that the adults attack any strange chick that comes near their own, even though it is of the same age, and that the chicks respond to their parents arriving with food, yet ignore other adults that come near them. Tinbergen (1953) believed that Herring Gull parents know their own young beginning about five days after hatching.

To the human observer, Herring Gull chicks and adults look so much alike as to be indistinguishable. To a Herring Gull, however, each bird undoubtedly has nuances in appearance—and probably voice, too—that give it individuality. Just what the nuances are in Herring Gulls remains to be determined, but studies made in colonies of Rockhopper and Adélie Penguins *(Eudyptes crestatus* and *Pygoscelis adeliae)* revealed that subtilties in voice almost certainly play a role in mutual recognition.

The chicks of both species of penguins, when left alone in their colonies after two or three weeks of age, customarily gather from their nests to huddle together in creches. Their parents, on returning from the sea to feed them, proceed directly to their respective nest sites and call. Immediately, their own chicks leave the creches, come to their parents, give begging calls, and are soon fed. If other chicks respond, as is sometimes the case, they are not only refused food, but are sometimes severely pecked and driven away.

To test the ability of the chick to recognize its parent by voice, Penney (1968) tape-recorded the calls given by several parent Adélie Penguins, then played them back to their chicks huddled with others in creches. In nine out of the ten playbacks, the chick, or the chicks, of the parent whose voice was played, "stood up, looked around the colony, and then moved to the edge of the creche in an alert posture. None of the other creche chicks changed position, although several raised their heads and looked briefly about. In seven of the nine cases the chicks kept moving and returned to their natal territory within one minute after the playback had started. These chicks remained on the territory for three to nine minutes after the sound stopped. On the territory they assumed an appearance best described as alert and anticipatory." Penney later analyzed by sound spectrograph the calls of the different parents and found that each had distinctive and consistent characters of voice sufficient to give that bird a recognizable individuality.

There is little evidence that mutual recognition occurs commonly in altricial birds although instances are reported. Lawrence (1967) found that woodpeckers refused to feed unrelated young of their own species. Probably mutual recognition has not developed widely in altricial birds to the extent that it has in precocial birds because it offers no advantage in species whose young are dependent for only a short period after they leave the nest and soon disperse. The fact that one altricial bird nest-helps another—that an adult will feed an unrelated nestling without attack from the parent of that nestling—is indirect evidence against mutual recognition. And the fact, too, that a young Brown-headed Cowbird *(Molothrus ater)* does not become imprinted on its foster parent, but instead, soon after dependency, seeks the companionship of its own species, is further indirect evidence.

DURATION OF FAMILY BOND

The bond of parents to young and young to parents persists for a varying length of time, depending on the species. In most passerine and other altricial birds, it weakens and soon disappears when the young have attained proficiency in flight and can forage successfully. There may be exceptions. Some families of permanent-resident titmice are reputed to stay together through the fall and winter months. Among precocial species the family bond is stronger for a longer time, owing to the young bird's need of brooding during its slow development of body-temperature control and also its need of protection during its prolonged flightless period. Families of swans, geese, and cranes maintain their integrity for many months, migrating south in family units to their wintering grounds and even returning north before the yearling birds disperse.

THE STUDY OF PARENTAL CARE

Select a nest containing eggs under incubation, and be prepared, as hatching approaches, to make detailed studies of the methods used by the birds in caring for their young. The study may be combined with the preceding study of young birds and their development. Before undertaking the study consult works by Kendeigh (1952) and Nice (1943, 1962) for important information already available.

Procedure

Make careful observations during the period of hatching from the pipping of the first egg. Watch the adults *continuously* for as long as possible. Note significant activities both at the nest and in the vicinity, any changes in behavior after incubation, the part taken by the sexes in nesting affairs, and the way they remove the eggshells.

Study the activity of the sexes during brooding, ignoring no detail of behavior either at the nest or in the vicinity. Determine the length of time *in days* involved in brooding. Consider the day of hatching as 0 day. Record specific data on the brooding rhythm (1) at different times of the day, (2) at different stages of brooding, and (3) under different weather conditions. Use the tabular form suggested by Pitelka (1941) for incubation. (See the directions for the study of incubation, page 362, in this book.) Attempt to determine when night brooding ceases.

In addition to studying activities connected with brooding and the brooding rhythm, study the activities associated with feeding. Determine the time of the first feeding after hatching and (if at all possible) the last feedings which occur long after the young have left the nest. Observe carefully the role of the sexes in feeding; the methods of feeding used at the beginning and later, when the food is gathered; the route followed when adults return to the nest; the responses of the young; and the adult's reactions to the young. Gather accurate data on the rate of feeding at different ages of the young. To do this, make at least one observation each day, not less than three hours long. Record the data in tabular form suggested in part by Pitelka (1941), *viz*:

>Date
>Age of young
>Time of day; total hours and minutes of observation
>Air temperature at beginning and end of observation
>Wind velocity
>Weather conditions
>1. Total feeding visits
>Number per hour
>Average
>Extremes
>Intervals
>Number per hour
>Average length

Maximum and minimum length
Number of visits in which only one young was fed
Number of visits in which only two were fed
(Repeat as above until all young are included)
Total number of visits in which young were fed
Average number of young fed per visit
Extremes
2. Total feeding visits of male
(Repeat as in 1)
3. Total feeding visits of female
(Repeat as in 1)

If there are size differences in the young or if there is anything peculiar about the number of young, location of nest, etc., take it into account when recording the data.

Observe the disposal of fecal sacs, recording the number removed or eaten per feeding visit, and the role of each sex in disposing of them.

Devote considerable attention to the methods of defense. Note how the adults react to humans at different times and under different circumstances. When possible, observe methods of defense in the presence of various birds and other animals. If feasible, test defense responses by deliberately placing animals (under control) near the birds. Use discretion so as not to interfere with the normal course of nesting events.

Make every effort to follow the family after nest-leaving, and strive to learn how long it takes before the family breaks up. Even meager information is valuable.

Photographs that show brooding attitudes and methods of feeding, nest sanitation, and defense will be useful in illustrating information obtained.

Presentation of Results

Prepare a report on the results obtained from a study of parental care. Prepare the manuscript as directed in Appendix B. An outline of suggested topics is given below.

Hatching. Number of eggs; dates of hatching; duration of hatching. Role and behavior of the sexes. Evident changes in behavior since incubation. Methods by which eggshells are removed (if removed). Number of unhatched eggs.

Brooding. **Start and Duration. Participation and Behavior of the Sexes:** Activities of the male and his relation to the nest; activities and behavior of the brooding bird on the nest and off; manner of approaching and of leaving the nest; relations of the sexes. **The Brooding Rhythm:** General description; frequency and length of attentive periods; relation of periods to climatic conditions, age of young, time of day and night.

Feeding. **Start and Duration. Participation of the Sexes:** Method by which the sexes take part, and variations throughout dependency of young. **Methods of Feeding:** Detailed account of the way food is delivered and received throughout brooding. **Time Involved in Feeding:** Number of days involved in feeding; time of first and last feedings in relation to age of young; rate of feeding at different ages of the young, using tables suggested and discussing the results in relation to age, time of day, and climatic conditions. Comment on automatic apportionment and any data on amount of food (e.g., items of food) brought per visit. **Type of Food:** Identify food when possible.

Nest Sanitation. Method of defecation; number of defecations in relation to feeding visits; part played by sexes in removal of fecal material; frequency of removal.

Nest Helpers and Parental Carrying. Give whatever information is obtained.

Defense. Different kinds of defense witnessed and their relation to nest and age of young; kinds of enemies stimulating defense; participation of sexes.

Young-Parent Recognition. Describe any instances and give supporting evidence.

Duration of Family Bond. Determine, if at all possible.

REFERENCES

Alley, R., and H. Boyd
1950   Parent-Young Recognition in the Coot, *Fulica atra*. Ibis, 92: 46-51.

Amadon, D.
1964   The Evolution of Low Reproductive Rates in Birds. Evolution, 18: 105-110. (The evolutionary trend among birds has been toward producing relatively few young, but bestowing on them protracted parental care. One evolutionary reason for this is that the complex physical skills of flight and food-getting require a certain time for maturation and perfection.)

Armstrong, E. A.
1949   Diversionary Display.—Part 1. Connotation and Terminology. Ibis, 91: 88-97.

Baskett, J. N.
1900   Sanitary Habits of Birds. Auk, 17: 299-300.

Blair, R. H., and B. W. Tucker
1941   Nest-sanitation. Brit. Birds, 34: 206-215; 226-235; 250-255.

Brackbill, H.
1944   Juvenile Cardinal Helping at a Nest. Wilson Bull., 56: 50.

Cogswell, H. L.
1949   Alternate Care of Two Nests in the Black-chinned Hummingbird. Condor, 51: 176-178.

Dexter, R. W.
1952   Extra-parental Cooperation in the Nesting of Chimney Swifts. Wilson Bull.. 64: 133-139.

Duffey, E., N. Creasey, and K. Williamson
1950   The "Rodent-Run" Distraction-behaviour of Certain Waders. Ibis, 92: 27-33.

Friedmann, H.
1934   The Instinctive Emotional Life of Birds. Psychoanal. Rev., 21: 1-57. (Contains a consideration of "injury-feigning" with the suggestion that it is a compromise between fear and reproductive emotions.)

Hann, H. W.
1937   Life History of the Oven-bird in Southern Michigan. Wilson Bull., 49: 145-237. (Extensive account of parental care.)

Herrick, F. H.
1900   Care of Nest and Young. Auk, 17: 100-103. (Nest sanitation.)
1935   Wild Birds at Home. D. Appleton-Century Company, New York. (Important information, particularly in Chapter 10.)

Hopcraft, J. B. D.
1968   Some Notes on the Chick-carrying Behavior in the African Jacana. Living Bird, 7: 85-88.

Howell, T. R., and W. R. Dawson
1954   Nest Temperatures and Attentiveness in the Anna Hummingbird. Condor, 56: 93-97.

Johnsgard, P. A., and J. Kear
1968   A Review of Parental Carrying of Young by Waterfowl. Living Bird, 7: 89-102.

Jourdain, F. C. R.
1936-37   The So-called "Injury-feigning" in Birds. Oologists' Record, 16: 25-37; 62-70; 17: 14-16. (An extensive treatise on the subject.)

Kendeigh, S. C.
1952   Parental Care and Its Evolution in Birds. Second corrected printing, 1955. Illinois Biol. Monogr., 22: i-x, 1-356.

Krause, H.
1965   Nesting of a Pair of Canada Warblers. Living Bird, 4: 5-11.

Lawrence, L. de K.
1967   A Comparative Life-history Study of Four Species of Woodpeckers. Amer. Ornith. Union, Ornith. Monogr. No. 5.

Nethersole-Thompson, C., and D. Nethersole-Thompson
1942   Egg-shell Disposal by Birds. Brit. Birds, 35: 162-169; 190-200; 214-223. (An important discussion of the egg-shell disposal of some 186 British birds.)

Nice, M. M.
1943   Studies in the Life History of the Song Sparrow, II. Trans. Linnaean Soc. New York, 6: i-viii; 1-328. (Chapters 19-21.)
1962   Development of Behavior in Precocial Birds. Trans. Linnaean Soc. New York, 8: i-xii; 1-211.

Nolan, V., Jr.
1958   Anticipatory Food-bringing in the Prairie Warbler. Auk, 75: 263-278.
1962   A Catbird Helper at a House Wren Nest. Wilson Bull., 74: 183-184.

Odum, E. P.
1941   Annual Cycle of the Black-capped Chickadee -2. Auk, 58: 518-535. (Care of the young.)

Penney, R. L.
1968   Territorial and Social Behavior in the Adélie Penguin. American Geo-

physical Union, Antarctic Res. Ser., 12: 83-131.

Pettingill, E. R.
1937 Grand Manan's Acadian Chickadee. Bird-Lore, 39: 277-282. (Instance of distraction display before a red squirrel but not before a human.)

Pettingill, O. S., Jr.
1931 An Analysis of a Series of Photographs of the Common Tern. Wilson Bull., 43: 165-172.
1936 The American Woodcock *Philohela minor* (Gmelin). Mem. Boston Soc. Nat. Hist., 9: 167-391. (Notes on parental care and methods of defense in a precocial species.)
1937 Behavior of Black Skimmers at Cardwell Island, Virginia. Auk, 54: 237-244. (Feeding and defense of young.)
1938 Intelligent Behavior in the Clapper Rail. Auk, 55: 411-415. (Instances of adults carrying young.)
1960 Créche Behavior and Individual Recognition in a Colony of Rockhopper Penguins. Wilson Bull., 72: 213-221.

Pitelka, F. A.
1940 Breeding Behavior of the Black-throated Green Warbler. Wilson Bull., 52: 3-18. (Detailed analysis of parental care, particularly feeding.)
1941 Presentation of Nesting Data. Auk, 58: 608-612.

Ramsay, A. O.
1951 Familial Recognition in Domestic Birds. Auk, 68: 1-16.

Schantz, W. E.
1939 A Detailed Study of a Family of Robins. Wilson Bull., 51: 157-169. (Parental care, departure from the nest, etc.)

Skutch, A. F.
1935 Helpers at the Nest. Auk, 52: 257-273.
1953 How the Male Bird Discovers the Nestlings. Ibis, 95: 1-37; 505-542.

1954- The Parental Stratagems of Birds.
55 Ibis, 96: 544-564; 97: 118-142. (Various maneuvers, including distraction displays, employed by birds in protecting their nests.)
1961 Helpers Among Birds. Condor, 63: 198-226.

Smith, S.
1941 The Instinctive Nature of Nest Sanitation. Brit. Birds, 35: 120-124.
1943 The Instinctive Nature of Nest Sanitation. Brit. Birds, 36: 186-188.

Spencer, O. R.
1943 Nesting Habits of the Black-billed Cuckoo. Wilson Bull., 55: 11-22. (Parental care.)

Stephens, T. C.
1917 The Feeding of Nestling Birds. Jour. Animal Behavior, 7: 191-206.

Stoddard, H. L.
1931 The Bobwhite Quail: Its Habits, Preservation and Increase. Charles Scribner's Sons, New York. (Parental care in Chapter 2.)

Tinbergen, N.
1953 The Herring Gull's World: A Study of the Social Behaviour of Birds. Collins, London.

Tinbergen, N., and others
1962 Egg Shell Removal by the Black-headed Gull, *Larus ridibundus* L.: A Behaviour Component of Camouflage. Behaviour, 19: 74-117.

Walkinshaw, L. H.
1937 The Virginia Rail in Michigan. Auk, 54: 464-475. (Includes description of parental carrying of young.)

White, W. W.
1941 Bird of First Brood of Swallow Assisting to Feed Second Brood. Brit. Birds, 34: 179.

Williamson, K.
1952 Regional Variation in the Distraction Displays of the Oystercatcher. Ibis, 94: 85-96.

LONGEVITY, NUMBERS, AND POPULATIONS

This book does not attempt a thorough presentation of the problems pertaining to the longevity, numbers, and populations of birds. The purpose here is merely to introduce the subjects and to suggest possible studies of populations.

LONGEVITY

The length of life or **longevity** of birds depends on many factors including their size. As a rule, the larger the bird, the longer it is likely to live. But this is a very broad generalization.

Some estimates of the average ages of wild birds, roughly calculated and computed from the time they are independent of their parents, give between one and two years for passerine birds and pigeons; two and three years and possibly longer for some herons, some shore birds, and gulls; and four and five years for some swifts (data from Lack, 1954). Some of the larger birds, especially those with a long pre-breeding period, have a much longer life expectancy. The two largest albatrosses, the Royal and Wandering *(Diomedea epomophora* and *D. exulans)*, that require eight or nine years to reach breeding maturity, may have a lifespan averaging between 30 and 40 years (based on Lack, 1954) and an occasional bird may live theoretically as long as 80 years (Westerskov, 1963; Tickell, 1968).

Do not confuse the average lifespan of a bird with its potential lifespan which is very much longer. A Herring Gull *(Larus argentatus)*, banded as a chick, was found dead exactly 36 years later (Pettingill, 1967) and a European Oystercatcher *(Haematopus ostralegus)*, banded as a chick, was trapped on its nest just as many years later (from Terres, 1968). Other maximum longevity records for wild birds are: Herring Gull, 31 years, 11 months; Eurasian Curlew *(Numenius arquata)*, 31 years, 6 months; Black-headed Gull *(Larus ridibundus)*, 30 years, 3 months; European Oystercatcher, 28 years, 5 months; and Arctic Tern *(Sterna paradisaea)*, 27 years. Rydzewski (1962), who assembled these records, lists the age for passerine birds between 10 and 15 years, with the maximum record—20 years—for a Starling *(Sturnus vulgaris)*.

No doubt, as time elapses, the returns coming in from the thousands upon thousands of birds banded will show ages surpassing all the existing longevity records. Captive birds, sheltered as they are from predation and other unfavorable aspects of the natural environment, have exceeded the normal life expectancy manyfold. Indeed, some captive birds have lived to astonishing ages—an Eagle Owl *(Bubo bubo)*, 68 years; Andean Condor *(Vultur gryphus)*, 65 years; one Siberian White Crane *(Grus leucogeranus)*, 62 years, another 59 years; Canada Goose *(Branta canadensis)*, 33 years; Common Pigeon *(Columba livia)*, 30 years; House Sparrow *(Passer domesticus)*, 23 years. For further listings, see Fisher and Peterson (1964) and Flower (1938).

NUMBERS

How many birds are there? Nobody knows with any certainty. Fisher (1951) has placed the number at 100 billion, but it could be higher. Peterson (1941, 1964) guess-estimated the number of breeding birds in conterminous United States to be "not less than" 6 billion and "probably closer" to 6 billion at the start of the breeding season and as many as 20 billion at the end. Nobody has ever disagreed in print with his figures.

About the only country with anything like an accurate estimation of its bird numbers is Finland which has an area nearly half the size of Texas. Using a strip census method through representative habitats in all parts of Finland from south to north, Merikallio (1958) and his associates found the number of birds to be about 64 million.

In Illinois, which is over a quarter the size of Texas, Graber and Graber (1963) estimated by the strip census method a statewide summer population of 61 million birds in 1957 and 54 million during the preceding winter.

Probably the Domestic Fowl *(Gallus gallus)* is the most numerous bird in the world, and it has been suggested that the House Sparrow and Starling, introduced from Europe to many parts of the world, may be almost as numerous. But excluding these birds, whose numbers have been aided and abetted by man, the most abundant birds are no doubt certain marine species inhabiting vast areas of sea where the supply of food is seemingly limitless.

The guano-producing cormorant, or Guanay *(Phalacrocorax bougainvillii)*, breeding on desert islands off the Peruvian coast, may be one of the most abundant. One island alone was estimated conservatively to have more than 5,600,000 adults and young, all of which required not less than a thousand tons of fish per day (Murphy, 1925). The several species of alcids of the North Atlantic and North Pacific must comprise many millions of individuals. Lockley (1953) put the breeding population of just one North Atlantic species, the Common Puffin *(Fratercula arctica)*, at 15 million. Fisher (1951) believes that the Wilson's Petrel *(Oceanites oceanicus)*, which breeds in the Antarctic area and wanders over all the oceans, is probably the most abundant species, marine or land, of any wild bird in the world.

Ornithologists are by no means certain as to the most abundant land bird in North America, but the Red-winged Blackbird *(Agelaius phoeniceus)* may lead all other native land birds. It breeds from the northern fringes of the Arctic tundra to Central America. Although primarily a marsh bird, it nests in various open, upland situations across the continent. Wherever one travels by car in the summer, he sees it on utility wires, seldom far from its nesting site in a grassy ditch, low shrub, or field. Possibly the Red-eyed Vireo *(Vireo olivaceus)* is the most common bird in the woodlands of eastern North America and the Horned Lark *(Eremophila alpestris)* the most common in the prairie country.

Accurate information on the absolute number of any one species is sparse. Generally, the species whose numbers have been determined accurately are those whose breeding areas are sharply restricted, completely known, and accessible for direct counting. Colonies of marine birds lend themselves particularly well to direct counting because they are confined to islands or headlands where their nests can be found and the number of breeding pairs computed therefrom. Fisher (1954), working with a group of collaborators in 1949, censused all the colonies of the Gannet *(Morus bassanus)* on the east side of the

North Atlantic and came up with a total count of 82,394 nests or 164,788 birds. More recently, Westerskov (1963) and Tickell (1968) provided figures for the world's populations, respectively, of the Royal and Wandering Albatrosses, both of which breed on islands exclusively in the far southern oceans. By collecting data on the number of breeding pairs in any one year on every island where the birds are known to nest and making calculations based on long-term knowledge of the species' breeding cycle and mortality rate, Westerskov determined the Royal Albatross population to consist of 18,960 birds, and Tickell, the Wandering Albatross population to consist of 58,760.

The most accurate figures available on absolute numbers are those for species with small populations. The Kirtland's Warbler *(Dendroica kirtlandii)*, which breeds only in a small section of the jack pine barrens in northern Lower Michigan, was systematically censused in 1951 and found to have 432 singing males or an adult population of fewer than one thousand (Mayfield, 1953). The remaining wild population of the Whooping Crane *(Grus americana)* approximates 55 birds, a number readily obtained each year on the Aransas National Wildlife Refuge, coastal Texas, where it winters exclusively.

For a statewide estimate in Illinois of numbers of individuals comprising some of the more common species, see Graber and Graber (1963).

POPULATIONS

A **population** of birds is the total number of individuals found in a given area. Seldom stable for any length of time, it fluctuates in numerical composition as a result of such factors as changes in climatic conditions, changes of season, epidemics, predation, and biotic community succession.

The study of a population attempts primarily to (1) count the number of individuals, (2) analyze composition, (3) determine fluctuations in number and composition from a given time to another, and (4) compare the number and composition in a given area with those in another.

Measurements of Populations

The proper measurement of a population is by a **census**—the enumeration of the individuals of a given area at a given time (modified from Leopold, 1933). Three methods are commonly employed: (1) Census by direct counting. (2) Census by sampling. (3) Census by the application of indices.

1. **Census by Direct Counting.** Direct counting of all individuals in an area is impractical because the majority of species are small and elusive. However, the populations of certain species—e.g., birds with colonial nesting habits or large birds living in open areas—are suited to this type of measurement. Aerial photographs of sea bird colonies will frequently yield fairly accurate data on the exact number of nesting pairs (see Kadlec and Drury, 1968a).

Birds which customarily roost in large communal gatherings may be counted, either as they gather or while they are roosting. The breeding population of colonial nesting species may be enumerated satisfactorily by counting all the occupied nests. If monogamous, as is usually the case, the number of breeding adult birds may be determined by multiplying the number of occupied nests by two. Ordinarily all communal roosting places and colonial nesting sites may be easily located in a particular area, thus allowing an accurate figure for the total population of the species.

Game birds of the upland variety can often be counted by persons, each with a bird dog, walking abreast over all parts of an area. Record is kept of all birds flushed. The error resulting from failure to flush individuals or from counting more than once individuals repeatedly flushed is considered negligible.

In a given area, trapping and banding all individuals of each species at or near the nests is sometimes used as a means of determining numerical composition.

2. **Census by Sampling.** Census by sampling permits the measurement of a total population in an area that cannot be covered either accurately or practically by direct counting. It is actually a direct census of a section of a large area, the section being considered representative—a **sample**—of the whole. The section may be a strip of measured width transecting the entire area or it may be a measured plot of ground. Only the birds observed within the measured boundaries are counted. Censuses of this sort are called, respectively, a **strip census** and a **sample-plot census**. The figures obtained provide basic data from which an estimate of the population of the entire area is possible.

3. **Census by Application of Indices.** The population of a large area may be censused by the use of indices. This method, though not a true census, is useful when direct counting or sampling is impracticable or when the information desired is not total population but the relative abundance of the different species composing the population.

Indices serve to measure a population indirectly. Briefly, they are any conditions which can be measured and which may be expected to vary in proportion to the population which cannot be measured (Leopold, 1933). In censusing bird populations, two indices commonly used are the *number of birds observed per time unit* (e.g., an hour or a day) and the *number observed per distance unit* (a mile).

Composition of Populations

A population of birds normally contains various diurnal and nocturnal species with both sexes represented. An analysis of its composition during the mid-seasons will reveal the following groups of individuals:

> **Mid-summer Population**
>> Summer residents
>>> Breeding adults a year or more of age
>>> Non-breeding adults a year or more of age
>>> Young-of-the-year
>>> Immature, non-breeding individuals a year or more of age
>> Permanent residents
>>> Breeding adults a year or more of age
>>> Non-breeding adults a year or more of age
>>> Young-of-the-year
>>> Immature, non-breeding individuals a year or more of age
>
> **Mid-winter Population**
>> Winter visitants
>> Permanent residents
>
> **Mid-spring and Mid-fall Populations**
>> Transients
>> Permanent residents
>> Summer residents

The determination of the above groups presupposes (1) a knowledge of the residential and transient status of the species in a given area, (2) a field recognition of differences between ages, and (3) a recognition and understanding of breeding behavior and habits.

An analysis of composition is facilitated by trapping and by banding and other marking (Appendix A). This allows accurate identification and recognition of individuals. When a summer population has been banded for several seasons in succession, one can determine the proportion of breeding adults new to the area, the proportion of young returning to breed, and so forth.

For an instructive paper on the composition of a population in a species, read Kadlec and Drury (1968b).

Fluctuations in Populations

Although subject to minor and somewhat irregular fluctuations, a population is normally quite stable—the rate of birth is in balance with the rate of death (i.e., the reproductive rate equaling the mortality rate). If the balance is upset, severe fluctuations occur. When possible, an attempt is made to determine the cause of the fluctuations. Listed below are ecological factors that affect, or are thought to affect, the numerical status of bird populations. The order of listing has no bearing on the importance.

Weather
Natural catastrophes (floods, volcanic eruptions)
Predation
Food supply
Diseases and parasites
Brood parasitism
Abnormal sex ratio
Territorial behavior
Competition
    Interspecific
    Intraspecific
Plant succession
Activities of man
    Land use
    Pollution of environment
    Use of pesticides and herbicides
    Introduction of exotic birds and other animals

The reproductive rate of birds is inherent. Thus, the clutch size is adapted to correspond with the largest number of young that the parents can feed and not to the mortality rate which is a consequence of a high reproductive rate (Lack, 1954). Considerable variation in reproductive rate exists within species as previously shown by the variation in clutch size (see the section, "Eggs, Egg-laying, and Incubation," page 352).

To maintain a stable population, a pair must produce many more young than will survive to adulthood since the mortality rate among young birds is very high. In a study of 100 nests (144 eggs) of the Arctic Tern *(Sterna paradisaea)*, only 15.9 percent of the chicks fledged (Pettingill, 1939). In another study of 428 nests (820 eggs) of the Least Tern *(Sterna albifrons)*, only 9 percent fledged (Austin, 1938). The failure of so many young birds to survive, bringing the mortality rate to about 85 percent and 90 percent for the two species of terns, is not unusual for nidicolous, precocial birds nesting close together in colonies. Anyone who visits a colony of gulls or terns, as fledging time approaches, is invariably appalled by the large number of unhatched or broken eggs and the carcasses of young at different stages of development. The normally aggressive action of adults toward one another, their nests and young, is one of the chief causes of mortality. For a description of this action as well as others in a colony, read the paper by Pettingill (1939) which traces the fate of 100 nests of the Arctic Tern.

The mortality rate of young among nidifugous species is nearly as high—about 75 percent (derived from Lack's summary, 1954)—and is not surprising when one considers the hazards to which chicks are subjected in their long out-of-the-nest period between hatching and fledging.

The mortality rate of young among altricial species is altogether much lower—roughly 45 percent—because their period between hatching and fledging is spent as nestlings with greater protection. Nice (1957) brought together data on nestling success (= young fledged) from 68 studies of altricial species, 35 of which build open nests and 33 of which nest in cavities. Her instructive summaries show considerable variation in nesting success in both groups, and, at the same time, reveal a greater success among cavity-nesters (66.8 percent) than open-nesters (46.9 percent). Her study emphasizes the point that the fewer the hazards to which young birds are subjected in their period of dependency, the lower their mortality rate.

Once young birds have attained independence and flight proficiency, their chances of survival rapidly improve, with the mortality rate decreasing sharply until it levels off at a low figure and remains so for a period dependent on the species' normal life expectancy.

For information on the many factors affecting populations, as well as on related problems, the student is referred to the work by Lack (1954) and its sequel (1966), also to the excellent review by Gibb (1961). For views contrary to Lack's—*viz.*, that the reproductive rate is inherent and thus adapted to correspond with the largest number of young which the parents can supply with food—the student should read Wynne-Edwards (1955) in the case of sea birds, and Skutch (1949) and Wagner (1957) in the case of tropical birds.

Comparison of Populations

A comparison of populations may have at least three objectives: (1) To show differences in their average densities. (2) To determine differences in their average composition. (3) To find differences in their seasonal fluctuations.

Populations with Like Habitat. Frequently, a comparison of populations in different areas having similar plant associations and physiographic features can meet all three of the objectives indicated. One must, however, measure the populations by the same method, at the same time of day and season, and in areas where the vegetation complex and physiographic features are as closely alike as the minimum factors of natural variation will allow.

Populations of Dissimilar Habitat. Sometimes it is worthwhile to compare populations with dissimilar habitat (e.g., beech-hemlock forest and oak-maple forest). The results serve mainly to show the differences in population density and composition. The populations must be measured by the same method at the same time of day and season.

Successional Populations. Biotic communities succeed one another in a given area until the dominant or climax community is established. (For a discussion of this phenomenon, see the section of this book, "Distribution," page 208.)

A study of changes in bird populations, associated with community succession in one area, is largely impracticable because of the great number of years involved. However, one may analyze bird populations in a region where different stages in a succession can be found in adjoining areas at one time. Such a study is valuable principally in demonstrating the changes in species composition. Comparison of the population densities and composition of different communities is generally difficult because of the merging of communities, but it can be done if one is careful to select typical stages of succession.

Seasonal Populations in the Same Habitat. A very common type of comparison is that of populations in the same habitat in different seasons of the year. If one selects an area of determined size, and studies its population by the same method in different seasons, he can determine changes in population density and composition from season to season.

Yearly Populations in the Same Habitat. The analysis of populations in the same habitat in the same season, or seasons, of successive years is useful for showing any fluctuations in density or changes in composition that may have occurred. Obviously, one must employ the same methods of measurement from year to year.

THE STUDY OF BIRD POPULATIONS

The following directions for four studies of bird populations will acquaint the student with several fundamental procedures. For the best results several students should work together.

1. Direct Counting

Visit a large colony of nesting birds, preferably ground-nesting species such as pelicans, cormorants, gulls, terns, or skimmers. Select the time of year when the nests are likely to contain either complete clutches of eggs, or newly hatched young.

Procedure. Members of the student group line up abreast, several feet apart. (The exact spacing depends on the particular circumstances of terrain and visibility of nests. Ordinarily ten feet apart is a workable distance.) Then move back and forth through the colony until the entire nesting area has been covered. The person on the right of the line is considered the leader; upon him rests the responsibility of making certain that no area is passed over twice and that no area is slighted. As the lines move—always straight abreast—each person notes the contents and counts the number of nests that occur between himself and the person to his left. Empty nests, destroyed eggs, and dead young should be separately recorded. If more than one species nests in the colony, keep separate records for each.

Tabulation of Results. At the conclusion of the census the figures obtained by the group are assembled by the leader. Later they may be tabulated on the suggested chart, "Census of a Nesting Colony," page 408. If previous censuses have been conducted in the same colony, it will be worthwhile to compare the results. Undoubtedly certain fluctuations will be noted. On the basis of the figures, observations, and other information, attempt to determine the causes of the fluctuations.

2. Plot Census of Breeding Birds

Select a square plot of woods (preferably with uniform plant association) of approximately 50 acres. Make at least five different counts at intervals during the breeding season between the time of territory establishment and the time when the young leave the nest. Conduct the counts early in the day when all males present are likely to be singing.

Procedure. Lay out a plot in grid form by first measuring off (either by pacing or by using a measured rope) a straight line 1,456 feet long. Use a compass to make certain that the line is straight. Then measure off another line of the same length, parallel to the first, 208 feet away. Continue measuring off additional 1,456-foot parallel lines until altogether eight such lines, 208 feet apart, have been marked. Make certain that the lines are permanently indicated by conspicuous markers (see Appendix A, page 432) conveniently spaced. Next lay out eight 1,456-foot lines 208 feet apart at right angles to the first eight lines, thus subdividing the plot into 49 units, each 208 feet square. Where the lines intersect, use numbered markers, conspicuously positioned, preferably on posts.

After the plot has been laid out, make a grid map. Number intersecting lines as on the plot. Indicate topographical features, peculiarities of vegetation, and various landmarks. Show the position of the plot in relation to the points of the compass. After the map is complete, make copies available to each student.

Students, when ready to make the first count, walk across the plot following different parallel lines. If there are enough students for all parallel lines in one direction, only one crossing a day is necessary. However, they must move at the same time and in the same direction. Each student is equipped with a copy of the map. Indicate on the map the places where species are found and their breeding status or activity. Use abbreviations and letters. Abbreviate bird names thus: Black-and-white Warbler = B-w War; White-throated Sparrow = W-t Sp; Red-eyed Vireo = R-e V. Indicate observations regarding status or activity by letters in parentheses after the abbreviations of names, thus: Singing male, pair, or bird not singing but obviously paired = (P); transient individual moving over or through plot but not residing = (T); occupied nest = (N); young bird = (Y).

When the second and succeeding counts of the plot are made, use a new map and follow the same procedure.

Tabulation of Results. When the five counts along the parallel lines have been completed, assemble the various maps and record the results for each count with crayons of different colors on a composite map. Tabulate the results using the chart suggested on page 408, "Plot Census of Breeding Bird Population." On this chart, list the species in the sequence of *The A. O. U. Check-list.* Record the number of pairs, transients, nests, and young in the adjoining columns. The following hypothetical examples demonstrate the method: One Great Blue Heron seen flying over the plot = 1 (T); three Red-eyed Vireos heard singing and one nest found = 3(P) 1 (N); one adult White-throated Sparrow and one young found = 1 (P) 1 (Y).

In estimating totals, include only pairs. Count singing males as one pair, a nest as another pair, etc. However, avoid counting a singing male and a nest as two pairs when there is actual evidence that they represent only one pair of breeding adults. The matter of knowing when to count, for example, a singing male and a nest as one pair or two is left to the judgment of the census-taker.

3. Strip Census

Make a census of an area that shows one or more habitats widely represented. Make it in the early morning when birds are normally more active. If possible, repeat the census on several successive mornings, thereby increasing the chances of including all the more inconspicuous birds.

Procedure. Lay out a route and measure its total length. Measure also the approximate length of the route through each habitat. Then pass along the route, recording all birds observed by sight and sound within 100 feet of each side of the route. Use a trail or road if it traverses a promising area. For details on making a census along a road, consult Howell (1951).

Tabulation of Results. Tabulate the results on the suggested chart, "Strip Census of a Bird Population," page 409. Do not attempt to compute the number of individuals of each species observed but rather record the number of species and total number of individuals observed.

4. Measurement of Relative Abundance.

While making trips, either for the purpose of census-taking or for other field work, it is worthwhile to record and compute the relative abundance of birds in various regions traversed. This information, obtained over a prolonged period of time, will yield a picture of bird distribution and fluctuations in numbers. Considering the value of the information, the amount of time and effort expended in obtaining it is relatively slight.

Relative abundance may be determined in numerous ways and under different conditions. The following two procedures are suggested, since they suit an average class trip, but they may be modified to meet particular circumstances.

First Procedure

While walking through an area, mark on a daily field check-list the species observed. Use a new list for each walk taken. Note the date, the length of time involved on the trip, and the biotic communities passed through. Devise a method for indicating on the check-list the communities in which the birds were observed.

Tabulation of Results. At the conclusion of the season compute the results in one of several ways. (1) To determine the relative abundance of all species observed, divide the number of trips (or days) a species is observed by the total number of trips (or days). Then list the species in the order of percentage figures from the highest to the lowest, thus giving a composite view of relative abundance. The suggested chart on page 409, "Relative Abundance of Birds in a Region," may be used for tabulating the results. (2) To determine the relative abundance of species in similar biotic communities, divide the number of times a species is observed in a community by the total number of times the community was visited. Then list the species for each community in the order of percentage figures on another chart of like design.

If it is desirable to use terms such as "abundant," "common," etc. for indicating relative abundance, the following method and symbols are suggested. Species with a frequency rating of from 90% to 100% = Abundant (A); from 65% to 89% = Common (C); from 31% to 64% = Moderately Common (MC); from 10% to 30% = Uncommon (UC); from 1% to 9% = Rare (R).

Second Procedure

When automobile transportation is necessary from the laboratory to the point of starting field work, one can frequently record birds observed along the highway. Use a check-list for marking off the species observed. Indicate the date and the particular region traversed. Note the mileage on the speedometer at the start and conclusion of the trip and compute the total number of miles covered.

Tabulation of Results. At the conclusion of the trip, divide the number of times a species was observed by the number of miles covered. The results will have obvious limitations; nevertheless they may be of value in giving some conception of the abundance of the more conspicuous species. If several trips are made through the same region during the course of the season, divide the total number of times a species is observed by the total number of miles covered. This will give a somewhat more accurate indication of the abundance of conspicuous species. Use the suggested chart, "Relative Abundance of Birds in a Region," page 409.

CENSUS OF A NESTING COLONY

	Species	Species	Species	Species						
Species in colony:										
Location of colony:										
Estimated size (in acres) of area containing nests:	Date of census									
Length of time involved in census	Number of persons making census:									
Total number of nests:										
Total number with eggs only:										
with one egg:										
with two eggs:										
with three eggs:										
with four eggs:										
with five eggs:										
Total number with one egg and one young:										
with three combination:										
with four combination:										
with five combination:										
Total number with young only:										
with one young:										
with two young:										
with three young:										
with four young:										
with five young										
Total number of empty nests:										
Total number of eggs and young:										
Average number of eggs and young per nest (empty nests not included):										
Estimated total of breeding adults:										
Estimated number of nests per acre:										
Approximate ratio of adults to eggs and young:										
Total number of eggs found destroyed:										
Total number of young found dead:										
Comments or notes:										

PLOT CENSUS OF BREEDING BIRD POPULATION

General location of plot:

Size of plot:

Classification of habitat:

General topography:

	First Trip	Second Trip	Third Trip	Fourth Trip	Fifth Trip
Date of census trip:					
Time of day:					
Temperature:					
Precipitation:					
Wind:					
Weather rating:					

Species	First Trip	Second Trip	Third Trip	Fourth Trip	Fifth Trip

Total number of pairs:

Number of pairs per 100 acres:

Estimated total number of pairs based on above censuses:

Final total of pairs per 100 acres:

RELATIVE ABUNDANCE OF BIRDS IN A REGION

Region covered:

How covered:

Dates of trips:

Number of observers:

Brief description of region covered:

Species in Order of Abundance	Number of Times seen	Percent of Abundance	Symbol of Abundance

STRIP CENSUS OF A BIRD POPULATION

General location of census:

General topography:

Total length and width of strip:

Date and time of day of census:

Number of persons participating:

Temperature: Precipitation:

Wind velocity: Weather rating:

Habitat	Length of habitat strip	Number of species	Estimated number of species per 100 acres	Estimated number adult individuals per 100 acres

Comments and notes:

REFERENCES

The student's attention is called to *Audubon Field Notes* (see Appendix G), which contains reports of the annual Christmas Bird Count and Breeding-bird Census in the April and December numbers, respectively. For an evaluation of the Christmas Bird Counts, see the paper by Stewart (1954). If the student is interested in making censuses, he will enjoy participating in these nation-wide projects. Current numbers of the journal give instructions on how to take part in them.

Listed below are important publications dealing with population problems and census-taking. The student desiring to adopt procedures more particularly suited to his purposes should consult them. For a comprehensive survey of the subject of census-taking, see Overton and Davis (1969).

Amman, G. D., and P. H. Baldwin
  1960 A Comparison of Methods for Censusing Woodpeckers in Spruce-Fir Forests of Colorado. Ecology, 41: 699-706.

Austin, O. L.
  1938 Some Results from Adult Tern Trapping in the Cape Cod Colonies. Bird-Banding, 9: 12-25.

Bergstrom, E. A.
  1956 Extreme Old Age in Birds. Bird-Banding, 27: 128-129.

Breckenridge, W. J.
  1935 A Bird Census Method. Wilson Bull., 47: 195-197. (A census of a square mile area. The author worked alone, traversing the area along compass lines.)

Clapp, R. B., and F. C. Sibley
  1966 Longevity Records of Some Central Pacific Seabirds. Bird-Banding, 37: 193-197.

Colquhoun, M. K.
  1940 The Density of Woodland Birds Determined by the Sample Count Method. Jour. Animal Ecol., 9: 53-67.

Davis, D. E.
  1951 The Analysis of Population by Banding. Bird-Banding, 22: 103-107.
  1960 A Chart for Estimation of Life Expectancy. Jour. Wildlife Management, 24: 344-348.

Dice, L. R.
  1930 Methods of Indicating Relative Abundance of Birds. Auk, 47: 22-24.

Eberhardt, L. L.
  1969 Population Analysis. In *Wildlife Management Techniques*. Third edition, revised. Edited by R. H. Giles, Jr. Wildlife Society. Washington, D. C.

Errington, P. L.
  1945 Some Contributions of a Fifteen-year Local Study of the Northern Bobwhite to a Knowledge of Population Phenomena. Ecol Monogr., 15: 1-34.

Farner, D. S.
  1955 Birdbanding in the Study of Population Dynamics. In *Recent Studies in Avian Biology*. Edited by A. Wolfson. University of Illinois Press, Urbana.

Fisher, J.
  1951 Watching Birds. Revised edition. Hammondsworth, Middlesex.
  1954 A History of Birds. Houghton Mifflin Company, Boston. (Chapters 8, "Absolute Numbers of Birds"; 9, "Comparative Numbers of Birds"; 10, "Changing Bird Populations"; 11, "Bird Numbers and Man.")

Fisher, J., and R. T. Peterson
  1964 The World of Birds. Doubleday and Company, Garden City, New York.

Flower, S. S.
  1938 Further Notes on the Duration of Life in Animals. Proc. Zool. Soc. London, 108 (Ser.A): 195-235.

Gibb, J. A.
  1961 Bird Populations. In *Biology and Comparative Physiology of Birds*. Volume 2. Edited by A. J. Marshall. Academic Press, New York.

Gibson, J. A.
  1950 Methods of Determining Breeding-cliff Populations of Guillemots and Razorbills. Brit. Birds, 43: 329-331.

Graber, R. R., and J. W. Graber
  1963 A Covarative Study of Bird Populations in Illinois, 1906-1909 and 1956-1958. Illinois Nat. Hist. Surv. Bull., 28: 377-528. (A classic example of how populations in the same areas may be compared after a lapse of many years —in this case fifty.)

Hensley, M. M., and J. B. Cope
  1951 Further Data on Removal and Repopulation of the Breeding Birds in a Spruce-Fir Forest Community. Auk, 68: 483-493.

Hiatt, R. W.
1942 A Frequency Distribution of Eastern and Western Kingbirds in Montana. Great Basin Nat., 3: 109-114.
1944 A Raptor Census in Montana. Amer. Midland Nat., 31: 684-688.

Hickey, J. J.
1943 A Guide to Bird Watching. Oxford University Press, New York. (Chapter 3, "Adventures in Bird Counting.")
1955 Some American Population Research on Gallinaceous Birds. In *Recent Studies in Avian Biology*. Edited by A. Wolfson. University of Illinois Press, Urbana.

Hosley, N. W., and others
1936 Forest Wildlife Census Methods Applicable to New England Conditions. Jour. Forest., 34: 467-471.

Howell, J. C.
1951 The Roadside Census as a Method of Measuring Bird Populations. Auk, 68: 334-357.

Kadlec, J. A., and W. H. Drury
1968a Aerial Estimation of the Size of Gull Breeding Colonies. Jour. Wildlife Management, 32: 287-293. (By sight and photography, but neither method "will reliably detect annual changes of less than about 25 percent.")
1968b Structure of the New England Herring Gull Population. Ecology, 49: 644-676. (An analysis of age structure, population increase, and mortality rate.)

Kashkarov, D. N.
1927 The Quantitative Method in the Field Study of Vertebrate Fauna and Analysis of Data Obtained. Acta Universitatis Asiae Mediae, Ser. 8a Fasc., 1: 1-24. (Written in Russian with an English summary.)

Kendeigh, S. C.
1944 Measurement of Bird Populations. Ecol. Monogr., 14: 67-106.
1947 Bird Population Studies in the Coniferous Forest Biome during a Spruce Budworm Outbreak. Dept. Lands and Forests, Ontario, Canada. Div. Res. Biol. Bull. No. 1.
1948 Bird Populations and Biotic Communities in Northern Lower Michigan. Ecology, 29: 101-114.

Kenoyer, L. A.
1927 A Study of Raunkaier's Law of Frequence. Ecology, 8: 341-349.

Kluijver, H. N.
1951 The Population Ecology of the Great

Tit, *Parus m. major* L. Ardea, 39: 1-135.

Lack, D.
1954 The Natural Regulation of Animal Numbers. Oxford University Press, London.
1966 Population. Studies of Birds. Oxford University Press, London.

Leopold, A.
1933 Game Management. Charles Scribner's Sons, New York. (Chapters 6 and 7 pertain to game censuses, surveys, etc.)

Lincoln, F. C.
1930 Calculating Waterfowl Abundance on the Basis of Banding Returns. U. S. Dept. Agric. Circ. 118.

Linsdale, J. M.
1928 A Method of Showing Relative Frequency of Occurrence of Birds. Condor, 30: 180-184.
1936 Frequency of Occurrence of Summer Birds in Northern Michigan. Wilson Bull., 48: 158-163.

Lockley, R. M.
1953 Puffins. Devin-Adair Company, New York.

Mayfield, H.
1953 A Census of the Kirtland's Warbler. Auk, 70: 17-20.

McClure, H. E.
1944 Censusing Pheasants by Detonations. Jour. Wildlife Management, 8: 61-65.

Merikallio, E.
1958 Finnish Birds: Their Distribution and Numbers. Fauna Fennica, 5.

Miller, A. H.
1943 Census of a Colony of Caspian Terns Condor, 45: 220-225.

Murphy, R. C.
1925 Bird Islands of Peru: The Record of a Sojourn on the West Coast. G. P. Putnam's Sons. New York.

Nice, M. M.
1927 Seasonal Fluctuations in Bird Life in Central Oklahoma. Condor, 29: 144-149.
1934 A Hawk Census from Arizona to Massachusetts. Wilson Bull., 46: 93-95.
1957 Nesting Success in Altricial Birds. Auk, 74: 305-321.

Nice, M. M., and L. B. Nice
1921 The Roadside Census. Wilson Bull., 33: 113-123.

Nicholson, E. M.
1931 The Art of Bird-watching: A Practical Guide to Field Observation. H. F. and G. Witherby, London. (Chapter 3, "The Bird Census Work.")

Odum, E. P.
1959 Fundamentals of Ecology. Second edition. W. B. Saunders Company, Philadelphia. (Chapters 5, "Introduction to Population and Community Ecology"; 6, "Principles and Concepts Pertaining to Organization at the Species Population Level.")

Overton, W. S., and D. E. Davis
1969 Estimating the Numbers of Animals in Wildlife Populations. In *Wildlife Management Techniques*. Third edition, revised. Edited by R. H. Giles, Jr. Wildlife Society, Washington, D. C.

Peterson, R. T.
1941 How Many Birds Are There? Audubon Mag., 43: 179-187.
1964 Birds over America. New and revised edition. Dodd, Mead and Company, New York.

Pettingill, O. S., Jr.
1939 History of One Hundred Nests of Arctic Tern. Auk, 56: 420-428.
1967 A 36-year-old Wild Herring Gull. Auk, 84: 123.

Pitelka, F. A.
1942 High Population of Breeding Birds within an Artificial Habitat. Condor, 44: 172-174.

Root, O. M.
1942 The Bog Birds of Cheboygan County, Michigan. Jack-Pine Warbler, 20: 39-44.

Rosene, W.
1951 Breeding Bird Populations of Upland Field Borders. Jour. Wildlife Management, 15: 434-436.

Rydzewski, W.
1962 Longevity of Ringed Birds. Ring, 3: 147-152.
1963 Longevity Records II. Ring, 3: 177-181.

Saunders, A. A.
1936 Ecology of the Birds of Quaker Run Valley, Allegany State Park, New York. New York State Mus. Handbook 16.

Shelford, V. E.
1954 An Experimental Approach to the Study of Bird Populations. Wilson Bull., 66: 253-258.

Skutch, A. F.
1949 Do Tropical Birds Rear as Many Young as They Can Nourish? Ibis, 91: 430-455.

Stamm, D. D., D. E. Davis, and C. S. Robbins
1960 A Method of Studying Wild Bird Populations by Mist-netting and Banding. Bird-Banding, 31: 115-130.

Stewart, P. A.
1954 The Value of the Christmas Bird Counts. Wilson Bull., 66: 184-195.

Stewart, R. E., and J. W. Aldrich
1949 Breeding Bird Populations in the Spruce Region of the Central Appalachians. Ecology, 30: 75-82.
1951 Removal and Repopulation of Breeding Birds in a Spruce-Fir Forest Community. Auk, 68: 471-482.

Studholme, A. T., and R. T. Norris
1942 Breeding Woodcock Populations. Auk, 59: 229-233.

Terres, J. K.
1968 Flashing Wings: The Drama of Bird Flight. Doubleday and Company, Garden City, New York.

Thompson, St. C.
1951 The Southeastern Cooperative Dove Study. Trans. 16th N. Amer. Wildlife Conf., pp. 296-306.

Tickell, W. L. N.
1968 The Biology of the Great Albatrosses, *Diomedea exulans* and *Diomedea epomophora*. American Geophysical Union, Antarctic Res. Ser., 12: 1-55.

Twomey, A. C.
1945 The Bird Population of an Elm-Maple Forest with Special Reference to Aspection, Territorialism, and Coactions. Ecol. Monogr., 15: 173-205.

Wagner, H. O.
1957 Variation in Clutch Size at Different Latitudes. Auk, 74: 243-250.

Westerskov, K.
1963 Ecological Factors Affecting Distribution of a Nesting Royal Albatross Population. Proc. XIIIth Internatl. Ornith. Congr., pp. 795-811.

White, K. A.
1942 Frequency of Occurrence of Summer Birds at the University of Michigan Biological Station. Wilson Bull., 54: 204-210.

Williams, A. B.
1936 The Composition and Dynamics of a Beech-Maple Climax Community. Ecol. Monogr., 6: 317-408.

Wynne-Edwards, V. C.
1955 Low Reproductive Rates in Birds, Especially Sea-birds. Proc. XIth Internatl. Ornith. Congr., pp. 540-547.

Young, H.
1949 A Comparative Study of Nesting Birds in a Five-acre Park. Wilson Bull., 61: 36-47.

ANCESTRY, EVOLUTION, AND DECREASE OF BIRDS

Despite the astronomical numbers of birds that have lived and died in the passing of millions of years, the fossil record is remarkably meager. The only logical explanation seems to be that bird bones are usually so small and fragile that they are completely destroyed through ingestion by predators, and to a lesser extent by scavengers or erosion. Most bones that by rare chance escaped destruction, became trapped in mud, silt, or some other soft substance, and then fossilized are at best a few from a leg or wing, occasional vertebrae, and fragments from a skull or sternum. The preservation of an entire skeleton is unusual.

Owing to this lack, the fossil record provides no direct indication of the ancestry of birds, no succession of forms from some primitive vertebrate animal. One can only speculate on the ancestry of birds, starting with the evidence at hand and searching backward.

SPECULATIONS ON ANCESTRY

Undeniably, birds arose from reptilian stock. In the first place, modern birds and reptiles show numerous physical features in common. For example, their skulls bear one rather than two occipital condyles for articulation with the vertebral column; their lower mandibles articulate with movable quadrate bones as in lizards and snakes instead of being hinged directly on the skull; their ears have a single bone, the columella, for conducting vibrations from the tympanum across the middle ear to the internal ear.

In the second place, there is the highly important evidence of reptilian ancestry from *Archaeopteryx lithographica*, an ancient winged creature whose fossil remains were found at four different times in lithographic slate near Solnhofen, Bavaria. The first discovery, in 1860, comprised the imprint of an undoubted feather, and the second, the next year, consisted of an incomplete skeleton (now in the British Museum) showing particularly forearm and leg bones, a long tail with many vertebrae and the impressions of feathers, some attached to the forearms, and a pair each to most of the tail vertebrae. The third discovery, in 1877, was a skeleton (now in a Berlin museum), virtually complete except for the lower mandible, the right foot, and some of the cervical vertebrae. The fourth discovery, in 1958, was another skeleton (now at the University of Erlangen), much less complete than the two found 80 years or so earlier.

A little larger than the Common Pigeon *(Columba livia)*, *Archaeopteryx* lived in the Upper Jurassic Period (see the Geologic Time Scale, next page), some 140 million years ago, at a time when reptiles were dominant vertebrate animals. Without the tell-tale feathers in the fossil remains, the discoverers might have identified it as just another reptile from that remote age. But now that numerous authorities (e.g., Heilmann, 1927; de Beer, 1954) have scrutinized and evaluated the 1861 and 1877 specimens and studied the 1861 specimen by direct and indirect lighting and by ultra-violet and X-rays, they consider *Archaeopteryx* more bird than reptile with nonetheless remarkable and—from and evolutionary viewpoint—significant features intermediate between the two.

GEOLOGIC TIME SCALE
(Starting with Permian Period)

Era	Period	Epoch	Age at Start (in years)
Cenozoic	Quaternary	Recent	15,000
		Pleistocene	1,000,000
	Tertiary	Pliocene	10,000,000
		Miocene	25,000,000
		Oligocene	40,000,000
		Eocene (including Palaeocene)	65,000,000
Mesozoic	Upper Cretaceous		100,000,000
	Lower Cretaceous		135,000,000
	Upper Jurassic		140,000,000
	Lower Jurassic		180,000,000
	Triassic		225,000,000
Palaeozoic	Permian		270,000,000

The skeleton in general is strongly avian in character, yet no bone reveals any evidence of being pneumatic.

The skull has such reptilian features as an overall heavier structure, large fossae in the facial region, no bill, and pointed teeth in sockets. Also, the brain-box shows that the cerebral hemispheres were elongated, the optic lobes dorsal in position, and the cerebellum small and not extended forward to overlap the posterior parts of the cerebral hemispheres. The orbit, however, is relatively large and contains a sclerotic ring.

The vertebral column, comprised of 10 cervical, 12 thoracic and lumbar, 6 sacral, and 20 caudal vertebrae, is distinctly reptilian, with little if any fusions of vertebrae. The vertebral centra are biconcave or amphicoelous—simpler than the saddle-shaped or heterocoelous type in modern birds. The caudal vertebrae are elongated and free, with no evidence of their fusing terminally to form a pygostyle. All the ribs are unjointed, lack uncinate processes, and do not reach the sternum. The sternum itself is short, broad, and unkeeled, while posterior and lateral to it are about 12 pairs of dermal ribs or gastralia which are present in the ventral abdominal wall of some reptiles.

The pectoral girdle shows, besides a scapula and coracoid, a clavicle fusing with its fellow of the opposite side to form a distinctly avian feature, the furcula or wishbone.

The forelimb is as long as the hindlimb. Its humerus is longer than the ulna, and the ulna is a little longer and stouter than the radius, but none of the three bones is very strongly developed. In the wrist are five carpals, two (the radiale and ulnare) proximal and three distal. The hand has three metacarpals. Whereas modern birds have only the radiale and ulnare separate and the other carpals and the metacarpals fused as one bone, the carpometacarpus, *Archaeopteryx* shows all the carpals and metacarpals separate with the possible exception of the outermost distal carpal being fused with the outermost metacarpal. Articulating with the metacarpals are three separate digits or fingers (the middle one longest) terminating in conspicuously long, curved, sharply pointed claws. Each digit is comprised of more phalanges than its homologue in modern birds.

The pelvic girdle is connected to, rather than fused with, the sacral vertebrae, and its three bones—the ilium, ischium, and pubis—are separate and individually identifiable instead of being fused into one bone. The pubis is distinctly avian by being long, slender, and directed posteriorly. At the distal end it fuses with its fellow of the opposite side to form a symphysis.

In the hindlimb the fibula is a separate bone alongside the tibia and extends all the way from the femur to the heel. The fibula and tibia are longer than the femur. The second, third, and fourth metatarsals remain separate and parallel for most of their length distally; only at their proximal ends are they fused with one another and certain tarsals to form the tarsometatarsus. As in the Common Pigeon and most other modern birds capable of perching, there are four digits or toes, the first (hallux) projecting backward. The fourth or outermost digit has four phalanges instead of five.

However strikingly bird-like some of the skeletal features of *Archaeopteryx* may be, none is quite so distinctly avian as the accompanying feathers. Careful studies of their impressions in lithographic slate show them to be in every way typical of modern flying birds. The forearm appears to have borne true primaries (as many as eight, according to Savile, 1957) and ten secondaries, each with overlapping coverts. Although the tail is more reptilian than avian in length and number of vertebrae, it supports rectrices—one pair with coverts attached to each of 15 caudal vertebrae beginning with the sixth. Besides the feathers of the forearms and tail are the impressions of numerous other feathers which, from their positions and size, clearly suggest that the body, neck, and legs had a feather covering.

Nobody knows precisely what *Archaeopteryx* looked like in life, how it was shaped and whether or not it was brightly or somberly colored. Numerous artists, however, have reconstructed it partly from anatomical evidence and the rest from supposition. See, for example, some very meticulous attempts in the paintings by Heilmann in the frontispiece to his book (1927), by Roger Tory Peterson (*in* Fisher and Peterson, 1964), and by Rudolf Freund in the frontispiece to this book.

While *Archaeopteryx* shows beyond a doubt that birds arose from reptilian stock, it is only to a limited extent helpful in providing clues to the ancestry of birds.

Existing contemporaneously with *Archaeopteryx* were two groups of reptiles—the pterosaurs (winged reptiles) and smaller bipedal (two-footed) dinosaurs—which shared many of the skeletal features with birds. The pterosaurs, while equipped with pneumatic bones as are modern birds, depended on wings with membranes and with bones of different number and proportions. The bipedal dinosaurs had long hindlimbs on which they could run while keeping their forelimbs off the ground. Their pectoral girdles, however, lacked certain features, including a furcula, that modern birds possess. Obviously, these two groups of reptiles were no more ancestral to modern birds than *Archaeopteryx*.

To find the ancestor common to birds (including *Archaeopteryx*), pterosaurs, and dinosaurs, one must search farther back in time for a more generalized form of reptile.

Speculation, based on current researches, points to the Pseudosuchia, a small group among the extinct thecodont (socket-toothed) reptiles which existed in the Old World during the early to middle Triassic Period some 200 million years ago.

The pseudosuchians were bipedal reptiles with hindlimbs that were longer than the forelimbs and used for walking, running, and jumping. Their skulls had many of the basic features found in birds. In one form, *Euparkeria capensis* from Africa, the skull had all the bony elements present in *Archaeopteryx*, although it was more heavily constructed with a higher profile and stronger jaws. The orbits and brain-box were smaller. In another form, *Ornithosuchus woodwardi* from Scotland, the skull was more lightly constructed and thus more closely resembled that of *Archaeopteryx*. The feet showed a tendency toward reduction in the fourth and fifth toes. As in other pseudosuchians, *Ornithosuchus* supported a paired row of scale-like epidermal plates on the dorsal surface of the back and tail.

Besides *Euparkeria* and *Ornithosuchus* there were many other similarly generalized forms of pseudosuchians existing at the time. It could have been from any one of these forms—or even from an older and still more generalized form which lived in the preceding Palaeozoic Era—that birds branched off, but more fossil evidence must come to light before there can be any certainty. The stock from which modern birds evolved has yet to be discovered.

EARLY EVOLUTION

If the numbers of early bipedal pseudosuchians were as vast as commonly believed, one can assume that they were subject to considerable competition for food as well as living space. Being carnivores by virtue of their socketed teeth, they were forced to seek small animals in every possible situation. At least one form of pseudosuchian and possibly more used their sharp claws to advantage in climbing trees where, through the continuing process of adaptive radiation, they developed physical attributes and techniques that enabled them to forage successfully. From an arboreal habit thus acquired, they proceeded (as suggested by Heilmann, 1927) to attain ever greater proficiency by jumping from branch to branch. This required the simultaneous development of more muscular power in their hindlimbs for leaping and of opposable first toes (halluces) for grasping.

The next probable step in development was the extension of jumping movements to gliding from higher to more distant lower branches and eventually from high in one tree down to the lower part of another. The evolutionary consequences of these feats were the gradual acquisition of sailing surface through the development of patagia between the brachia and antebrachia on the leading edge of the forelimbs, the expansion and elongation of scales from the trailing edge of the forelimbs, and (for balance as well as sailing surface) the expansion and lateral projection of the scales along the tail. In due course, the scales became lengthened and elaborated to form contour feathers.

Archaeopteryx well represents that stage in the evolution toward flight where the subject became adapted to an arboreal life and capable of extensive gliding. Its feet were suited to grasping and perching; its forelimbs, while retaining some grasping ability as evidenced by the three-clawed fingers of the hands, were elongated and more suited to supporting the body in the air. The great extent to which *Archaeopteryx* depended on gliding is reflected by a number of anatomical features—for example, the lightening of the skull; the increase in size of the orbits to accommodate larger eyes for greater distance perception and visual acuity; the larger brain-box to accommodate the enlargement of those brain centers that control and coordinate the more complicated neuro-muscular mechanisms for gliding.

With true contour feathers already formed in *Archaeopteryx*, and with no evidence in the few fossil remains of early reptiles that even remotely suggest how they formed, the origin of feathers is a fruitful field for speculation (see a brilliant paper on this subject by Parkes, 1966). The general consensus is that feathers developed first on the forelimbs and tail of an early arboreal reptile by the lengthening, broadening, and overlapping of those scales that would increase sailing surface and at the same time be lighter without sacrificing strength and durability. Since the trend toward lightness is of selective advantage to any airborne animal, it was quite natural for the modifying process to involve the remaining scales on the forelimbs and tail and eventually all the scaly covering of the body, neck, and so on. A secondary advantage to feathers was no doubt to streamline the body by smoothing over angular surfaces and folds from head to tail. All of the feathers were contour feathers and developed initially as an adaptation to gliding. All other kinds of feathers—down feathers (both teleoptilian and neossoptilian) and filoplumes—found in modern birds presumably evolved from the contour feathers of the type in *Archaeopteryx*.

After *Archaeopteryx* comes a wide gap in the fossil record until the Cretaceous Period some 20 million years later. In this long interval the evolution of birds must have accelerated, for by the Cretaceous, if one may judge by the several species discovered, birds were not only well established and globally distributed but had taken divergent paths of development, many becoming aquatic and many others remaining terrestrial. Among the species were the first known representatives of the Pelecaniformes and Ciconiiformes and representatives of other orders, now completely extinct, that are known by fossil material from sedimentary rocks laid down by the great inland sea which occupied the west-central portion of North America for some 40 million years. Nearly complete skeletons have been found for some of these birds, notably *Ichthyornis victor* and *Hesperornis regalis*.

Ichthyornis, a superficially gull-like bird about the size of a Common Pigeon, was obviously capable of sustained flight, for it had well-developed wings with a carpometacarpus fully formed, strong pectoral girdles amply supported by a furcula, and a keeled sternum for the attachment of large muscles to control the wing strokes. *Hesperornis*, a huge bird five to six feet long, was highly specialized for diving—perhaps the equal of loons in diving ability—and totally flightless. Far back on its elongated body the hindlimbs, with their paddle-like feet, projected at right angles to the body axis. Probably it could maneuver on land only to the extent of reaching its nest. The wings were vestigial, with no bones distal to a slender humerus; the pectoral girdles were not only weak but the clavicles failed to meet at the midline; and the sternum was flat and unkeeled. Whatever powers of flight the bird once had were long since lost.

The skeletal structure of *Ichthyornis* and *Hesperornis*, though widely divergent for purposes correlated with locomotion, shared certain similarities such as a typical avian tarsometatarsus, a small skull with bones well fused, and the reduction in the number, and the fusing of, the terminal caudal vertebrae to form a pygostyle. At the same time they showed certain differences that cannot be explained. Thus in *Hesperornis* both mandibles retained reptilian teeth, fewer in the upper mandible than the lower, while in *Ichthyornis* apparently no teeth persisted. *Hesperornis* had heterocoelous vertebrae characteristic of modern birds, but *Ichthyornis* had amphicoelous (bi-concave) vertebrae that were peculiarly enjoined as they are in fish. This accounts for the name *Ichthyornis* —fish bird.

Ichthyornis was undoubtedly able to fly in the manner of modern birds and, according to the fossil record of other Cretaceous birds, was not alone in having achieved this ability. But just how it flew is a matter of conjecture.

Presumably, during the merging of the Upper Jurassic and Lower Cretaceous Periods, the feat of gliding, as exemplified by *Archaeopteryx*, was rather quickly extended for increasingly longer distances by flapping the forelimbs, thus sustaining the weight of the body against gravity. This gave the performer such selective advantages as capturing insects on the wing and more readily escaping from enemies and unfavorable aspects of the environment. In the development and achievement of flight, the long tail that was advantageous only in sailing was foreshortened and its feathers directed posteriorly and spread fan-like to provide a mechanism for steering and braking; the eyes were enlarged further to give still greater visual acuity; and the teeth played a steadily decreasing role and ultimately disappeared, allowing the mandibles to be lighter in weight.

The attainment of flight was unquestionably accompanied by a heightening of body metabolism through modifications in the organ systems. Cretaceous birds such as *Ichthyornis* and *Hesperornis* must have been homoiothermous (warm-blooded) but how early in their development birds acquired a thermoregulatory mechanism is problematical. Some authorities contend that *Archaeopteryx* was at least somewhat homoiothermous, as were its pseudosuchian ancestors and the pterosaurs, and that the feather covering of *Archaeopteryx* developed to hold its body heat as it took to an arboreal life in which there was more exposure to lower air temperature. But this argument is hardly valid, if, as generally believed, the climate at the time of *Archaeopteryx* was uniformly warm, possibly tropical, the year round. The pterosaurs were able to fly successfully without any such special covering. It is more likely that the feather covering developed to facilitate gliding and later flight and that, as warm-bloodedness increased with the attainment of flight, the feather covering eventually assumed an insulating function.

LATER EVOLUTION

Although the fossil record of the Cretaceous is poor, advances in the avifauna must have been great for with the coming of the Tertiary Period toothed birds had disappeared and a host of new forms began to emerge. Among them are the earliest known representatives of a great number of modern families or orders that include the penguins (Spheniscidae), rheas (Rheidae), loons (Gaviidae), tropicbirds (Phaëthontidae), anhingas (Anhingidae), cormorants (Phalacrocoracidae), herons (Ardeidae), ducks (Anatidae), vultures (Cathartidae), hawks (Accipitridae), grouse (Tetraonidae), cranes (Gruidae), rails (Rallidae), sandpipers (Scolopacidae), Auks (Alcidae), cuckoos (Cuculidae), owls (Strigidae), swifts (Apodidae), trogons (Trogonidae), and the first few forms of Passeriformes.

The rate of avian evolution in the Eocene probably reached the all-time maximum in the history of birds. From their origin in Eurasia or Africa, birds had spread all over the world. One reason may have been that the earthly scene had changed. Gone were the dinosaurs, pterosaurs, and other early reptiles that had dominated the Cretaceous, leaving habitats more than ever available to birds. Besides the proliferation of many families of modern birds and many forms obviously ancestral to modern species, there also appeared several gigantic flightless birds which seem to have replaced the giant reptiles of the Cretaceous. They thrived until the emergence of the giant mammals, and then died out, leaving no known descendants. A notable example is *Diatryma steini* found in Wyoming. Specialized for terrestrial life, with massive, powerful legs, a huge head with an enormous bill, and small degenerate wings, it stood nearly seven feet tall.

The fossil deposits of the Oligocene in the mid-Tertiary yield, among other forms, the first known grebes (Podicipediformes), albatrosses (Diomedeidae), shearwaters (Procellariidae), storks (Ciconiidae), turkeys (Meleagrididae), limpkins (Aramidae), plovers (Charadriidae), stilts (Recurvirostridae), gulls (Laridae), pigeons (Columbidae), parrots (Psittacidae), kingfishers (Alcedinidae), and woodpeckers (Picidae) together with a few more passerine birds.

In the Miocene and Pliocene deposits, toward the close of the Tertiary, the fossil remains of birds are much more numerous and varied. While the majority represent forms already known, others are the first recorded evidences of ostriches (Struthionidae), tinamous (Tinamidae), falcons (Falconidae), oystercatchers (Haematopodidae), goatsuckers (Caprimulgidae), and nearly a dozen families of passerine birds.

From the Pleistocene Epoch or Ice Age come the first records of the cassowaries (Casuariidae), emus (Dromiceidae), elephant birds (Aepyornithidae), moas (Dinornithidae), kiwis (Apterygidae), ospreys (Pandionidae), jacanas (Jacanidae), phalaropes (Phalaropodidae), skuas (Stercorariidae), barn owls (Tytonidae), hummingbirds (Trochilidae), motmots (Momotidae), and many passerine species. Undoubtedly, these birds evolved much earlier, in the Tertiary, since most all species living today and those recently extinct are believed to have been in existence at the beginning of the Pleistocene (Wetmore, 1959) and well established in the Pleistocene (Howard, 1950). This may not have been the case with some of the "higher" passerines which are thought to have acquired their species distinctions as a result of the Pleistocene. For example, see Mengel (1964) or a partial summation of his paper in this book, page 139.

In its abundance and variety of forms, the fossil record of the Pleistocene is the richest of all the epochs. Besides 732 living species, represented by fossils, there are 270 extinct species. One, a condor-like bird, *Teratornis incredibilis*, with an estimated wingspread of 16 to 17 feet, may have been the largest bird ever to fly. Several Pleistocene deposits have yielded many hundreds of bird bones, but the most productive of any in the world are at Rancho La Brea near the center of Los Angeles, California, where during the Ice Age countless numbers of birds became entrapped in asphalt and their bones preserved in beds of tar.

THE NUMBER OF BIRDS IN THE PAST

From the fossil evidence and theoretical knowledge of environmental conditions that existed during the long history of birds, one can easily surmise that there were far greater numbers of species and individuals in past ages than at the present time; but when he attempts to determine numbers he is frustrated by the meagerness of the record. Speculation is his only recourse.

The rapid multiplication of bird species in the Eocene was favored not only by the decline of reptiles and the consequent availability of their habitats, but also by the persisting warm climate and the increasing development of seed plants (angiosperms) which formed immense forests, creating many new niches for occupancy. By the time of the Oligocene, mountains had begun to rise, and, as the land dried in their lee, forests soon decreased and grasslands formed. Here again were new habitats—in the mountains and on the plains—for avian radiation. Through the Miocene and into the Pliocene there was a slight cooling of the climate although it stayed warm or temperate much farther north than at the present time and there were no seasonal changes. But in the

Pleistocene and its succession of four glacial stages, with the long and warm interglacial periods, the relative uniformity of the climate ended and its effects on bird life were catastrophic. With each invading ice sheet, the prevailing temperature lowered. Birds were forced to shift and sometimes compress their ranges southward. Failing to cope with such radical changes, some species died out.

How many species of birds have been identified from fossil remains? Roughly 1,700, according to Austin (1961). Of this number, about 800 are still in existence and 900 are extinct. Adding the 900 extinct species to the 8,600 species known to be living in the world today, the total is 9,500.

Brodkorb (1960) theorized at great length on the numbers of bird species in past epochs and arrived at the maximum figure of 1,634,000 for all species evolved. This number may be excessively high or at best too speculative on the basis of the meager evidence at hand (e.g., see Moreau, 1966, for an evaluation of Brodkorb's calculations), but if it is a good approximation, the 9,500 species known today is about one-half of one percent or a mere fraction of the species that have existed.

Brodkorb correlated his figures for the different epochs with the changes in environment resulting from plant evolution and expansion of habitats, shifts in climate, fluctuating ocean levels, and so on. He suggested, for instance, that in the Miocene, when grasslands were widespread and ecologically important, grass-seed-eaters became a significant part of the avifauna and that it was in this period of time that the Fringillidae, the last family to develop, underwent its principal radiation.

LOSS OF FLIGHT

In the long history of birds, various forms have been characteristically flightless—that is, unable to fly because either they lacked wings altogether or they did not have wings large enough and powerful enough for sustained locomotion in the air. Most students of paleontology agree that this condition developed secondarily in all instances from stock at one time capable of flight. All birds, whether they can fly or not, are recognizable structurally as such because they evolved essentially as flying creatures.

Hesperornis, Diatryma, and a dozen or more other genera of birds, long since extinct, were flightless. Among modern birds and some only recently extinct are a considerable number of flightless birds, two well-known groups being the ratites and the penguins.

The ratites, so called from their unkeeled and consequently raft-like sternum, comprise the ostriches (Africa), rheas (South America), emus (Australia), cassowaries (New Guinea and Australia), kiwis (New Zealand), and the recently extinct moas (New Zealand) and elephant birds (Madagascar). All are terrestrial with strong legs and feet and all, except the kiwis, are large and heavy bodied. Many ornithologists feel confident that the birds arose independently in their respectively isolated areas and became similar through convergence; but a few (e.g., Bock, 1963) feel differently, believing that the birds arose together and then dispersed to remote areas, perhaps before losing their ability to fly. Regardless of how they arose, one may postulate that their ancestors took to living in treeless country where they became grazers. The cursorial habit of walking about for food and running to escape enemies became increasingly adequate and flight decreasingly essential; hence, their legs and feet grew stronger while their wings and keeled sternum degenerated. In some instances, birds, isolated on islands and free from predation by carnivores, grew to gigantic size. In New Zealand one of the 27 species of moas, *Dinornis maximus,* stood 12 feet tall and may have weighed about 500 pounds. Although the chicken-sized kiwis were similarly isolated in New Zealand, they never tended toward giantism; instead they reverted to a forest habitat where great size would have been only a handicap.

Unlike the ratites, penguins retained the keeled sternum and flight muscles because, in a sense, they simply re-adapted their locomotion from flying in the air to flying under water. Presumably, penguins evolved from flying aquatic birds—possibly stock ancestral also to the Procellariiformes (Simpson, 1946)—that reached the cool water ringing the Southern Hemisphere. Besides short, blade-like wings, or "flippers," suitable for propulsion, penguins acquired correlated structural features such as a streamlined or torpedo-shaped body with legs so far back that they must stand upright when walking; feathering almost scale-like in aspect, without apteria; and a thick skin over a heavy layer of fat that provides insulation against the cool environment. A few species of penguins eventually adjusted to the more frigid water adjacent to Antarctica and soon nested exclusively on its periphery and outlying islands.

The auks which tend to fill the niche in the Northern Hemisphere, occupied by penguins in the Southern, struck a compromise in their adaptation (Storer, 1960a). From presumed gull-like ancestors, all species, with one exception, derived wings small enough for under-water swimming, yet large enough for aerial flight. The exception was the Great Auk *(Alca impennis);* like the penguins, it forsook the air altogether for submarine flight.

THE DECREASE OF BIRDS

Over the ages since life began, species of animals and plants have arisen, flourished, and died out. This applies as much to birds as to other forms.

Taking a backward look at the long history of birds, one can readily see an overall increase in the number of species from the Upper Jurassic to the late Tertiary. In this great period of time the rate of increase exceeded the rate of extinction. If there was an Age of Birds, it must have been from the Miocene through the Pleistocene, a period of some 20,000,000 years. Since the Pleistocene, the rate of extinction is presumed to have exceeded the rate of increase. Brodkorb (1960) assembled data to show a reduction in species of at least 25 percent. Later Moreau (1966) strongly challenged this figure as groundless although he did not refute the concept of a decrease.

The extent to which prehistoric man contributed to the decrease was probably negligible. As Greenway (1958) fancifully stated, "man and birds arranged a means of living together to the end that no birds were extirpated." But there is no denying that in historic times man has played an awesome role in the extinction of birds.

The first species actually known to be eliminated by man is the Dodo *(Raphus cucullatus)* on the island of Mauritius in the Indian Ocean. In the 174 years following the discovery of the island by the Portuguese in 1507, men from European ships, and the cats, pigs, monkeys, and rats which they brought with them, succeeded in destroying the entire population. It was nothing less than miraculous that this flightless, ground-nesting species survived as long as it did.

The Great Auk was the first species on the coast of North America that man annihilated. Breeding, probably in colonies, on rocky, coastal islands in the North Atlantic, this flightless bird was readily accessible to roving sailors and fishermen. They took its eggs for food and slaughtered it for meat, feathers, oil, and codfish bait. The last two specimens of the Great Auk were taken on Eldey, a volcanic rock off the southwest coast of Iceland, on June 3, 1844.

At about this time on mainland North America two species, the Carolina Parakeet *(Conuropsis carolinensis)* and the Passenger Pigeon *(Ectopistes migratorius),* were on their way to extinction through excessive killing, but nobody yet realized it. Both species lived east of the Great Plains; both were gregarious, commonly existing in large flocks. The Carolina Parakeet, though perhaps never very abundant, was especially fond of fruit and consequently much despised by farmers who could kill large numbers easily. When a flock raided an apple tree, it was possible to shoot every bird because those individuals escaping the first blast from the gun hovered over those killed until they too were shot. The last specimen was killed in the wild on April 18, 1901. The Passenger Pigeon, at the time of the white man's arrival in North America, may have numbered 3,000,000,000 individuals—a population never attained by any other bird species known—and constituted between 25 and 40 percent of the bird population in the United States (Schorger, 1955). Despite this apparent security in numbers, the species was wiped out in the course of a century, the last wild specimen being recorded with certainty bètween September 9 and 15, 1899. By a remarkable coincidence, both the last captive Carolina Parakeet and the last captive Passenger Pigeon died in the same place in the same month and year—in the Cincinnati Zoological Garden in September, 1914 (Greenway, 1958).

The dramatic decline of the Carolina Parakeet and Passenger Pigeon in the last century eclipses the demise of another North American species, the Labrador Duck *(Camptorhynchus labradorius)*, the last recorded specimen of which was taken in the fall of 1875. It was never an abundant bird and it was never hunted extensively. Just what caused its extinction will never be known with any certainty.

In this century, the one remaining Heath Hen *(Tympanuchus cupido cupido)*, the eastern subspecies of the Greater Prairie Chicken, was last seen on March 11, 1932. Once prevalent along the Atlantic seaboard from New Hampshire south to Virginia, the Heath Hen became confined after 1869 to Martha's Vineyard, an island off the coast of Massachusetts. Here it survived in varying numbers, reaching a population close to 2,000 by 1916. But in the spring of that year a severe fire swept its breeding grounds, no doubt destroying many nests and nesting sites (Gross, 1928). In any case, its final decline began soon thereafter and was accelerated in the few remaining years by predation from cats and rats, diseases acquired from poultry, and, toward the end, an excessive ratio of males to females.

All told since 1681, the last year when the Dodo was alive, no less than 78 species and 49 well-marked subspecies have become extinct over the world. Fisher (1964), who compiled these figures, found "fairly strong direct proof" that man destroyed nearly half of the 78 species. Although Fisher makes no statement as to man's role in the destruction of the 49 subspecies, it was undoubtedly as great, if not even greater.

Fisher attributes the causes of extinction by man to be primarily (1) direct killing, (2) destruction of habitat, and (3) predation by cats, rats, and other human symbiotes. Another cause, but difficult to prove, is man's introduction of competing bird species.

Worthy of note from Fisher's figures is that only nine of the 78 extinct species and two of the 49 extinct subspecies were continental; all the others lived on islands. This clearly reflects the fact that insular birds, normally with small populations, are particularly sensitive to changes. Sometimes flightless, often quite tame, and usually with quite specialized feeding habits, they tend to lack the versatility to escape from, compete with, or adjust to man-imposed modifications in their natural environment.

Besides those species and subspecies known to be extinct there are a far greater number verging on extinction and, in some cases, may actually be extinct already. Listed on the following page are North American species and subspecies whose populations are under 100.

Species
 California Condor, *Gymnogyps californianus*
 Whooping Crane, *Grus americana*
 Eskimo Curlew, *Numenius borealis*
 Bachman's Warbler, *Vermivora bachmanii*
Subspecies
 Florida Everglade Kite, *Rostrhamus sociabilis plumbeus*
 American Ivory-billed Woodpecker, *Campephilus p. principalis*

In the following list are those North American species and subspecies with populations estimated to be not more than 1,000 and no less than 100.

Species
 Kirtland's Warbler, *Dendroica kirtlandii*
 Dusky Seaside Sparrow, *Ammospiza nigrescens*
 Cape Sable Seaside Sparrow, *Ammospiza mirabilis*
Subspecies
 Aleutian Canada Goose, *Branta canadensis leucopareia*
 Attwater's Greater Prairie Chicken, *Tympanuchus cupido attwateri*
 Yuma Clapper Rail, *Rallus longirostris yumanensis*

For bird species elsewhere in the world that are near or in danger of extinction, consult Fisher *et al.* (1969).

Now and then a bird, believed extinct, has been found still existing in some unexplored area or has simply been overlooked. For example, the Takahe, *Notornis mantelli*, a heavy, flightless gallinule which once lived over much of New Zealand's South Island was not reported after 1898 until rediscovered exactly fifty years later in a remote mountain valley (Williams, 1960). The rare Puerto Rican Whip-poor-will, *Caprimulgus noctitherus*, described from bones found in prehistoric cave deposits and from a single specimen collected in 1888, and seen alive only once—in 1911—thereafter went unreported by ornithologists until fifty years later, when it was rediscovered by a tape-recording of its call and from a collected specimen (Reynard, 1962). The giant race of the Canada Goose, *Branta canadensis maxima*, which breeds in the northern Great Plains, was rediscovered in 1962 after being considered extinct for three decades (Hanson, 1965).

While it is not impossible that even a few other supposedly extinct birds will some day show up, it is only wishful thinking to expect that many are still extant. The truth must be faced: Most of the species declared extinct are in fact gone forever.

Will other birds become extinct? Undoubtedly. Species have apparently been decreasing since the Pleistocene and no species can live on indefinitely. Indeed, every species has a normal life expectancy—perhaps as short as 16,000 years (calculation by Fisher, 1964; based on Brodkorb, 1960). The disturbing problem confronting ornithologists and other people interested in birds is how to blunt the human threat that may shorten life expectancy and hasten extinction.

The surging human population, now estimated at 3 billion and expected to be doubled by the year 2000, is not in itself a threat to the longevity of bird species. The threat comes in man's thoughtless abuse of the earthly environment. Promiscuous dissemination of poisons, pollution of air and water, total destruction rather than the selective use of habitats, indiscriminate creation of hazards to bird migration—all such actions, if they increase as the human population expands, can in a short time eliminate bird species by the score.

The main hope for birds lies in more aggressive conservation of the natural environment. Every student of ornithology should become a *militant* conservationist with a special objective—to apply his newly acquired knowledge about birds toward countering the threat of their early extinction.

REFERENCES

Austin, O. L., Jr.
1961 Birds of the World. Golden Press, New York.

Bock, W. J.
1963 The Cranial Evidence of Ratite Affinities. Proc. XIIIth Internatl. Ornith. Congr., pp. 39-54.

Brodkorb, P.
1960 How Many Species of Birds Have Existed? Bull. Florida State Mus., Biol. Sci., 5: 41-53.

1963 Catalogue of Fossil Birds. Part 1 (Archaeopterygiformes through Ardeiformes). Bull. Florida State Mus., Biol. Sci., 7: 179-293.

1964 Catalogue of Fossil Birds. Part 2 (Anseriformes through Galliformes). Bull. Florida State Mus., Biol. Sci., 8: 195-335.

1967 Catalogue of Fossil Birds. Part 3 (Ralliformes, Ichthyornithiformes, Charadriiformes). Bull. Florida State Mus., Biol. Sci., 11: 99-220.

de Beer, G.
1954 *Archaeopteryx lithographica:* A Study Based upon the British Museum Specimen. British Museum (Natural History), London.

1964 Archaeopteryx. In *A New Dictionary of Birds.* Edited by A. L. Thomson. McGraw-Hill Book Company, New York.

Fisher, J.
1964 Extinct Birds. In *A New Dictionary of Birds.* Edited by A. L. Thomson. McGraw-Hill Book Company, New York.

Fisher, J., and R. T. Peterson
[1964] The World of Birds. Doubleday and Company, Garden City, New York.

Fisher, J., and others
1969 Wildlife in Danger. Viking Press, New York.

Greenway, J. C., Jr.
1958 Extinct and Vanishing Birds of the World. Special Publ. No. 13, Amer. Committee Internatl. Wild Life Protection. (Dover reprint available.)

Gross, A. O.
1928 The Heath Hen. Mem. Boston Soc. Nat. Hist., 6: 487-588.

Hanson, H. C.
1965 The Giant Canada Goose. Southern Illinois University Press, Carbondale.

Heilmann, G.
1927 The Origin of Birds. D. Appleton and Company, New York.

Howard, H.
1950 Fossil Evidence of Avian Evolution. Ibis, 92: 1-21.

Mengel, R. M.
[1964] The Probable History of Species Formation in Some Northern Wood Warblers (Parulidae). Living bird, 3: 9-43.

Moreau, R. E.
1966 On Estimates of the Past Numbers and of the Average Longevity of Avian Species. Auk, 83: 403-415.

Parkes, K. C.
1966 Speculations on the Origin of Feathers. Living Bird, 5: 77-86.

Reynard, G. B.
1962 The Rediscovery of the Puerto Rican Whip-poor-will. Living Bird, 1: 51-60.

Savile, D. B. O.
1957 The Primaries of *Archaeopteryx.* Auk, 74: 99-101.

Schorger, A. W.
1955 The Passenger Pigeon: Its Natural History and Extinction. University of Wisconsin Press, Madison.

Simpson, G. G.
1946 Fossil Penguins. Bull. Amer. Mus. Nat. Hist., 87: 1-99.

Storer, R. W.
1960a Evolution in the Diving Birds. Proc. XIIth Internatl. Ornith. Congress, pp. 694-707.

1960b Adaptive Radiation in Birds. In *Biology and Comparative Physiology of Birds.* Volume 1. Edited by A. J. Marshall. Academic Press, New York.

Swinton, W. E.
1960 The Origin of Birds. In *Biology and Comparative Physiology of Birds.* Volume 1. Edited by A. J. Marshall. Academic Press, New York.

1964 Fossil Birds. In *A New Dictionary of Birds.* Edited by A. L. Thomson. McGraw-Hill Book Company, New York.

1965 Fossil Birds. British Museum (Natural History), London.

Wetmore, A.
1931 Birds. Smithsonian Scientific Series. Volume 9. Washington, D. C.
1955 Paleontology. In *Recent Studies in Avian Biology*. Edited by A. Wolfson. University of Illinois Press, Urbana.
1956 A Check-list of the Fossil and Prehistoric Birds of North America and the West Indies. Smithsonian Misc. Coll., 131 (5): 1-105.

1959 Birds of the Pleistocene in North America. Smithsonian Misc. Coll., 138: 1-24.
Williams, G. R.
1960 The Takahe (*Notornis mantelli*, Owen, 1848): A General Survey. Trans. Royal Soc. New Zealand, 88: 235-258.

BRECKENRIDGE

APPENDIX A
ORNITHOLOGICAL FIELD METHODS

Numerous methods are used for facilitating the study and observation of birds in the field. Some of the more common are briefly outlined below.

BLINDS FOR OBSERVATION AND PHOTOGRAPHY

Blinds, or hides, are frequently necessary for close-up observation and photography. While they often must be constructed to suit particular locations, the following specifications will accommodate a wide variety of situations.

Framework. For a **ground blind,** the framework should be of metal, preferably piping three-quarters of an inch in diameter. The piping should be cut, threaded, and screwed together by "elbows" and "cross" to form a frame as in Plate XXIX, Figure 2. The upright pieces of piping are each six feet long (five feet to accommodate the covering and one foot to be driven into the ground for support). The framework is four feet square at the top and base. For an **elevated blind,** the framework should be of two-by-four inch boards. The boards should be cut and nailed together to form a frame, as in Plate XXIX, Figure 1. The total height is twenty-five feet. The frame is four feet square at the top and six feet square at the base. A floor for the blind, constructed of boards one inch thick in nailed to two-by-four cross-pieces which are in turn bolted to the inside of the upright supports. The bolts allow the cross-pieces to be easily detached, so that the floor can be raised or lowered to the desired height.

Covering. The covering should be a tough-fibered fabric (preferably canvas) that is water-proof, non-translucent, and colored dark green or brown. It should be cut and sewed together in one piece in such a manner as to fit over the framework of the ground blind and tie tightly at the side-corners, as in Plate XXIX, Figure 3. The front (i.e., the side facing the direction of observation) contains several zippered openings. The same covering may be used for the elevated blind framework. When the floor is only five feet from the top the covering may be fitted over the top of the framework and tied. (See Plate XXIX, Figure 1.) But when the floor is lower than five feet from the top the covering may be suspended from within the framework so as to meet the floor. The covering should possess "eyes" at the top corners through which guys may be inserted.

When placing blinds in the vicinity of nests for observation and photography certain precautions must be taken. Always put the framework together far from the nests in order to prevent disturbance. For *small birds 12 inches in length or less,* quickly place the blind framework about 10 feet from the nest, put on the covering and tie well so as to prevent excessive flapping in the wind, attach guys, than leave the vicinity until the next day. If the blind appears to be "accepted" after this lapse of time, move the blind

Plate XXIX

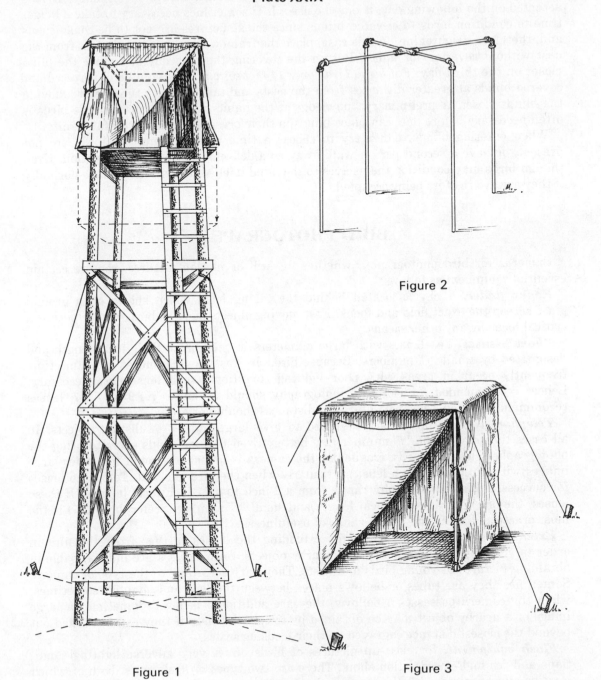

Figure 1

Figure 2

Figure 3

OBSERVATION BLINDS

closer to the nest (seldom nearer than five feet), and again leave the vicinity. It it is accepted on the following day, it may be used. It is sometimes necessary to take a longer time to condition birds to elevated blinds since the structures are apt to be conspicuous and, therefore, frightening. In this case, place the framework in position 10 feet from the nest without covering for a day. Put on the covering the next day and move the blind closer on the third day. For *large birds over 12 inches*, place the completely erected and covered blinds at greater distances from the nests and take a longer time before moving the blinds closer. A preliminary knowledge of the habits and reactions of large birds is often necessary before one can place blinds in their presence with successful results.

When entering blinds, either try to choose a time when the nesting birds are not present, or have a second person walk away as a deflector of the birds' attention. Even though birds may condition themselves to the blind itself, they will not behave normally if they are aware of its being occupied.

BIRD PHOTOGRAPHY

Cameras for bird photography, whether for still or motion pictures, require certain essential equipment as follows:

Reflex finder, a device located behind the taking lens, which shows on a ground glass screen the exact field and focus. Fast-moving objects such as birds do not permit a critical focus by any other means.

"Fast" lenses, i.e., lenses with large diameters in relation to their focal length and designated by small "f" numbers. Because birds are fast-moving and at the same time frequently occur in areas with poor lighting conditions, such lenses are necessary. Lenses recommended for close-up photography should be either f 1.9 or f 2.7; lenses recommended for photographing birds at distances should not exceed f 5.6.

Telephoto lenses, i.e., lenses designed to give large images of distant objects. In all cases the smaller the "f" number, the better. Even though blinds serve to bring the photographer closer to birds, frequently the camera is not near enough to render large images with a regular 1-inch lens, particularly when the bird is small. Telephoto lenses for successful bird photography range from a 2-inch to a 17-inch focal length. Of these lenses, the one with a 4-inch focal length and small "f" number (preferably f 2.7) is the most practical, having the widest range of usefulness.

Extension devices, i.e., devices for extending the lenses farther from the film in order to get large, detailed images of nearby objects. Such devices are indispensable in obtaining pictures of young birds and nests. They vary according to the type of camera. Sometimes they are tubes or bellows placed between the lens and adapter; sometimes, when the camera possesses a bellows, they are additional bellows. Sometimes a large image of a nearby object may be obtained by telephoto lenses if they can be racked out beyond the closest distance engraved on their focusing scales.

Flash equipment, for close-up pictures of birds under very adverse lighting conditions and for high-speed action shots. There are two types of flash units, both of which may be synchronized with the camera shutter. One is the photoflash, with expendable flash lamps (i.e., each lamp good for one flash only) and the other is the electronic flash with repeating flash tubes (each tube good for as many as 10,000 flashes). The electronic flash, sometimes referred to by other names, such as high-speed flash and strobe flash, is ideal for bird photography, since it enables one to take shots at 1/1,000 to 1/20,000 of a second. One disadvantage in using flash equipment for close-up shots is that it makes the background of the pictures dark, due to underexposure. Some photographers overcome this by using colored cards with mat finish as backgrounds. Blue blotters give the most natural effects.

Tripod, for supporting the camera when photographing from a blind, when using a telephoto lens of long focal length, or when taking pictures of nearby objects at slow speed.

Gunstock mount, for holding the camera when obtaining high-speed shots of birds in action, such as flight. With this type of mount a photographer can follow bird movement with greater facility. Under most circumstances, a camera on a tripod cannot be moved rapidly enough to keep up with fast action.

The references given below will aid the student in understanding numerous problems involved in photographing birds and in using equipment.

Allen, A. A.
  1963a Bird Photography at Sapsucker Woods. Living Bird, 2:93-120.
  1963b Stalking Birds with Color Camera. Second edition. National Geographic Society, Washington, D. C.
Bailey, A. M.
  1951   Nature Photography with Miniature Cameras. Mus. Pictorial No. 1, Denver Museum of Natural History, Colorado.
Cruickshank, A. D., and others
  1957   Hunting with the Camera. Harper and Brothers, New York.
Fischer, R. B.
  1952   Bird Photography for Bird-banders. Bird-Banding, 23: 63-72.
Hosking, E., and C. Newberry
  1961   Bird Photography as a Hobby. Stanley Paul, London.
Kinne, Russ
  1962   The Complete Book of Nature Photography. A. S. Barnes and Company, New York.
Reichert, R. J., and E. Reichert
  [1961] Binoculars and Scopes and Their Uses in Photography. Chilton Company, Philadelphia.
Van Riper, W., R. J. Niedrach, and A. M. Bailey
  1952   Nature Photography with the High-speed Flash. Mus. Pictorial No. 5, Denver Museum of Natural History, Colorado.
Warham, J.
  1956   The Technique of Photographing Birds. Focal Press, London.
  1966   The Technique of Wildlife Cinematography. Focal Press, London.

RECORDING BIRD VOCALIZATIONS

Recording bird songs and other vocalizations suitable for spectrographic analysis requires careful choice of equipment together with a few simple techniques in its application.

The equipment must have a higher quality and fidelity than one can expect of a home recorder for music and human speech and it must be designed to respond *uniformly* to a wider range of sound frequencies (sound waves) in cycles per second. Furthermore, it must not only be portable and rugged for use in the field but include special accessories for picking up sounds efficiently at a distance with a minimum of distortion and environmental noises.

The following information on equipment and methods comes primarily from Peter Paul Kellogg who, over the years at Cornell University, has been eminently successful in recording bird vocalizations.

Microphone. This instrument, responsible for converting acoustical energy into electrical energy—i.e., sound waves into electrical impulses—should be able to stand rough use, be responsive to sound waves ranging from 2,000 to 10,000 cycles per second, have an electrical output high enough and an impedance low enough to transmit impulses over a long cable to the recorder without substantial loss in strength. The so-called dynamic microphone with low impedance meets these specifications more closely than any other type and is therefore recommended for recording most bird sounds. A useful accessory that may be fitted like a cap over the receiving end of the microphone is a windscreen which prevents the wind from striking the microphone directly and consequently causing a low-frequency fluttering in the recording. During a moderate to strong wind, however, the windscreen is of no value in keeping the wind from causing serious interference.

Parabolic Reflector ("Parabola"). This is a metal, dish-like device which, when fitted to the microphone, is indispensable for collecting many times more sound energy than the microphone alone and thus serves to reduce the distance between the source of the sound and the microphone. Once the sound waves are picked up by the parabola, they are concentrated or reflected at a focal point out from the center of the dish. The parabola must be large enough to collect, from a distance, frequencies as low as 2,000 cps. Obviously then, the larger the parabola, the greater is its efficiency—and unwieldiness. A parabola 40 inches in diameter will collect about 1,600 times more sound energy from the point at which it is directed than would the microphone alone at the same distance. The compromise between effectiveness and convenience in the size of a parabola is one with a 36-inch diameter and a 12-inch focal length. If the focal length is much less than 12 inches, the microphone has to be set deeper into the dish and this will result in its picking up more resonance from the sides; conversely, if it is much greater than 12 inches, the microphone must be set inconveniently far out in front of the dish. On a parabola with a 36-inch diameter and 12-inch focal length the microphone is best fitted a few inches farther out than the rim of the dish. It must be set precisely at the focal point with its receiving end *facing* the center of the parabola.

A parabola of aluminum has the desirable lightness for manual use, but it is noisy and will amplify vibrations from the slightest touch. The handle should be wrapped with rubber or soft cloth and the back should be coated with a "sound-deadener" used on car bodies. To reduce the shininess of the front, lest it startle the birds whose voices are being recorded, a light coat of dull-colored, non-glossy paint may be applied. While a parabola is usually hand-held, it should nonetheless bear a bracket so that it can be set on a camera tripod for occasional uses in a stationary position.

Recorder. The ideal machine is a portable magnetic tape recorder, powered by dry-cell ("flashlight") batteries and weighing between eight and 20 pounds. The recorder should have the following features and accessories: (1) Besides the recording head and amplifier, a set of earphones for testing the sounds before recording them and for monitoring the sounds almost at the instant when they are being recorded. (2) A playback head and amplifier for determining whether or not the recordings are meeting the desired standards and an erase head for wiping the recordings if they are unsatisfactory. (3) Recording speeds at 7.5 and 15 inches per second. The slower speed is satisfactory for sounds of low frequency, but the higher speed is better for the sounds of high frequency and is strongly recommended for all recordings. Sounds recorded at the higher speed have a finer quality that is particularly desirable if they are to be reproduced later on discs or analyzed on spectrograms. (4) A cable to the microphone of the best quality and at least 10 feet long. It should contain two wires in a woven-wire shield that are grounded in the recorder and covered with rubber or plastic. (5) A durable carrying case with an over-the-shoulder strap and pockets in the sides for keeping the microphone and cable when not in use.

When recording, the operator ordinarily carries the recorder slung from the shoulder, wears earphones plugged into the recorder, manages the recorder's controls with one hand, and holds the parabola with the other. Through the earphones he can determine when the microphone is in focus on the vocalizing bird and whether or not he should correct the quality of the sound being received by manipulating the controls on the recorder.

One of the serious problems in recording is to avoid extraneous sounds such as wind and all the noises emanating from human activities. The early morning when there is apt to be less wind and less human activity is usually the best time for recording bird songs with a minimum of interference. If conditions are very quiet, it is possible, with the above-specified equipment including a 36-inch parabola, to record songs at distances of 300 feet or more.

For further information and reading on the subject of bird-sound recording, the student is referred to the following:

Kellogg, P. P.
  1960 Considerations and Techniques in Recording Sound for Bio-acoustics Studies. In *Animal Sounds and Communication*. Edited by W. E. Lanyon and W. N. Tavolga. American Institute of Biological Sciences, Washington, D. C.
  1961 Sound Recording as an Aid to Bird Study, Atlantic Nat., 16: 85-92
  1962 Bird-sound Studies at Cornell. Living Bird, 1: 37-48.
North, M. E. W.
  1956 Tape-recording for the Field Ornithologist. In *The Ornithologists' Guide*. Edited by H. P. W. Hutson. British Ornithologists' Union.
Stillwell, N.
  1964 Bird Songs: Adventures and Techniques in Recording the Songs of American Birds. Doubleday and Company, Garden City, New York.

RECORDING COLORS OF BIRDS AND BIRD EGGS

Frequently, in describing birds and bird eggs, or in recording the colors of a particular specimen, it is necessary to have some color standard to follow so that the names of different hues will have a precise meaning. The best color standard available is *Atlas de los Colores* by C. Villalobos-Domínguez and J. Villalobos, English text by A. M. Homer (El Ateneo, Florida 340, Buenos Aires, Argentina, 1947; obtainable from Stechert-Hafner, Inc., 31 East 10th Street, New York, New York 10003). This work contains a table converting the symbols, which the authors use, to the nomenclature in R. Ridgway's *Color Standards and Nomenclature* (1912), once the accepted standard for ornithologists but now out of print.

Lacking a copy of Villalobos, the student may follow the more simplified version of the Villalobos standard, with a double-page color chart prepared under the supervision of J. Villalobos, in the introduction to the *Handbook of North American Birds*, Volume 1, edited by R. S. Palmer (Yale University Press, New Haven, 1962). The choice of color terms available is sufficiently wide for most descriptions.

MEASURING ELEVATION OF NESTS AND FLIGHT PATHS

When nests are in trees or other situations too high for accurate determination with measuring stick or tape, a hand-held and hand-operated finder may be useful. Variously called rangefinder, clinometer, altimeter, or height meter, depending on the model, it can be obtained from any one of several companies specializing in equipment for work in forestry.

This same instrument, usually available at modest cost, can also be applied to determining within 100 to 150 feet the elevation of flight paths taken by birds above land or water in their local maneuvers. It can also be applied to measuring straight line distances such as the extent of a bird's territory, the width of a river, and so on.

MARKING NEST SITES, BOUNDARIES OF TERRITORIES, AND CENSUS LINES

Useful for marking nest sites, territorial boundaries, and lines to be followed in census work is a vinyl plastic ribbon called flagging, available in an assortment of bright colors. This can be obtained in rolls from companies that supply equipment for foresters. It is easy to tie and easy to see; it lasts indefinitely and its color is permanent. The fact that it lasts so long should prompt the student to remove all flagging when he is through with his study lest it become litter.

CAPTURING WILD BIRDS

Frequently, it is necessary to capture wild adult birds for banding, marking, weighing, or examination. There are several devices and techniques commonly used. See page 438 for permits required.

Traps for Nesting Birds. Birds nesting **on the ground** may be captured in a "pull-string" trap which consists of a wooden frame, two feet square and a foot high, with a covering of No. 2 galvanized hardware cloth on the sides and top. It is placed over the nest and tilted upward at a 45-degree angle by a trip-stick to which is attached a string leading to the trapper, who is either hidden in a near-by blind or behind some object farther away. When the bird enters the nest, the trip-stick is quickly pulled from under the trap, which falls over the nest and imprisons the bird. The bird is removed by reaching under the trap. Birds nesting **in trees** may also be captured by a trap, specially constructed to suit the particular nesting site. It is made entirely of hardware cloth, cut to fit around and below the nest and supporting branch, and held in place either by the branch or a prop coming from the ground below. The top, which is several inches above the nest rim, is hinged, weighted, and held partially open by a trip-stick to which a string is attached. When the bird settles on the nest, after entering through the top (the only opening), the trip-stick is pulled out and the weighted top falls to hold the bird a prisoner. A more simple method of capturing tree-nesting adults is a net made of several hair nets sewed together and attached to a wire hoop. This is placed over the nest at a 45-degree angle and held in place by strings. After settling on the nest the bird is abruptly flushed, and it flies into the net where it becomes entangled. Birds nesting **in cavities** may be captured by a mist net (see below) placed in such a manner in front of the entrance to the cavity that the bird flies into it when leaving the nest or returning.

Mist Nets. Made of fine nylon or silk and dyed black to make them practically invisible against almost any dark background, mist nets are excellent for capturing flying birds in quantity. The nets, about 30 to 38 feet long and 3 to 7 feet high, are available in different mesh sizes for birds ranging from kinglets and warblers to grouse, large shore birds, and medium-sized ducks and hawks. Each net consists of five horizontal lines or "trammels," which are stretched taut between two vertical poles, and the netting itself which hangs loose from the trammels to form bags. The nets are positioned strategically across flight paths commonly taken by birds. Birds fail to see the netting, strike it, and fall into the bags where they become entangled and held. The person operating the nets should be in constant attendance, and not let the birds be captive longer than the time necessary to remove them. When birds become badly entangled, as is sometimes the case, great patience and skill is required to remove them.

Bait Traps. Birds which normally feed on the ground may be captured in traps suitably baited. The variety of traps is infinite, ranging from single-cell types, which automatically capture one bird at a time, to large "decoy" enclosures which can imprison hundreds at one setting. The simple trap described for capturing ground-nesting birds (see above) can be easily adapted by setting it up over bait instead of a nest. Another simple trap is a wire-cloth, bottomless box placed flat on the ground over the bait with an entrance consisting of a funnel, its big opening flush with the side of the box and its small opening far inside. Birds, after entering through the funnel, fail to exit since they overlook the small inner opening of the funnel as they strive futilely to get through the wire-cloth sides of the trap. The birds are removed through a covered opening in the top of the trap.

Cannon Nets. Gregarious species such as blackbirds, doves, gulls, and waterfowl, that are easily attracted to bait, may be captured effectively by large cannon or rocket nets. Each net is attached to three or more heavy projectiles which are inserted in "cannons," or mortars, loaded with a propellant, usually black powder. When simultaneously detonated electrically by an operator in a blind some distance away, the projectiles instantly carry the net over the feeding birds. As there is an element of danger involved in setting up and firing cannon nets, the student is advised to obtain prior instruction from an experienced operator before using them.

Corral Nets. Waterfowl, while flightless during the simultaneous molt of their remiges, and young of a few colony-nesting species, may be driven into enclosures or corrals, constructed either of wire cloth or netting. Long fences or "leads" flare from the entrance to form a funnel into which the birds are first herded and ultimately guided into the trap.

Hand Nets Used with Spotlights. Shore birds, such as woodcock, and many aquatic birds may be hand-netted at night after first being "blinded" by a powerful spotlight. The operation usually requires at least two persons side by side, one to direct the spotlight and the other to wield a large circular net on a long pole. Dark, overcast nights are best. If the bird is on the ground, the night-lighters walk quietly toward it, all the while holding the beam on the bird until it can be reached with the net. Should the bird flush, it can sometimes be brought down by "dazzling" it with the spotlight. If the bird is on the water, the night-lighters can follow the same procedures from a boat by quietly maneuvering their craft toward the bird until they can take it in the net.

Chemicals. Birds may be captured by bait treated with a tranquilizing or stupefying agent. Since most such chemicals are invariably lethal in heavy doses, no student should apply any agent of this sort without first being thoroughly familiar with its potentialities, knowing precisely how to administer it, and understanding how to handle doped birds and effect their recovery.

The references given below will provide much detailed information on methods of capturing wild birds. For a listing of the agencies supplying traps and nets, the student should write to one of the five regional bird-banding associations—Northeastern, Eastern, Inland, Western, and Ontario—whose addresses are given in the current issue of *Bird-Banding* (see Appendix G).

Brownlow, H. G.
  1952 The Design, Construction and Operation of Heligoland Traps. Brit. Birds, 45: 387-399.
     (Consists of a "tapering wire netting enclosure open at the wide end, and closed at the

narrow end by a collecting box with a transparent back, which appears to birds driven into the trap as a way of escape, and induces them to enter the box.")

Crider, E. D., and J. C. McDaniel
  1967 Alpha-Chloralose Used to Capture Canada Geese. Jour. Wildlife Management, 31: 258-264. (A sugar compound of chloral hydrate taken in the food, anesthetizes the birds in 40 to 80 minutes. Can be lethal unless used properly.)
Cummings, G. E., and O. H. Hewitt
  1964 Capturing Waterfowl and Marsh Birds at Night with Light and Sound. Jour. Wildlife Management, 28: 120-126.
Dill, H. H., and W. H. Thornsberry
  1950 A Cannon-projected Net Trap for Capturing Waterfowl. Jour. Wildlife Management, 14: 132-137.
Drewien, R. D., and others
  1967 Back-pack Unit for Capturing Waterfowl and Upland Game by Night-lighting. Jour. Wildlife Management, 31: 778-783.
Edgar, R. L.
  1968 Catching Colonial Seabirds for Banding. Bird-Banding, 39: 41-43. (Using an adjustable noose on a pole.)
Edwards, M. G
  1961 New Use of Funnel Trap for Ruffed Grouse Broods. Jour. Wildlife Management, 25: 89.
Greenlaw, J. S., and J. Swinebroad
  1967 A Method for Constructing and Erecting Aerial-nets in a Forest. Bird-Banding, 38: 114-119.
Gullion, G. W.
  1965 Improvements in Methods for Trapping and Marking Ruffed Grouse. Jour. Wildlife Management, 29: 109-116.
Heimerdinger, M. A., and R. C. Leberman
  1966 The Comparative Efficiency of 30 and 36 MM. Mesh in Mist Nets. Bird-Banding, 37: 280-285.
Humphrey, P. S., D. Bridge, and T. E. Lovejoy
  1968 A Technique for Mist-netting in the Forest Canopy. Bird-Banding, 39: 43-50.
Johns, J. E.
  1963 A New Method of Capture Utilizing the Mist Net. Bird-Banding, 34: 209-213. (Net stretched horizontally just above shallow water and shore, then dropped by release string, captures shore birds successfully.)
Johnston, D. W.
  1965 An Effective Method for Trapping Territorial Male Indigo Buntings. Bird-Banding, 36: 80-83. (Equipment: a mist net, a stuffed male in breeding plumage, and a recording of species' song. Set-up must be *in* an occupied territory or between contiguous territories.)
Hussell, D. J. T., and J. Woodford
  1961 The Use of a Heligoland Trap and Mist-nets at Long Point, Ontario. Bird-Banding, 32: 115-125.
Labisky, R. F.
  1968 Nightlighting: Its Use in Capturing Pheasants, Prairie Chickens, Bobwhites, and Cottontails. Illinois Nat. Hist. Surv. Biol. Notes No. 62.
Lacher, J. R., and D. D. Lacher
  1964 A Mobile Cannon Net Trap. Jour. Wildlife Management, 28: 595-597. (Cannons mounted on the front of a jeep.)
Lindmeier, J. P., and R. J. Jessen
  1961 Results of Capturing Waterfowl in Minnesota by Spotlighting. Jour. Wildlife Management, 25: 430-431. (Success averaged 1.3 ducks per man-hour.)
Liscinsky, S. A., and W. J. Bailey, Jr.
  1955 A Modified Shorebird Trap for Capturing Woodcock and Grouse. Jour. Wildlife Management, 19: 405-408. (Using funnel nets strategically placed in feeding and resting areas.)
Lockley, R. M., and R. Russell
  1953 Bird-Ringing: The Art of Bird Study by Individual Marking. Crosby Lockwood and Son, London. (Contains a well-illustrated account of bird traps; how they are constructed and operated.)
Loftin, H.
  1960 Use of Decoys in Netting Shorebirds. Bird-Banding, 31: 89-90. ("Decoys used are simply

profiles cut from plywood or even pasteboard, with a heavy wire rod attached for sticking into the sand.")

Lord, R. D. Jr., and W. W. Cochran
  1963 Techniques in Radiotracking Wild Animals. In *Bio-Telemetry*. Edited by L. Sclater. Pergamon Press, New York.

Low, S. H.
  1935 Methods of Trapping Shore Birds. Bird-Banding, 6: 16-22.
  1957 Banding with Mist Nets. Bird-Banding, 28: 115-128. (The classic paper on the subject.)

McCamey, F.
  1961 The Chickadee Trap. Bird-Banding, 32: 51-55.

McClure, H. E.
  1966 An Asian Bird-banders' Manual. Migratory Animal Pathological Survey, Box 3443, Hong Kong, B. B. C. (Extensive information on trapping and netting; profusely illustrated with photographs and drawings of different types of traps.)

Nolan, V., Jr.
  1961 A Method of Netting Birds at Open Nests in Trees. Auk, 78: 643-645.

Rogers, J. P.
  1964 A Decoy Trap for Male Lesser Scaups. Jour. Wildlife Management, 28: 408-410.

Salyer, J. W.
  1962 A Bow-net Trap for Ducks. Jour. Wildlife Management, 26: 219-221.

Serventy, D. L., and others.
  1962 Trapping and Maintaining Shore Birds in Captivity. Bird-Banding, 33: 123-130.

Sheldon, W. G.
  1960 A Method of Mist Netting Woodcocks in Summer. Bird-Banding, 31: 130-135. (Nets placed on spring singing fields and operated for a few minutes after sunset.)

Silvy, N. J., and R. J. Robel
  1967 Recordings Used to Help Trap Booming Greater Prairie Chickens. Jour. Wildlife Management, 31: 370-373.
  1968 Mist Nets and Cannon Nets Compared for Capturing Prairie Chickens on Booming Grounds. Jour. Wildlife Management, 32: 175-178. Mist nets have more advantages than disadvantages.)

Smith, N. G.
  1967 Capturing Seabirds with Avertin. Jour. Wildlife Management, 31: 479-483. (A narcotizing agent put in food, stupefies the birds in a few minutes. Can be lethal unless used properly.)

Spencer, A. W., and J. W. De Grazio
  1962 Capturing Blackbirds and Starlings in Marsh Roosts with Dip Nets. Bird-Banding, 33: 42-43.

Taber, R. D., and I. McT. Cowan
  1969 Capturing and Marking Wild Animals. In *Wildlife Management Techniques*. Third edition, revised. Edited by R. H. Giles, Jr. Wildlife Society, Washington, D. C. (Probably the most comprehensive coverage of the subject, with details on equipment.)

Thompson, M. C., and R. L. DeLong
  1967 The Use of Cannon and Rocket-projected Nets for Trapping Shorebirds. Bird-Banding, 38: 214-218.

Vandenbergh, J. G.
  1960 A Bird Holding Cage. Bird-Banding, 31: 221-222.

Woodford, J., and D. J. T. Hussell
  1961 Construction and Use of Heligoland Traps. Bird-Banding, 32: 125-141. (Contains a wealth of detailed information and very useful suggestions.)

Zwickel, F. C., and J. F. Bendell
  1967 A Snare for Capturing Blue Grouse. Jour. Wildlife Management, 31: 202-204.

BANDING WILD BIRDS

There are many schemes for banding (called ringing in Europe) wild birds throughout the world. All involve the use of metal leg bands, systematically numbered and carrying an address to which the recoverer of the banded bird may report.

In North America, the United States and Canadian Governments (i.e., the U. S. Bureau of Sport Fisheries and Wildlife and the Canadian Wildlife Service) cooperate in a single, continent-wide program. Each band bears a number and directions to notify the Fish and Wildlife Service, Washington, D.C. After the band is attached to the bird's leg, a record is sent to the Bird Banding Laboratory at the Migratory Bird Populations Station, Laurel, Maryland, where it is placed on file. If at some future time a person kills the bird, finds it dead, or captures it, he reports the number to the Fish and Wildlife Service, thus revealing its whereabouts. In North America, well over 20,000,000 birds have been banded and approximately 1,700,000 more are banded each year. Nearly 2,000,000 have been "recovered" and reported to the Bird Banding Laboratory.

Any person wishing to band birds must first obtain the necessary permits (see page 438). The Bird Banding Laboratory than provides, free of charge, all bands, forms for reporting, postage-free envelopes, and a *Bird Banding Manual.* The bander is expected to provide his own traps, nets, and other equipment.

The data from recoveries sent to the Bird Banding Laboratory are computerized, and then made available to students and other persons qualified to undertake research. In order to protect the interests of banders, many of whom are engaged in long-range research projects and who have invested significant amounts of money in the accumulation of data, the Bird Banding Laboratory places certain limitations on the publication of information from its files. Requests for data from the files should be forwarded to the Director, Migratory Bird Populations Laboratory, Laurel, Maryland 20810.

MARKING WILD BIRDS

It is sometimes desirable to mark birds for individual recognition. See page 438 for necessary permits.

Colored Plastic Leg Bands. Various combinations of colored plastic bands enable the student to identify individual birds visually without having to recapture them. By devising a coding system (e.g., green band on the *left* leg of one bird, green band on the *right* leg of another, green and yellow bands on the *left* leg of still another, etc.) it is possible to recognize individuality in a great number of birds. Caution: While it is all right to place two or more plastic bands on the same leg, never use two or more metal bands on the same leg, as their constant tapping together produces a sharp "flange" that can seriously injure or even sever the leg.

Tags, Streamers, and Collars. A wide variety of markers have been devised for large birds to enable their recognition from a distance. Most such markers consist of plastic material in different colors attached to the bird by loops or by non-corrosive metal clips. Caution: Any marker can be a handicap to a bird, even a cause of its death, unless properly designed and applied to suit the bird's habits and habitat. No marker should (1) impair flight, (2) be so loose as to become entangled in vegetation, or (3) be heavy enough to erode the feathers or to chafe the skin.

Bill Markers. Several types of disks may be used on the bills of larger birds, primarily waterfowl. Made of plastic in different colors, the markers are attached on either side of the bird's bill by monofilament nylon through the nostrils. Their chief detriment is their tendency to become entangled in vegetation or debris while the bird is feeding.

Feather Marking and Coloring. Birds may be "feather-marked" by cementing to their upper tail coverts white feathers that have been dyed in different colors. Birds may also have their flight feathers or other parts of their plumage colored by applying either an aniline dye or a quick-drying lacquer (e.g., the so-called "airplane dope" used for model airplanes). Considerable care must be taken, however, to apply the substance lightly and to make sure that it is dry and that the feathers are not stuck together when the birds are released.

When it is desirable to mark birds which cannot for one reason or another be captured, it is sometimes possible to spray them with a quick-drying dye or lacquer shot from a water pistol, or to place fresh enamel on objects (e.g., the edge of a nest-opening) with which the birds will come in contact. Either method will give the birds random markings sufficient for individual identification.

The Bird Banding Laboratory maintains a central file containing records of all color-marking schemes throughout North America. This prevents two or more research workers from duplicating each other's schemes and enables the Laboratory to respond to reports of sightings of color-marked birds and to notify the research workers of the birds' whereabouts.

Use of Radiotelemetry. When a bird cannot be followed visually, it may be equipped with a miniaturized radio transmitter, held in place by a harness slipped over the bird's neck or wings or by being stuck with an adhesive to the skin of the back between the wings. The transmitter is then monitored by a receiver and the movements of the bird charted.

The references below contain further information with regard to marking wild birds.

Ballou, R. M., and F. W. Martin
   1964   Rigid Plastic Collars for Marking Geese. Jour. Wildlife Management, 28: 846-847.
Bartonek, J. C., and C. W. Dane
   1964   Numbered Nasal Discs for Waterfowl. Jour. Wildlife Management, 28: 688-692.
Baumgartner, A. M.
   1938   Experiments in Feather Marking Eastern Tree Sparrows for Territory Studies. Bird-Banding, 9: 124-135. (Consult for methods of marking by dyeing and attaching feathers.)
Brander, R. B., and W. W. Cochran
   1969   Radio-location Telemetry. In *Wildlife Management Techniques*. Third edition, revised. Edited by R. H. Giles, Jr. Wildlife Society, Washington, D. C.
Cottam, C.
   1956   Use of Marking Animals for Ecological Studies: Marking Birds for Scientific Purposes. Ecology, 37: 675-681.
Emlen, J. T., Jr.
   1954   Territory, Nest Building, and Pair Formation in the Cliff Swallow. Auk, 71: 16-35. (Consult for methods of marking birds that cannot be captured.)
Fankhauser, D.
   1964   Plastic Adhesive Tape for Color-marking Birds. Jour. Wildlife Management, 28: 594.
Gullion, G. W., R. L. Eng, and J. J. Kupa
   1962   Three Methods for Individually Marking Ruffed Grouse. Jour. Wildlife Management, 26: 404-407. (Color banding, dyeing, and back-tagging.)
Hamerstrom, F. N., Jr., and O. E. Mattson
   1964   A Numbered, Metal Color-band for Game Birds. Jour. Wildlife Management, 28: 850-852.
Havlin, J.
   1968   Wing-tagging Ducklings in Pipped Eggs. Jour. Wildlife Management, 32: 172-174.
Hester, A. E.
   1963   A Plastic Wing Tag for Individual Identification of Passerine Birds. Bird-Banding, 34: 213-217.
Hewitt, O. H., and P. J. Austin-Smith
   1966   A Simple Wing Tag for Field-marking Birds. Jour. Wildlife Management, 30: 625-627. (For use with small birds.)
Kennard, J. H.
   1961   Dyes for Color-marking. Bird-Banding, 32: 228-229. (Successful use of Drimark, a commercial product made in Japan.)
Kozicky, E. L., and H. G. Weston, Jr.
   1952   A Marking Technique for Ring-necked Pheasants. Jour. Wildlife Management, 16: 223. (Tails shortened by scissors, coated with Duco cement, and then painted.)

Lindmeier, J. P.
 1960   Experimental Marking and Dyeing of Ring-necked Pheasants. Flicker, 32: 43-45.
Martin, F. R.
 1963   Colored Vinylite Bands for Waterfowl. Jour. Wildlife Management, 27: 288-290.
Sherwood, G. A.
 1966   Flexible Plastic Collars Compared to Nasal Discs for Marking Geese. Jour. Wildlife Management, 30: 853-855. (Collars superior in visibility, retention, and ease of placement.)
Southern, W. E.
 1965   Biotelemetry: A New Technique for Wildlife Research. Living Bird, 4: 45-58.
Swank, W. G.
 1952   Trapping and Marking of Adult Nesting Doves. Jour. Wildlife Management, 16: 87-90. (Large feathers on wings and tail the best surfaces for marking.)
Taber, R. D., and I. McT. Cowan
 1969   Capturing and Marking Wild Animals. In *Wildlife Management Techniques*. Third edition, revised. Edited by R. H. Giles, Jr. Wildlife Society, Washington, D. C. (Many methods of marking wild birds are suggested.)
Thomas, J. W., and R. G. Marburger
 1964   Colored Leg Markers for Wild Turkeys. Jour. Wildlife Management, 28: 552-555.
Wadkins, L. A.
 1948   Dyeing Birds for Identification. Jour. Wildlife Management, 12: 388-391.

COLLECTING BIRDS

If specimens cannot be captured by a net or trap, they may be collected with a gun. The weapon commonly used is either a 12-gauge, or 16-gauge, double-barreled shot-gun equipped with an auxiliary .410 barrel that may be inserted in one of the regular barrels. The regular barrel is used for large birds, the auxiliary barrel for small birds. The shells and desired loads may be secured through commercial channels. Usually the .410 shell is loaded with No. 12 ("dust") shot for birds under 10 inches. The 12, or 16, shells are loaded with No. 10 shot for birds of medium size and with Nos. 7½ and 6 shot for large birds.

Permits are required for collecting (see below). Any student holding them should regard himself as specially privileged and personally responsible to the granting agencies and to the institution he represents. Callous or indiscriminate killing of birds is not only unwarranted but can create bad public relations. To avoid misunderstandings and objections, heed these admonitions: (1) Advise local conservation officers and constabulary about collecting activities contemplated. (2) Obtain the permission of land owners before collecting on their property. (3) Never, *never* carry firearms or shoot birds in the presence of persons uninformed about the purpose or value of the procedure.

PERMITS FOR CAPTURING, BANDING, MARKING, AND COLLECTING BIRDS

Most species of wild, migratory North American birds are fully protected by federal regulations under the provisos of treaties between the governments of the United States, Canada, and other countries. In some states and Canadian provinces, additional species in their avifauna are fully protected by regulations not afforded by the federal governments.

In accordance with federal regulations, it is therefore illegal to capture, hold, transport, or kill any protected species and to take or otherwise destroy its nests or eggs. This means that the following acts involving *a protected species* are illegal.

(1) To be in the possession of any live specimen, even one that is sick or injured.

(2) To pick up or otherwise be in the possession of a dead specimen or the parts (including the feathers) of a specimen.

(3) To be in the possession of the eggs or nests, including *old* nests long since abandoned by the builders.

Federal regulations in the United States and Canada provide for the issuance of various permits which authorize qualified students and *bona fide* research workers to capture, mark, and hold live specimens, to "salvage" dead specimens, and to collect specimens, their nests and eggs, of protected species. Special restrictions, however, are placed on any study activities involving eagles or other species designated "endangered," or involving the use of mist nets or chemicals.

Students or research workers who wish to capture wild migratory birds for the purpose of banding or marking must first obtain a federal scientific marking permit. Each applicant must be at least 18 years of age, show evidence of a sound ornithological background, and provide the names of at least three licensed banders or recognized ornithologists who will vouch for the applicant's knowledge and ability.

Applications for United States banding permits should be forwarded to Chief, Bird Banding Laboratory, Migratory Bird Populations Station, Laurel, Maryland 20810; for Canadian permits to Director, Canadian Wildlife Service, 400 Laurier Avenue West, Ottawa 4, Ontario, Canada.

The federal permit is invalid unless the permittee also possesses any required state or provincial permits or licenses. Therefore, *after* obtaining the federal permit, the permittee must apply for a state or provincial license. Applications for state permits should be forwarded to the appropriate agencies listed at the end of this section. Applicants for provincial licenses, if needed, may obtain the necessary advice and addresses from the Chief, Canadian Wildlife Service.

Students or research workers who wish to collect (live or dead), propagate, or hold in captivity for "special use" one or more specimens of wild migratory birds, or to take the nests and eggs, may obtain the necessary permits. The collecting of protected birds, their nests and eggs, is warranted only when particularly significant problems are being studied. Permits to collect a variety of species are not generally granted.

Applications for permits to collect, propagate, or hold wild migratory birds in the United States must first be directed to the Bureau of Sport Fisheries and Wildlife through the appropriate regional office listed below. Each applicant must be at least 18 years of age, give a careful description of the nature and ultimate purpose of the study to be undertaken, and, later, if he has collected specimens, render a report stating the exact number taken and the manner of their disposal.

Region I:	Regional Director P. O. Box 3737 Portland, Oregon 97208	Alaska, California, Hawaii, Idaho, Montana, Nevada, Oregon, Washington
Region II:	Regional Director P. O. Box 1306 Albuquerque, New Mexico 87103	Arizona, Colorado, Kansas, New Mexico, Oklahoma, Texas, Utah, Wyoming
Region III:	Regional Director 1006 West Lake Street Minneapolis, Minnesota 55408	Illinois, Indiana, Iowa, Michigan, Minnesota, Missouri, Nebraska, North Dakota, Ohio, South Dakota, Wisconsin

Region IV: Regional Director
Peachtree-Seventh Building
Atlanta, Georgia 30323

Alabama, Arkansas, Florida, Georgia,
Kentucky, Louisiana, Maryland, Mississippi,
North Carolina, South Carolina, Tennessee,
Virginia, Puerto Rico

Region V: Regional Director
U. S. Post Office and Courthouse
Boston, Massachusetts 02109

Connecticut, Delaware, Maine,
Massachusetts, New Hampshire, New
Jersey, New York, Pennsylvania, Rhode
Island, Vermont, West Virginia

If a student or any other person picks up a sick or injured wild migratory bird and wants to nurse it back to health, he may obtain temporary legal possession by writing the appropriate regional office (above), requesting a permit to retain it. The permittee will be required to release the bird when it has recovered.

As in the case of federal permits to capture and mark wild migratory birds, no permit to collect, propagate, or hold (except sick or injured birds) is valid until the permittee obtains another permit or license from the state agency (below) or provincial office (through the Canadian Wildlife Service) which has jurisdiction over the area where the project is undertaken.

Applications for state permits should be addressed to the officer of the appropriate agency listed below.

Alabama: Director, Department of Conservation, 64 North Union Street, Montgomery 36104
Alaska: Commissioner, Department of Fish and Game, Subport Building, Juneau 99801
Arizona: Director, Game and Fish Department, 1688 West Adams, Phoenix 85007
Arkansas: Director, Game and Fish Commission, State Capitol, Little Rock 72201
California: Director, Department of Fish and Game, 1416 9th Street, Sacramento 95814
Colorado: Director, Game, Fish and Parks Department, 6060 Broadway, Denver 80216
Connecticut: Director, Board of Fisheries and Game, State Office Building, Hartford 06106
Delaware: Director, Board of Game and Fish Commissioners, Box 457, Dover 19901
District of Columbia: Chief of Police, Metropolitan Police, 300 Indiana Avenue, Washington 20001
Florida: Director, State Game and Fish Commission, 620 South Meridian, Tallahassee 32304
Georgia: Director, State Game and Fish Commission, 401 State Capitol, Atlanta 30334
Hawaii: Director, Division of Fish and Game, Department of Land and Natural Resources, 400 South Beretania Street, Honolulu 96813
Idaho: Director, Fish and Game Department, 600 South Walnut, Box 25, Boise 83707
Illinois: Director, Department of Conservation, 102 State Office Building, Springfield 62706
Indiana: Head, Division of Fish and Game, Department of Natural Resources, 603 State Office Building, Indianapolis 46209
Iowa: Director, State Conservation Commission, East 7th and Court Avenue, Des Moines 50309
Kansas: Director, Forestry, Fish and Game Commission, Box F, Pratt 67124
Kentucky: Commissioner, Department of Fish and Wildlife Resources, State Office Annex, Frankfort 40601
Louisiana: Director, Wild Life and Fisheries Commission, 400 Royal Street, New Orleans 70130
Maine: Commissioner, Department of Inland Fisheries and Game, State House, Augusta 04430
Maryland: Director, Department of Game and Inland Fish, State Office Building, Box 231, Annapolis 21404
Massachusetts: Director, Division of Fisheries and Game, 100 Cambridge Street, Boston 02202
Michigan: Director, Department of Natural Resources, Lansing 48926
Minnesota: Commissioner, Department of Conservation, 301 Centennial Building, 658 Cedar Street, St. Paul 55101
Mississippi: Executive Director, Game and Fish Commission, Game and Fish Building, 402 High Street, Jackson 39205
Missouri: Director, State Conservation Commission, P. O. Box 180, Jefferson City 65101
Montana: Director, Fish and Game Department, Helena 59601
Nebraska: Director, Game, Forestation and Park Commission, State Capitol Building, Lincoln 68509
Nevada: Director, Fish and Game Commission, Box 678, Reno 89504

New Hampshire: Director, Fish and Game Department, 34 Bridge Street, Concord 03301

New Jersey: Director, Division of Fish and Game, Department of Conservation and Economic Development, Labor and Industry Building, Box 1809, Trenton 08625

New Mexico: Director, Department of Game and Fish, State Capitol, Santa Fe 87501

New York: Director, Fish and Game, Conservation Department, State Office Buildings, Campus, Albany 12226

North Carolina: Executive Director, Wildlife Resources Commission, P. O. Box 2919, Raleigh 27602

North Dakota: Commissioner, State Game and Fish Department, Bismark 58501

Ohio: Chief, Division of Wildlife, Department of Natural Resources, 1500 Dublin Road, Columbus 43212

Oklahoma: Director, Department of Wildlife Conservation, P. O. Box 53098, Oklahoma City 73105

Oregon: Director, State Game Commission, Box 3503, Portland 97208

Pennsylvania: Executive Director, Game Commission, P. O. Box 1567, Harrisburg 17120

Rhode Island: Chief, Division of Conservation, Department of Natural Resources, 83 Park Street, Providence 02903

South Carolina: Director, Division of Game and Boating, Wildlife Resources Department, 1015 Main Street, Box 167, Columbia 29209

South Dakota: Director, Department of Game, Fish, and Parks, State Office Building, Pierre 57501.

Tennessee: Director, Game and Fish Commission, Room 600, Doctors Building, Nashville 37219

Texas: Executive Director, Parks and Wildlife Commission, Austin 78701

Utah: Director, State Department of Fish and Game, 1596 West North Temple, Salt Lake City 84116

Vermont: Commissioner, Fish and Game Board, Montpelier 05602

Virginia: Executive Director, Commission of Game and Inland Fisheries, 7 North 2nd Street, Box 1642, Richmond 23213

Washington: Director, Department of Game, 600 North Capitol Way, Olympia 98501

West Virginia: Director, Department of Natural Resources, State Office Building 3, Charleston 25305

Wisconsin: Director, Conservation Department, Box 450, Madison 53701

Wyoming: Director, Game and Fish Commission, Box 1589, Cheyenne 82001

PREPARING AND STORING SPECIMENS

Study skins of birds have great value in research on birds in the field as well as in taxonomic investigations. Read *The Contribution of Museum Collections to Knowledge of Living Birds* by K. C. Parkes (Living Bird, 2: 121-130, 1963) for highly instructional information along this line.

The preparation of birds as study skins requires experience that can be gained only through practice. For a description of procedures in bird-skinning, cataloguing, and labeling, consult the following: *Perserving Birds for Study* by E. R. Blake (Fieldiana: Technique No. 7, Chicago Nat. Hist. Mus., 1949); *The Preparation of Birds for Study* by J. P. Chapin (Science Guide No. 58, Amer. Mus. Nat. Hist., 1946); "Making Birdskins" in *Handbook of Birds of Eastern North America* by F. M. Chapman (D. Appleton and Company, New York, 1932; Dover reprint available); *Collecting and Preparing Study Specimens of Vertebrates* by E. R. Hall (Misc. Publ. No. 30, Univ. Kansas Mus. Nat. Hist., 1962). See also *A Bird Skin Drying Form for Field Use* by G. E. Watson (Bird-Banding, 33: 95-96, 1962).

If the student is to prepare study skins, he should bear in mind that it is important not only to make skins that neatly and properly preserve each bird's general form and feather arrangement, but also to record all necessary data pertaining to each specimen. This means placing on each bird skin label full information as to locality and date of collection, name of collector, sex, age, and weight of specimen, amount of fat, condition of gonads, presence or absence of incubation patch, and the perishable colors of the soft parts. A skin, no matter how good it is, has little value for study purposes without data of this sort. For details as to just what information should be placed on labels and a discussion of the importance of full data, the student is urged to read *Principles and Practices in Collecting and Taxonomic Work* by J. Van Tyne (Auk, 69: 27-33, 1952). The student will also find instructive *An Aspect of Collectors' Technique* by T. T. McCabe (Auk, 60: 550-558, 1943) in which suggestions are given for recording cyclic conditions of gonads, fat, and incubation patch.

For comprehensive information on the preservation and storage of bird skins and birds in the flesh (including embryos), as well as illustrated descriptions of the preparation of bird skins and bird eggs, the student is referred to *The Preservation of Natural History Specimens*, Volume 2, by R. Wagstaffe and J. H. Fidler (Philosophical Library, New York, 1968).

In preserving small birds in quantity—e.g., large kills from television towers—it may not be feasible to make regular skins of them; yet the specimens may be necessary for later study and analysis. See *A New Method for Preserving Bird Specimens* by R. A. Norris (Auk, 78: 436-440, 1961) wherein skins may be mounted flat on large cards with ample room for data. The procedure is time-saving and the space required for storage is economical.

The preparation of skeletal material, as well as specimens to be kept in the flesh, requires special care. Read *Suggestions Regarding Alcoholic Specimens and Skeletons of Birds* by A. J. Berger (Auk, 72: 300-303, 1955) for helpful information. Skeletons may be cleaned by putting them in a colony of dermestid beetles maintained for the purpose. For details on the method, read *Dermestid Beetles for Cleaning Skulls and Skeletons in Small Quantities* by R. Hardy (Turtox News, 23: 69-70, 1945) and *Defleshing of Skulls by Beetles* by E. J. Coleman and J. R. Zbijewska (Turtox News, 46: 204-205, 1968). Or skeletons may be cleaned by macerating them in a solution of ammonium hydroxide. For details, read *Method for the Preparation and Preservation of Articulated Skeletons* by C. P. Egerton (Turtox News, 46: 156-157, 1968).

DETERMINING THE SEX AND AGE OF LIVE BIRDS

When coloration or other features of a captured bird's exterior do not reveal its sex or age, its cloaca may be examined for the presence, absence, or state of development of the following structures which will provide the information desired. (Before examining a cloaca the student should first understand its general anatomy by reading the directions for the dissection of the pigeon's cloaca, page 90.)

Penis

The ventral wall of the male proctodeum may be modified to form a penis. In waterfowl (swans, geese, and ducks) the penis is a thickening which, when erected and protruded through the vent, acts as an intromittent organ during copulation. A groove on its dorsal surface conveys the spermatozoa into the cloaca of the female. In adult ducks —and presumably all waterfowl—the penis, except when protruded, is covered by a sheath which appears as a grayish white fold on the left wall of the proctodeum. But in immature ducks—i.e., ducks that are no older than six months—the penis is merely a short prominence just inside the lip of the vent and has no sheath.

By examining the cloaca of a duck it is possible to determine the sex and, in case the individual is a male, the age. The technique used in making the examination is briefly: Have the bird held with the posterior abdomen up and the tail pointing away from the examiner; force the tail back with the forefingers; with the thumbs, part the feathers over the vent and press down and away from the two sides of the vent simultaneously, spreading the opening of the vent and exposing the cloaca within. If the duck is a male, a penis will protrude; if the male is an adult, the penis will be sheathed, but if it is an immature bird, the penis will be small and unsheathed.

A penis like the waterfowl's occurs in ostriches, rheas, cassowaries, emus, kiwis, tinamous, and guans. A rudimentary form of the same structure also persists in herons and flamingos. Males of a few other kinds of birds often show a slight modification of the cloacal wall for copulatory purposes. For example, the Domestic Fowl *(Gallus)*, Turkey *(Meleagris)*, and Ring-necked Pheasant *(Phasianus colchicus)* have paired thickenings with erectile tissue on the ventral wall. Males of most kinds of birds, however, lack a penis.

Cloacal Protuberance

The posterior wall of the male proctodeum may swell and protrude posteriorly, bearing the encircled vent with it. Called the cloacal protuberance, it contains chiefly two bodies of compactly coiled tubules (vasa deferentia), which are believed to act as storage reservoirs for spermatozoa. So far as known, the protuberance occurs in certain passerine species only.

The protuberance increases in size as the breeding season advances, reaching its maximum size at the height of the season. After the breeding season it is very small, if not entirely absent. The presence of the protuberance is an unquestionable indicator of the male sex, and its size shows the breeding condition. A protuberance is easily looked for by holding the bird with posterior abdomen up and parting the feathers over the vent. If there is any pronounced swelling around the vent, the bird is a male; if the area around the vent is greatly distended, the male is very near or in breeding condition, but if the swelling is slight, the male has not neared or has definitely passed breeding condition.

Bursa of Fabricius

This is an outpocketing of the dorsal wall of the proctodeum and is connected to the proctodeum by an orifice. It is present in all young birds, but as birds approach maturity it disappears by involution and its orifice at the same time becomes smaller and finally closes. An orifice in the dorsal cloacal wall means that there is a bursa behind it and is thus positive proof of the individual's immaturity. With larger species of birds a cloacal examination is practicable, for the orifice is big enough to be readily seen, but with smaller birds an examination requires too great precision to be recommended. The examination procedure is, briefly: Under strong light, have the bird held with the posterior abdomen up and the tail pointing toward the examiner; with the outer fingers of one hand force back the tail; with the thumb and index finger of the same hand part the feathers over the vent, grasp the dorsal lip of the vent and pull it toward the tail; with the thumb and index finger of the other hand grasp the ventral lip of the vent and pull it away from the tail. The cloaca will now be sufficiently open to view. Look for the orifice in the dorsal wall of the cloaca. Sometimes the orifice, even though present, may be covered by a membranous fold of the adjacent cloacal wall. Therefore, probe the area gently with a blunt instrument in order to uncover the orifice, should it be there.

Certain investigators, by measuring the depth of bursae with scaled instruments, have been able to determine the ages of young birds by the stages of bursal involution.

Further information on determining the sex and age of live birds, including details on techniques, may be obtained in the following references:

Boag, D. A.
   1965   Indicators of Sex, Age, and Breeding Phenology in Blue Grouse. Jour. Wildlife Management, 29: 103-108.
Campbell, H., and R. E. Tomlinson
   1962   Some Observations on the Bursa of Fabricius in Chukars. Jour. Wildlife Management, 26: 324.
Carney, S. M., and A. D. Geis
   1960   Mallard Age and Sex Determination from Wings. Jour. Wildlife Management, 24: 372-381.
Dane, C. W.
   1968   Age Determinations of Blue-winged Teal. Jour. Wildlife Management, 22: 267-274. (Three means were used but none was 100 percent reliable.)
Davis, D. E.
   1947   Size of Bursa of Fabricius Compared with Ossification of Skull and Maturity of Gonads. Jour. Wildlife Management, 11: 244-251, (Includes information on many species of tropical birds.)
Dorney, R. S.
   1966   A New Method for Sexing Ruffed Grouse in Late Summer. Jour. Wildlife Management, 30: 623-625. (By measuring the length of a single barb of the central tail feather.)
Ellison, L. N.
   1968   Sexing and Aging Alaskan Spruce Grouse by Plumage. Jour. Wildlife Management, 32: 12-16.
Eng, R. L.
   1955   A Method for Obtaining Sage Grouse Age and Sex Ratios from Wings. Jour. Wildlife Management, 19: 267-272.
Fredrickson, L. H.
   1968   Measurements of Coots Related to Sex and Age. Jour. Wildlife Management, 32: 409-411.
Gower, W. C.
   1939   The Use of the Bursa of Fabricius as an Indication of Age in Game Birds. Trans. 4th N. Amer. Wildlife Conf., pp. 426-430.
Henderson, F. R., F. W. Brooks, R. E. Wood, and R. B. Dahlgren
   1967   Sexing of Prairie Grouse by Crown Feather Patterns. Jour. Wildlife Management, 31: 764-769.
Hochbaum, H. A.
   1942   Sex and Age Determination of Waterfowl by Cloacal Examination. Trans. 7th N. Amer. Wildlife Conf., pp. 299-307.
Horwich, R. H.
   1966   Feather Development as a Means of Aging Young Mockingbirds *(Mimus polyglottos)*. Bird-Banding, 37: 257-267.
Kessel, B.
   1951   Criteria for Sexing and Aging European Starlings *(Sturnus vulgaris)*. Bird-Banding, 22: 16-23.
Kirkpatrick, C. M.
   1944   The Bursa of Fabricius in Ring-necked Pheasants. Jour. Wildlife Management, 8:118-129. (Includes a description of the technique for determining ages by the depth of bursae.)
Martin, F. W.
   1964   Woodcock Age and Sex Determination from Wings. Jour. Wildlife Management, 28: 287-293.
Miller, A. H.
   1946   A Method of Determining the Age of Live Passerine Birds. Bird-Banding, 17: 33-35. (In dead birds it is possible to tell age by the condition of the skull. Thus if the specimen is immature, its skull is uniformly pinkish; if it is an adult, its skull is whitish and finely speckled. The author suggests a simple operation on a live bird that will reveal the condition of the skull.)
Miller, W. J., and F. H. Wagner
   1955   Sexing Mature Columbiformes by Cloacal Characters. Auk, 72: 279-285.
Norris, R. A.
   1961   A Modification of the Miller Method of Aging Live Passerine Birds. Bird-Banding, 32: 55-57.

Palmer, W. L.
1959   Sexing Live-trapped Juvenile Ruffed Grouse. Jour. Wildlife Management, 23: 111-112.
Parks, G. H.
1962   A Convenient Method of Sexing and Aging the Starling. Bird-Banding, 33: 148-151. (Use of the Kessel, 1951, method.)
Petrides, G. A.
1950   Notes on Determination of Sex and Age in the Woodcock and Mourning Dove. Auk, 67: 357-360. (The bursa is useful for age determination in the Woodcock only by dissection, and in the live Mourning Dove during the breeding season.)
Roseberry, J. L., and W. D. Klimstra
1965   A Guide to Age Determination of Bobwhite Quail Embryos. Illinois Nat. Hist. Surv. Biol. Notes No. 55.
Salt, W. R.
1954   The Structure of the Cloacal Protuberance of the Vesper Sparrow *(Pooecetes gramineus)* and Certain Other Passerine Birds. Auk, 71: 64-73.
Scott, D. M.
1967   Postjuvenal Molt and Determination of Age of the Cardinal. Bird-Banding, 38: 37-51.
Silovsky, G. D., and others
1968   Methods for Determining Age of Band-tailed Pigeons. Jour. Wildlife Management, 32: 421-424. (By wings and cloacal characters.)
Swank, W. G.
1955   Feather Molt as an Ageing Technique for Mourning Doves. Jour. Wildlife Management, 19: 412-414.
Taber, R. D.
1969   Criteria of Sex and Age. In *Wildlife Management Techniques*. Third edition, revised. Edited by R. H. Giles, Jr. Wildlife Society, Washington, D. C. (Methods of determining sex and age, particularly in game birds.)
Thompson, D. R., and C. Kabat
1950   The Wing Molt of the Bob-white. Wilson Bull., 62: 20-31. (Contains information on how age may be determined by the postjuvenal molt of the primaries.)
Weaver, H. R., and W. L. Haskell
1968   Age and Sex Determination of the Chukar Partridge. Jour. Wildlife Management, 32: 46-50. (By wings during the fall hunting season.)
Weeden, R. B.
1961   Outer Primaries as Indicators of Age among Rock Ptarmigan. Jour. Wildlife Management, 25: 337-339.
Weeden, R. B., and A. Watson
1967   Determining the Age of Rock Ptarmigan in Alaska and Scotland. Jour. Wildlife Management, 31: 825-826. (Distinguished by primary coloration.)
Wentworth, B. C., E. M. Pollack, and J. R. Smyth, Jr.
1967   Sexing Day-old Pheasants by Sex-linked Down Color. Jour. Wildlife Management, 31: 741-745.
Wight, H. M.
1956   A Field Technique for Bursal Inspection of Mourning Doves. Jour. Wildlife Management, 20: 94-95.
Wight, H. M., L. H. Blankenship, and R. E. Tomlinson
1967   Aging Mourning Doves by Outer Primary Wear. Jour. Wildlife Management, 31: 832-835.
Wolfson, A.
1952   The Cloacal Protuberance—A Means for Determining Breeding Condition in Live Male Passerines. Bird-Banding, 23: 159-165.
1954   Notes on the Cloacal Protuberance, Seminal Vesicles, and a Possible Copulatory Organ in Male Passerine Birds. Bull. Chicago Acad. Sci., 10(1): 1-23.
Wood, M.
[1969]   A Bird-Bander's Guide to Determination of Age and Sex of Selected Species. College of Agriculture, Pennsylvania State University, University Park. (Spiral binding. Includes Passeriformes and "a few other species commonly handled by bird-banders" in northeastern United States.)

Zwickel, F. C., and A. N. Lance
  1966   Determining the Age of Young Blue Grouse. Jour. Wildlife Management, 30: 712-717.
           (Based on molt and development of primary feathers.)
Zwickel, F. C., and C. F. Martinsen
  1967   Determining Age and Sex of Franklin Spruce Grouse by Tails Alone. Jour. Wildlife
           Management, 31: 760-763.

WEIGHING BIRDS

Weights of birds have three main uses: (1) To aid systematic and physiological studies within species; (2) to ascertain differences or similarities between sexes; (3) to record the growth of young.

The study of bird weights reveals a wide variety of significant problems such as: seasonal fluctuations in weight within a species, or in individuals; weight in relation to migration; weight in relation to time of day or night; weight in relation to age; weight in relation to the type and amount of food consumed. See *The Biological significance of Bird Weights* by M. M. Nice (Bird-Banding, 9: 1-11, 1938) for a discussion of some of these problems; also *Avian Anatomy, 1925-1950, and Some Suggested Problems* by H. I. Fisher (in *Recent Studies in Avian Biology,* edited by A. Wolfson, University of Illinois Press, Urbana, 1955) for a review of some of the uses for bird weights.

Weights are taken in grams on scales sensitive to one-tenth of a gram. Scales recommended for field and laboratory use should have the following specifications: For **small birds,** triple-beam, agate-bearing scales with weighing capacity of 0.01 to 111 grams without separate, auxiliary weights and up to 201 grams with auxiliary weights; for **large birds,** triple-beam, agate-bearing trip scales with weighing capacity of 610 grams without separate, auxiliary weights and up to 2,610 grams with auxiliary weights. When the scales are used in the field they should be housed in specially constructed carrying cases with handles. The cases should be so designed that they may be entirely opened on one side by hinged doors. Birds may thus be weighed within the case where, if the opened side faces leeward, there is a minimum of interference from wind.

When birds are too active to stay on the scales while being weighed, place them in a cloth sack or paper bag. The weight of the sack or bag should be determined separately and deducted from the total. If the sack or bag is to be used over a period of days, the weight should be determined each day, since it is liable to fluctuate under different atmospheric conditions.

MEASURING BIRDS

The measurements of birds have three primary uses: (1) To aid systematic studies (i.e., studies of differences or similarities between families, genera, species, or subspecies, or variations within the limits of species); (2) to determine the differences or similarities between sexes; (3) to record the growth of young.

Innumerable measurements of birds may be taken. For the above purposes, the following measurements are most generally used, ordinarily in this order.

Length (L), the distance from the tip of the bill to the tip of the longest rectrix. The specimen is placed flat on its back and is gently stretched. The commissure of the bill is brought parallel to the ruler. (See Plate XXX, Figure 4.)

Extent (Ex), the distance from tip to tip of the longest primaries of the outstretched wings. The specimen is placed flat on its back, and the wings are grasped at the wrist joints.

Plate XXX

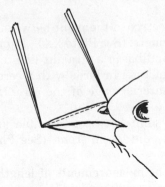

Figure 1

Figure 2

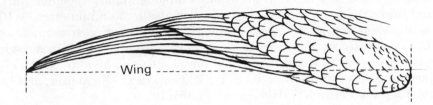

Figure 3

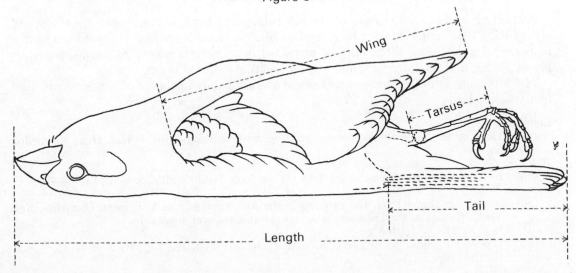

Figure 4

MEASUREMENTS OF BIRDS

Wing (W), the distance from the bend of the wing to the tip of the longest primary. The curvature is *not* straightened and the measurement is made with dividers from the bend *directly* to the tip. (See Plate XXX, Figures 3 and 4.)

Tail (T), the distance from the tip of the longest rectrix to the point between the middle rectrices where they emerge from the skin. Use dividers. (See Plate XXX, Figure 4).

Bill (B), the distance from the tip of the upper mandible in a straight line to the base of the feathers on the forehead. (See Plate XXX, Figure 1.) In birds with a cere, or similar structure, the measurement is made from the anterior edge of the cere. (See Plate XXX, Figure 2.)

Tarsus (Tar), the distance from the point of the joint between the tibia and metatarsus to the point of the joint at the base of the middle toe in front. (See Plate XXX, Figure 4.) Always use dividers.

In young birds, either naked or with neossoptiles, the measurements of length, extent, wing, and tail are taken from *the tips of the developing teleoptiles*. If neossoptiles are present on the tips of the teleoptiles, they are not included in the measurements.

Measurements of length and extent can be accurately taken only on birds in the flesh. The other measurements may be taken on bird skins.

Measurements are best taken in millimeters although many "popular" bird books use inches and decimal fractions of an inch. A ruler about 250 millimeters (= 10 inches) in length is sufficient for general use. For birds that exceed such a ruler, the extreme points may be marked on a flat surface and the distance between them measured.

If a student desires to obtain additional measurements, he may secure directions by consulting *Measurements of Birds*, by S. P. Baldwin, H. C. Oberholser, and L. G. Worley (Sci. Publ. Cleveland Mus. Nat. Hist., 2: i-ix; 1-165, 1931).

ATTRACTING BIRDS

Attracting birds to the vicinity of the laboratory and home permits close-up study and observation that cannot always be gained by other means. Moreover, it provides a source of personal enjoyment. Birds may be attracted in numerous ways—by furnishing nest-boxes, water, and food, by planting, and by creating watered areas.

The following list of references will prove useful in showing some of the methods used in attracting birds.

Bellrose, F. C.
  1955 Housing for Wood Ducks. Second printing, with revisions. Illinois Nat. Hist. Surv. Circ. 45.
Bolen, E. G.
  1967 Nesting Boxes for Black-bellied Tree Ducks. Jour. Wildlife Management, 31: 794-797.
Davison, V. E.
  1967 Attracting Birds: From the Prairies to the Atlantic. Thomas Y. Crowell Company, New York. (Particular emphasis on the food preferences of over 400 species.)
Kalmbach, E. R., and W. L. McAtee
  1957 Homes for Birds. Revised edition. U. S. Dept. Int., Fish and Wildlife Service, Conservation Bull. No. 14.
Lemmon, R. S.
  1948 How to Attract the Birds. American Garden Guild and Doubleday and Company, New York.
McAtee, W. L.
  1939 Wildfowl Food Plants: Their Value, Propagation, and Management. Collegiate Press, Ames, Iowa.
  1941 Plants Useful in Upland Wildlife Management. U. S. Dept. Int., Fish and Wildlife Service, Conservation Bull. No. 7.

1947   Attracting Birds. Revised edition. U. S. Dept. Int., Fish and Wildlife Service, Conservation Bull. No. 1.

McElroy, T. P., Jr.
1960   The New Handbook of Attracting Birds. Second edition, revised and enlarged. Alfred A. Knopf, New York.

McKenny, M.
1939   Birds in the Garden and How to Attract Them. University of Minnesota Press, Minneapolis. (Reprinted by Grossett and Dunlap, New York.)

Martin, A. C., H. S. Zim, and A. L. Nelson
1951   American Wildlife and Plants. A Guide to Wildlife Food Habits: The Use of Trees, Shrubs, Weeds, and Herbs by Birds and Mammals of the United States. McGraw-Hill Book Company, New York. (Dover reprint available.).

Mason, C. R.
1955   Invite Hummingbirds to Your Garden. Bull. Massachusetts Audubon Soc., 39: 217-221. (Much information on the kinds of flowering plants that are especially attractive to hummingbirds.)

Pettit, T. S.
1949   Birds in Your Back Yard. Harper and Brothers, New York.

Sawyer, E. J.
1944   Bird Houses, Baths, and Feeding Shelters. How to Make and Where to Place Them. Fourth edition. Cranbrook Inst. Sci. Bull. No. 1, Bloomfield Hills, Michigan.

Schutz, W. E.
1963   Bird Watching, Housing and Feeding. Bruce Publishing Company, Milwaukee.

Terres, J. K.
1968   Songbirds in Your Garden. New expanded edition. Thomas Y. Crowell Company, New York.

CONTROLLING OBJECTIONABLE BIRDS

A few bird species, in becoming exceptionally numerous, sometimes interfere with man's activities, including agricultural interests. The following publications introduce some of the problems and the means undertaken to cope with them.

Anderson, T. E.
1969   Identifying, Evaluating, and Controlling Wildlife Damage. In *Wildlife Management Techniques*. Edited by R. H. Giles, Jr. Wildlife Society, Washington, D. C.

Boudreau, G. W.
1967   Red-winged Blackbird Responses to Acoustic Stimuli. U. S. Air Force Office of Scientific Research Documentary Technical Report.
1968   Alarm Sounds and Responses of Birds and Their Application in Controlling Problem Species. Living Bird, 7: 27-46.

Frings, H., and J. Jumber
1954   Preliminary Studies on the Use of Specific Sound to Repel Starlings *(Sturnus vulgaris)* from Objectionable Roosts. Science, 119: 318-319.

Murton, R. K., and E. N. Wright, Editors
1968   The Problems of Birds as Pests. Symposia of the Institute of Biology No. 17. Academic Press, London. (Includes 12 papers grouped under "Birds and Aircraft" and "Birds and Agriculture.")

APPENDIX B
PREPARATION OF A PAPER

The student may have occasion to prepare a paper to meet a class requirement or for publication. The instructions below may aid in the preparation.

PREPARATION OF THE MANUSCRIPT

The manuscript must be typewritten on paper of good quality, preferably bond (size 8½ x 11 inches), with lines double-spaced. On each page allow a left-hand margin 1 to 1½ inches wide. Never fold the manuscript. Make and keep one carbon copy.

Number the pages consecutively, beginning with the first page *after* the table of contents. Fasten the pages together with clips—do not staple them. Place tables, figures, plates, and maps on separate sheets of paper and number them, but do not include them in the pagination.

Underscore all words to be printed in italics.

Do not abbreviate geographical names or names of days and months except in lists and tables. Do not abbreviate "Figure," "Plate," etc., when referring to illustrations in the paper. Thus: "Figure 2," not "Fig. 2."

Abbreviate designations of measurements after the first time given. Thus: "... millimeters" the first time, "... mm" the second time, without a period after the abbreviation.

Write out ordinal numbers except in tables: "First, second," etc., not "1st, 2nd."

Write out all numbers below 10 except for pages, tables, plates, etc., or when giving numbers in a series.

Use capitals for designating various items. Thus: "Bird A"; "Nest 3"; "Plate 6."

Capitalize all common names of birds when designating a species, but not otherwise. Thus: "Song Sparrow," but "sparrow"; "Eastern Meadowlark," but "meadowlark." Follow *The A. O. U. Check-list* (latest edition and its supplements) for the correct scientific and common names of North American birds.

Do not capitalize common names of species other than of birds.

Underscore all technical names of genera, species, and subspecies, but not technical names of other taxonomic categories.

Underscore the phonetic representations of bird sounds.

Give the technical name of a species the first time mentioned in the paper, but not again unless in the summary.

Use parentheses for technical names after their common names and for citations to literature. Otherwise keep parenthetical interruptions to a minimum.

Follow the latest edition of one standard dictionary for spelling, use of hyphens, etc.

For further information on the preparation of a manuscript, consult *Style Manual for Biological Journals* (American Institute of Biological Sciences, 2000 P Street NW, Washington, D. C. 20036; second edition, 1964).

If the paper is for publication, consult the editorial directions inside the back cover of the journal to which the paper will be submitted.

ORGANIZATION

Title Page

The title page includes (1) the title of the study in as few words as possible, (2) the name and address of the author, and (3) the date when the paper is submitted. If the paper is for publication, omit the title page and incorporate the information on the first and/or last pages of the paper in accordance with the custom of the journal to receive the paper for consideration.

Table of Contents

A table of contents helps to guide the instructor as well as an editor who may review the paper for publication. The table follows the title page, thus:

TABLE OF CONTENTS

Text

Introduction. Set forth the scope and purposes of the paper, including the common and technical names of the species studied, dates of the study, the precise political and geographical location of the study, the number of birds studied, the number of days involved in field and laboratory work, the number of hours involved in observations. Include under a subheading the methods and techniques used.

Observations or Results. This is the body of the paper. Here correlate all the essential information and present it in a logical sequence. Sift out data irrelevant to the main theme. Always keep the reader oriented as the theme unfolds, keeping him posted on dates and places. If this part of the paper is long, use subheadings so that the reader will not lose track of the theme. Make sure that these subheadings are given in the Table of Contents. Avoid long lists of numbers. If many must be included, incorporate them in tables or graphs.

Discussion. Here discuss certain facts and suppositions, particularly when they are at variance with known information, or need explanation, interpretation, and analysis. While it may be appropriate to discuss some of the results in those sections of the paper where the facts were given, it is usually better to reserve the principal discussion for this part of the paper. The discussion is second in importance only to the summary and is often the part of the paper of greatest interest to the reader.

Summary. Repeat significant facts and/or main points brought out in the paper. Do not repeat methods, techniques, and descriptions. Since the summary is often the first part of the paper read by the instructor or editor, it is frequently the basis for first judging the value of the paper. Therefore, put the facts and/or points in short paragraphs, concisely worded, for quick perusal.

Acknowledgments. Give full credit to the persons and organizations contributing directly or indirectly to the study. When in doubt about acknowledging someone's assistance, give it rather than risk the loss of his good will.

References. The student must not prepare a paper without first being conversant with the literature on the subject treated. Whenever necessary in his paper he should refer to statements in the literature that (1) corroborate, (2) differ from, (3) fill gaps in, or (4) serve to explain his own findings.

If the student refers to more than 10 publications, he must list them by authors alphabetically under the caption "Literature Cited." All titles must be quoted verbatim and the words spelled exactly as in the original; all words in titles except articles, prepositions, and conjunctions should be capitalized unless the journal to receive the paper for consideration requires another style of capitalization. In case the student quotes a citation without having seen the original himself, he must state "Original publication not examined" or "Title quoted from (name or source)." Do not include under "Literature Cited" titles which are not referred to in the text.

Below are cited several types of publications to show style.

Burroughs, J.
  1887 Birds and Bees. Riverside Press, Cambridge, Massachusetts. (Note how the author's name is reversed and placed by itself on one line. There is no period after the date.)
Forbush, E. H.
  1927 Birds of Massachusetts and Other New England States. Volume 2. Massachusetts Department of Agriculture, Boston. (Note the Arabic numeral followed by a period. Do not use Roman numerals in references to volumes, numbers, etc.)
Hann, H. W.
  1937 Life History of the Oven-bird in Southern Michigan. Wilson Bull., 49: 45-237. (Note the punctuation, also the abbreviation for Bulletin. Consult the *Style Manual for Biological Journals* when in doubt about the proper abbreviation.)
Odum, E. P.
  1941a Annual Cycle of the Black-capped Chickadee—1. Auk, 58: 314-333.
  1941b Winter Homing Behavior of the Chickadee. Bird-Banding, 12: 113-119. (When two or more articles are published by the same author in the same year, the dates of publication are distinguished by small letters in series. Page numbers include the entire article, not the part cited.)
Taverner, P. A., and G. M. Sutton
  1934 The Birds of Churchill, Manitoba. Annals Carnegie Mus., 23: 1-83. (When a paper is authored by more than one person, only the name of the first author is reversed.)

All citations in the text refer to the publications under "Literature Cited." Note the following variety of citations:

"Hann (1937: 148) stated that...."

"One authority (Forbush, 1927: 152) has this to say...."

"At Churchill, Manitoba, Taverner and Sutton (1934: 154) reported...."

In the above, the numbers after the colon refer to the exact page of the quotation from the papers. If the papers are mentioned in a general way, give only their dates. Example: "The study of the Ovenbird by Hann (1937) is a good example...."

Some journals require the following style of listing references:

Burroughs, J. 1877. Birds and Bees. Riverside Press, Cambridge, Massachusetts.
Forbush, E. H. 1927. Birds of Massachusetts and Other New England States. Volume 1. Massachusetts Department of Agriculture, Boston.

The style has the advantage of saving space and the great disadvantage of obscuring the dates of publication from quick reference. It is not recommended for typewritten papers.

If the student refers to not more than 10 publications in his paper, he may cite them parenthetically in the body of the paper. For example: (J. Burroughs, *Birds and Bees*, Riverside Press, Cambridge, Massachusetts, 1887); H. W. Hann (Wilson Bull., 49: 45-237, 1937). Note that the authors' names are not reversed and the title of the book is underscored. Also note the different punctuation.

Tables, Figures, Plates, and Maps. Tables are generally lists of data or related items arranged for ease of reference or comparison, often in parallel columns. Figures are generally graphs, diagrams, and drawings; plates are generally photographs and paintings.

All tables, figures, plates, and maps must be separately titled, numbered consecutively in separate series (e.g., Table 1, Table 2, etc.; Figure 1, Figure 2, etc.) with their captions on attached separate sheets of paper. All photographs not taken by the author must be credited, either by a credit line or in "Acknowledgments."

Tables, figures, plates, and maps should *supplement* the text, not take the place of it. Refer to each in the text by number; if necessary, indicate its significance.

STYLE OF WRITING

Use a simple style. Avoid wordiness. Make precise statements. Whenever possible, *use the first person* in giving personal observations and opinions—the most effective means of indicating the source. Never use contractions—e.g., write "the bird was not," instead of "the bird wasn't."

After carefully organizing all data in a logical sequence and writing the paper, review it for style, keeping in mind some of the common pitfalls described below that all too frequently show up in scientific reports.

The deadly passive voice: Example: "The worms *were eaten* by the Robin." Make the verb active and the paper live. Revise: "The Robin ate the worms."

The ambiguous observer: Example: "Fifteen Evening Grosbeaks were seen on January 1." Always let the reader know who made the observation. "[ ] saw 15 Evening Grosbeaks on January 1."

Meaningless adjectives: "An *interesting* behavior, a *beautiful* forest, an *exciting* experience." The adjectives are lazy words and tell nothing. Describe the behavior, the forest, and the experience and let the reader judge the quality of the behavior, the forest, and the experience for himself.

Unnecessary use of "would": "When the rain came, the Robin *would* cover her eggs." "The hawk *would* take off at the slightest disturbance." Write instead: "When the rain came, the Robin covered her eggs." "The hawk took off at the slightest disturbance." Avoid the past perfect tense unless an imprecise statement is intended.

The superfluous "located": Example: "The nest *located* at the edge of the road." Strike out "located" and the statement conveys the same meaning.

Too many adjectival nouns: Examples: "Bird population decline, pesticide identification techniques, forest habitat preference." The use of words in this manner, applied in the name of economy of space or to suggest the scientific approach, impairs rather than facilitates readability. "The decline in the population of birds, the techniques for identifying pesticides, the preference for habitat in a forest" takes little more space and detracts nothing whatsoever from the stature of the writer.

The unidentified "it": Question every "it" in the paper. Is "it" necessary? If so, make sure that "it" has a definite antecedent. Many times one can avoid "it." Examples: "It is possible that...." "It is the opinion of the writer that...." Write instead: "The possibility that...." "My opinion is that...." Or "I believe that...."

"Utilization" and other pompous words: Example: "Utilization of habitat." This says nothing more than "the uses of habitat," but it has a high sounding ring, chosen to impress the reader. Utilization is synonymous with use. When given a choice, always use (not utilize!) the simpler word.

APPENDIX C
BIBLIOGRAPHIES PERTAINING TO ORNITHOLOGY

Listed below are bibliographies and library catalogues that will assist the student in his search through ornithological literature. In addition to these bibliographies the student will find of great assistance the reviews and notices of various publications, appearing regularly in *Bird-Banding* (highly recommeded), *The Auk, The Ibis, The Wilson Bulletin,* and *The Atlantic Naturalist.*

Allen, F. P.
    1935   A Check List of Periodical Literature and Publications of Learned Societies of Interest to Zoologists in the University of Michigan Libraries. Univ. Michigan Mus. Zool. Circ. No. 2.

Allouse, B. E.
    1954   A Bibliography of the Vertebrate Fauna of Iraq and Neighbouring Countries. II Birds. Iraq Nat. Hist. Mus. Publ. 5. (Includes Iran, Jordan, Palestine, Sinai, Syria, Lebanon, and Turkey.)

Anker, J.
    1938   Bird Books and Bird Art: An Outline of the Literary History and Iconography of Descriptive Ornithology, Based Principally on the Collection of Books Containing Plates with Figures of Birds and Their Eggs Now in the University Library at Copenhagen and Including a Catalogue of These Works. Levin and Munksgaard, Copenhagen.

Aves in *Biological Abstracts.* BioSciences Information Service of Biological Abstracts, Philadelphia. Volumes 1-49 (1926-68). Others to follow. Issues appear semi-monthly.

Aves in *Zoological Record.* Zoological Society of London. Volumes 1-102 (1864-1968). Others to follow. Annual volumes appear 12 to 18 months after year closes.

Bailey, H. B.
    1881   "Forest and Stream" Bird Notes. An Index and Summary of all the Ornithological Matter Contained in *Forest and Stream,* Vols. 1-12. Forest and Stream Publishing Company, New York.

British Museum
    1903-   Catalogue of the Books, Manuscripts, Maps and Drawings in the British Museum (Natural History). Five volumes. British Museum, London.
    15
    1922-   Supplement. Three Volumes.
    40

Burns, F. L.
    1915   A Bibliography of Scarce of Out of Print North American Amateur and Trade Periodicals Devoted More or Less to Ornithology. Supplement to Oologist, 32: 1-32.

Carnegie Library
    1929   A List of Books on Ornithology Contained in the Carnegie Library of Pittsburgh. Sewickley, Pennsylvania.

Coues, E.
    1878-   American Ornithological Bibliography. Part 1 (1878), Misc. Publ. U. S. Geol. Surv. Terr.
    80   No. 11: 567-784; Parts 2-3 (1879-80), Bull. U. S. Geol. Geog. Surv. Terr. No. 5: 239-330, 521-1066; Part 4 (1881), Proc. U. S. Nat. Mus., 2: 359-482.

Crispens, C. G., Jr.
  1960 Quails and Partridges of North America: A Bibliography. University of Washington Press, Seattle.
Ellis, J. C.
  1949 Nature and Its Applications: Over 200,000 Selected References to Nature Forms and Illustrations of Nature as Used in Every Way. F. W. Faxon Company, Boston.
Fisher, H. I.
  1947 Bibliography of Hawaiian Birds since 1890. Auk, 64: 78-97.
Gallatin, F., Jr.
  1908 A Catalogue of a Collection of Books on Ornithology in the Library of Frederic Gallatin, Jr. New York. (Privately printed.)
Grinnell, J.
  1909 A Bibiliography of California Ornithology (1797-1907). Pacific Coast Avifauna No. 5, Cooper Ornithological Club, Santa Clara.
  1924 Bibliography of California Ornithology (1908-1923). Pacific Coast Avifauna No. 16, Cooper Ornithological Club, Berkeley.
  1939 Bibliography of California Ornithology (1924-1938). Pacific Coast Avifauna No. 26, Cooper Ornithological Club, Berkeley.
Gurney, J. H.
  1921 Early Annals of Ornithology. H. F. and G. Witherby, London.
Hachisuka, M.
  1952 Bibliography of Chinese Birds. Quart. Jour. Taiwan Mus., 5: 71-209.
Harting, J. E.
  1891 Catalogue of Books Ancient and Modern Relating to Falconry with Notes, Glossary, and Vocabulary. Bernard Quaritch, London.
Kuroda, M. D.
  1942 A Bibliography of the Duck Tribe: Anatidae, Mostly from 1926 to 1940, Exclusive of That of Dr. Phillips' Work. Herald Press, Tokyo.
Legendre, M.
  1926- Bibliographie des Faunes Ornithologiques des Regions Françaises. Rev. Franc. Ornith.,
  27   10: 80-87, 182-191, 264-279, 372-382; 11: 60-71, 153-161.
Lincoln, F. C.
  1928 A Bibliography of Bird Banding in America. Supplement to Auk, 4 : 1-73.
Lovell, H. B., and M. Slack
  1949 Bibliography of Kentucky Ornithology. Kentucky Ornith. Soc., Occas. Papers No. 1, Bowling Green.
Low, G. C.
  1931 The Literature of the Charadriiformes from 1894-1928, with a Classification of the Order, and Lists of the Genera, Species and Subspecies. Second edition. H. F. and G. Witherby, London.
Marelli, C. A.
  1934 Contribuciones al Estudio de la Fauna Argentina. Bibliographia Relativa a la Ornitolpgoa. Mem. del Jardin Zool., 5: 37-106.
Mathews, G. M.
  1953 Die illustrierten Vogelbücher. Hiersemann Verlag, Stuttgart, Germany. (A bibliography of illustrated bird books published up to 1951.)
Mathews, G. M.
  1925 Bibliography of the Birds of Australia. Supplements 4 and 5 to *The Birds of Australia.* H. F. and G. Witherby, London.
McAtee, W. L.
  1913 Index to Papers Relating to the Food of Birds by Members of the Biological Survey in Publications of the United States Department of Agriculture 1885-1911. U. S. Dept. Agric., Bur. Biol. Surv. Bull. No. 43: 1-69.
Meisel, M.
  1924- A Bibliography of American Natural History. Three volumes. Premier Publishing Com-
  29   pany, Brooklyn, New York.
Minnesota, University of
  1925 Check List of Periodicals and Serials in Biology and Allied Sciences Available in Libraries of the University of Minnesota and Vicinity. Minneapolis.

Mullens, W. H., and H. K. Swann
  1916- A Bibliography of British Ornithology from the Earliest Times to the End of 1912. (Pub-
  17   lished in six parts.) Macmillan and Company, London.
Mullens, W. H., H. K. Swann, and F. R. C. Jourdain
  1919- A Geographical Bibliography of British Ornithology from the Earliest Times to the End
  20   of 1918 Arranged by Counties. (Published in six parts.) Witherby and Company, London.
Nissen, C.
  1953   Die Illustrierten Vogelbücher. Hiersemann Verlag, Stuttgart, Germany. (A bibliography
          of illustrated bird books published up to 1951.)
Oliver, H. C.
  1968   Annotated Index to Some Early New Zealand Bird Literature. Wildlife Publication No.
          106, Department of Internal Affairs, Wellington, New Zealand.
Osborn, H.
  1930   Bibliography of Ohio Zoology. Ohio Biol. Surv. Bull. 23, Ohio State University Press,
          Columbus.
Phillips, J. C.
  1926   Bibliography of A Natural History of the Ducks. Riverside Press, Cambridge, Massa-
          chusetts. (Reprinted, with one plate, from *A Natural History of the Ducks.* Volume 4,
          1926.)
  1930   American Game Mammals and Birds. A Catalogue of Books, 1582-1925. Houghton Mifflin
          Company, Boston.
Porter, C. E.
  1912   Bibliografía Ornitolójica de Chile. Universidad de Santiago.
Reichenow, A.
  1894   Bibliographia Ornithologiae Aethiopicae. Jour. f. Ornith., 42: 172-226.
Riley, J. H., and C. W. Richmond
  1922- A Partial Bibliography of Chinese Birds. Jour. North China Branch Royal Asiatic Soc.,
  23   53: 196-237; 54: 225-226.
Ripley, S. D., and L. L. Scribner
  1961   Ornithological Books in the Yale University Library Including the Library of William
          Robertson Coe. Yale University Press, New Haven.
Roberts, B.
  1941   A Bibliography of Antarctic Ornithology. British Museum (Natural History), London.
Ronsil, R.
  1948- Bibliographie Ornithologique Francaise. In *Encyclopédie Ornithologique*, Vols. 8 and 9.
  49   (Described in Auk, 67: 262-263, 1950.)
Schalow, H.
  1906   Beitrage zu Einer Ornithologischen Bibliographie des Atlas Gebietes. Jour. f. Ornith.,
          54: 100-143.
Schlegel, R.
  1925   "Ornis Saxonica." Ein Beitrag zur Bibliographie des Gebiets. Jour. f. Ornith., 73: 247-255.
Shaver, J. M.
  1931   A Bibliography of Tennessee Ornithology. Jour. Tennessee Acad. Sci., 6: 179-190.
Sitwell, S., H. Buchanan, and J. Fisher
  1953   Fine Bird Books: 1700-1900. Collins and Van Nostrand, London and New York.
Smith, R. C.
  1958   Guide to the Literature of the Zoological Sciences. Fifth edition. Burgess Publishing
          Company, Minneapolis.
Stephens, T. C.
  1945   An Annotated Bibliography of South Dakota Ornithology. Privately printed by the
          author, Sioux City, Iowa.
  1956   An Annotated Bibliography of North Dakota Ornithology. Nebraska Ornith. Union
          Occas. Papers No. 1.
Stone, W.
  1933   American Ornithological Literature 1883-1933. In *Fifty Years' Progress of American
          Ornithology* published by the American Ornithologists' Union.
Strong, R. M.
  1939- A Bibliography of Birds. Field Mus. Nat. Hist. Zool. Ser., 25. Parts 1-4. (Part 3, pp. 11-17,
  59   gives a number of bibliographies and library catalogues not included in this list.)

Thayer, E., and V. Keyes
   1913   Catalogue of a Collection of Books on Ornithology in the Library of John E. Thayer, Boston. (Privately printed.)
Underwood, M. H.
   1954   Bibliography of North American Minor Natural History Serials in the University of Michigan Libraries. University of Michigan Press, Ann Arbor.
Van Tyne, J., and A. J. Berger
   1959   Fundamentals of Ornithology. John Wiley and Sons, New York. ("Ornithological Sources," pp. 553-558, gives a number of bibliographies not included in this list.)
Whittell, H. M.
   1954   The Literature of Australian Birds: A History and a Bibliography of Australian Ornithology. Paterson Brokensha Proprietary, Perth.
Wildlife Abstracts 1935-60. Fish and Wildlife Service, U. S. Department of the Interior, Washington, D. C. (A bibliography and index of the abstracts in Wildlife Review; see below.)
Wildlife Review. Fish and Wildlife Service, U. S. Department of the Interior, Washington, D. C. Numbers 1-130 (1935-68). Others to follow.
Wistar Institute
   1909   Biological Serials, Exclusive of Botany, in the Libraries of Philadelphia. Wistar Inst. Bull. No. 2.
Wood, C. A.
   1931   An Introduction to the Literature of Vertebrate Zoology. Oxford University Press, London.
Zimmer, J. T.
   1926   Catalogue of the Edward E. Ayer Ornithological Library. Two volumes. Field Mus. Nat. Hist. Zool. Ser., 16.

APPENDIX D
BIBLIOGRAPHY OF LIFE
HISTORY STUDIES

The following bibliography has been compiled in order to lead the student directly to some of the more important life history treatises not found in the regular ornithological journals. Part I presents some of the general works dealing extensively with life histories. Part II presents some of the works dealing with particular groups and species. In neither part is completeness attempted. Space limits the bibliography primarily to North American birds.

PART I

Bailey, A. M., and R. J. Niedrach
  1965 Birds of Colorado. Two volumes, Denver Museum of Natural History.
Behle, W. H.
  1958 The Bird Life of Great Salt Lake. University of Utah Press, Great Salt Lake. (Extensive information on the White Pelican, Double-crested Cormorant, Treganza Great Blue Heron, and California Gull.)
Belopol'skii, L. O.
  1957 Ecology of Sea Colony Birds of the Barents Sea. Translated from Russian and published in 1961 by Israel Program for Scientific Translations, Jerusalem. (Available from Office of Technical Services, U. S. Department of Commerce, Washington, D. C.)
Bendire, C.
  1892 Life Histories of North American Birds. Smithsonian Inst. Spec. Bull. No. 1. (Includes Tetraonidae, Phasianidae, Cracidae, Columbidae, Cathartidae, Falconidae, Strigidae, Bubonidae.)
  1895 Life Histories of North American Birds. Smithsonian Inst. Spec. Bull. (Includes Psittacidae, Cuculidae, Trogonidae, Picidae, Caprimulgidae, Micropodidae, Trochilidae, Cotingidae, Tyrannidae, Alaudidae, Corvidae, Sturnidae, Icteridae.)
Bent, A. C.
  1919 Life Histories of North American Diving Birds. U. S. Natl. Mus. Bull. 107. (Reprinted 1946, Dodd, Mead and Company, New York.)
  1921 Life Histories of North American Gulls and Terns. U. S. Natl. Mus. Bull. 113. (Reprinted 1947, Dodd, Mead and Company, New York.)
  1922 Life Histories of North American Petrels and Pelicans and Their Allies. U. S. Natl. Mus. Bull. 121.
  1923 Life Histories of North American Wild Fowl. Part 1. U. S. Natl. Mus. Bull. 126.
  1925 Life Histories of North American Wild Fowl. Part 2. U. S. Natl. Mus. Bull. 130.
  1926 Life Histories of North American Marsh Birds. U. S. Natl. Mus. Bull. 135.
  1927 Life Histories of North American Shore Birds. Part 1. U. S. Natl. Mus. Bull. 142.
  1929 Life Histories of North American Shore Birds. Part 2. U. S. Natl. Mus. Bull. 146.
  1932 Life Histories of North American Gallinaceous Birds. U. S. Natl. Mus. Bull. 162.
  1937 Life Histories of North American Birds of Prey. Part 1. U. S. Natl. Mus. Bull. 167.

1938   Life Histories of North American Birds of Prey. Part 2. U. S. Natl. Mus. Bull. 170.

1939   Life Histories of North American Woodpeckers. U. S. Natl. Mus. Bull. 174.

1940   Life Histories of North American Cuckoos, Goatsuckers, Hummingbirds, and their Allies. U. S. Natl. Mus. Bull. 176.

1942   Life Histories of North American Flycatchers, Larks, Swallows, and Their Allies. U. S. Natl. Mus. Bull. 179.

1946   Life Histories of North American Jays, Crows, and Titmice. U. S. Natl. Mus. Bull. 191.

1948   Life Histories of North American Nuthatches, Wrens, Thrashers, and Their Allies. U. S. Natl. Mus. Bull. 195.

1949   Life Histories of North American Thrushes, Kinglets, and Their Allies. U. S. Natl. Mus. Bull. 196.

1950   Life Histories of North American Wagtails, Shrikes, Vireos, and Their Allies. U. S. Natl. Mus. Bull. 197.

1953   Life Histories of North American Wood Warblers. U. S. Natl. Mus. Bull. 203.

1958   Life Histories of North American Blackbirds, Orioles, Tanagers, and Allies. U. S. Natl. Mus. Bull. 211.

1968   Life Histories of North American Cardinals, Grosbeaks, Buntings, Towhees, Finches, Sparrows, and Allies. Parts 1, 2, and 3. Compiled and edited by O. L. Austin, Jr. U. S. Natl. Mus. Bull. 237.

Dover reprints of the entire series (above) by Bent are available.

Brewster, W.

1906   The Birds of the Cambridge Region of Massachusetts. Mem. Nuttall Ornith. Club No. 4. Cambridge, Massachusetts.

1924-   The Birds of the Lake Umbagog Region of Maine. Bull. Mus. Comp. Zool., 66, Parts 1-3.
37   (Part 4 of this paper was compiled by Ludlow Griscom. See below.)

Craighead, J. J., and F. C. Craighead, Jr.

1956   Hawks, Owls and Wildlife. Stackpole Company, Harrisburg, Pennsylvania, and Wildlife Management Institute, Washington, D. C.

Fisher, J., and R. M. Lockley

1954   Sea-Birds. Houghton Mifflin Company, Boston.

Forbush, E. H.

1916   A History of the Game Birds, Wild-Fowl and Shore Birds of Massachusetts and Adjacent States. Second edition. Massachusetts State Board of Agriculture, Boston.

1925-   Birds of Massachusetts and Other New England States. Three volumes. Massachusetts
29   Department of Agriculture, Boston.

Grinnell, J., H. C. Bryant, and T. I. Storer

1918   The Game Birds of California. University of California Press, Berkeley. (Includes ducks, geese, swans, spoonbills, ibises, cranes, rails, gallinules, coots, shore birds, quail, grouse, and pigeons.)

Grinnell, J., J. Dixon, and J. M. Linsdale

1930   Vertebrate Natural History of a Section of Northern California through the Lassen Peak Region. University of California Press, Berkeley.

Grinnell, J., and T. I. Storer

1924   Animal Life in the Yosemite: An Account of the Mammals, Birds, Reptiles, and Amphibians in a Cross-section of the Sierra Nevada. University of California Press, Berkeley.

Griscom, L.

1938   The Birds of the Lake Umbagog Region of Maine. (Part 4.) Bull. Mus. Comp. Zool., 66, Part 4.

Kendeigh, S. C.

1952   Parental Care and Its Evolution in Birds. Second corrected printing, 1955. Illinois Biol. Monogr., 22: i-x; 1-356.

Linsdale, J. M.

1938   Environmental Responses of Vertebrates in the Great Basin. Amer. Midland Nat., 19: 1-206.

Murphy, R. C.

1936   Oceanic Birds of South America. Two volumes. American Museum of Natural History, New York.

Palmer, R. S.
  1949   Maine Birds. Bull. Mus. Comp. Zool., 102: 1-656.
Palmer, R. S., Editor
  1962   Handbook of North American Birds. Volume 1. Loons through Flamingos. Yale University Press, New Haven, Connecticut.
Parmelee, D. F., H. A. Stephens, and R. H. Schmidt
  1967   The Birds of Southeastern Victoria Island and Adjacent Small Islands. Natl. Mus. Canada Bull. 222.
Reilly, E. M., Jr.
  1968   The Audubon Illustrated Handbook of American Birds. Edited by O. S. Pettingill, Jr. McGraw-Hill Book Company, New York.
Roberts, T. S.
  1936   The Birds of Minnesota. Two volumes. Second revised edition. University of Minnesota Press, Minneapolis.
Sherman, A. R.
  1952   Birds of an Iowa Dooryard. Edited by F. J. Pierce. Christopher Publishing House, Boston. (A posthumous work of a pioneer student of life histories. Contains both new and reprinted material.)
Skutch, A. F.
  1954   Life Histories of Central American Birds: Families Fringillidae, Thraupidae, Icteridae, Parulidae, and Coerebidae. Pacific Coast Avifauna No. 31, Cooper Ornithological Society, Berkeley, California.
  1960   Life Histories of Central American Birds II: Families Vireonidae, Sylviidae, Turdidae, Troglodytidae, Paridae, Corvidae, Hirundinidae, and Tyrannidae. Pacific Coast Avifauna No. 34, Cooper Ornithological Society, Berkeley, California.
  1967   Life Histories of Central American Highland Birds. Publ. Nuttall Ornith. Club No. 7, Cambridge, Massachusetts.
Stone, W.
  1937   Bird Studies at Old Cape May. Two volumes. Academy of Natural Sciences of Philadelphia. (Dover reprint available.)
Stoner, D.
  1932   Ornithology of the Oneida Lake Region: With Reference to the Late Spring and Summer Seasons. Roosevelt Wild Life Annals, 4: 277-764.
Sutton, G. M.
  1932   The Birds of Southampton Island. Mem. Carnegie Mus., 12 (pt. 2, sec. 2): 1-275.
Todd, W. E. C.
  1940   Birds of Western Pennsylvania. University of Pittsburgh Press, Pittsburgh.
  1963   Birds of the Labrador Peninsula and Adjacent Areas. University of Toronto Press.
Townsend, C. W.
  1905   The Birds of Essex County, Massachusetts. Mem. Nutthall Ornith. Club No. 3, Cambridge, Massachusetts.
  1920   Supplement to the Birds of Essex County, Massachusetts, Mem. Nuttall Ornith. Club No.5, Cambridge, Massachusetts.

PART II

Gaviiformes

Olson, S. T., and W. H. Marshall
  1952   The Common Loon In Minnesota. Minnesota Mus. Nat. Hist. Occas. Papers No. 5.
Rankin, N.
  1947   Haunts of British Divers. Collins, London. (Black-throated Diver, *Colymbus arcticus*, and Red-throated Diver, *Colymbus stellatus.*)

Podicipediformes

Munro, J. A.
  1941   The Grebes. Studies of Waterfowl in British Columbia. Brit. Columbia Prov. Mus. Occas. Papers No. 3.
See Rankin, under Gaviiformes, for Great-crested Grebe *(Podiceps cristatus).*

Procellariiformes

Fisher, J.
1952 The Fulmar. Collins, London.
Fisher, J., and G. Waterson
1941 The Breeding, Distribution, History and Population of the Fulmar *(Fulmarus glacialis)* in the British Isles. Jour. Animal Ecol., 10: 204-272.
Lockley, R. M.
1942 Shearwaters. J. M. Dent and Sons, London.
Roberts, B.
1940 The Life Cycle of Wilson's Petrel *Oceanites oceanicus* (Kuhl). Brit. Graham Land 1934-37 Sci. Repts., 1 (2): 141-194.

Pelecaniformes

Gurney, J. H.
1913 The Gannet: A Bird with a History. Witherby and Company, London.
Lewis, H. F.
1929 The Natural History of the Double-crested Cormorant *Phalacrocorax auritus auritus* (Lesson). Ru-Mi-Lou Books, Ottawa, Canada.
Mendall, H. L.
1936 The Home-life and Economic Status of the Double-crested Cormorant *Phalacrocorax auritus auritus* (Lesson). The Maine Bulletin, 39 (No. 3). Orono.

Ciconiiformes

Allen, R. P.
1942 The Roseate Spoonbill. Natl. Audubon Soc. Res. Rept. No. 2, New York. (Dover reprint available.)
1956 The Flamingos: Their Life History and Survival. With Special Reference to the American or West Indian Flamingo *(Phoenicopterus ruber)*. Natl. Audubon Soc. Res. Rept. No. 5, New York.
Allen, R. P., and F. P. Mangels
1940 Studies of the Nesting Behavior of the Black-crowned Night Heron. Proc. Linnaean Soc. New York, Nos. 50-51: 1-28.
Bouet, G.
1950 La Vie des Cigognes. Les Éditions Braun et Cie, Paris.
Chapman, F. M.
1905 A Contribution to the Life History of the American Flamingo *(Phoenicopterus ruber)*, with Remarks upon Specimens. Bull. Amer. Mus. Nat. Hist., 21: 53-77.
Cottrille, W. P., and B. D. Cottrille
1958 Great Blue Heron: Behavior at the Nest. Univ. of Michigan Mus. Zool. Misc. Publ. No. 102.
Gallet, É.
1950 The Flamingos of the Camargue. (Translated from the French by Sumner Austin.) Basil Blackwell, Oxford, England.
Haverschmidt, F.
1949 The Life of the White Stork. E. J. Brill, Leiden.
Lowe, F. A.
1955 The Heron. Collins, London. (Concerns *Ardea cinerea.*)
Meyerriecks, A. J.
1960 Comparative Breeding Behavior of Four Species of North American Herons. Publ. Nuttall Ornith. Club No. 2, Cambridge, Massachusetts.
Rooth, J.
1965 The Flamingos on Bonaire (Netherlands Antilles): Habitat, Diet and Reproduction of *Phoenicopterus ruber ruber*. Uitgaven "Natuurwetenschappelijke Studiekring voor Suriname en den Nederlandse Antillen," Utrecht, No. 41.

Anseriformes

Banko, W. E.
1960 The Trumpeter Swan: Its History, Habits, and Population in the United States. N. Amer. Fauna No. 63: i-x; 1-214, U. S. Fish and Wildlife Service, Washington, D. C.

Bennett, L. J.
1938   The Blue-winged Teal: Its Ecology and Management. Collegiate Press, Ames, Iowa.
Coulter, M. W., and W. R. Miller
1968   Nesting Biology of Black Ducks and Mallards in Northern New England. Vermont Fish and Game Dept. Bull. No. 68-2.
Delacour, J.
1954- The Waterfowl of the World. Four volumes. Country Life, London.
64
Einarsen, A. S.
1965   Black Brant: Sea Goose of the Pacific Coast. University of Washington Press, Seattle.
Grice, D., and J. P. Rogers
1965   The Wood Duck in Massachusetts. Massachusetts Division of Fisheries and Game, Boston.
Hanson, H. C.
1965   The Giant Canada Goose. Southern Illinois University Press, Carbondale.
Hochbaum, H. A.
1959   The Canvasback on a Prairie Marsh. Second edition. Stackpole Company, Harrisburg, Pennsylvania, and Wildlife Management Institute, Washington, D. C.
Kortright, F. H.
1942   The Ducks, Geese and Swans of North America. American Wildlife Institute, Washington, D. C.
Kossack, C. W.
1950   Breeding Habits of Canada Geese under Refuge Conditions. Amer. Midland Nat., 43: 627-649.
Low, J. B.
1945   Ecology and Management of the Redhead, *Nyroca americana*, in Iowa. Ecol. Monogr., 15: 35-69.
Mendall, H. L.
1958   The Ring-necked Duck in the Northeast. Univ. Maine Bull., 60: i-xvi; 1-317.
Millais, J. G.
1902   The Natural History of the British Surface-feeding Ducks. Longmans, Green and Company, London.
1918   British Diving Ducks. Two volumes. Longmans, Green and Company, London.
Munro, J. A.
1939   Studies of Water-fowl in British Columbia. Barrow's Golden-eye, American Golden-eye. Trans. Royal Canadian Inst., 22: 259-318.
1941   Studies of Waterfowl in British Columbia. Greater Scaup Duck, Lesser Scaup Duck. Canadian Jour. Res., 19: 113-138.
1942   Studies of Waterfowl in British Columbia. Buffle-head. Canadian Jour. Res., 20: 133-160.
1943   Studies of Waterfowl in British Columbia. Mallard. Canadian Jour. Res., 21: 223-260.
1944   Studies of Waterfowl in British Columbia. Pintail. Canadian Jour. Res., 22: 60-86.
1949   Studies of Waterfowl in British Columbia. Green-winged Teal. Canadian Jour. Res., 27: 149-178.
Phillips, J. C.
1922- A Natural History of the Ducks. Four volumes. Houghton Mifflin Company, Boston.
26
Ryder, J. P.
1967   The Breeding Biology of Ross' Goose in the Perry River Region, Northwest Territories. Canadian Wildlife Serv. Rept. Ser., No. 3.
Savage, C.
1952   The Mandarin Duck. Adam and Charles Black, London.
Soper, J. D.
1930   The Blue Goose. Department of the Interior, Ottawa, Canada.
1942   Life History of the Blue Goose, *Chen caerulescens* (Linnaeus). Proc. Boston Nat. Hist., 42: 121-225.
Sowls, L. K.
1955   Prairie Ducks: A Study of Their Behavior, Ecology, and Management. Stackpole Company, Harrisburg, Pennsylvania, and Wildlife Management Institute, Washington, D. C.

Williams, C. S.
   1967   Honker: A Discussion of the Habits and Needs of the Largest of Our Canada Geese. D. Van Nostrand Company, Princeton, New Jersey.
Wright, B. S.
   1954   High Tide and East Wind: The Story of the Black Duck. Stackpole Company, Harrisburg, Pennsylvania, and Wildlife Management Institute, Washington, D. C.

Falconiformes

Brown, L., and D. Amadon
   1968   Eagles, Hawks and Falcons of the World. Two volumes. McGraw-Hill Book Company, New York.
Fitch, H. S.
   1963   Observations on the Mississippi Kite in Southwestern Kansas. Univ. Kansas Publ. Mus. Nat. Hist., 12: 503-519.
Gordon, S.
   1955   The Golden Eagle: King of Birds. Collins, London.
Grossman, M. L., and J. Hamlet
   1964   Birds of Prey of the World. Clarkson N. Potter, New York.
Herrick, F. H.
   1934   The American Eagle. D. Appleton-Century Company, New York.
Hickey, J. J., Editor
   1969   Peregrine Falcon Populations: Their Biology and Decline. University of Wisconsin Press, Madison.
Koford, C. B.
   1953   The California Condor. Natl. Audubon Soc. Res. Rept. No. 4, New York. (Dover reprint available.)
MacPherson, H. B.
   1909   The Home-life of a Golden Eagle. Witherby and Company, London.
Miller, A. H., I. I. McMillan, and E. McMillan
   1965   The Current Status and Welfare of the California Condor. Natl. Audubon Soc. Res. Rept. No. 6, New York.
Robinson, T. S.
   1957   Notes on the Development of a Brood of Mississippi Kites in Barber County, Kansas. Trans. Kansas Acad. Sci., 60: 174-180.
Slevin, J. R.
   1929   A Contribution to Our Knowledge of the Nesting Habits of the Golden Eagle. Proc. California Acad. Sci., 18: 45-71.
Willgohs, J. F.
   1961   The White-tailed Eagle *Haliaëtus alibicilla albicilla* (Linné) in Norway. Norwegian Universities Press, Bergen.

Galliformes—Tetraonidae

Ammann, G. A.
   1957   The Prairie Grouse of Michigan. Game Division, Department of Conservation, Lansing.
Baker, M. F.
   1953   Prairie Chickens of Kansas. Univ. Kansas Mus. Nat. Hist. Misc. Publ. No. 5.
Bendell, J. F., and R. W. Elliott
   1967   Behaviour and the Regulation of Numbers in Blue Grouse. Canadian Wildlife Serv. Rept. Ser., No. 4.
Bump, G., and others
   1947   The Ruffed Grouse: Life History, Propagation, Management. New York State Conservation Department, Albany.
Edminster, F. C.
   1947   The Ruffed Grouse. Its Life History, Ecology and Management. Macmillan Company, New York.
Fisher, L. W.
   1939   Studies of the Eastern Ruffed Grouse in Michigan *(Bonasa umbellus umbellus)*. Michigan State College Agric. Exp. Sta. Tech. Bull. 166.

Girard, G. I.
  1937   Life History, Habits, and Food of the Sage Grouse, *Centrocercus urophasianus* Bonaparte. Univ. Wyoming Publ., 3: 1-56.

Gross, A. O.
  1928   The Heath Hen. Mem. Boston Soc. Nat. Hist., 6: 487-588.
  1930   Progress Report of the Wisconsin Prairie Chicken Investigation. Wisconsin Conservation Commission, Madison.

Lehmann, V. W.
  1941   Attwater's Prairie Chicken: Its Life History and Management. N. Amer. Fauna No. 57: i-iv; 1-65, U. S. Fish and Wildlife Service, Washington, D. C.

Patterson, R. L.
  1952   The Sage Grouse in Wyoming. Sage Books, Denver.

Schwartz, C. W.
  1945   The Ecology of the Prairie Chicken in Missouri. Univ. Missouri Stud., 20: 1-99.

Galliformes—Phasianidae

Allen, D. L.
  1956   Pheasants in North America. Stackpole Company, Harrisburg, Pennsylvania, and Wildlife Management Institute, Washington, D. C.

Baskett, T. S.
  1947   Nesting and Production of the Ring-necked Pheasant in North-central Iowa. Ecol. Monogr., 17: 1-30.

Beebe, W.
  1926   Pheasants, Their Lives and Homes. Two volumes. Doubleday, Page and Company, Garden City, New York.

Buss, I. O.
  1946   Wisconsin Pheasant Populations. Wisconsin Conservation Department, Madison.

Delacour, J.
  1951   The Pheasants of the World. Country Life, London.

Glading, B.
  1938   Studies on the Nesting Cycle of the California Valley Quail in 1937. California Fish and Game, 24: 318-340.

Gorsuch, D. M.
  1934   Life History of the Gambel Quail in Arizona. Univ. Arizona Bull., 5: 1-89.

Hamerstrom, F. N., Jr.
  1936   A Study of the Nesting Habits of the Ring-necked Pheasant in Northwest Iowa. Iowa State College Jour. Sci., 10: 173-203.

McCabe, R. A., and A. S. Hawkins
  1946   The Hungarian Partridge in Wisconsin. Amer. Midland Nat., 36: 1-75.

Robertson, W. B., Jr.
  1958   Investigations of Ring-necked Pheasants in Illinois. Illinois Dept. Cons. Div. Game Management Tech. Bull. No. 1.

Robinson, T. S.
  1957   The Ecology of Bobwhites in South-central Kansas. State Biological Survey and Museum of Natural History, University of Kansas.

Rosene, W., Jr.
  1969   The Bobwhite Quail: Its Life and Management. Rutgers University Press, New Brunswick, New Jersey.

Schick, C.
  1952   A Study of Pheasants on the 9,000-acre Prairie Farm, Saginaw County, Michigan. Game Division, Michigan Department of Conservation, Lansing.

Stoddard, H. L.
  1931   The Bobwhite Quail: Its Habits, Preservation and Increase. Charles Scribner's Sons, New York.

Sumner, E. L., Jr.
  1935   A Life History Study of the California Quail, with Recommendations for Conservation and Management. California Fish and Game, 21: 165-256, 277-342.

Yeatter, R. E.
  1934 The Hungarian Partridge in the Great Lakes Region. Univ. Michigan School of Forestry and Cons., Bull. No. 5, Ann Arbor.
  1943 The Prairie Chicken in Illinois. Bull. Illinois Nat. Hist. Surv., 22: 377-416.
Yocom, C. F.
  1943 The Hungarian Partridge, *Perdix perdix* Linn., in the Palouse Region, Washington. Ecol. Monogr., 13: 167-202.

Galliformes—Meleagrididae

Dalke, P. D., A. S. Leopold, and D. L. Spencer
  1946 The Ecology and Management of the Wild Turkey in Missouri. Missouri Cons. Comm. Tech. Bull. No. 1.
Hewitt, O. H., Editor
  1967 The Wild Turkey and Its Management. Wildlife Society, Washington, D. C.
Latham, R. M.
  1956 Complete Book of the Wild Turkey. Stackpole Company, Harrisburg, Pennsylvania.
Ligon, J. S.
  1946 History and Management of Merriam's Wild Turkey. Univ. New Mexico Publ. in Biol., No. 1.
Mosby, H. S., and C. O. Handley
  1943 The Wild Turkey in Virginia: Its Status, Life History and Management. Commission of Game and Inland Fisheries, Richmond.
Schorger, A. W.
  1966 The Wild Turkey: Its History and Domestication. University of Oklahoma Press, Norman.
Wheeler, R. J., Jr.
  1948 The Wild Turkey in Alabama. Alabama Department of Conservation, Montgomery.

Charadriiformes—Charadrii

Ennion, E. A. R.
  1949 The Lapwing. Methuen and Company, London.
Haverschmidt. F.
  1963 The Black-tailed Godwit. E. J. Brill, Leiden.
Mendall, H. L., and C. M. Aldous
  1943 The Ecology and Management of the American Woodcock. Maine Cooperative Wildlife Research Unit, Orono.
Nethersole-Thompson, D.
  1951 The Greenshank. Collins, London. (Concerns *Tringa nebularia.*)
Palmer, R. S.
  1967 Species Accounts. In *The Shorebirds of North America.* Edited by G. D. Stout. Viking Press, New York.
Pettingill, O. S., Jr.
  1936 The American Woodcock *Philohela minor* (Gmelin). Mem. Boston Soc. Nat. Hist., 9: 167-391.
Sauer, E. G. F.
  1962 Ethology and Ecology of Golden Plovers on St. Lawrence Island, Bering Sea. Psychol. Forsch., 26: 399-470. (Text in English.)
Sheldon, W. G.
  1967 The Book of the American Woodcock. University of Massachusetts Press, Amherst.
Spencer, K. G.
  1953 The Lapwing in Britain. A. Brown and Sons, London.
Vogt, W.
  1938 Preliminary Notes on the Behavior and Ecology of the Eastern Willet. Proc. Linnaean Soc. New York. No. 49: 8-42.

Charadriiformes—Lari

Beck, D. E.
  1942 Life History Notes on the California Gull. No. 1. Great Basin Nat., 3: 91-108.

Darling, F. F.
    1938   Bird Flocks and the Breeding Cycle. A Contribution to the Study of Avian Sociality. University Press, Cambridge, England. (Text based in large part on the study of four species of gulls, *viz: Larus argentatus, L. canus, L. marinus,* and *L. fuscus.*)

Kirkman, F. B.
    1937   Bird Behaviour. A Contribution Based Chiefly on a Study of the Black-headed Gull. T. Nelson and Sons, London.

Lashley, K. S.
    1915   Notes on the Nesting Activities of the Noddy and Sooty Terns. Carnegie Inst. Washington Publ., 211: 61-83.

Marples, G., and A. Marples
    1934   Sea Terns or Sea Swallows, Their Habits, Language, Arrival and Departure. Country Life, London.

Noble, G. K., and M. Wurm
    1943   The Social Behavior of the Laughing Gull. Annals New York Acad. Sci., 45: 179-220.

Palmer, R. S.
    1941   A Behavior Study of the Common Tern *(Sterna hirundo hirundo* L.*).* Proc. Boston Soc. Nat. Hist., 42: 1-119.

Paludan, K.
    1951   Contributions to the Breeding Biology of *Larus argentatus* and *L. fuscus.* Ejnar Munksgaard, Copenhagen.

Tinbergen, N.
    1953   The Herring Gull's World: A Study of the Social Behaviour of Birds. Collins, London.

Vermeer, K.
    1963   The Breeding Ecology of the Glaucous-winged Gull *(Larus glaucescens)* on Mandarte Island, B. C. Occas. Papers British Columbia Prov. Mus. No. 13.

Watson, J. B.
    1908   The Behavior of Noddy and Sooty Terns. Carnegie Inst. Washington Publ., 103: 187-255.

Ytreberg, N. J.
    1956   Contribution to the Breeding Biology of the Black-headed Gull *(Larus ridibundus* L.*)* in Norway. Nytt Magasin for Zool., 4: 5-106.

Charadriiformes—Alcae

Bedard, J.
    1969   Histoire Naturelle du Gode, *Alca torda,* L., dans le Golfe Saint-Laurent, Province de Quebec, Canada. Canadian Wildlife Serv. Rept. Ser., No. 7. (English edition to follow.)

Lockley, R. M.
    1953   Puffins. Devin-Adair Company, New York.

Paludan, K.
    1947   Alken: Dens Ynglebiologi og dens Forekomst i Danmark. Ejnar Munksgaard, Copenhagen. (The breeding biology and occurrence of the Razor-bill, *Alca torda.* See review in Wilson Bull., 59: 215, 1947.)

Perry, R.
    1940   Lundy: Isle of Puffins. Lindsay Drummond, London. (Contains valuable information on the Razor-bill, *Alca torda,* Guillemot, *Uria aalge,* and Puffin, *Fratercula arctica.*)

Storer, R. W.
    1952   A Comparison of Variation, Behavior, and Evolution in the Sea Bird Genera Uria and Cepphus. Univ. California Publ. in Zool., 52: 121-222.

Thoresen, A. C., and E. S. Booth
    1958   Breeding Activities of the Pigeon Guillemot, *Cepphus columba columba* (Pallas). Publ. Dept. Biol. Sci. Walla Walla College and Biol. Sta. No. 23.

Tuck, L. M.
    1960   The Murres: Their Distribution, Populations and Biology: A Study of the Genus *Uria.* Canadian Wildlife Series, 1. Canadian Wildlife Service, Ottawa.

Gruiformes

Allen, R. P.
    1952   The Whooping Crane. Natl. Audubon Soc. Res. Rept. No. 3, New York.

1956   A Report on the Whooping Crane's Northern Breeding Grounds. Supplement to Natl. Audubon Soc. Res. Rept. No. 3.

Howard, E.
1940   A Waterhen's Worlds. University Press, Cambridge. (Concerns *Gallinula chloropus*.)

Meanley, B.
1969   Natural History of the King Rail. N. Amer. Fauna No. 67: i-viii; 1-108, U. S. Fish and Wildlife Service, Washington, D. C.

Walkinshaw, L. H.
1949   The Sandhill Crane. Cranbrook Inst. Sci. Bull. No. 29, Bloomfield Hills, Michigan.

Columbiformes

Cottam, C., and J. B. Trefethen, Editors
1968   Whitewings: The Life History, Status and Management of the White-winged Dove. D. Van Nostrand Company, Princeton, New Jersey.

Cowan, J. B.
1952   Life History and Productivity of a Population of Western Mourning Doves in California. California Fish and Game, 38: 505-521.

Goodwin, D.
1967   Pigeons and Doves of the World. British Museum (Natural History), London.

Hanson, H. C., and C. W. Kossack
1963   The Mourning Dove in Illinois. Illinois Dept. Cons. Tech. Bull. No. 2.

McClure, H. E.
1943   Ecology and Management of the Mourning Dove, *Zenaidura macroura* (Linn.), in Cass County, Iowa. Iowa State College Agric. Exp. Sta. Res. Bull. 310.

Mitchell, M. H.
1935   The Passenger Pigeon in Ontario. Royal Ontario Mus. Zool. Contr. No. 7: 1-181.

Murton, R. K.
1965   The Wood Pigeon. Collins, London.

Neff, J. A.
1947   Habits, Food, and Economic Status of the Band-tailed Pigeon. N. Amer. Fauna No. 58: 1-76, U. S. Fish and Wildlife Service, Washington, D. C.

Quay, T. L.
1951   Mourning Dove Studies in North Carolina. Game Division, North Carolina Wildlife Resources Commission, Raleigh.

Schorger, A. W.
1955   The Passenger Pigeon: Its Natural History and Extinction. University of Wisconsin Press, Madison.

Cuculiformes

Baker, E. C. S.
1942   Cuckoo Problems. H. F. and G. Witherby, London.

Chance, E. P.
1940   The Truth about the Cuckoo. Country Life, London.

Friedmann, H.
1948   The Parasitic Cuckoos of Africa. Washington Acad. Sci. Monogr. No. 1.

Herrick, F. H.
1910   Life and Behavior of the Cuckoo. Jour. Exp. Zool., 9: 169-233.

Strigiformes

Ligon, J. D.
1968   The Biology of the Elf Owl, *Micrathene whitneyi*. Univ. Michigan Mus. Zool. Misc. Publ. No. 136.

Wallace, G. J.
1948   The Barn Owl in Michigan: Its Distribution, Natural History and Food Habits. Michigan State College Agric. Exp. Sta. (Section on Zool.) Tech. Bull. 208.

Apodiformes

Berlioz, J.
1944   La Vie des Colibris. Histoires Naturelles 4, Gallimard, Paris.

Fischer, R. B.
1958 The Breeding Biology of the Chimney Swift. New York State Mus. and Sci. Service Bull. 368.
Lack, D.
1956 Swifts in a Tower. Methuen and Company, London.
Martin, A., and A. Musy
1959 La Vie des Colibris: Les Trochilidés. Éditions Delachaux et Niestlé, Neuchâtel, Switzerland.
Scheithauer, W.
1966 Hummingbirds. Thomas Y. Crowell Company, New York. (Although concerned with species in captivity, the book contains considerable information on breeding behavior.)
Wagner, H. O.
1946 Observaciones sobre la Vida de *Calothorax lucifer*. Anales del Inst. Biol. Mexico, 17: 283-299.

Coraciiformes
White, H. C.
1953 The Eastern Belted Kingfisher in the Maritime Provinces. Fisheries Res. Board Canada Bull. No. 97.

Piciformes
Lawrence, L. de K.
1967 A Comparative Life-history Study of Four Species of Woodpeckers. Amer. Ornith. Union, Ornith. Monogr. No. 5.
Ritter, W. E.
1938 The California Woodpecker and I. University of California Press, Berkeley.
Sielmann, H.
1958 My Year with the Woodpeckers. Barrie and Rockliff, London.
Tanner, J. T.
1942 The Ivory-billed Woodpecker. Natl. Audubon Soc. Res. Rept. No. 1, New York. (Dover reprint available.)

Passeriformes—Tyrannidae through Troglodytidae
Allen, R. W., and M. M. Nice
1952 A Study of the Breeding Biology of the Purple Martin *(Progne subis)*. Amer. Midland Nat., 47: 606-665.
Amadon, D.
1944 A Preliminary Life History Study of the Florida Jay, *Cyanocitta c. coerulescens*. Amer. Mus. Novitates No. 1252.
Armstrong, E. A.
1955 The Wren. Collins, London. (Concerns *Troglodytes troglodytes*.)
Crossin, R. S.
1967 The Breeding Biology of the Tufted Jay. Proc. Western Found. Vert. Zool., 1 (5): 265-299.
Erickson, M. M.
1938 Territory, Annual Cycle, and Numbers in a Population of Wren-tits *(Chamaea fasciata)*. Univ. California Publ. in Zool., 42: 247-334.
Hinde, R. A.
1952 The Behaviour of the Great Tit *(Parus major)* and Some Other Related Species. E. J. Brill, Leiden.
Kale, H. W., II
1965 Ecology and Bioenergetics of the Long-billed Marsh Wren *Troglodytes palustris griseus* (Brewster) in Georgia Salt Marshes. Publ. Nuttall Ornith. Club No. 5, Cambridge, Massachusetts.
Kendeigh, S. C.
1952 Parental Care and Its Evolution. Second corrected printing, 1955. Illinois Biol. Monogr., 22: i-x; 1-356. (Contains detailed treatment of the House Wren, *Troglodytes aedon*.)
Kuerzi, R. G.
1941 Life History Studies of the Tree Swallow. Proc. Linnaean Soc. New York. Nos. 52-53: 1-52.

Linsdale, J. M.
   1937   The Natural History of Magpies. Pacific Coast Avifauna No. 25, Cooper Ornithological
          Club, Berkeley.
Lunk, W. A.
   1962   The Rough-winged Swallow *Stelgidopteryx ruficollis* (Vieillot): A Study Based on Its
          Breeding Biology in Michigan. Publ. Nuttall Ornith. Club No. 4, Cambridge, Massachusetts.
Mumford, R. E.
   1964   The Breeding Biology of the Acadian Flycatcher. Univ. Michigan Mus. Zool. Misc. Publ.
          No. 125.
Norris, R. A.
   1958   Comparative Biosystematics and Life History of the Nuthatches *Sitta pygmaea* and *Sitta
          pusilla.* Univ. California Publ. in Zool., 56: 119-300.
Pickwell, G. B.
   1931   The Prairie Horned Lark. Trans. Acad. Sci. St. Louis, 27: 1-153.
Stoner, D.
   1936   Studies on the Bank Swallow, *Riparia riparia riparia* (Linnaeus), in the Oneida Lake
          Region. Roosevelt Wild Life Annals, 9: 126-233.
   1939   Temperature, Growth, and Other Studies on the Eastern Phoebe. New York State Mus.
          Circ. 22.
Yeates, G. K.
   1934   The Life of the Rook. Philip Allen, London.

Passeriformes—Mimidae through Parulidae

Barlow, J. C.
   1962   Natural History of the Bell Vireo, *Vireo belli* Audubon. Univ. Kansas Publ. Mus. Nat.
          Hist., 12: 241-296.
Biaggi, V., Jr.
   1955   The Life History of the Puerto Rican Honeycreeper. Special Publication, University of
          Puerto Rico Agricultural Experiment Station.
Buxton, J.
   1950   The Redstart. Collins, London. (The subject is *Phoenicurus phoenicurus.*)
Chapman, F. M.
   1907   The Warblers of North America. D. Appleton and Company, New York. (Dover reprint
          available.)
Eaton, S. W.
   1957   A Life History Study of *Seiurus noveborancensis* (with Notes on *Seiurus aurocapillus*
          and the Species of *Seiurus* Compared). St. Bonaventure Univ. Sci. Stud., 19: 7-36.
Erwin, W. G.
   1935   Some Nesting Habits of the Brown Thrasher. Jour. Tennessee Acad. Sci., 10: 179-204.
Graber, J. W.
   1961   Distribution, Habitat Requirements, and Life History of the Black-capped Vireo *(Vireo
          atricapilla).* Ecol. Monogr., 31: 313-336.
Griscom, L., A. Sprunt, Jr., and others
   1957   The Warblers of America. Devin-Adair Company, New York.
Hillstead, A. F. C.
   1945   The Blackbird: A Contribution to the Study of a Single Avian Species. Faber and Faber,
          London. (Concerns *Turdus merula.*)
Hofslund, P. B.
   1959   A Life History of the Yellowthroat, *Geothlypis trichas.* Proc. Minnesota Acad. Sci., 27:
          144-174.
Howell, J. C.
   1942   Notes on the Nesting Habits of the American Robin *(Turdus migratorius* L. *).* Amer. Mid-
          land Nat., 28: 529-603.
Kessel, B.
   1957   A Study of the Breeding Biology of the European Starling (*Sturnus vulgaris* L.) in North
          America. Amer. Midland Nat., 58: 257-331.

Lack, D.
1939   The. Behaviour of the Robin. Part I. The Life History with Special Reference to Aggressive Behaviour and Territory. Part II. A Partial Analysis of Aggressiveness and Recognitional Behaviour. Proc. Zool. Soc. London, 109: 169-219. (Concerns *Erithacus rubecula.*)
1965   The Life of the Robin. Fourth edition. H. F. and G. Witherby, London.
Mayfield, H.
1960   The Kirtland's Warbler. Cranbrook Inst. Sci. Bull. No. 40, Bloomfield Hills, Michigan.
Miller, A. H.
1931   Systematic Revision and Natural History of the American Shrikes *(Lanius).* Univ. California Publ. in Zool., 38: 11-242.
Nickell, W. P.
1965   Habitats, Territory, and Nesting of the Catbird. Amer. Midland Nat., 73: 433-478.
Shaver, N. E.
1918   A Nest Study of the Maryland Yellow-throat. Univ. Iowa Stud., Stud. in Nat. Hist., 8: 1-12.
Smith, S.
1950   The Yellow Wagtail. Collins, London. (Concerns *Motacilla flava.*)
Snow, D. W.
1958   A Study of Blackbirds. George Allen and Unwin, London. (Concerns *Turdus merula.*)
Stoner, D.
1920   Nesting Habits of the Hermit Thrush in Northern Michigan. Univ. Iowa Stud., Stud. in Nat. Hist., 9: 1-21.
Sutton, G. M.
1949   Studies of the Nesting Birds of the Edwin S. George Reserve. Part 1. The Vireos. Univ. Michigan Mus. Zool. Misc. Publ. No. 74.
Wallace, G. J.
1939   Bicknell's Thrush, Its Taxonomy, Distribution and Life History. Proc. Boston Nat. Hist., 41: 211-402.
Young, H.
1955   Breeding Behavior and Nesting of the Eastern Robin. Amer. Midland Nat., 53: 329-352.

Passeriformes—Ploceidae through Fringillidae

Allen, A. A.
1914   The Red-winged Blackbird: A Study in the Ecology of a Cat-tail Marsh. Proc. Linnaean Soc. New York, Nos. 24-25: 43-128.
Bailey, A. M., R. J. Niedrach, and A. L. Baily
1953   The Red Crossbills of Colorado. (Part I by Bailey and Niedrach; Part II by Baily.) Mus. Pictorial No. 9, Denver Museum of Natural History.
Barbour, R. W.
1951   Observations on the Breeding Habits of the Red-eyed Towhee. Amer. Midland Nat., 45: 672-678.
Blanchard, B. D.
1941   The White-crowned Sparrows *(Zonotrichia leucophrys)* of the Pacific Seaboard: Environment and Annual Cycle. Univ. California Publ. in Zool., 46: 1-178.
Blanchard, B. D., and M. M. Erickson
1949   The Cycle in the Gambel Sparrow. Univ. California Publ. in Zool., 47: 255-318.
Cartwright, B. W., T. M. Shortt, and R. D. Harris
1937   Baird's Sparrow. Trans. Royal Canadian Inst., 21: 153-197.
Chapman, F. M.
1928   The Nesting Habits of Wagler's Oropendola *(Zarhynchus wagleri)* on Barro Colorado Island. Bull. Amer. Mus. Nat. Hist., 58: 123-166.
Dwight, J., Jr.
1895   The Ipswich Sparrow (*Ammodramus princeps* Maynard) and Its Summer Home. Mem. Nuttall Ornith. Club No. 2, Cambridge, Massachusetts.
Esten, S. R.
1925   A Comparative Nest Life of the Towhee, Meadow Lark and Rose-breasted Grosbeak. Proc. Indiana Acad. Sci., 34: 397-401.

Friedmann, H.
  1929   The Cowbirds. A Study in the Biology of Social Parasitism. Charles C. Thomas, Spring-
         field, Illinois.
Hyde, A. S.
  1939   The Life History of Henslow's Sparrow, *Passerherbulus henslowi* (Audubon). Univ.
         Michigan Mus. Zool. Misc. Publ. No. 41.
Jones, V. E., and E. Fichter
  1961   Nesting of the House Finch at Pocatello, Idaho. Tebiwa: Jour. Idaho State College Mus.,
         4 (2): 1-9.
Lanyon, W. E.
  1957   The Comparative Biology of the Meadowlarks *(Sturnella)* in Wisconsin. Publ. Nuttall
         Ornith. Club No. 1, Cambridge, Massachusetts.
Linsdale, J. M.
  1928   Variations in the Fox Sparrow *(Passerella iliaca)* with Reference to Natural History and
         Osteology. Univ. California Publ. in Zool., 30: 251-392.
  1957   Goldfinches on the Hastings Natural History Reservation. Amer. Midland Nat., 57: 1-119.
Mountfort, G.
  1957   The Hawfinch. Collins, London.
Nice, M. M.
  1937   Studies in the Life History of the Song Sparrow, I. Trans. Linnaean Soc. New York, 4:
         i-vi; 1-247. (Dover reprint available.)
  1943   Studies in the Life History of the Song Sparrow, II. Trans. Linnaean Soc. New York, 6:
         i-viii; 1-328. (Dover reprint available.)
Nethersole-Thompson, D.
  1966   The Snow Bunting. Oliver and Boyd, Edinburgh.
Prescott, K. W.
  1965   Studies in the Life History of the Scarlet Tanager, *Piranga olivacea.* New Jersey State
         Mus., Investigations No. 2.
Shaver, J. M., and M. B. Roberts
  1933   A Brief Study of the Courtship of the Eastern Cardinal *(Richmondena cardinalis car-
         dinalis* (Linnaeus)). Jour. Tennessee Acad. Sci., 8: 116-123.
Summers-Smith, J. D.
  1963   The House Sparrow. Collins, London.
Tinbergen, N.
  1939   The Behavior of the Snow Bunting in Spring. Trans. Linnaean Soc. New York, 5: 1-94.
Woolfenden, G. E.
  1956   Comparative Breeding Behavior of *Ammospiza caudacuta* and *A. maritima.* Univ. Kansas
         Publ. Mus. Natl. Hist., 10: 45-75.

APPENDIX E
SELECTED BIBLIOGRAPHY OF
REGIONAL WORKS

The following bibliography will introduce the student to the more recent works covering birds known to occur in different parts or political divisions of the world. Older works, although out of print, are included when they supply especially useful information or are still the best treatises available.

UNITED STATES

General
Reilly, E. M., Jr.
  1968   The Audubon Illustrated Handbook of American Birds. Edited by O. S. Pettingill, Jr. McGraw-Hill Book Company, New York.
See also several guides for identification listed on pages 230-231 of this book.

Alabama
Imhof, T. A.
  1962   Alabama Birds. University of Alabama Press, University.

Alaska
Cahalane, V. H.
  1959   A Biological Survey of Katmai National Monument. Smithsonian Misc. Coll., 138 (5): i-iv; 1-246. (Near the base of the Alaska Peninsula in southwestern Alaska; birds, pp. 83-155.)
Childs, H. E., Jr.
  1969   Birds and Mammals of the Pitmegea River Region, Cape Sabine, Northwestern Alaska. Univ. Alaska Biol. Papers No. 10.
Gabrielson, I. N., and F. C. Lincoln
  1959   The Birds of Alaska. Stackpole Company, Harrisburg, Pennsylvania, and Wildlife Management Institute, Washington, D. C.
Kenyon, K. W., and R. E. Phillips
  1965   Birds from the Pribilof Islands and Vicinity. Auk, 82:624.635.
Kessel, B., and T. J. Cade
  1958   Birds of the Colville River, Northern Alaska. Univ. Alaska Biol. Papers No. 2.
Kessel, B., and G. B. Schaller
  1960   Birds of the Upper Sheenjek Valley, Northeastern Alaska. Univ. Alaska Biol. Papers No. 4.
Murie, A.
  1963   Birds of Mount McKinley National Park, Alaska. Mount McKinley Natural History Association.
Murie, O. J.
  1959   Fauna of the Aleutian Islands and Alaska Peninsula. N. Amer. Fauna No. 61: i-xiv; 1-406, U. S. Fish and Wildlife Service, Washington, D. C. (Birds, pp. 27-261.)
Williamson, F. S. L., and L. J. Peyton
  1962   Faunal Relationships of Birds in the Iliamna Lake Area, Alaska. Univ. Alaska Biol. Papers No. 5. (Iliamna Lake is at the base of the Alaska Peninsula.)

Arizona

Monson, G., and A. R. Phillips
  1964   A Checklist of the Birds of Arizona. University of Arizona Press, Tucson.
Phillips, A., J. Marshall, and G. Monson
  1964   The Birds of Arizona. University of Arizona Press, Tucson.

Arkansas

Baerg, W. J.
  1951   Birds of Arkansas. Univ. Arkansas Agric. Exp. Sta. Bull. No. 258 (revised): 1-188.

California

Grinnell, J., and A. H. Miller
  1944   The Distribution of the Birds of California. Pacific Coast Avifauna No. 27, Cooper Orni-
         thological Club, Berkeley, California.
Johnson, D. H., M. D. Bryant, and A. H. Miller
  1948   Vertebrate Animals of the Providence Mountains Area of California. Univ. California
         Publ. in Zool., 48: 221-376. (Covers birds and other vertebrates of a section of eastern San
         Bernardino County, in the Mohave Desert.)
McCaskie, R. G., and R. C. Banks
  1966   Supplemental List of Birds of San Diego County, California. Trans. San Diego Soc. Nat.
         Hist., 14: 157-168.
Sumner, L., and J. S. Dixon
  1953   Birds and Mammals of the Sierra Nevada. University of California Press, Berkeley.

Colorado

Bailey, A. M., and R. J. Niedrach
  1965   Birds of Colorado. Two volumes. Denver Museum of Natural History.
  1967   A Pictorial Checklist of Colorado Birds, with Brief Notes on the Status of Each Species in
         Neighboring States of Nebraska, Kansas, New Mexico, Utah, and Wyoming. Denver Mu-
         seum of Natural History, Denver.
Knorr, O. A.
  1959   The Birds of El Paso County, Colorado. Univ. Colorado Stud. Biol. Ser., 5: 1-48.

Connecticut

Mackenzie, L.
  1961   The Birds of Guilford, Connecticut. Peabody Museum of Natural History, Yale University,
         New Haven, Connecticut.
Manter, J. A.
  1965   Birds of Storrs, Connecticut, and Vicinity. Natchaug Ornithological Society, Storrs.
Sage, J. H., L. B. Bishop, and W. P. Bliss
  1913   The Birds of Connecticut. Connecticut State Geol. and Nat. Hist. Surv. Bull. No. 20: 1-370.

Delaware

Rhoads, S. N., and C. J. Pennock
  1905   Birds of Delaware: A Preliminary List. Auk, 22: 194-205.

District of Columbia

Aldrich, J. W., and others
  1947   A Field List of Birds of the District of Columbia Region. Audubon Society of the District
         of Columbia, Washington, D. C.
See also Maryland.

Florida

Greene, E. R.
  1945   Birds of the Lower Florida Keys. Quart. Jour. Florida Acad. Sci., 8: 199-265.
Sprunt, A., Jr.
  1954   Florida Bird Life. Coward-McCann, New York.

Stevenson, H. M.
  1960   A Key to Florida Birds. Peninsular Publishing Company, Tallahassee. (Includes an annotated list, giving descriptions of species and their status.)
Weston, F. M.
  1965   A Survey of the Birdlife of Northwestern Florida. Tall Timbers Res. Sta. Bull. No. 5.

Georgia
Burleigh, T. D.
  1958   Georgia Birds. University of Oklahoma Press, Norman.

Hawaii
Bailey, A. M.
  1956   Birds of Midway and Laysan Islands. Mus. Pictorial No. 12, Denver Museum of Natural History, Colorado.
Clapp, R. B., and P. W. Woodward
  1968   New Records of Birds from the Hawaiian Leeward Islands. Proc. U. S. Natl. Mus., 124 (3640): 1-39.
Fisher, H. I.
  1951   The Avifauna of Niihau Island, Hawaiian Archipelago. Condor, 53: 31-42.
Hawaii Audubon Society
  1967   Hawaii's Birds. Hawaii Audubon Society, Honolulu.
Kenyon, K. W., and D. W. Rice
  1958   Birds of Kure Atoll, Hawaii. Condor, 60: 188-190.
Munro, G. C.
  1960   Birds of Hawaii. Revised edition. Charles E. Tuttle Company, Rutland, Vermont.
Peterson, R. T.
  1961   A Field Guide to Western Birds. Second edition. Houghton Mifflin Company, Boston. (Part 2 is on the birds of the Hawaiian Islands.)
Richardson, F., and J. Bowles
  1964   A Survey of the Birds of Kauai, Hawaii. Bernice P. Bishop Mus. Bull. 227.

Idaho
Arvey, M. D.
  1947   A Check-list of the Birds of Idaho. Univ. Kansas Publ. Mus. Nat. Hist., 1: 193-216. (Additions and corrections were published in 1950 by Arvey, Condor, 52: 275. See the comments on Arvey's work published in 1952 by M. Jollie, Condor, 54: 172-173.
Hand, R. L.
  1941   Birds of the St. Joe National Forest, Idaho. Condor, 43: 220-232.
Johnston, D. W.
  1949   Populations and Distribution of Summer Birds of Latah County, Idaho. Condor, 51: 140-149.
Levy, S. H.
  1950   Summer Birds in Southern Idaho. Murrelet, 31: 2-8.

Illinois
Ford, E. R.
  1956   Birds of the Chicago Region. Chicago Acad. Sci. Special Publ. No. 12
George, W. G.
  1968   Check List of Birds of Southern Illinois. (Mimeographed; available from Department of Zoology, Southern Illinois University, Carbondale.)
Smith, H. R., and P. W. Parmalee
  1955   A Distributional Check List of the Birds of Illinois. Illinois State Mus. Popular Sci. Ser., 4.

Indiana
Brooks, E.
  1945   Common Birds of Indiana. Blatchley Nature Study Club. Noblesville, Indiana. (Concerns mainly the birds of Hamilton County, which is centrally located in the state.)

Butler, A. W.
  1898   The Birds of Indiana. Report of the State Geologist of Indiana for 1897, pp. 515-1187. Indianapolis.

Iowa

Anderson, R. M.
  1907   The Birds of Iowa. Proc. Davenport Acad. Sci., 11: 125-417.
DuMont, P. A.
  1933   A Revised List of the Birds of Iowa. Univ. Iowa Stud. in Nat. Hist., 15: 1-171.
Musgrove, J. W., and M. R. Musgrove
  1947   Waterfowl in Iowa. Second edition. State Conservation Commission, Des Moines.

Kansas

Johnston, R. F.
  1964   The Breeding Birds of Kansas. Univ. Kansas Publ. Mus. Nat. Hist., 12: 575-655.
  1965   A Directory to the Birds of Kansas. Univ. of Kansas, Mus. Nat. Hist. Misc. Publ. No. 41.

Kentucky

Mengel, R. M.
  1965   The Birds of Kentucky. Amer. Ornith. Union, Ornith. Monogr. No. 3.

Louisiana

Lowery, G. H., Jr.
  1960   Louisiana Birds. Second edition. Louisiana State University Press, Baton Rouge.

Maine

Cruickshank, A. D.
  [1950] Summer Birds of Lincoln County, Maine. National Audubon Society, New York.
Palmer, R. S.
  1949   Maine Birds. Bull. Mus. Comp. Zool., 102: 1-656.
Tyson, C., and J. Bond
  1941   Birds of Mt. Desert Island, Acadia National Park, Maine. Academy of Natural Sciences of Philadelphia.

Maryland

Manville, R. H.
  1968   Natural History of Plummers Island, Maryland. XX. Annotated List of the Vertebrates. Special Publication of the Washington Biologists' Field Club.
Stewart, R. E., and C. S. Robbins
  1958   Birds of Maryland and the District of Columbia. N. Amer. Fauna No. 62: i-vi; 1-401, U. S. Fish and Wildlife Service, Washington, D. C.

Massachusetts

Bagg, A. C., and S. A. Eliot, Jr.
  1937   Birds of the Connecticut Valley in Massachusetts. The Hampshire Bookshop, Northampton.
Forbush, E. H.
  1925- Birds of Massachusetts and Other New England States. Three volumes. Massachusetts
  29   Department of Agriculture, Boston.
Griscom, L.
  1949   The Birds of Concord. Harvard University Press, Cambridge, Massachusetts.
Griscom, L., and G. Emerson
  1959   Birds of Martha's Vineyard with an Annotated Check List. Privately printed. (Available from the Massachusetts Audubon Society, South Lincoln.)
Griscom, L., and E. V. Folger
  1948   The Birds of Nantucket. Harvard University Press, Cambridge, Massachusetts.
Griscom, L., and D. E. Snyder
  1955   The Birds of Massachusetts: An Annotated and Revised Check List. Peabody Museum, Salem.

Hill, N. P.
  1965   The Birds of Cape Cod, Massachusetts. William Morrow and Company, New York.
Johnson, G. P.
  1946   Birds of Springfield, Massachusetts, and Vicinity. Seventh edition. Museum of Natural History, Springfield.

Michigan

Cuthbert, N. L.
  [1962]   The Birds of Isabella County, Michigan. Mount Pleasant.
Hatt, R. T., and others
  1948   Island Life: A Study of the Land Vertebrates of the Islands of Eastern Lake Michigan. Cranbrook Inst. Sci. Bull. 27, Bloomfield Hills, Michigan.
Kelley, A. H., and others
  1963   Birds of the Detroit-Windsor Area: A Ten-year Survey. Cranbrook Inst. Sci. Bull. 45, Bloomfield Hills, Michigan.
Wood, N. A.
  1951   The Birds of Michigan. Univ. Michigan Mus. Zool. Misc. Publ. No. 75.
Zimmerman, D. A., and J. Van Tyne
  1959   A Distributional Check-list of the Birds of Michigan. Univ. Michigan Mus. Zool. Occas. Papers No. 608.

Minnesota

Roberts, T. S.
  1936   The Birds of Minnesota. Two volumes. Second revised edition. University of Minnesota Press, Minneapolis.

Mississippi

Burleigh, T. D.
  1944   The Bird Life of the Gulf Coast Region of Mississippi. Louisiana State Univ. Mus. Zool. Occas. Papers No. 20: 329-490.
Coffey, B. B., Jr.
  1936   A Preliminary List of the Birds of Mississippi. Mimeographed and distributed by the author.

Missouri

Bennitt, R.
  1932   Check-list of the Birds of Missouri. Univ. Missouri Stud., 7: 1-81.
Widmann, O.
  1907   A Preliminary Catalog of the Birds of Missouri. Trans. Acad. Sci. St. Louis, 17: 1-288.

Montana

Hoffmann, R. S., R. L. Hand, and P. L. Wright
  1959   Recent Bird Records from Western Montana. Condor, 61: 147-151.
Saunders, A. A.
  1921   A Distributional List of the Birds of Montana. Pacific Coast Avifauna No. 14, Cooper Ornithological Club, Berkeley, California.

Nebraska

Rapp, W. F., Jr., J. L. C. Rapp, H. E. Baumgarten, and R. A. Moser
  1958   Revised Check-list of Nebraska Birds. Nebraska Ornith. Union Occas. Papers No. 5.
Tout, W.
  1947   Lincoln County Birds. Published by the author, North Platte, Nebraska.

Nevada

Gullion, G. W., W. M. Pulich, and F. G. Evenden
  1959   Notes on the Occurrence of Birds in Southern Nevada. Condor, 61: 278-297.

Linsdale, J. M.
  1936 The Birds of Nevada. Pacific Coast Avifauna No. 23, Cooper Ornithological Club,
        Berkeley, California. (A supplement to this work was published in 1951 by the same
        author, Condor, 53: 228-249.)
Van Rossem, A. J.
  1936 Birds of the Charleston Mountains, Nevada. Pacific Coast Avifauna No. 24, Cooper Orni-
        thological Club, Berkeley, California.

New Hampshire
Richards, T.
  1958 A List of the Birds of New Hampshire. Audubon Society of New Hampshire.

New Jersey
Fables, D., Jr.
  1955 Annotated List of New Jersey Birds. Urner Ornithological Club, Newark.
Stone, W.
  1937 Bird Studies at Old Cape May. Two volumes. Academy of Natural Sciences of Philadel-
        phia. (Dover reprint available.)

New Mexico
Bailey, F. M.
  1928 Birds of New Mexico. New Mexico Department of Fish and Game, Santa Fe.
Ligon, J. S.
  1961 New Mexico Birds and Where to Find Them. University of New Mexico Press, Albuquerque.

New York
Beardslee, C. S.
  1965 Birds of the Niagara Frontier Region: An Annotated Check-list. Bull. Buffalo Soc. Nat.
        Sci., 22.
Bull, J.
  1964 Birds of the New York [City] Area. Harper and Row, New York.
Eaton, E. H.
  1923 Birds of New York. Two volumes. Second edition. New York State Museum, Albany.
Eaton, S. W.
  1953 Birds of the Olean and Salamanca Quadrangles. St. Bonaventure Univ. Sci. Stud., 15:
        1-27.
Hyde, A. S.
  1939 The Ecology and Economics of the Birds along the Northern Boundary of New York State.
        Roosevelt Wild Life Bull., 7: 67-215.
Saunders, A. A.
  1929 The Summer Birds of the Northern Adirondack Mountains. Roosevelt Wild Life Bull.,
        5: 327-499.
Spiker, C. J.
  1935 A Popular Account of the Bird Life of the Finger Lakes Section of New York with Main
        Reference to the Summer Season. Roosevelt Wild Life Bull., 6: 391-551.
Stoner, D.
  1932 Ornithology of the Oneida Lake Region: With Reference to the Late Spring and Summer
        Seasons. Roosevelt Wild Life Annals, 2: 277-764.
Stoner, D., and L. C. Stoner
  1952 Birds of Washington Park, Albany, New York. New York State Mus. Bull. No. 344: 1-268.

North Carolina
Pearson, T. G., C. S. Brimley, and H. H. Brimley
  1942 Birds of North Carolina. Revised in 1959 by D. L. Wray and H. T. Davis. North Carolina
        Department of Agriculture, Raleigh.

North Dakota
Wood, N. A.
　1923　A Preliminary Survey of the Bird Life of North Dakota. Univ. Michigan Mus. Zool. Misc. Publ. No. 10.

Ohio
Borror, D. J.
　1950　A Check-list of the Birds of Ohio, with the Migration Dates for the Birds of Central Ohio. Ohio Jour. Sci., 50: 1-32.
Campbell, L.
　1968　Birds of the Toledo Area. The Blade, Toledo.
Kemsies, E., and W. Randle
　1953　Birds of Southwestern Ohio. Published by the authors, Cincinnati, Ohio.
Trautman, M. B.
　1935　Second Revised List of the Birds of Ohio. Ohio Dept. Agric. Bull. Bur. Sci. Res. Div. Cons., 1: 3-16.
　1940　The Birds of Buckeye Lake, Ohio. Univ. Michigan Mus. Zool. Misc. Publ. No. 44.
Williams, A. B.
　1950　Birds of the Cleveland Region. Sci. Publ. Cleveland Mus. Nat. Hist., 10: 1-215.

Oklahoma
Sutton, G. M.
　1967　Oklahoma Birds: Their Ecology and Distribution, with Comments on the Avifauna of the Southern Great Plains. University of Oklahoma Press, Norman.

Oregon
Farner, D. S.
　1952　The Birds of Crater Lake National Park. University of Kansas Press, Lawrence.
Gabrielson, I. N., and S. G. Jewett
　1940　Birds of Oregon. Oregon State College, Corvallis.
Gullion, G. W.
　1951　Birds of the Southern Willamette Valley, Oregon. Condor, 53: 129-149.

Pennsylvania
Arnett, J. H., Jr., and others
　1954　A Field List of Birds of the Philadelphia Region. Delaware Valley Ornithological Club, Academy of Natural Sciences of Philadelphia.
Poole, E. L.
　1947　A Half Century of Bird Life in Berks County, Pennsylvania. Reading Publ. Mus. and Art Gallery Bull. No. 19: 3-133.
　1964　Pennsylvania Birds: An Annotated List. Delaware Valley Ornithological Club, Philadelphia.
Street, P. B.
　1956　Birds of the Pocono Mountains, Pennsylvania. Delaware Valley Ornithological Club, Philadelphia.
Todd, W. E. C.
　1940　Birds of Western Pennsylvania. University of Pittsburgh Press, Pittsburgh.
Wood, M.
　1952　Birds of the State College Region, Pennsylvania. Pennsylvania State College Agric. Exp. Sta. Bull. No. 558.

Rhode Island
Howe, R. H., Jr., and E. Sturtevant
　1899　The Birds of Rhode Island. Privately published.
　1903　A Supplement to the Birds of Rhode Island. Middletown.

South Carolina
Norris, R. A.
　1963　Birds of the AEC Savannah River Plant Area. Contributions from the Charleston Museum, 14.

Sprunt, A., Jr., and E. B. Chamberlain
  1949   South Carolina Bird Life. University of South Carolina Press, Columbia.

South Dakota

Over, W. H., and C. S. Thoms
  1946   Birds of South Dakota. Revised edition. Univ. South Dakota Mus. Nat. Hist. Stud. No. 1.
Pettingill, O. S., Jr., and N. R. Whitney, Jr.
  1965   Birds of the Black Hills. Cornell Laboratory of Ornithology Special Publ. No. 1.
Stephens, T. C., W. G. Youngworth, and W. R. Felton, Jr.
  1955   The Birds of Union County, South Dakota. Nebraska Ornith. Union Occas. Papers No. 1.

Tennessee

Ganier, A. F.
  1933   A Distributional List of the Birds of Tennessee. Tennessee Ornithological Society, Nashville.
Howell, J. C., and M. B. Monroe
  1957   The Birds of Knox County, Tennessee. Jour. Tennessee Acad. Sci., 32: 247-322.
  1958   The Birds of Knox County, Tennessee. Migrant, 29: 17-24.
Stupka, A.
  1963   Notes on the Birds of the Great Smoky Mountains National Park. University of Tennessee Press, Knoxville.
Wetmore, A.
  1939   Notes on the Birds of Tennessee. Proc. U. S. Natl. Mus., 86: 175-243.

Texas

Peterson, R. T.
  1960   A Field Guide to the Birds of Texas. Houghton Mifflin Company, Boston.
Wolfe, L. R.
  1956   Check-list of the Birds of Texas. Published by the author, Kerrville, Texas.

Utah

Behle, W. H.
  1943   Birds of Pine Valley Mountain Region, Southwestern Utah. Bull. Univ. Utah, 34 (2): 1-85.
  1955   The Birds of the Deep Creek Mountains of Central Western Utah. Univ. Utah Biol. Ser., 11 (4): 1-34.
  1958a The Bird Life of Great Salt Lake. University of Utah Press, Salt Lake City.
  1958b The Birds of the Raft River Mountains, Northwestern Utah. Univ. Biol. Ser., 11 (6): 1-40.
  1960   The Birds of Southeastern Utah. Univ. Utah Biol. Ser., 12 (1): 1-56.
Behle, W. H., J. B. Bushman, and C. M. Greenhalgh
  1958   Birds of the Kanab Area and Adjacent High Plateaus of Southern Utah. Univ. Utah Biol. Ser., 11 (7): 1-92.
Hayward, C. L.
  1967   Birds of the Upper Colorado River Basin. Brigham Young Univ. Sci. Bull., Biol. Ser., 9 (2): 1-64. (Mainly in the Utah portion, with general observations in other parts.)
Twomey, A. C.
  1942   The Birds of the Uinta Basin. Annals Carnegie Mus., 28: 341-490.
Wauer, R. H., and D. L. Carter
  1965   Birds of Zion National Park and Vicinity. Zion Natural History Association, Springvale, Utah.
Woodbury, A. M., C. Cottam, and J. W. Sugden
  1949   Annotated Check-list of the Birds of Utah. Bull. Univ. Utah, 39 (16): 1-40.
Woodbury, A. M., and H. N. Russell, Jr.
  1945   Birds of the Navajo Country. Bull. Univ. Utah, 35 (14): 1-160. (Southeastern Utah.)

Vermont

Fortner, H. C., W. P. Smith, and E. J. Dole
  1933   A List of Vermont Birds. Bull. No. 41, Department of Agriculture, Montpelier.

Virginia

Murray, J. J.
  1952   A Check-list of the Birds of Virginia. Virginia Society of Ornithology, Lexington.
  1957   The Birds of Rockbridge County, Virginia. Virginia Society of Ornithology, Sweet Briar.
Wetmore, A.
  1950   The List of Birds of the Shenandoah National Park (Third Revision). Shenandoah Nat. Hist. Assoc. Bull. No. 1, Luray, Virginia.

Washington

Jewett, S. G., and others
  1953   Birds of Washington State. University of Washington Press, Seattle.
Larrison, E. J., and K. G. Sonnenberg
  1968   Washington Birds: Their Location and Identification. Seattle Audubon Society.

West Virginia

Brooks, M.
  1944   A Check-list of West Virginia Birds. West Virginia Univ. Agric. Exp. Sta. Bull. 316.
Seeber, E. L., and R. M. Edeburn
  1952   A Preliminary Report of the Birds in the Ohio River Valley in West Virginia between the Great Kanawha and Big Sandy Rivers. Marshall College, Huntington, West Virginia. (Mimeographed.)
Wetmore, A.
  1937   Observations on the Birds of West Virginia. Proc. U. S. Natl. Mus., 84: 401-441.

Wisconsin

Gromme, O. J.
  1963   Birds of Wisconsin. University of Wisconsin Press, Madison.
Kumlien, L., and N. Hollister
  1951   The Birds of Wisconsin. With Revisions by A. W. Schorger. Wisconsin Society for Ornithology, Madison.

Wyoming

McCreary, O.
  1939   Wyoming Bird Life. Revised edition. Burgess Publishing Company, Minneapolis. (Mimeographed: long out of print.)
Skinner, M. P.
  1925   The Birds of the Yellowstone National Park. Roosevelt Wild Life Bull., 3: 11-189.

CANADA

General

Godfrey, W. E.
  1966   The Birds of Canada. Natl. Mus. Canada Bull. No. 203 (Biol. Ser. 73): 1-428.
Reilly, E. M., Jr.
  1968   The Audubon Illustrated Handbook of American Birds. Edited by O. S. Pettingill, Jr. McGraw-Hill Book Company, New York.

Sectional or Provincial

Austin, O. L., Jr.
  1932   The Birds of Newfoundland Labrador. Mem. Nuttall Ornith. Club No. 7, Cambridge, Massachusetts.
Belcher, M.
  1961   Birds of Regina. Saskatchewan Nat. Hist. Soc. Spec. Publ. No. 3.
Devitt, O. E.
  1967   The Birds of Simcoe County, Ontario. Brereton Field Naturalists' Club, Barrie, Ontario.
Drent, R. H., and C. J. Guiguet
  1961   A Catalogue of British Columbia Sea-bird Colonies. Occas. Papers British Columbia Prov. Mus. No. 12.

Godfrey, W. E.
  1951   Notes on the Birds of Southern Yukon Territory. Natl. Mus. Canada Bull. No. 123: 88-115.
  1954   Birds of Prince Edward Island. Natl. Mus. Canada Bull. No. 132: 155-213.
  1968   Notes on Birds of the Amos Region, Quebec, Natl. Mus. Canada, Natural History Papers No. 44.
Harper, F.
  1958   Birds of the Ungava Peninsula. Univ. Kansas Mus. Nat. Hist. Misc. Publ. No. 17.
Houston, C. S., and M. G. Street
  1959   The Birds of the Saskatchewan River: Carlton to Cumberland. Saskatchewan Natural History Society, Special Publ. No. 2.
MacPherson, A. H., and T. H. Manning
  1959   The Birds and Mammals of Adelaide Peninsula, N. W. T. Natl. Mus. Canada Bull. No. 161: 1-73.
Manning, T. H.
  1952   Birds of the West James Bay and Southern Hudson Bay Coasts. Natl. Mus. Canada Bull. No. 125: 1-114.
Manning, T. H., E. O. Höhn, and A. H. MacPherson
  1956   The Birds of Banks Island. Natl. Mus. Canada Bull. No. 143: 1-136.
Munro, J. A., and I. McT. Cowan
  1947   A Review of the Bird Fauna of British Columbia. British Columbia Prov. Mus. Special Publ. No. 2: 1-285.
Nero, R. W.
  1963   Birds of the Lake Athabasca Region, Saskatchewan. Saskatchewan Natural History Society, Special Publ. No. 5.
  1967   The Birds of Northeastern Saskatchewan. Saskatchewan Natural History Society, Special Publ. No. 6.
Parmelee, D. F., H. A. Stephens, and R. H. Schmidt
  1967   The Birds of Southeastern Victoria Island and Adjacent Small Islands. Natl. Mus. Canada Bull. No. 222: i-x; 1-229.
Peters, H. S., and T. D. Burleigh
  1951   The Birds of Newfoundland. Department of Natural Resources, Province of Newfoundland, St. John's.
Quilliam, H. R.
  1965   History of the Birds of Kingston, Ontario. (An extraordinarily thorough coverage of an area "circumscribed by a circle of thirty-miles radius" centered on MacDonald Park, Kingston, from historic times to the present; privately printed and obtainable from the author, Mrs. C. D. Quilliam, R. R. 1, Kingston.)
Rand, A. L.
  1946   List of Yukon Birds and Those of the Canol Road. Natl. Mus. Canada Bull. No. 105: 1-76.
  1943   Birds of Southern Alberta. Natl. Mus. Canada Bull. No. 111: 1-105.
Salt, W. R., and A. L. Wilk
  1966   The Birds of Alberta. Second (revised) edition. Department of Industry and Development, Government of Alberta, Edmonton.
Snyder, L. L.
  1957   Arctic Birds of Canada. University of Toronto Press, Toronto.
Squires, W. A.
  1952   The Birds of New Brunswick. New Brunswick Mus. Monogr. Ser. No. 4.
Todd, W. E. C.
  1963   Birds of the Labrador Peninsula and Adjacent Areas. University of Toronto Press, Toronto.
Tufts, R. W.
  1961   The Birds of Nova Scotia. Nova Scotia Museum, Halifax.

GREENLAND

Salomonsen, F.
  1951   The Birds of Greenland. Ejnar Munksgaard, Copenhagen, Denmark.
  1967   Fuglene på Grønland. Rhodos, Copenhagen, Denmark. (In Danish.)

MEXICO

Alden, P.
1969 Finding the Birds in Western Mexico: A Guide to the States of Sonora, Sinaloa, and Nayarit. University of Arizona Press, Tucson.
Blake, E. R.
1953 Birds of Mexico. University of Chicago Press, Chicago.
Edwards, E. P.
1968 Finding Birds in Mexico. Second edition, revised and enlarged. Published by the author, Sweet Briar, Virginia.
Friedmann, H., L. Griscom, and R. T. Moore
1950 Distributional Check-list of the Birds of Mexico. Pacific Coast Avifauna No. 29, Cooper Ornithological Club, Berkeley, California. (Covers Tinamidae through Trochilidae. For Part 2, see below under Miller, Friedmann, Griscom, and Moore.)
Leopold, A. S.
1959 Wildlife of Mexico: The Game Birds and Mammals. University of California Press, Berkeley.
Miller, A. H., H. Friedmann, L. Griscom, and R. T. Moore
1957 Distributional Check-list of the Birds of Mexico. Part 2. Pacific Coast Avifauna No. 33, Cooper Ornithological Society, Berkeley. (Covers Trogonidae through Fringillidae. For Part 1, see above under Friedmann, Griscom, and Moore.)
Sutton, G. M.
1951 Mexican Birds: First Impressions Based upon an Ornithological Expedition to Tamaulipas, Neuvo Leon, and Coahuila. With an Appendix Briefly Describing All Mexican Birds. University of Oklahoma Press, Norman.

CENTRAL AMERICA

Eisenmann, E.
1952 Annotated List of Birds of Barro Colorado Island, Panama, Canal Zone. Smithsonian Misc. Coll., 117 (5): 1-62.
1955 The Species of Middle American Birds. Trans. Linnaean Soc. New York, 7: 1-128.
Carriker, M. A., Jr.
1910 An Annotated List of the Birds of Costa Rica Including Cocos Island. Annals Carnegie Mus., 6: 314-915.
Dickey, D. R., and A. J. van Rossem
1938 The Birds of El Salvador. Field Mus. Nat. Hist. Zool. Ser., 23: 1-609.
Griscom, L.
1932 The Distribution of Bird-life in Guatemala: A Contribution to the Study of the Origin of Central American Bird-life. Bull. Amer. Mus. Nat. Hist., 64: i-x; 1-439.
1935 The Ornithology of the Republic of Panama. Bull. Amer. Mus. Comp. Zool., 78: 261-382.
Monroe, B. L., Jr.
1968 A Distributional Survey of the Birds of Honduras. Amer. Ornith. Union, Ornith. Monogr. No. 7.
Russell, S. M.
1964 A Distributional Study of the Birds of British Honduras. Amer. Ornith. Union, Ornith. Monogr. No. 1.
Slud, P.
1964 The Birds of Costa Rica: Distribution and Ecology. Bull. Amer. Mus. Nat. Hist., 128: 1-430.
Smithe, F. B.
1966 The Birds of Tikal [Guatemala]. Natural History Press, Garden City, New York.
Sturgis, B. B.
1928 Field Book of Birds of the Panama Canal Zone. G. P. Putnam's Sons, New York.
Wetmore, A.
1946 The Birds of San José and Pedro González Islands, Republic of Panama. Smithsonian Misc. Coll., 106 (1): 1-60.
1957 The Birds of Isla Coiba, Panama. Smithsonian Misc. Coll., 134 (9): 1-105.
1965 The Birds of the Republic of Panama. Part 1: Tinamidae to Rynchopidae. Smithsonian Misc. Coll., 150.
1968 The Birds of the Republic of Panama. Part 2: Columbidae to Picidae. Smithsonian Misc. Coll., 150.

WEST INDIES

Allen, R. P.
1961 Birds of the Caribbean. Viking Press, New York.
Bond, J.
1961 Birds of the West Indies: A Guide to the Species of Birds That Inhabit the Greater Antilles, Lesser Antilles, and Bahama Islands. Houghton Mifflin Company, Boston.
Herklots, G. A. C.
1961 The Birds of Trinidad and Tobago. Collins, London.
Leopold, N. F.
1963 Checklist of Birds of Puerto Rico and the Virgin Islands. University of Puerto Rico Agricultural Experiment Station, Rio Piedras.
Voous, K. H.
1955 The Birds of St. Martin, Saba, and St. Eustatius. Studies on the Fauna of Curacao and Other Caribbean Islands, No. 25. (Obtainable from the Zoological Laboratory of the State University, Utrecht, Holland.)
1957 The Birds of Aruba, Curacao, and Bonaire. Studies on the Fauna of Curacao, and Other Caribbean Islands, No. 29. (Obtainable from the Zoological Laboratory of the State University, Utrecht, Holland.)
Wetmore, A.
1927 The Birds of Puerto Rico and the Virgin Islands. New York Acad. Sci., Sci. Surv. Puerto Rico and Virgin Islands, Vol. 9, Pts. 3 and 4.
Wetmore, A., and B. H. Swales
1931 The Birds of Haiti and the Dominican Republic. U. S. Natl. Mus. Bull. 155.

SOUTH AMERICA

General

Meyer de Schauensee, R.
1966 The Species of Birds of South America and Their Distribution. Academy of Natural Sciences of Philadelphia.
Olrog, C. C.
1968 Las Aves Sudamericanas: Una Guia de Campo. Volume 1 (Penguins through Woodpeckers). Fundacion-Instituto Miguel Lillo, Universidad de Tucumán, Tucumán Argentina. (Second and final volume to follow.)

Countries or Regions

Cawkell, E. M., and J. E. Hamilton
1961 The Birds of the Falkland Islands. Ibis, 103a: 1-27.
Chapman, F. M.
1917 The Distribution of Bird-life in Colombia: A Contribution to a Biological Survey of South America. Bull. Amer. Mus. Nat. Hist., 36: i-x; 1-729.
1926 The Distribution of Bird-life in Ecuador: A Contribution to a Study of the Origin of Andean Bird-life. Bull. Amer. Mus. Nat. Hist., 55: i-xiv; 1-784.
Friedmann, H., and F. D. Smith, Jr.
1950 A Contribution to the Ornithology of Northeastern Venezuela. Proc. U. S. Natl. Mus., 100: 411-538.
1955 A Further Contribution to the Ornithology of Northeastern Venezuela. Proc. U. S. Natl. Mus., 104: 463-524.
Haverschmidt, F.
1968 Birds of Surinam. Oliver and Boyd, Edinburgh.
Hellmayr, C. E.
1929 A Contribution to the Ornithology of Northeastern Brazil. Field Mus. Nat. Hist. Zool. Ser., 12 (18): 235-501.
Hudson, W. H.
1920 Birds of La Plata. Volumes 1 and 2. E. P. Dutton and Company, New York.
Johnson, A. W.
1965- The Birds of Chile and Adjacent Regions of Argentina, Bolivia and Peru. Volumes 1 and
67 2. Buenos Aires. (Available in United States from Pierce Book Company, Winthrop, Iowa 50682.)

Koepcke, M.
  1964 Las Aves del Departmento de Lima. Casilla 5129, Miraflores, Lima.
Meyer de Schauensee, R.
  1964 The Birds of Colombia and Adjacent Areas of South and Central America. Livingston
        Publishing Company, Narberth, Pennsylvania.
Mitchell, M. H.
  1957 Observations on Birds of Southeastern Brazil. University of Toronto Press.
Murphy, R. C.
  1936 Oceanic Birds of South America. Two volumes. American Museum of Natural History,
        New York.
Naumberg, E. M. B.
  1930 The Birds of Matto Grosso, Brazil. A Report on the Birds Secured by the Roosevelt-
        Rondon Expedition. Bull. Amer. Mus. Nat. Hist., 60: i-ix; 1-432.
Nelson, B.
  1968 Galapagos: Islands of Birds. William Morrow and Company, New York.
Olrog, C. C.
  1959 Las Aves Argentinas: Una Guia de Campo. Instituto Miguel Lillo, Universidad de Tucu-
        mán.
  1963 Lista y Distribución de las Aves Argentinas. Instituto Miguel Lillo, Universidad de Tucu-
        mán.
Phelps, W. H., and W. H. Phelps, Jr.
  1963 Lista de las Aves de Venezuela con su Distribución. Volume 1, Part 2. Passeriformes.
        Second edition. Bol. Soc. Venezolana Cien. Nat., 24 (104-105): 1-498.
  1963 Lista de las Aves de Venezuela con su Distribución. Volume 1, Part 2. Passeriformes.
        Second edition. Bol. Soc. Venezolana Cien. Nat., 24 (104-10): 1-498.
Snyder, D. E.
  1966 The Birds of Guyana. Peabody Museum, Salem, Massachusetts.
Wetmore, A.
  1926 Observations on the Birds of Argentina, Paraguay, Uruguay, and Chile. U. S. Natl. Mus.
        Bull. 133.
Zimmer, J. T.
  1931- Studies of Peruvian Birds. (Altogether 66 studies published in American Museum of
  55   Natural History Novitates, the first being Novitates No. 500, the last No. 1723.)

ATLANTIC ISLANDS

Bannerman, D. A.
  1963 Birds of the Atlantic Islands. Volume 1. A History of the Birds of the Canary Islands and
        of the Salvages. Oliver and Boyd, Edinburgh.
Bannerman, D. A., and W. M. Bannerman
  1965 Birds of the Atlantic Islands. Volume 2. A History of the Birds of the Madeira, the Des-
        ertas, and the Porto Santo Islands. Oliver and Boyd, Edinburgh.
  1966 Birds of the Atlantic Islands. Volume 3. A History of the Birds of the Azores. Oliver and
        Boyd, Edinburgh.
  1968 Birds of the Atlantic Islands. Volume 4. History of the Birds of the Cape Verde Islands.
        Oliver and Boyd, Edinburgh.
Bourne, W. R. P.
  1957 The Breeding Birds of Bermuda. Ibis, 99: 94-105.
Elliott, H. F. I.
  1957 A Contribution to the Ornithology of the Tristan da Cunha Group. Ibis, 99: 545-586.

EUROPE AND ASIA

General

Vaurie, C.
  1959 The Birds of the Palearctic Fauna: A Systematic Reference. Order Passeriformes. H. F.
        and G. Witherby, London.
  1965 The Birds of the Palearctic Fauna: A Systematic Reference. Non Passeriformes. H. F. and
        G. Witherby, London.

EUROPE

General

Peterson, R. T., G. Mountfort, and P. A. D. Hollom
  1966   A Field Guide to the Birds of Britain and Europe. Revised and enlarged edition. Houghton Mifflin Company, Boston. (Includes Iceland.)
Voous, K. H.
  1960   Atlas of European Birds. Thomas Nelson and Sons, London.

British Isles

Bannerman, D. A.
  1953-63   The Birds of the British Isles. Twelve volumes. Oliver and Boyd, Edinburgh.
Baxter, E. V., and L. J. Rintoul
  1953   The Birds of Scotland: Their History, Distribution, and Migration. Two volumes. Oliver and Boyd, Edinburgh.
Hollom, P. A. D.
  1960   The Popular Handbook of Rarer British Birds. H. F. and G. Witherby, London.
  1962   The Popular Handbook of British Birds. Revised edition. H. F. and G. Witherby, London.
London Natural History Society
  1964   The Birds of the London Area. A new revised edition. Rupert Hart-Davis, London.
Ruttledge, R. F.
  1966   Ireland's Birds: Their Distribution and Migrations. H. F. and G. Witherby, London.
Witherby, H. F., and others
  1938-   The Handbook of British Birds. Five volumes. Reprinted with revisions, 1943-44. H. F. and
  41   G. Witherby, London.

Other Islands

Robert, E. L.
  1954   The Birds of Malta. Progress Press, Malta.
Timmermann, G.
  1938-   Die Vögel Islands. Visindafélag Íslendinga Nos. 21, 24, and 28. Reykjavík (Iceland; in
  49   German.)

Continental Europe

Bauer, K. M., and U. N. Glutz von Blotzheim, Editors
  1966-   Handbuch der Vögel Mitteleuropas. Volume 1 (Gaviiformes-Phoenicopteriformes), Vol-
  69   ume 2, Parts 1 and 2 (Anseriformes). Akademische Verlagsgesellschaft, Frankfurt am Main, Germany. (Other volumes to follow.)
Curry-Lindahl, K.
  1959-   Våra Fåglar i Norden. Four volumes. Second edition. Bokforläget Natur och Kultur, Stock-
  63   holm. (Includes all of the Scandinavian Peninsula.)
Dement'ev, G. P., and others
  1966-   Birds of the Soviet Union. Volumes 1, 2, 4, and 6. Translated from the Russian. Isreal Pro-
  68   gram for Scientific Translation, Jerusalem. Available from U. S. Department of Commerce, Clearinghouse for Federal Scientific and Technical Information, Springfield, Virginia.
Glutz von Blotzheim, U.
  1961   Die Brutvögel der Schweiz. Schweizerische Vogelwarte Sempach, Aarau. (Breeding birds of Switzerland; in German.)
Jespersen, P.
  1946   The Breeding Birds of Denmark, with Special Reference to Changes during the Last Century. Ejnar Munksgaard, Copenhagen. (Includes an annotated list of breeding species; in English.)
Kanellis, A., Editor
  1969   Catalogus Faunae Graeciae. Pars 2. Aves. Thessaloniki. (In German; privately printed.)
Keve, A.
  1960   Nomenclator Avium Hungariae. Hungarian Ornithological Institute, Budapest. (In German and Hungarian.)

Lambert, A.
   1957   A Specific Check List of the Birds of Greece. Ibis, 99: 43-68.
Lletget, A. G.
   1945   Sinopsis de las Aves de España y Portugal. Trab. Inst. Cien. Nat. Madrid, 2: 1-346.
Løppenthin, B.
   1950   Birds [in Denmark]. In *List of Danish Vertebrates*. Dansk Videnskabs Forlag, Copen-
          hagen. (In English.)
Løvenskiöld, H. L.
   1947-50   Handbok over Norges Fugler. Gyldendal, Oslo.
Matvejev, S. D.
   1950   La Distribution et la Vie des Oiseaux en Serbie. Acad. Serb. Sci. Monogr., 161 (3): 1-363.
          (Jugoslavia; in French.)
Mayaud, N.
   1953   Liste des Oiseaux de France. Alauda, 21: 1-63.
Merikallio, E.
   1958   Finnish Birds: Their Distribution and Numbers. Fauna Fennica 5. (In English.)
Moltoni, E.
   1945   Elenco degli Uccelli Italiani. Rivista Italiana di Ornitologia, Milano. (An annotated
          check-list.)
Mountfort, G., and I. J. Ferguson-Lees
   1961   Observations on the Birds of Bulgaria. Ibis, 103a: 443-471.
Niethammer, G., H. Kramer, and H. E. Wolters
   1964   Die Vögel Deutschland: Artenliste. Akademische Verlagsgesellschaft, Frankfurt am
          Main, Germany. (Check-list.)
Pateff, P.
   1950   The Birds of Bulgaria. Bulgarian Academy of Sciences, Sofia. (Summary in English.)
Salomonsen, F.
   1963   Oversigt over Danmarks Fugle. Ejnar Munksgaard, Copenhagen. (Check-list; partly in
          English.)
van Ijzendoorn, A. L. J.
   1950   The Breeding Birds of the Netherlands. E. J. Brill, Leiden. (In English.)
Verheyen, R.
   1943-   Les Oiseaux de Belgique. Eight parts. Institut Royal des Sciences Naturelles de Belgique,
   52   Brussels.

ASIA

Northern and Central Asia

Austin, O. L., Jr.
   1948   The Birds of Korea. Bull. Mus. Comp. Zool., 101: 1-301.
Austin, O. L., Jr., and N. Kuroda
   1953   The Birds of Japan: Their Status and Distribution. Bull. Mus. Comp. Zool., 109: 277-637.
Caldwell, H. R., and J. C. Caldwell
   1931   South China Birds: A Complete, Popular and Scientific Account of Nearly Five Hundred
          and Fifty Forms of Birds Found in Fukien, Kwangtung, Kiangsi, Kiangsu, and Chekiang
          Provinces. Hester May Venderburgh, Shanghai.
Dement'ev, G. P., and others
   1966-68   Birds of the Soviet Union. (See under Europe.)
MacFarlane, A. M., and A. D. Macdonald
   1966   An Annotated Checklist of the Birds of Hong Kong. Hong Kong Bird Watching Society.
          (Obtainable from the Secretary of the Society, care of The Chartered Bank, Hong Kong.)
Vaurie, C.
   1964   A Survey of the Birds of Mongolia. Bull. Amer. Mus. Nat. Hist., 127: 105-143.
Yamashina, Y.
   1961   Birds in Japan: A Field Guide. Tokyo News Service.

Southeastern Asia

Ali, S.
   1949   Indian Hill Birds. Oxford University Press, London.

1961 The Book of Indian Birds. Sixth edition (revised and enlarged). Bombay Natural History Society.

1962 The Birds of Sikkim. Oxford University Press, London.

Bates, R. S. P., and E. H. N. Lowther
1952 Breeding Birds of Kashmir. Oxford University Press, London.

Chasen, F. N.
1935 A Handlist of Malaysian Birds. Bull. Raffles Mus. No. 11, Singapore.

Deignan, H. G.
1963 Checklist of the Birds of Thailand. U. S. Natl. Mus. Bull. 226.

Delacour, J.
1947 Birds of Malaysia. Macmillan Company, New York.

Glenister, A. G.
1951 The Birds of the Malay Peninsula, Singapore and Penang: An Account of All the Malayan Species, with a Note on Their Occurrence in Sumatra, Borneo, and Java and a List of Birds of Those Islands. Oxford University Press, London.

Henry, G. M.
1955 A Guide to the Birds of Ceylon. Oxford University Press, London.

Lekagul, B.
1968 Bird Guide of Thailand. Privately published. (Available from the author, care of Association for the Conservation of Nature, 4 Old Custom House, Bangkok, Thailand.)

Ripley, S. D., II
1961 A Synopsis of the Birds of India and Pakistan Together with Those of Nepal, Sikkim, Bhutan, and Ceylon. Bombay Natural History Society.

Smythies, B. E.
1953 The Birds of Burma. Second (revised) edition. Oliver and Boyd, Edinburgh.

Whistler, H.
1949 Popular Handbook of Indian Birds. Revised by N. B. Kinnear. Oliver and Boyd, Edinburgh.

Wildash, P.
1968 Birds of South Vietnam. Charles E. Tuttle Company, Rutland, Vermont.

Southwestern Asia

Allouse, B. E.
1960 Birds of Iraq. Ar-Rabitta Press, Baghdad.

Arnold, P.
1962 Birds of Israel. Shalit Publishers, Haifa.

Bannerman, D. A., and W. M. Bannerman
1958 Birds of Cyprus. Oliver and Boyd, Edinburgh.

Meinertzhagen, R.
1954 Birds of Arabia. Oliver and Boyd, Edinburgh.

Paludan, K.
1959 On the Birds of Afghanistan. Vidensk. Medd. Dansk Naturh. For., 122: 1-332. (In English.)

AFRICA

Northern Africa

Archer, G. F., and E. M. Godman
1937- The Birds of British Somaliland and the Gulf of Aden. Four volumes. Oliver and Boyd,
61 Edinburgh.

Cave, F. O., and J. D. Macdonald
1955 Birds of the Sudan: Their Identification and Distribution. Oliver and Boyd, Edinburgh.

Etchécopar, R. D., and F. Hüe
1967 The Birds of North Africa from the Canary Islands to the Red Sea. Oliver and Boyd, Edinburgh.

Heim de Balsac, H., and N. Mayaud
1962 Les Oiseaux du Nord-ouest de l'Afrique. Lechevalier, Paris.

Central and Southern Africa

Bannerman, D. A.
  1953   The Birds of West and Equatorial Africa. Two volumes. Oliver and Boyd, Edinburgh.

Benson, C. W., and M. P. S. Irwin
  1967   A Contribution to the Ornithology of Zambia. Zambia Mus. Paper No. 1.

Chapin, J. P.
  1932- The Birds of the Belgian Congo. Four parts. Bulletin of the American Museum of Natural
  54   History, Volume 65 (1932); 75 (1939); 75A (1953); 75B (1954).

Clancey, P. A.
  1964   The Birds of Natal and Zululand. Oliver and Boyd, Edinburgh.

Dekeyser, P. O., and J. H. Derivot
  1966   Les Oiseaux de l'Ouest Africain. Institut Fondamental d'Afrique Noire, Université de
         Dakar. (A guide to identification, illustrating 1,160 species.)

Jackson, F. J., and W. L. Sclater
  1938   The Birds of Kenya Colony and the Uganda Protectorate. Three volumes. Gurney and
         Jackson, London.

McLachlan, G. R., and R. Liversidge
  1957   Roberts' Birds of South Africa. Central News Agency, Cape Town.

Mackworth-Praed, C. W., and C. H. B. Grant
  1952- Birds of Eastern and North Eastern Africa. Two volumes (second edition of first volume
  55   published in 1957). Longmans, Green and Company, London.
  1962- Birds of the Southern Third of Africa. Two volumes. Longmans, Green and Company,
  63   London.

Rand, A. L.
  1936   The Distribution and Habits of Madagascar Birds: A Summary of Field Notes of the Mis-
         sion Zoologique Franco-Anglo-Americaine 'a Madagascar. Bull. Amer. Mus. Nat. Hist.,
         72: 143-499.

Smithers, R. H. N.
  1964   A Checklist of the Birds of Bechuanaland Protectorate and the Caprivi Strip: With Data
         on Ecology and Breeding. Trustees National Museums Southern Rhodesia, Bulawayo.

Smithers, R. H. N., M.P. S. Irwin, and M. L. Paterson
  1957   A Check List of the Birds of Southern Rhodesia. Rhodesia Ornithological Society.

Traylor, M. A.
  1963   Check-list of Angolan Birds. Publ. Culturais Comp. Diamantes de Angola, No. 61.

Williams, J. G.
  1964   A Field Guide to the Birds of East and Central Africa. Houghton Mifflin Company, Boston.
  1968   A Field Guide to the National Parks of East Africa. Houghton Mifflin Company, Boston.
         (Part 3, pp. 230-341, concerns the rarer birds.)

Winterbottom, J. M.
  1968   A Check List of the Land and Fresh Water Birds of the Western Cape Province. Ann. S.
         Africa Mus., 53: 1-276.

AUSTRALIA, NEW ZEALAND, AND ISLANDS
OF THE WESTERN PACIFIC

Australia

Cayley, N. W.
  1966   What Bird Is That? A Guide to the Birds of Australia. Fourth edition. Angus and Robert-
         son, Sydney.

Condon. H. T.
  1969   A Handlist of the Birds of South Australia. Third edition, revised and enlarged. South
         Australian Ornithological Association.

Hill, R.
  1967   Australian Birds. Thomas Nelson (Australia), Melbourne.

Serventy, D. L., and H. M. Whittell
  1967   Birds of Western Australia. Fourth edition. Lambert Publications, Perth, Western
         Australia.

Sharland, M.
   1958   Tasmanian Birds: A Field Guide to the Birds Inhabiting Tasmania and Adjacent Islands, Including Sea Birds. Third edition. Angus and Robertson, Sydney.
Wheeler, W. R.
   1967   A Handlist of the Birds of Victoria. Victorian Ornithological Research Group, Melbourne.

New Zealand

Falla, R. A., R. B. Sibson, and E. G. Turbott
   1967   A Field Guide to the Birds of New Zealand. Houghton Mifflin Company, Boston.
Oliver, W. R. B.
   1955   New Zealand Birds. Second edition. A. H. and A. W. Reed, Wellington.
Westerskov, K. E.
   1967   Know Your New Zealand Birds. Whitcomb and Combs, Christchurch.

Islands of Western Pacific

Amerson, A. B., Jr.
   1969   Ornithology of the Marshall and Gilbert Islands. Atoll Res. Bull. No. 127, Smithsonian Institution, Washington, D. C.
Baker, R. H.
   1951   The Avifauna of Micronesia, Its Origin, Evolution, and Distribution. Univ. Kansas Publ. Mus. Nat. Hist., 3: 1-359.
Delacour, J.
   1966   Guide des Oiseaux de la Nouvelle-Calédonie et de Ses Dépendances. Éditions Delachaux et Niestlé, Neuchâtel, Switzerland.
Delacour, J., and E. Mayr
   1946   Birds of the Philippines. Macmillan Company, New York.
Mayr, E.
   1945   Birds of the Southwest Pacific: A Field Guide to the Birds of the Area between Samoa, New Caledonia, and Micronesia. Macmillan Company, New York.
Mercer, R.
   1966   A Field Guide to Fiji Birds. Fiji Mus. Special Publ. Ser. No. 1.
Murphy, R. C., A. M. Bailey, and R. J. Neidrach
   1954   Canton Island. Denver Mus. Nat. Hist., Mus. Pictorial No. 10. (Extensive information on birds.)
Rand, A. L., and E. T. Gilliard
   1967   Handbook of New Guinea Birds. Weidenfeld and Nicholson, London.
Rand, A. L., and D. S. Rabor
   1960   Birds of the Philippine Islands: Siquijor, Mount Malindang, Bohol, and Samar. Chicago Nat. Hist. Mus., Fieldana: Zool., 35: 221-441.
Ripley, S. D.
   1944   The Bird Fauna of the West Sumatra Islands. Bull. Mus. Comp. Zool., 94: 307-430.
Smythies, B. E.
   1960   The Birds of Borneo. Oliver and Boyd, Edinburgh.

MARINE BIRDS

Alexander, W. B.
   1963   Birds of the Ocean: Containing Descriptions of All the Sea-birds of the World, with Notes on Their Habits and Guides to Their Identification. New and revised edition. G. P. Putnam's Sons, New York.
King, W. B.
   1967   Seabirds of the Tropical Pacific Ocean: Preliminary Smithsonian Identification Manual. Smithsonian Institution, Washington, D. C.
Watson, G. E.
   1966   Seabirds of the Tropical Atlantic Ocean. Smithsonian Identification Manual. Smithsonian Institution, Washington, D. C.
Watson, G. E., R. L. Zusi, and R. W. Storer
   1963   Preliminary Field Guide to the Birds of the Indian Ocean. Smithsonian Institution, Washington, D. C.

APPENDIX F
BOOKS FOR GENERAL INFORMATION AND RECREATIONAL READING

PART I

The books listed below provide a source of general information on birds and bird life.

Allen, A. A.
1961   The Book of Bird Life. Second edition. D. Van Nostrand Company, New York.
Allen, G. M.
1925   Birds and Their Attributes. Marshall Jones Company, Boston. (Dover reprint available.)
Amadon, D.
1966   Birds around the World: A Geographical Look at Evolution and Birds. Natural History Press, Garden City, New York.
Austin, O. L., Jr.
1961   Birds of the World: A Survey of the Twenty-seven Orders and One Hundred and Fifty-five Families. Golden Press, New York.
Aymar, G. C.
1935   Bird Flight. Dodd, Mead, New York.
Barruel, P.
1954   Birds of the World: Their Life and Habits. Oxford University Press, New York.
Beebe, C. W.
1906   The Bird: Its Form and Function. Henry Holt and Company, New York. (Dover reprint available.)
Berger, A. J.
1961   Bird Study. John Wiley and Sons, New York.
Coues, E.
1903   Key to North American Birds. Fifth edition. Two volumes. Dana Estes and Company, Boston.
Darling, L. and L.
1962   Bird. Houghton Mifflin Company, Boston.
Fisher, J.
1954   A History of Birds. Houghton Mifflin Company, Boston.
Fisher, J., and R. T. Peterson
[1964] The World of Birds. Doubleday and Company, Garden City, New York.
Gilliard, E. T.
1958   Living Birds of the World. Doubleday and Company, Garden City, New York.
Grassé, P. -P., Editor
1950   Traité de Zoologie: Anatomie, Systématique, Biologie. Volume 15. Oiseaux. Masson et Cie, Paris.
Heinroth, O., and K. Heinroth
1958   The Birds. University of Michigan Press, Ann Arbor.
Hess, G.
1951   The Bird: Its Life and Structure. Greenberg, New York.
Hutson, H. P. W., Editor
1956   The Ornithologists' Guide. British Ornithologists' Union, London. (Distributed by H. F.

and G. Witherby, 5 Warwick Court, London, W. C. 1. Contains instructions for undertaking various aspects of bird study.)

Jack, A.
  1953   Feathered Wings. A Study of the Flight of Birds. Methuen and Company, London.

Knowlton, F. H.
  1909   Birds of the World. Henry Holt and Company, New York.

Lanyon, W. E.
  1963   Biology of Birds. Natural History Press, Garden City, New York.

Lister, M.
  1956   The Bird Watcher's Reference Book. Phoenix House, London.

Marshall, A. J., Editor
  1960-61   Biology and Comparative Physiology of Birds. Two volumes. Academic Press, New York.

Meinertzhagen, R.
  1959   Pirates and Predators: The Practical and Predatory Habits of Birds. Oliver and Boyd, Edinburgh.

Murphy, R. C., and D. Amadon
  1953   Land Birds of America. McGraw-Hill Company, New York.

Newton, A.
  1896   A Dictionary of Birds. Adam and Charles Black, London.

Pickwell, G. B.
  1939   Birds. McGraw-Hill Book Company, New York.

Pycraft, W. P.
  1910   A History of Birds. Methuen and Company, London.

Rand, A. L.
  1956   American Water and Game Birds. E. P. Dutton and Company, New York.
  1967   Ornithology: An Introduction. W. W. Norton and Company, New York.

Saunders, A. A.
  1954   The Lives of Wild Birds. Doubleday and Company, Garden City, New York.

Sweney, F.
  1959   Techniques for Drawing and Painting Wildlife. Reinhold Publishing Corporation, New York. (Includes birds; helpful instructions for a beginner.)

Thomson, A. L., Editor
  1964   A New Dictionary of Birds. McGraw-Hill Book Company, New York.

Thomson, J. A.
  1923   The Biology of Birds. Macmillan Company, New York.

Tunnicliffe, C. F.
  1945   Bird Portraiture. Studio Publications, New York. (Instructions for painting birds.)

Van Tyne, J., and A. J. Berger
  1959   Fundamentals of Ornithology. John Wiley and Sons, New York.

Vaucher, C.
  1960   Sea Birds. Oliver and Boyd, Edinburgh.

Wallace, G. J.
  1963   An Introduction to Ornithology. Second edition. Macmillan Company, New York.

Welty, J. C.
  1963   The Life of Birds. W. B. Saunders Company, Philadelphia.

Wetmore, A.
  1931   Birds. Smithsonian Scientific Series. Volume 9. Washington, D. C.

Wing, L. W.
  1956   Natural History of Birds: A Guide to Ornithology. Ronald Press Company, New York.

Wolfson, A., Editor
  1955   Recent Studies in Avian Biology. University of Illinois Press, Urbana.

Young, J. Z.
  1962   The Life of Vertebrates. Second edition. Oxford University Press, New York. (Chapters 16, 17, and 18 concern birds.)

PART II

The books listed below pertain more or less to ornithology and ornithological matters. All are recommended for recreational reading.

Allen, E. G.
1951   The History of American Ornithology before Audubon. Trans. Amer. Phil. Soc., 41: 387-591. (Republished by Russell and Russell, New York, 1969.)

Allen, R. P.
1957   On the Trail of Vanishing Birds. McGraw-Hill Book Company, New York.

Armstrong, E. A.
1940   Birds of the Grey Wind. Oxford University Press, New York. (Reminiscences about birds and Ireland.)
1958   The Folklore of Birds: An Enquiry into the Origin and Distribution of Some Magico-Religious Traditions. Collins, London.

Austin, E. S., Editor
1967   Frank M. Chapman in Florida: His Journals and Letters. University of Florida Press, Gainesville. (Competently organized and woven together with biographical notes by the editor.)

Barton, R.
1955   How to Watch Birds. McGraw-Hill Book Company, New York.

Bodsworth, F.
1955   Last of the Curlews. Illustrated by T. M. Shortt. Dodd, Mead and Company, New York. (A skillful narrative about the Eskimo Curlew by one of Canada's eminent naturalist-writers.)

Borland, H.
1965   Our Natural World: The Land and Wildlife of America as Seen and Described by Writers since the Country's Discovery. Doubleday and Company, Garden City, New York. (Many articles by well-known names in ornithology.)

Boynton, M. F., Editor
1956   Louis Agassiz Fuertes: His Life Briefly Told and His Correspondence Edited. Oxford University Press, New York. (The editor, Mrs. Boynton, is the daughter of Fuertes.)

Brewster, W.
1937a October Farm. From the Concord Journals and Diaries of William Brewster. With an Introduction by Daniel Chester French. Harvard University Press, Cambridge, Massachusetts.
1937b Concord River. Selections from the Journals of William Brewster. Edited by Smith O. Dexter. Harvard University Press, Cambridge, Massachusetts.

Broley, M. J.
1952   Eagle Man. Farrar, Straus and Young, New York. (About Charles L. Broley's extensive banding of Bald Eagles in Florida.)

Chapman, F. M.
1929   My Tropical Air Castle. Nature Studies in Panama. D. Appleton and Company, New York.
1933   Autobiography of a Bird Lover. D. Appleton and Company, New York.
1938   Life in an Air Castle. Nature Studies in the Tropics. D. Appleton-Century Company, New York.

Cherry-Garrard, A.
1939   The Worst Journey in the World: Antarctic 1910-1913. Chatto and Windus, London. First published in 1922. (An enduring classic that includes the incredible expedition in the Antarctic night to find the eggs of the Emperor Penguin.)

Cowles, R. B.
1959   Zulu Journal: Field Notes of a Naturalist in South Africa. University of California Press, Berkeley and Los Angeles. (Impressions and experiences obtained over a span of some fifty years. Contains considerable information about birds.)

Cruickshank, A., and H. Cruickshank
1958   1001 Questions about Birds. Dodd, Mead and Company, New York.

Cruickshank, H. G.
1941   Bird Islands down East. Macmillan Company, New York. (Experiences with birds on islands off the coast of Maine.)
1948   Flight into Sunshine. Bird experiences in Florida. Macmillan Company, New York.
1968   A Paradise of Birds. Dodd, Mead and Company, New York. (Adventures of the author and her photographer husband with birds in Texas.)

Cruickshank, H., Compiler
1964   Thoreau on Birds. McGraw-Hill Book Company, New York. (Selections from Thoreau's writings with authoritative commentary by the compiler.)

Darling, F. F.
   1940   Island Years. G. Bell and Sons, London. (Adventures of an ornithologist and his family on several uninhabited Scottish islands.)
Delacour, J.
   1966   The Living Air: The Memoirs of an Ornithologist. Country Life, London. (The autobiography of Jean Delacour.)
Eifert, V. S.
   1962   Men, Birds, and Adventures: The Thrilling Story of the Discovery of American Birds. Dodd, Mead and Company, New York.
Fisher, J.
   1951   Watching Birds. Revised edition. Penguin Books, Harmondsworth, England.
Gallico, P.
   1941   The Snow Goose. Illustrations by P. Scott. Michael Joseph, London. (A moving short story with the Snow Goose central to the theme.)
Green, R.
   1951   How I Draw Birds. A Practical Guide for the Bird-Watcher. Adam and Charles Black, London. (Highly recommended to anyone wishing to learn how to draw birds "in pencil, pen and ink, wash, or a combination of all three." Handsomely illustrated.)
Greenewalt, C. H.
   1960   Hummingbirds. American Museum of Natural History, New York. (A portfolio of ultra-high-speed photographs in full color.)
Herrick, F. H.
   1938   Audubon the Naturalist: A History of His Life and Time. Second edition. D. Appleton-Century Company, New York. (The definitive biography of John James Audubon.)
Hickey, J. J.
   1943   A Guide to Bird Watching. Oxford University Press, New York. (A stimulating introduction to the study of birds in the field.)
Hoover, H.
   1963   The Long-shadowed Forest. Thomas Y. Crowell Company, New York. (Life with her artist husband on the shores of a lake on the Canadian border of Minnesota.)
Howard, L.
   1952   Birds as Individuals. Collins, London. (Intimate studies of birds which have lost their fear of the human observer.)
Jaques, F. P.
   1942   Birds Across the Sky. Harper and Brothers, New York. (The author states in the foreword: "This is a book for people who like birds, but even more it is for those who would like to like them. I am not a born ornithologist; I only married one.")
   1951   As Far as the Yukon. Harper and Brothers, New York. (An account of a trip taken by the author and her bird-artist husband.)
Krutch, J. W., and P. S. Eriksson, Editors
   1962   A Treasury of Birdlore. Doubleday and Company, Garden City, New York. (Selections from the writings of many well-known ornithologists and naturalists.)
Lack, D.
   1965   Enjoying Ornithology. Methuen and Company, London. (Popular articles, broadcast talks, and a few more specialized studies by a professional ornithologist who enjoys bird watching.)
Lawrence, L. de K.
   1968   The Lovely and the Wild. McGraw-Hill Book Company, New York. (Studies of birds in the Pimisi Bay region of Ontario; delightfully written.)
Leopold, A.
   1949   A Sand County Almanac and Sketches Here and There. Oxford University Press, New York. (Classic essays.)
Livingston, J. A.
   1966   Birds of the Northern Forest. Paintings by J. F. Lansdowne. Houghton Mifflin Company, Boston.
   1968   Birds of the Eastern Forest. Paintings by J. F. Lansdowne. Houghton Mifflin Company, Boston.
Lorenz, K. Z.
   1952   King Solomon's Ring. New Light on Animal Ways. Thomas Y. Crowell Company, New

York. (Anecdotes about birds and other animals in captivity, by a noted German animal psychologist. Paperback edition available.)

Matthiessen, P.
  1959 Wildlife in America. The Viking Press, New York. (A historical review, handsomely illustrated, of birds and other vertebrate animals that have been depleted by man.)

McCoy, J. J.
  1966 The Hunt for the Whooping Cranes: A Natural History Detective Story. Lothrop, Lee, and Shepard Company, New York.

McMillan, I.
  1968 Man and the California Condor: The Embattled History and Uncertain Future of North America's Largest Free-living Bird. E. P. Dutton and Company, New York.

McNulty, F.
  1966 The Whooping Crane: The Bird That Defies Extinction. E. P. Dutton and Company, New York.

Murie, A.
  1961 A Naturalist in Alaska. Devin-Adair Company, New York.

Murie, M. E.
  1962 Two in the Far North. Alfred A. Knopf, New York. (About numerous expeditions with her biologist husband, Olaus J. Murie, in Alaska's wilderness.)

Murphy, R.
  1964 The Peregrine Falcon. Houghton Mifflin Company, Boston. (A charming story of the first year in the life of a Peregrine Falcon.)

Murphy, R. C.
  1947 Logbook for Grace. Macmillan Company, New York. (The author's day-to-day experiences as a young ornithologist aboard the whaling brig *Daisy* during its cruise to Antarctic waters in 1912-13.)

Nice, M. M.
  1939 The Watcher at the Nest. Macmillan Company, New York. (Dover reprint available. The author's experiences studying Song Sparrows and a few other birds; written on a popular level.)

Peterson, R. T.
  1957 The Bird Watcher's Anthology. Harcourt, Brace and Company, New York. (Selections from the writings of many ornithologists, amateur and professional, with graceful introductions by the author.)
  1964 Birds over America. New and revised edition. Dodd, Mead and Company, New York. (The author's experiences and observations during his travels through the United States.)

Peterson, R. T., and the Editors of *Life*
  1963 The Birds. Time Incorporated, New York.

Peterson, R. T., and J. Fisher
  1955 Wild America: The Record of a 30,000-mile Journey around the Continent by a Distinguished Naturalist and His British Colleague. Houghton Mifflin Company, Boston.

Pettingill, E. R.
  1960 Penguin Summer: An Adventure with the Birds of the Falkland Islands. Clarkson N. Potter, New York.

Pettingill, O. S., Jr., Editor
  1965 The Bird Watcher's America. McGraw-Hill Book Company, New York. (Forty-four leading naturalists write about the best places for birds in the United States and Canada.)

Plate, R.
  1966 Alexander Wilson: Wanderer in the Wilderness. David McKay Company, New York. (The most recent biography of "The Father of American Ornithology.")

Rand, A. L.
  1955 Stray Feathers from a Bird Man's Desk. Doubleday and Company, Garden City, New York. (Consists of 60 short chapters about aspects of bird life, many of which are unusual.)

Rankin, N.
  1951 Antarctic Isle: Wild Life in South Georgia. Collins, London. (About penguins and other birds, seals, and whales; based largely on the author's personal experiences. Illustrated with superb photographs.)

Ripley, S. D.
  1942 Trail of the Money Bird. Harper and Brothers, New York. (A trip to the South Seas and New Guinea. The "money bird" is a small species of bird-of-paradise.)
  1952 Search for the Spiny Babbler: An Adventure in Nepal. Houghton Mifflin Company, Boston.
  1957 A Paddling of Ducks. Harcourt, Brace, and Company, New York. (Experiences and techniques in raising waterfowl, and some related problems informally explained.)
Roberts, T. S.
  1960 Bird Portraits in Color. Revised by W. J. Breckenridge, D. W. Warner, and R. W. Dickerman. University of Minnesota Press, Minneapolis. (Contains the 92 full-page color plates from Roberts' *The Birds of Minnesota*, long out of print.)
Roberts, B., Editor
  1968 Edward Wilson's Birds of the Antarctic. Humanities Press, New York. (A selection of over 300 drawings and paintings by the ornithologist and medical officer on Robert Falcon Scott's two Antarctic expeditions. See also Wilson, 1967, and Seaver, 1933.)
Scott, P.
  1951 Wild Geese and Eskimos. Country Life, London. (A journal of the Perry River Expedition of 1949 in search of the nesting grounds of Ross's Geese.)
Scott, P., and J. Fisher
  1954 A Thousand Geese. Houghton Mifflin Company, Boston. (The story of an expedition to Iceland to find the principal breeding ground of the Pink-footed Goose.)
Seaver, G.
  1933 Edward Wilson of the Antarctic: Naturalist and Friend. John Murray, London. (See also B. Roberts, 1968, and Wilson, 1967.)
Stefferud, A., Editor
  1966 Birds in Our Lives. United States Department of the Interior, Washington, D. C.
Stoddard, H. L., Sr.
  1969 Memoirs of a Naturalist. University of Oklahoma Press, Norman. (Autobiography of a pioneer in wildlife and forest management.)
Stonehouse, B.
  1960 Wideawake Island: The Story of the B. O. U. Centenary Expedition to Ascension. Hutchinson and Company, London.
Sutton, G. M.
  1934 Eskimo Year. A Naturalist's Adventures in the Far North. Macmillan Company, New York.
  1936 Birds in the Wilderness. Adventures of an Ornithologist. Macmillan Company, New York.
  1961 Iceland Summer: Adventures of a Bird Painter. University of Oklahoma Press, Norman.
Terres, J. K.
  1968 Flashing Wings: The Drama of Bird Flight. Doubleday and Company, Garden City, New York.
Terres, J. K., Editor
  1958 The Audubon Book of True Nature Stories. Thomas Y. Crowell Company, New York. (A collection of writings about birds and other subjects by many well-known naturalists.)
  1961 Discovery: Great Moments in the Lives of Outstanding Naturalists. J. P. Lippincott Company, Philadelphia. (Entertainingly written, first-person accounts by many well known in ornithology.)
Thomas, R.
  1952 Crip, Come Home. Harper and Brothers, New York. (The story of a Brown Thrasher, *Toxostoma rufum*, that lived for ten years in the author's garden in Arkansas.)
Tinbergen, N.
  1953 The Herring Gull's World. A Study of the Social Behaviour of Birds. Collins, London.
  1958 Curious Naturalists. Country Life, London. (Twenty-five years of experiences in field studies by a naturalist and animal behaviorist. Available in paperback.)
Welker, R. H.
  1955 Birds and Men: American Birds in Science, Art, Literature, and Conservation, 1800-1900. The Belknap Press of Harvard University Press, Cambridge, Massachusetts. (A scholarly yet very readable book.)

Wetmore, A., and others
  1964   Song and Garden Birds of North America. National Geographic Society, Washington, D.C.
  1965   Water, Prey, and Game Birds of North America. National Geographic Society, Washington, D. C.
Williamson, K., and J. M. Boyd
  1960   St. Kilda Summer. Hutchinson and Company, London. (Studying birds on a fascinating island 100 miles west of the Scottish mainland.)
Wilson, E.
  1967   Diary of the "Discovery" Expedition to the Antarctic Regions, 1901-1904. Edited by A. Savours. Humanities Press, New York. (The diaries of Edward Wilson, pioneer Antarctic ornithologist, illustrated by many of his drawings and paintings.)

BRECKENRIDGE

APPENDIX G
CURRENT ORNITHOLOGICAL JOURNALS

Listed below are the titles of current journals; many are strictly ornithological while others frequently contain ornithological information. The name of the journal is followed by the name of the supporting organization, if any, and either the name and address of the editor or a more permanent address if available.

NORTH AMERICAN

The principal journals appearing in the United States and Canada are given in the following two lists. The first contains the four leading journals—publications which are essential for any student of American birds.

The Auk (quarterly). American Ornithologists' Union. Permanent address: Museum of Natural History, Smithsonian Institution, Washington, D. C. 20560.

Bird-Banding (quarterly). A Journal of Ornithological Investigation. Northeastern Bird-Banding Association. E. Alexander Bergstrom, 37 Old Brook Road, West Hartford, Connecticut 06117.

The Condor (quarterly). Cooper Ornithological Society. Ralph J. Raitt, Department of Biology, New Mexico State University, Las Cruces, New Mexico 88001.

The Wilson Bulletin (quarterly). Wilson Ornithological Society. Permanent address: Division of Birds, Museum of Zoology, University of Michigan, Ann Arbor, Michigan 48104.

Alabama Birdlife (quarterly). Alabama Ornithological Society. Julian L. Dusi, P. O. Box 742, Auburn 36830.

The American Midland Naturalist (quarterly). University of Notre Dame, Notre Dame, Indiana 46556.

American Zoologist (quarterly). American Society of Zoologists. Editorial office: Department of Biology, University of Virginia, Charlottesville 22903.

Animal Kingdom (bi-monthly). New York Zoological Society. Editorial office: Zoological Park, Bronx, New York 10460.

Arkansas Audubon Newsletter (quarterly). Arkansas Audubon Society. Edith M. (Mrs. Henry N.) Halberg, 5809 North Country Club Boulevard, Little Rock 72207.

Atlantic Naturalist (quarterly). Audubon Naturalist Society of the Central Atlantic States. Editorial office: 1621 Wisconsin Avenue N. W., Washington, D. C. 20007.

Audubon (bi-monthly). National Audubon Society. Editorial office: 1130 Fifth Avenue, New York, New York 10028.

The Audubon Bulletin (quarterly). Illinois Audubon Society. Permanent address: Field Museum of Natural History, Roosevelt Road and Lake Shore Drive, Chicago 60605.

Audubon Field Notes (bi-monthly). National Audubon Society in collaboration with United States Fish and Wildlife Service. Editorial office: 1130 Fifth Avenue, New York, New York 10028.

Audubon Warbler (issued monthly except July, August, September). Oregon Audubon Society. Permanent address: Pittock Bird Sanctuary, 5151 Northwest Cornell Road, Portland 97210.

The Bluebird (quarterly). The Audubon Society of Missouri. James P. Jackson, 105 Terry Lane, Washington, Missouri 63090.

The Blue Jay (quarterly). Saskatchewan Natural History Society. George F. Ledingham, 2335 Athol Street, Regina.

The Bulletin of the Connecticut Audubon Society (bi-monthly). Audubon Society for the State of Connecticut. Permanent address: Larsen Sanctuary, 2325 Burr Street, Fairfield 06430.

Bulletin of the Oklahoma Ornithological Society (quarterly). Mrs. Sophia C. Mery, 345 S. E. Boston Avenue, Bartlesville 74002.

Bulletin of the Texas Ornithological Society (quarterly). Michael Kent Rylander, Department of Biology, Texas Technological College, Lubbock 79409.

Canadian Audubon (bi-monthly). The Canadian Audubon Society. Editorial office: 46 St. Clair Avenue West, Toronto 7.

The Canadian Field-Naturalist (quarterly). Ottawa Field-Naturalists' Club. Theodore Mosquin, Plant Research Institute, Central Experimental Farm, Ottawa.

Cassinia (annual). Delaware Valley Ornithological Club. Permanent address: Academy of Natural Sciences of Philadelphia, 19th and the Parkway, Philadelphia, Pennsylvania 19103.

The Chat (quarterly). Carolina Bird Club. Permanent address: Shuford Memorial Sanctuary, P. O. Box 1220, Tryon, North Carolina 28782.

EBBA News (issued six times a year). Eastern Bird-Banding Association. Frank P. Frazier, Jr., P. O. Box 13, Long Valley, New Jersey 07853.

Ecological Monographs (quarterly). Ecological Society of America. Zoology editor: Alan E. Stiven, Department of Zoology, University of North Carolina, Chapel Hill 27514.

Ecology (bi-monthly). Ecological Society of America. Zoology editor: Monte Lloyd, Department of Biology, University of Chicago, Chicago, Illinois 60637.

The Elepaio (monthly; mimeographed). Journal of the Hawaii Audubon Society. Permanent address: P. O. Box 5032, Honolulu, Hawaii 96814.

Evolution: International Journal of Organic Evolution (quarterly). Society for the Study of Evolution. Editorial office: Department of Geophysical Sciences, University of Chicago, Chicago, Illinois 60637.

The Florida Naturalist (quarterly). Florida Audubon Society. Editorial office: South Lake Sybelia Drive, Maitland 32751.

The Great Basin Naturalist (irregular). Editorial office: Department of Zoology, Brigham Young University, Provo, Utah 84601.

The Gull (monthly). Golden Gate Audubon Society. Permanent address: P. O. Box 103, Berkeley, California 94701.

The Hermit Thrush (bi-monthly). Green Mountain Audubon Society. Mr. and Mrs. Robert N. Spear, Jr., RFD 4, Winooski, Vermont 05404.

The Indiana Audubon Quarterly. Indiana Audubon Society. Henry C. West, 4660 East 42nd Street, Indianapolis 46226.

Inland Bird Banding News (bi-monthly). Inland Bird-Banding Association. Larry L. Hood, Box 478, Laurel, Maryland 20810.

Iowa Bird Life (quarterly). Iowa Ornithologists' Union. Peter C. Petersen, Jr., 235 McClellan Boulevard, Davenport 52803.

The Jack-Pine Warbler (quarterly). Michigan Audubon Society. Permanent address: 7000 North Westnedge, Kalamazoo 49001.

The Journal of Wildlife Management (quarterly). The Wildlife Society. Permanent address: 3900 Wisconsin Avenue N. W., Washington, D. C. 20016.

Kansas Ornithological Society Bulletin (quarterly). The Kansas Ornithological Society. Richard F. Johnston, Museum of Natural History, University of Kansas, Lawrence 66044.

The Kentucky Warbler (quarterly). Kentucky Ornithological Society. Anne L. (Mrs. F. W.) Stamm, 9101 Spokane Way, Louisville 40222.

The Kingbird (quarterly). Federation of New York State Bird Clubs. Permanent address: Cornell Laboratory of Ornithology, 159 Sapsucker Woods Road, Ithaca 14850.

The Living Bird (annual). Cornell Laboratory of Ornithology, 159 Sapsucker Woods Road, Ithaca, New York 14850.

The Loon (formerly The Flicker; quarterly). Minnesota Ornithologists' Union. Permanent address: James Ford Bell Museum of Natural History, University of Minnesota, Minneapolis 55455.

Maine Field Naturalist (quarterly). Maine Audubon Society and Portland Society of Natural History. Permanent address: 22 Elm Street, Portland 04111.

Maryland Birdlife (quarterly). Maryland Ornithological Society. Permanent address: Cylburn Mansion, 4915 Greenspring Avenue, Baltimore 21209.

Massachusetts Audubon (quarterly). Massachusetts Audubon Society. Permanent address: South Great Road, Lincoln 01773.

The Migrant (quarterly). Tennessee Ornithological Society. Lee R. Herndon, Route 6, Elizabethton 37643.

The Murrelet (tri-annually). Pacific Northwest Bird and Mammal Society. Dr. Gordon D. Alcorn, University of Puget Sound, Tacoma, Washington 98406.

The Narragansett Naturalist (quarterly). Audubon Society of Rhode Island. Permanent address: 40 Bowen Street, Providence 02903.

National Wildlife (bi-monthly). National Wildlife Federation. Editorial office: 534 North Broadway, Milwaukee, Wisconsin 53202.

Natural History (monthly October through May; bi-monthly June to September). Editorial office: American Museum of Natural History, Central Park West at 79th Street, New York, New York 10024.

The Nebraska Bird Review (quarterly). Nebraska Ornithologists' Union. Permanent address: University of Nebraska State Museum, Lincoln 68508.

The New Hampshire Audubon Quarterly. Audubon Society of New Hampshire. Permanent address: 63 North Main Street, Concord, New Hampshire 03301.

New Jersey Nature News (quarterly). New Jersey Audubon Society. Editorial office: 902 Westwood Avenue, River Vale P. O., Westwood 07675..

Ontario Bird Banding (quarterly). Ontario Bird Banding Association. Douglas D. Dow, Department of Zoology, University of Western Ontario, London.

The Oriole (quarterly). Georgia Ornithological Society. Leslie B. Davenport, Jr., Biology Department, Armstrong State College, Savannah 31406.

Pacific Discovery (bi-monthly). California Academy of Sciences. Editorial office: California Academy of Sciences, Golden Gate Park, San Francisco 94118.

The Passenger Pigeon (quarterly). Wisconsin Society for Ornithology. Permanent address: 821 Williamson Street, Madison 53703.

Physiological Zoology (quarterly). Editorial office: University of Chicago Press, Chicago, Illinois 60637.

The Raven (quarterly). Virginia Society of Ornithology. Mailing address: Box 57, Charlottesville 22902.

The Redstart (quarterly). Brooks Bird Club. Permanent address: 707 Warwood Avenue, Wheeling, West Virginia 26003.

Science (weekly). American Association for the Advancement of Science. Permanent address: 1515 Massachusetts Avenue N. W., Washington, D. C. 20005.

The Scissortail (quarterly; mimeographed). Oklahoma Ornithological Society. Mrs. Sophia C. Mery, 345 S. E. Boston Avenue, Bartlesville 74002.

South Dakota Bird Notes (quarterly). South Dakota Ornithologists' Union. J. W. Johnson, 1421 Utah Avenue S. E., Huron 57350.

The Southwestern Naturalist (quarterly). Southwestern Association of Naturalists. Robert L. Packard, Department of Biology, Texas Technological College, Lubbock 79409.

Systematic Zoology (quarterly). The Society of Systematic Zoology. Editorial office: Museum of Natural History, University of Kansas, Lawrence 66044.

Utah Audubon News (monthly; mimeographed). Utah Audubon Society. Vera Cassel, 374 Sixth Avenue. Salt Lake City 84112.

Western Bird Bander (quarterly). Western Bird-Banding Association. Mrs. Eleanor L. Radke, P. O. Box 446, Cave Creek, Arizona 85331.

Zoologica (quarterly). New York Zoological Society. Editorial office: New York Zoological Park, Bronx Park, New York 10460.

FOREIGN

The journals of greatest importance appearing regularly outside the United States and Canada are in the following list.

Alauda (quarterly). La Société d'Études Ornithologiques. École Normale Supérieure, Laboratoire de Zoologie, 24, rue Lhomond, Paris 5 Ve, **France.**

Angewandte Ornithologie (issued one to two times a year). Internationalen Union für Angewandte Ornithologie. Dr. H. Bruns, D-62 Wieshaden, Postfach 169, Germany.

Animal Behaviour (quarterly). Association for the Study of Animal Behaviour. Editors: P. P. Bateson, Sub-department of Animal Behaviour, High Street, Madingley, Cambridge, England, and Dr. J. Hirsch, Department of Psychology, University of Illinois, Urbana 61801.

Animals (monthly). The International Wildlife Magazine. Nigel Sitwell, 21-22 Great Castle Street, London W.1, England. (American office: 133 East 55th Street, New York, New York 10022.)

Anzeiger der Ornithologischen Gesellschaft in Bayern. Gymnasialprofessor Dr. W. Wust, Hohenlohestr. 61, München 19, Germany.

Aquila (annual). Instituti Ornithologici Hungarici. Dr. A. Vertse, Madártani Intézet, Garas #U. 14, Budapest, II., Hungary.

Ardea (quarterly). Tijdschrift der Nederlandse Ornithologische Unie. Prof. Dr. K. H. Voous, Zöologisch Museum, Plantage Middenlaan 53, Amsterdam, The Netherlands.

Ardeola (annual). Sociedad Espanõla de Ornitologiá. Editorial office: Museo de Ciencias Naturales, Castellana, 80, Madrid-6, Spain.

Australian Natural History (quarterly). Editorial office: Australian Museum, College Street, Sydney.

Avicultural Magazine (bi-monthly). The Avicultural Society. Editor: Miss Phyllis Barclay-Smith, 51 Warwick Avenue, London, W.9, England.

Behaviour (issued four times a year). An International Journal of Comparative Ethology. Published by E. J. Brill, Leiden, The Netherlands.

Beiträge zur Vogelkunde (quarterly). Akademische Verlagsgesellschaft. Prof. Dr. Heinrich Dathe, Am Tierpark 41, DDR-1136 Berlin, Germany.

Bird Study (quarterly). British Trust for Ornithology. Editorial office: Beech Grove, Tring, Hertfordshire, England.

Birds (formerly Bird Notes; bi-monthly). The Royal Society for the Protection of Birds. Editorial office: The Lodge, Sandy, Bedfordshire, England.

Birds Illustrated (monthly). Editorial office: The Butts, Half Acre, Brentford, Middlesex, England.

Bokmakierie (quarterly). South African Ornithological Society. Editorial office: Percy Fitz-Patrick Institute of African Ornithology, University of Cape Town, Rondebosch, Cape Province, South Africa.

British Birds (monthly). Editorial office: 10 Merton Road, Bedford, England.

Bulletin of the British Ornithologists' Club (bi-monthly). C. W. Benson, University Museum, Department of Zoology, Downing Street, Cambridge, England.

Dansk Ornithologisk Forenings Tidsskrift (quarterly). Dansk Ornithologisk Forening. Finn Salomonsen, Zoologisk Museum, University of Copenhagen, Universitetsparken 15, 2100 Copenhagen Ø, Denmark.

The Emu (quarterly). Royal Australasian Ornithologists Union. Permanent address: National Museum of Victoria, 285-321 Russell Street, Melbourne 3000.

Vår Fågelvärld (quarterly). Sveriges Ornitologiska Förening. Permanent address: Östermalmsgat. 65, 114 50 Stockholm Ö. Postgiro 19 94 99, Sweden.

Die Gefiederte Welt (monthly). Dr. Joachim Steinbacher, 638 Bad Homburg v.d.H., Kinzigstr. 47, Germany.

Le Gerfaut (quarterly). L'Institut Royal des Sciences Naturelles de Belgique, 31, rue Vautier, Bruxelles 4-C.C.P. 916.81, Belgium.

Handbuch der Oologie (annual). Dr. Wilhelm Meise, Zoologisches Staatsinstitut und Zoologisches Museum, Hamburg, Germany.

El Hornero (annual). La Asociación Ornitólogica del Plata. Editorial office: Avenida Añgel Gallardo 470, Buenos Aires, Argentina.

The Ibis (quarterly). British Ornithologists' Union. Permanent address: The Bird Room, British Museum (Natural History), Cromwell Road, London, S. W. 7, England.

Irish Bird Report (annual). Irish Ornithologists' Club. Major R. F. Ruttledge, Doon, Newcastle, County Wicklow, Ireland.

The Journal of Animal Ecology (three times annually). The British Ecological Society. H. N. Southern, Animal Ecology Research Group, Department of Zoology, Botanic Garden, High Street, Oxford, England.

Journal of the Bombay Natural History Society (quarterly). Editorial office: Hornbill House, Opposite Lion Gate, Apollo Street, Fort, Bombay 1-BR, India.

The Journal of Ecology (three times annually). The British Ecological Society. P. Greig-Smith, School of Plant Biology, University College of North Wales, Bangor, Caerns, Wales.

Journal fur Ornithologie (quarterly). Deutschen Ornithologen-Gesellschaft. Prof. Dr. E. Stresemann, Zoologisches Museum, Invalidenstr. 43, Berlin 104, Germany.

Journal of Zoology (three volumes annually). Zoological Society of London. Editorial office: Regent's Park, London, N. W. 1, England.

Larus (annual). Ornitološki Zavoda u Zabrebu. The Department of Ornithology, Institute of Biology, University, Zagreb, Jugoslavia.

Limosa (quarterly). Nederlandse Ornithologishe Unie. Dr. C. G. B. ten Kate, Fernhoutstraat 13, Kampen, The Netherlands.

Nature (weekly). MacMillan (Journals) Limited, 4 Little Essex Street, London WC2, England.

Nos Oiseaux (bi-monthly). La Société Romande pour l'Etude et la Protection des Oiseaux. Paul Géroudet, Avenue de Champel 37, 1206 Genéve, Switzerland.

Notornis (quarterly). Ornithological Society of New Zealand. R. B. Sibson, 26 Entrican Avenue, Remuera, S. E. 2, New Zealand.

L'Oiseau et la Revue Francaise d'Ornithologie (quarterly). La Société Ornithologique de France. Museum d' Historie Naturelle, 55, rue de Buffon, Paris (Ve), France.

Oiseaux de France. Bulletin Trimestriel du Groupe des Jeunes Ornithologistes. 129, Boulevard Saint-Germain, Paris 6e, France.

The Oologists' Record (quarterly). Editorial office: Five Magpies, Elton, Newnham-on-Severn, Gloucestershire. England.

Ornis Fennica (quarterly). Ornitologiska Föreningen i Finland. Editorial office: Department of Zoology, University of Turku, Turku 2, Finland.

Der Ornothologische Beobachter (bi-monthly). Schweizerische Gesellschaft für Vogelkunde und Vogelschutz. Dr. E. Sutter, Naturhistorisches Museum, Augustinerstrgasse 2, 4051 Basel, Switzerland.

Ornithologische Mitteilungen (monthly). Dr. H. Bruns, Institut für Biologie und Lebensschutz, D-6229 Schlangenbad-Georgenborn, Schlosspark Hohenbuchau, Weiherallee 29, Germany.

The Ostrich (quarterly). South African Ornithological Society. Editorial office: Percy Fitz-Patrick Institute of African Ornithology, University of Cape Town, Rondebosch, Cape Town, South Africa.

Pavo (two times annually). The Indian Journal of Ornithology. Professor J. C. George, M. S. University Department of Zoology, Faculty of Science, Baroda 2, India.

The Ring (quarterly). Polish Zoological Society. Dr. W. Rydzewski, Laboratory of Ornithology, Sienkiewicza 21, Wroclaw, Poland.

Rivista Italiana di Ornitologia (quarterly). Dr. E. Moltoni, Museo Civico di Storia Naturale, Corso Venezia 55, Milano, Italy.

Scottish Birds (quarterly). Scottish Ornithologists' Club. A. T. Macmillan, 12 Abinger Gardens, Edinburgh EH12 6DE, Scotland.

The South Australian Ornithologist (two times annually). South Australian Ornithological Association. Brian Glover, 14 Kauri Road, Hawthorndene, South Australia 5051.

Sterna (quarterly). Tidsskrift utgitt av Norsk Ornitologisk Forening og Stavanger Museum. Editorial office: Stavanger Museum, 4000 Stavanger, Norway.

Tori (annual). Ornithological Society of Japan. Editorial office: Yamashina Institute for Ornithology, 49 Nampeidai-machi, Shibuya-ku, Tokyo.

Wildfowl: Annual Journal of The Wildfowl Trust. Editorial office: New Grounds, Slimbridge, Gloucestershire, England.

Die Vogelwarte (quarterly) Vogelwarten Helgoland und Radolfzell. Dr. R. Drost, Vogelwarte Helgoland, 294 Wilhelmshaven-Rüstersiel, Germany.

Die Zeitschrift für Tierpsychologie (eight times annually). Dr. O. Koehler, 78 Freiburg/Br., Zoologisches Institut, Katharinenstr. 20, Germany.

APPENDIX H
CLUTCH SIZES, INCUBATION PERIODS, AND AGES AT FLEDGING

The list below gives the normal range in clutch sizes, incubation periods, and ages at first flight in 68 families of birds. The figures pertain only to species representing the families in North America north of Mexico and have been derived from various sources including the data compiled by Edgar M. Reilly, Jr. in *The Audubon Illustrated Handbook of American Birds* (McGraw-Hill Book Company, New York, 1968) and by David Lack in *Ecological Adaptations for Breeding in Birds* (Methuen and Company, London, 1968).

The student must bear in mind that the figures for incubation periods and ages at first flight are mainly estimates and serve only as a basis for comparing the different familial groups and for further investigation.

Family	Clutch Size	Incubation Period (in days)	Age at First Flight (in days)
Gaviidae	2	28-30	70-80
Podicipedidae	3-5	20-25	?
Diomedeidae, *Diomedea immutabilis* and *D. nigripes* only	1	64-65	140-165
Procellariidae, *Puffinus puffinus, P. griseus,* and *P. tenuirostris*	1	51-53	70-97
Hydrobatidae, *Oceanodroma leucorhoa*	1	41-42	63-70
Phaëthontidae *Phaëthon lepturus*	1	41-42	70-80
Pelecanidae	2-3	28-36	60-65
Sulidae, *Morus*	1	43-45	95-107
Phalacrocoracidae	3-6	28-31	46-53
Anhingidae	3-4	25-28	?
Fregatidae	1	44-55	170-190
Ardeidae	3-5	17-28	30-60
Ciconiidae	3-4	28-32	50-55
Threskiornithidae	2-4	21-24	50-56
Phoenicopteridae	1-2	28-32	75-78
Anatidae			
Cygninae, *Cygnus buccinator*	5-6	33-37	100-108
Anserinae	4-6	23-29	35-70
Dendrocygninae	12-14	?	?
Anatinae	7-12	22-28	37-56

Family	Clutch Size	Incubation Period (in days)	Age at First Flight (in days)
Aythyinae	3-12	22-29	49-77
Oxyurinae	6-10	23-24	52-56
Merginae	9-12	29-31	?
Cathartidae, *Cathartes aura* and *Coragyps atratus*	2	38-41	70-80
Accipitridae			
Except eagles	2-5	28-38	23-45
Eagles	2	43-45	70-84
Pandionidae	2-3	32-35	51-59
Falconidae, except *Caracara*	2-5	28-32	25-42
Tetraonidae	9-12	22-24	7-10
Phasianidae			
Quails	10-15	22-23	10-18
Phasianus colchicus	7-10	23-25	7-8
Meleagrididae	11-13	27-29	11-17
Gruidae	2	28-36	60-70
Aramidae	5-7	?	?
Rallidae	5-12	18-24	?
Haematopodidae	3	24-27	34-37
Charadriidae	3 or 4	21-27	27-40
Scolopacidae	4	17-28	14-21
Recurvirostridae	3-5	22-25	25-28?
Phalaropodidae	4	18-20	?
Stercorariidae	2 or 2-3	25-28	35-45
Laridae			
Larinae	1-5	20-29	35-50
Sterninae	1-3	20-35	28-35
Rynchopidae	4-5	30-32	38-42
Alcidae	1 or 2	21-39	38-40
Columbidae	1 or 2	13-19	14-28
Cuculidae	2-5	14-18	15-16?
Tytonidae	5-7	32-34	42-50
Strigidae	2-8	21-34?	23-60
Caprimulgidae	2	19-20	20-25
Apodidae, *Chaetura pelagica*	4-5	19-21	29-31
Trochilidae	2	16-17	19-22
Alcedinidae, *Megaceryle alcyon*	5-7	23-24	29-35?
Picidae, except *Campephilus*	3-10	11-14	24-28
Tyrannidae	3-5	12-16	13-16
Alaudidae	3-5	11-14	10-12
Hirundinidae	4-6	12-16	18-26
Corvidae			
Corvus	4-6	16-21	26-40
Nucifraga	2-3	17-18	24-28
All other	4-7	16-18	15-21
Paridae	4-8	11-15	14-18

Family	Clutch Size	Incubation Period (in days)	Age at First Flight (in days)
Sittidae	5-8	12-16	18-21
Certhiidae	5-6	14-15	14-15
Chamaeidae	3-5	15-16	15-16
Cinclidae	4-5	15-17	24-25
Troglodytidae	5-8	12-16	13-22
Mimidae	3-5	12-15	11-18
Turdidae	3-6	11-16	11-18
Sylviidae	4-9	13-15	10-14
Motacillidae	4-5	13-14	12-13
Bombycillidae	3-6	12-16	14-18
Ptilogonatidae	2-3	14-16	18-19
Laniidae	4-6	11-16	19-20
Sturnidae	4-6	11-13	19-22
Vireonidae	4	12-14	11-12
Parulidae	3-6	11-14	8-12
Ploceidae, *Passer*	4-6	11-14	12-15
Icteridae	3-5	11-14	9-20
Thraupidae	4-5	13-14	10-12
Fringillidae	3-5	11-15	9-14

APPENDIX I
ECTOPARASITES OF BIRDS:
A BRIEF REVIEW

by Robert E. Beer

Mallophagan Parasites of Birds

Lice of the insect order Mallophaga feed on fragments of feathers, hairs, and flakes of epidermal material which they bite off and chew with their strong, sharp mandibles. Unlike the sucking lice in the order Anoplura, Mallophaga imbibe blood only in rare instances, such as from a bleeding surface wound of the host. Only a few species attack mammals. The vast majority, the bird or biting lice, live chiefly as ectoparasites on birds.

Mallophagans rarely leave their avian hosts. Careful searchings in aggregations of birds heavily infested with lice reveal the absence of ectoparasites from the ground or nesting material that, moments before, accommodated large crowds of birds. Wild birds, raised in captivity, support few Mallophagan parasites.

These ectoparasites are reluctant to leave a host, even a dead host, and do so only when a live host actually contacts the dead body, permitting them to transfer without crossing a host-free zone. Usually, the bird lice die on the host and rather soon—within a few hours after the death of the host, though, in rare cases, workers have found live Mallophagans on drying bird skins as long as a week after the death of the bird. Because of these habits many Mallophagans in collections are specimens taken dead from the skins or carcasses of their hosts.

Most entomologists believe that the bird lice evolved from species in the order Psocoptera, the bark and book lice, a group of free-living insects that generally feed on fungus spores and fragments of organic debris; they also believe that the Mallophagan parasites of mammals evolved from the Mallophagan parasites of birds and that lice in the order Anoplura, which includes only mammal ectoparasites, evolved from the species of Mallophaga that parasitize mammals. Since the selective pressures relating to feeding habits strongly influenced the evolutionary process that produced the many species of bird lice, one can expect considerable diversity in the relationship of the various species of bird lice to their hosts.

In the order Mallophaga, the family Gyropidae contains species assigned to four genera; all are ectoparasitic on mammals most of which are neotropical. All species in the three genera of the family Boopidae parasitize Australian kangaroos and wallabies. All of the family Trimenoponidae parasitize neotropical rodents. The species of Menoponidae are widespread on birds, with members of the genera *Chapinia*, *Bucerophagus*, and *Bucerocolpocephalum* restricted to hornbills. The families Lemobothriidae, Ricinidae, and Philopteridae are also widespread on birds, with members of the philopterid genus *Aptericola* confined to New Zealand bird hosts. The Nesiotinidae are restricted to penguins and the species of Trochilophagidae to hummingbirds.

On the basis of gross morphology of the body we can sort most bird-parasitic Mallophagans into two groups: those with large heads, large mandibles, and fat bodies which typically inhabit the head and neck regions of the host; the slender, flat, small-headed species, capable of rapid and agile movements, which live on the backs and wings. The slender forms readily escape destruction by the preening actions of hosts; the fat, sluggish forms must reside in areas that the host cannot reach with the bill.

In the host association several relationships are worthy of note. The bird family Formicariidae, which includes the antbirds, is large with over 50 genera and more than 600 species and subspecies. The Mallophagan ectoparasites, recorded from antbirds, include 15 species in the genus *Formicaphagus* with hosts distributed among several genera of formacariids. The louse genus *Formicaricola* has seven species collected from a single antbird species. The two louse species in the genus *Machaerilaemus* and one member of the widespread genus *Furnaricola* occur only on antbirds.

Strong host specificity is obvious in the louse genera *Dennyus* and *Eureum* with all species ectoparasitic on swifts. Members of the genus *Odontophorus* live only on hosts in a single genus of Phasianidae; lice in the genus *Epipectus* only on Picidae; and species of *Rallicola* only on Rallidae. *Corvicola* species have appeared, thus far, only on Corvidae peculiar to the island of Guam.

At the opposite pole of host specificity are such louse genera as the *Furnaricola*, species of which occur on five families of passerines. A single species of tinamou (Tinamidae) entertains twelve species of Mallophagans in eight genera, distributed among three families. Another species of tinamou is host to 15 Mallophagan species in twelve genera, distributed among three families.

The relationship of bird lice to hosts can provide information for phyletic and zoogeographic studies. Important items germaine to the evolution of both birds and their arthropod parasites continue to emerge from the researches of ornithologists and arthropodologists, attentive to host-parasite relationships.

Listed below are references selected as an introduction to the literature on Mallophaga.

Ash, J. S.
  1960   A Study of the Mallophaga of Birds with Particular Reference to Their Ecology. Ibis, 102: 93-110.
Bergstrand, J. L., and W. D. Klimstra
  1964   Ectoparasites of the Bobwhite Quail in Southern Illinois. Amer. Midland Nat., 72: 490-498.
Carriker, M. A., Jr.
  1944   Studies in Neotropical Mallophaga (III). Proc. U. S. Natl. Mus., 95 (3180): 81-233.
  1957   Studies in Neotropical Mallophaga, XVI: Bird Lice of the Suborder Ischnocera. Proc. U. S. Natl. Mus., 196 (3375): 409-439.
  1966   A Revision of the Genus *Furnaricola* (Mallophaga) with Descriptions of New Species. Proc. U. S. Natl. Mus., 118 (3532): 405-432.
  1966   New Species and Records of Mallophaga (Insecta) from Neotropical Owls (Strigiformes). Amer. Midland Nat., 76: 74-99.
Clay, T.
  1949   Some Problems in the Evolution of a Group of Ectoparasites. Evolution, 3: 279-299.
  1951   The Mallophaga as an Aid to the Classification of Birds with Special Reference to the Structure of the Feathers. Proc. Xth Internatl. Ornith. Congr., pp. 207-215.
Elbel, R. E.
  1967   Amblyceran Mallophaga (Biting Lice) Found on the Bucerotidae (Hornbills). Proc. U. S. Natl. Mus., 120 (3558): 1-76.
Emerson, K. C.
  1954   Review of the Genus *Menopon* Nitsch, 1818 (Mallophaga). Ann. Mag. Nat. Hist., Ser. 12, No. 7: 225-232.
Hopkins, G. H. E.
  1942   The Mallophaga as an Aid to the Classification of Birds. Ibis, 84: 94-106.

Hopkins, G. H. E., and T. Clay
  1952   A Check List of the Genera and Species of Mallophaga. British Museum (Natural History),
         London.
Malcomson, R. O.
  1960   Mallophaga from Birds of North America. Wilson Bull., 72: 182-197. (According to the
         author, there are 2,600 recorded species of living Mallophaga. This paper lists 800 of these
         species and about 500 species of birds from which they have been recorded.)
Zlotorzycka, J.
  1962   Mallophaga Parasitizing with the Bird Families Columbidae and Phasianidae in Poland.
         Acta Zoologica Cracoviensia, 7 (5): 63-86.

Flies (Diptera) and Fleas (Siphonaptera) Associated with Birds

Blood-sucking flies of certain groups contain some species that show strong selectivity for bird hosts, and others that opportunistically attack birds as well as other vertebrates. Mosquitoes (family Culicidae), blackflies (Simuliidae), and punkies (Ceratopogonidae) of various species take blood meals from birds occasionally, with little predisposition to feed on particular species. Attacks of this sort are not unimportant, however, for microbial pathogens are transmitted in this manner.

Perhaps the most notorious bird-parasitic Dipterans are the louse flies (family Hippoboscidae). These flat-bodied flies, with a leathery appearance, show notable relationships with their bird and mammal hosts. More than 400 species have been described, and of these nearly 100 have been taken from hosts in 18 bird orders, with about half of the parasites exhibiting host restriction to a significant degree. Feeding by these strange flies occurs only in the adult stage and is accomplished by sucking blood from their hosts through the extensible proboscis.

Evidence seems to suggest that host specificity is often the result of habitat selection by the flies rather than selection of a particular species as a food source, though a high level monophagy is certainly present in some species. An example of habitat specificity is seen in the case of one hippoboscid species that attacks gannets, petrels, and terns indiscriminately. On the other hand, *Hippobosca struthionis* is host specific on the Ostrich *(Struthio camelus)* in Africa.

Two genera of bluebottle flies (family Calliphoridae) contain species parasitic on birds. In their larval stages the species in the genus *Apaulina* are obligatory, intermittent, nocturnal feeders on nestlings. Although birds in other orders are often attacked, most *Apaulina* species prefer passerine hosts, especially those birds that nest in holes. Almost every family in the order Passeriformes has had one or more species reported as hosting an *Apaulina*. There appears to be little, if any, monophagy in this group of parasites although *Apaulina metallica* is regarded as the most injurious ectoparasite of Bank Swallows *(Riparia riparia)*. Several species of *Protocalliphora* also spend their larval stages in bird nests, intermittently emerging from the nest material to suck blood from the nestlings.

Fleas are often encountered in bird nests and are seen less frequently on the birds themselves. Larval fleas are non-parasitic, subsisting on organic detritus of various sorts. Adult fleas feed exclusively on blood sucked from bird and mammal hosts. About 100 species and subspecies of fleas are now recognized as bird fleas, with about 20 of these occurring in North America.

Associations with particular bird species are undoubtedly governed by habitat restriction in some species, by geographic distribution in other species, and by host predilection in some of the fleas. Often, mammal fleas are found on birds that prey on mammals, for, when the prey is consumed, its parasites frequently move to the body of the predator. Thus, hawks and owls are found serving as temporary hosts for typical mammal fleas.

The true bird fleas are most often associated with birds that nest in burrows, or birds that build closely knit and protected nests. Such closed abodes characteristically provide an environment with relatively high humidity which is apparently optimum for flea development. Burrowing Owls *(Speotyto cunicularia)* and swallows are frequent hosts of fleas.

Some species of fleas, such as *Ceratophyllus garei*, parasitize ground-nesting birds regardless of the relationship of the bird species. *C. idius* is primarily a parasite of Tree Swallows *(Iridoprocne bicolor)*, but it also attacks Purple Martins *(Progne subis)*, and a few other hosts. *C. scopulorum*, like several other species in the genus, seems to be an oligophagous species, parasitizing Barn and Cliff Swallows *(Hirundo rustica* and *Petrochelidon pyrrhonota)*. Several fleas, such as *C. diffinis* and *C. gallinae*, are nonspecific in their host selection. The common sticktight flea, *Echidnophaga gallinacea*, native to the Old World but well established elsewhere, is parasitic on many mammal as well as bird species.

Listed below are references on flies and fleas commonly associated with birds. For additional information, with references, see "Nest Fauna" in this book, page 344.

Benton, A. H., and V. Shatrau
1965   The Bird Fleas of Eastern North America. Wilson Bull., 77: 76-81.
Curran, C. H.
1934   The Families and Genera of North American Diptera. Ballou Press, New York.
Hall, D. G.
1948   The Blowflies of North America. Thomas Say Foundation (Monumental Printing Company, Baltimore).
Holland, G. P.
1949   The Siphonaptera of Canada. Publ. 817, Tech. Bull. 70. Department of Agriculture, Ottawa.
1951   Notes on Some Bird Fleas with the Description of a New Species of *Ceratophyllus* and a Key to the Bird Fleas Known from Canada (Siphonaptera: Ceratophyllidae). Canadian Ent., 83: 281-289.
Hubbard, C. A.
1947   Fleas of Western North America: Their Relation to Public Health. Iowa State College Press, Ames.
Oldroyd, H.
1964   The Natural History of Flies. Weidenfeld and Nicholson, London.

Mites Associated with Birds

An astonishing number and considerable variety of mites occur often in the nests of birds. Many other mites live as parasites on and in avian hosts.

The systematics of mites (class Arachnida: subclass Acari) is in such a state of flux today that it is difficult to identify, with some degree of confidence, the taxonomic groups to which various species of bird parasites may be permanently assigned. In the discussion that follows an attempt is made to use the latest revisionary works of recent authors for the systematic alignments.

Since the animal group to which mites belong appears to have a diversity equaled only by insects, one is not surprised to find mites as successful inhabitants of a wide variety of habitats. As bird parasites, they occur as vagrant, external, and ephemeral associates of their hosts or as highly specialized internal parasites, with all degrees of host-parasite relationships between these extremes.

The most recent classification of mites separates the Acari into three orders, namely, the Parasitiformes, the Opilioacariformes, and the Acariformes. The Opilioacariformes, a small group that includes rather large, free-living tropical and subtropical species, feed on dead arthropods and possibly prey on live arthropods as well. In the remaining orders, Parasitiformes and Acariformes, large numbers of species are associated with birds in relationships ranging from casual to extremely intimate.

Of the four suborders of Parasitiformes, the most primitive is the Holothyrina. For this group, the only species known to affect birds was reported in 1922. On the island of Mauritius ducks and geese that swallowed large numbers of a species of *Holothyrus* died from poisonous secretions of the mites. Two other suborders, the Ixodides and the Argasides, embrace the hard and soft ticks respectively. Several species in the families Argasidae and Ixodidae frequently parasitize birds, and many of these species not only have a highly developed monophagy, but also a quite precise site of attachment—behind the ears, at the base of the wings, in the rump area, and so on.

Some of the species of mites in the suborder Mesostigmata are the most obvious of bird parasites, more because of their large size than because of their greater abundance. Within this group one can observe the entire range of host-parasite associations and reconstruct the probable evolution of parasitic feeding with some confidence. From the non-specialized, free-living predators in the families Macrochelidae and Parasitidae that occupy the nest niche and subsist on various arthropods also living in the nests, have evolved forms that occasionally take meals of blood exuding from the abraded skin of bird occupants. From this point it was a short step to facultative parasitism—i.e., when an organism can adapt to a parasitic or free life as opposed to obligate parasitism which insists on a parasitic existence. In such forms, represented in the mite families Laelapidae and Dermanyssidae, structural differences from the free-living types are very slight. In these forms one finds many species that are intermittent feeders, some of them attacking the host at night and retreating quickly, after engorging with blood, to take refuge in the nest material. From such facultative parasites arose the obligatory parasites displaying astounding modifications of body design, developmental processes, and behavior. Strong sclerotization of the body gives way to thin cuticles that offer little restraint during intermittent, heavy feeding which tends to balloon the body. Short, chelate appendages in the mouth parts become elongated to form highly efficient piercing stylets that in turn coalesce, producing a feeding tube. Some of the postembryonic stages, present in free-living, predaceous forms, are often omitted from the developmental cycle, and the stages that are retained move through their stadia in much shortened time periods. Some groups moved from the external surfaces of their hosts to internal niches, residing in the nares, air passages, lungs, and other tissues and organs.

In *Haemogamasus* (family Laelapidae), while undoubtedly a genus developed primarily in association with mammals, a few species parasitize birds. The structure of the mouth-parts in this genus demands that feeding consist of sucking blood or body fluids from the hosts. Most species are probably facultative parasites and able to adjust to various hosts, yet the ubiquitous and very common *Haemogamasus glasgowi*, now known from many mammal and birds hosts, may actually be a complex of several distinct species or subspecies in which monophagy is highly developed.

In the family Dermanyssidae, members of the genus *Liponyssoides* are commonly mammal parasites, but birds serve as hosts for some species. In the only other genus of the family, *Dermanyssus*, all species are bird parasites. All mites of the family are nidicolous, emerging from the shelter of the nest material or adjacent sites for brief periods of attachment to the host where they engorge with blood. The well known chicken mite, *Dermanyssus gallinae*, is a pest of poultry but occasionally parasitizes wild bird species that associate with poultry. *D. prognephilus* is commonly called the purple martin mite for its preferred host appears to be *Progne subis*, though other cavity-dwelling birds are occasionally attacked.

Several species of *Ornithonyssus* (family Macronyssidae) attack only mammals; others show strong preferences for birds. One often encounters *O. sylviarum* and *O. bursa*, two bird parasites, on many unrelated hosts. In this same family of mites is the genus *Pellonyssus*, containing species parasitic on birds only and only on birds in the orders Piciformes, Caprimulgiformes, and Passeriformes.

The large group of endoparasites of the respiratory tract, now placed in the family Rhinonyssidae, probably evolved from the external bird parasites of the family Macronyssidae. Several genera illustrate varying degrees of host specificity. The genus *Cas* is restricted to the Hirundinidae; species of *Rallinyssus* are found only on gruiform hosts; the large genera *Ptilonyssus* and *Paraneonyssus*, thus far, have been taken only from the Passeriformes. *Rhinoecius* species occur only in the nasal passages of owls, and the species of *Larinyssus* in charadriiform birds. On the other hand, species of the genus *Neonyssus* occur through the bird orders Tinamiformes, Ciconiiformes, Falconiformes, Columbiformes and Coraciiformes. Moreover, species in this genus range from monophagous, to oligophagous, and on to polyphagous. As in *Neonyssus*, one finds species of *Rhinonyssus* choosing hosts from several orders of birds.

Three groups of mites in the order Acariformes contain important bird parasites. Those in the suborder Parasitengona belong to the family Trombiculidae, the well known chiggers. Almost without exception, chiggers are parasitic on vertebrate hosts during their larval stage, but the nymphs and adults are free-living predators. Among the nearly 50 genera of trombiculids are found many that exhibit strong host affinities. *Neoschongastia* species occur almost exclusively on birds, attaching externally on the host while feeding on internal tissue fluids. Several species in this genus are monophagous; others show a tendency to attack large numbers of bird species. The members of *Toritrombicula* are known only from shore birds and the unusual *Womersia strandtmani* encysts in dermal layers of pelicans only.

Other chiggers in scattered genera frequently live on birds, and many species select their vertebrate hosts readily and opportunistically, whether the host is warm or cold blooded, feathered, furred, or scaled. Such is the case in various species of *Neotrombicula*, *Trombicula*, *Eutrombicula*, and others.

The suborder Eleutherengona of the order Acariformes contains four families of avian parasites. One family, the Syringophilidae or quill mites, invades the flight feathers of the host's wings and tail. All species of this family, which now includes only two genera, show a high degree of host specificity. Unlike *Syringophilus*, members of the genus *Picobia* reside subcutaneously in their hosts.

A second family of the Eleutherengona, the Harpyrhynchidae, also displays a highly developed monophagy. These mites, distributed in the two-genera complement of the family, are also exclusively parasites of birds. In general, they cluster externally in the head region, in groups around the eyelids, ears, and upper neck, appearing as tiny, spheroid, yellowish or whitish, sluggish creatures, with all stages of development usually represented in the cluster.

The family Speleognathidae includes species—pale-colored creatures—large enough to be seen with the naked eye, yet seldom seen because they run very rapidly on the slimy secretions that coat the air passages and air sacs. Occasionally, they dart in and out of the nares, but they never stay outside the host's body for very long. Apparently, all species in the family (which some authors combine with the family Ereynetidae) inhabit respiratory passages of the hosts, all show a highly developed host specificity, all seem to feed on mucous secretions and other tissue fluid, and most parasitize birds.

The ornithologist may also encounter a fourth family of the Eleutherengona—the Cheyletidae. All Cheyletidae, associated with birds, as far as known, appear in nests and on birds only to prey on the mites and insects often abundant in these places. Some cheyletid species, with strangely elongated body form, invade quills where they feed on such mites as the Syringophilidae. Though the vast majority of cheyletid species are free-living predators, some species are truly ectoparasitic on mammals. Therefore, truly parasitic cheyletids may be found on birds at some future time. Many of the species associated with birds are apparently so habitat specific in their requirements that they are found only in association with particular species of birds, or only in the nests of particular species.

The suborder Acaridei (= Astigmata, order Acariformes) is a large group of mites, many of which ornithologists have seen in extravagant numbers. Some of the largest populations of nidicolous species often involve members of this group. Nests sometimes seem to quiver as a result of the bustling commotion caused by thousands of individuals prowling through the nest material, scavenging on all kinds of detritus.

In addition to such free-living scavengers, largely represented in the families Acaridae, Glyciphagidae, and others, many of the Acaridei are clearly true parasites. Most ornithologists know well several families of feather mites. Habitat selection, often developed to a very precise degree, frequently isolates species, groups of species, or even particular families on certain areas of the bird host. The members of the family Analgesidae are ordinarily associated with flight feathers rather than with body feathers, and within this family one finds the genera *Analges* and *Rivoltasia* on the downy areas of the feathers, and other genera on the more distal sections of the feathers.

The families Proctophyllodidae and Dermoglyphidae are much like the Analgesidae in that they are exclusively ectoparasitic on birds and show the similar ranges of microhabitat selection on their hosts. Closely related to these groups are the Pterolichidae, often the most common feather mites on shore birds, but occurring on other birds as well; the Freyanidae, feather mites on ducks; the Epidermoptidae which attack both birds and mammals, feeding on skin as well as fur and feather fragments; and the Turbinoptidae which appear to be epidermoptids that have moved internally to occupy the respiratory passages of their hosts.

Unrelated to the analgesoid families mentioned above are some sarcoptoid bird parasites. Most of the families of sarcoptoid mites, also of the suborder Acaridei, are associated with mammals. However, the genus *Knemidokoptes* in the family Sarcoptidae contains at least three species of skin-burrowing bird parasites. The families Cytoditidae and Laminosioptidae contain endoparasitic species that encyst in subcutaneous tissues or burrow or wander freely on or in the linings of air sacs, thoracic cavities, in the lungs, on the liver, heart, and kidneys, or just about anywhere in the body cavity. Small numbers of these mites apparently do not affect the health of their hosts; occasionally, fantastic invasions kill rapidly. Species from both families occur in wild birds as well as in domestic fowl.

Listed below in decreasing order of priority are the principal references to the mites commonly associated with birds.

Evans, G. O., J. G. Sheals, and D. Macfarlane
    1961   The Terrestrial Acari of the British Isles. Volume 1, Introduction and Biology. British Museum (Natural History), London.

Strandtmann, R. W., and G. W. Wharton
    1958   A Manual of Mesostigmatid Mites Parasitic on Vertebrates. Contribution No. 4, Institute of Acarology, University of Maryland, College Park.

Baker, E. W., and others
    1956   A Manual of Parasitic Mites of Medical or Economic Importance. National Pest Control Association, New York.

Wharton, G. W., and H. S. Fuller
    1952   A Manual of the Chiggers. Mem. Ent. Soc. Washington, No. 4.

Woodrooffe, J. G.
    1953   An Ecological Study of the Insects and Mites in the Nests of Certain Birds in Britain. Bull. Ent. Res., 44: 739-772.

Cooley, R. A., and G. M. Kohls
    1944   The Argasidae of North America, Central America and Cuba. Amer. Midland Nat. Mongr. No. 1. Notre Dame University Press. (Includes a classified list of hosts.)

Arthur, D. R.
    1963   British Ticks. Butterworth and Company, London.

Fairchild, G. B., G. M. Kohls, and V. J. Tipton
    1966   The Ticks of Panama (Acarina: Ixodoidae). In *Ectoparasites of Panama*. Edited by R. L. Wenzel and V. J. Tipton. Field Museum Natural History, Chicago.

Arthur, D. R.
    1965   Ticks of the Genus *Ixodes* in Africa. Oxford University Press, New York.

Brennan, J. M., and C. E. Yunker
    1966   The Chiggers of Panama (Acarina: Trombiculidae). In *Ectoparasites of Panama*. Edited by R. L. Wenzel and V. J. Tipton. Field Museum Natural History, Chicago.

Nordberg, S.
    1936   Biologisch-okologische Untersuchungen über die Vogelnidicolen. Acta Zool. Fenn., 21: 1-168.

Dubinin, V. B.
    1951   Feather Mites (Analgesoidea). Part 1. Introduction to their study. Fauna U. S. S. R., Arachnida, 6 (5): 1-363. (In Russian.)
    1953   Feather Mites (Analgesoidea). Part 2. Families Epidermoptidae and Freyanidae. Fauna U. S. S. R., Arachnida, 6 (6): 1-411. (In Russian.)
    1956   Feather Mites (Analgesoidea). Part 3. Family Pterolichidae. Fauna U. S. S. R., Arachnida, 6 (7): 1-813. (In Russian.)

Radovsky, F. J.
    1968   Evolution and Adaptive Radiation of Gamasina Parasitic on Vertebrates (Acarina: Mesostigmata). Parasitology (Leningrad), 2 (2): 124-136. (In Russian.)

INDEX

Jay, Blue: niche, 5; coloration, 49;
    migration, 275; flight lanes, 293
Jay, Brown: air sacs, 111
Jay, Gray: salivary glands, 106
Jay, Piñon: monotypic species, 135
Jay, Scrub: range, 200
Journals, ornithological: 498
Jugulum, 11
Junco, Slate-colored:
    erythrocytes, 113; gonads, 123
Juncos: speciation, 139
Juvenal plumage, 188
Juvenile:
    wandering in migration, 278;
    definition, 371

Kea: playing, 254
Key: orders, 175; families, 177
Kidneys: function, 119
Killdeer: visual field, 129
Kinesis: bones of head, 66
Kingbird, Eastern: ecotone, 217;
    flight speed, 279; song, 321;
    nest, 344
Kingbird, Western:
    ecotone, 217;
    migration route, 284
Kingfisher, Belted:
    classification, 141;
    nomenclature, 145;
    toes, 160; community, 218
Kinglet, Ruby-crowned:
    number of feathers, 54
Kite, Everglade: niche, 5;
    numbers, 423
Kiwi: sense of smell, 126;
    egg size, 351;
    incubation period, 357
Knee, 26

Laboratory identification, 174
Lapwing, Eurasian:
    migration, 271; flight speed, 280
Lark, Horned: habitat, 5;
    tarsus, 158; toe nails, 159;
    plumage, 162; numbers of, 400
Learned behavior, 245, 247
Learning: habituation, 247
Legs: bones, 25; muscles, 76
"Lek," 331
Lice: feather, 55
Life histories: bibliography, 459
Life zones, 220
Lift: wing in flight, 22
Liver: anatomy, 86; function, 108
Locality: type, 145
Locomotion:
    identification clues, 232
Longevity, 7, 399;
    references, 410
Longspur, Chestnut-collared:
    hybridization, 137
Longspur, Lapland:
    mortality, 281

Longspur, McCown's:
    hybridization, 137
Lore, 11
Lotus-bird: carrying young, 392
Lovebird, Peach-faced:
    behavior, 260
Lumbar vertebrae, 61
Lungs: anatomy, 84
Lyrebird, Superb: rectrices, 37

Major biotic communities:
    list, 208; comparison with life
    zones, 223
Malar region, 11
Male songs, 319
Mallard: respiration, 111;
    classification, 141;
    nomenclature, 145;
    behavior, 246; learning, 248;
    orientation, 299
Mallee Fowl:
    incubation period, 357;
    development of young, 381
Mallophaga: 55; taxonomy, 140;
    of birds, 507; references, 508
Mandibles, 15
Mandibular ramus, 15
Mandibular tomia, 15
Manus: of wing, 18
Marine distribution, 206
Marine regions, 206
Martin, Purple: homing, 279;
    ectoparasites, 510
Mating: definition, 329;
    duration of, kinds of, 330;
    displays, 331; study of,
    references, 333
Meadowlark, Eastern:
    number of feathers, 54;
    sympatry, 136
Meadowlark, Western:
    sympatry, 136
Measuring: elevation of nests
    and flight paths, 431;
    specimens, 446
Megapodes: nest, 337
Melanins: in feathers, 47
Melanism, 193
Membrane, nictitating: 10
Mesquite:
    subclimax community, 215
Metacarpals: of wing, 16
Metatarsals: accessory, leg, 26
Migration: introduction, 6;
    definition, daily, seasonal, 267;
    causes of, 268; diurnal,
    nocturnal, preparation for,
    stimulus for, 269; effects of
    weather, 270; regularity of, 271;
    waves, 274; invasion, irregular,
    irruption, 275; juvenile
    wandering, reverse, speed of
    flight, 278; fat reserve,
    mortality in, 281; routes, 284;

altitudinal, 285; distance, 292;
    flight lanes,
    concentrations, 293; flocking, 294;
    direction-finding, 296;
    study of, 301; references, 304;
    song, 320
Migratory flight: altitude, 282
Migratory restlessness:
    navigation, 269, 298
Mimicry: song, 322
Mites: feather, 55;
    description, 510; references, 514
Mobbing, 256, 393
Mockingbird: ecotone, 217;
    song, 321; mimicry, 323
Moist Coniferous Forest:
    biotic community, 215
Molt: terminology, 188, 190;
    sequence, 188
Molting, 190
Monocularity, 128
Monotypic species, 135
Mouth: 15; parts of, 81;
    physiology, modifications of, 106
Murre, Common: color phases, 193
Muscles: dissection, 72; wing, 74;
    legs, pelvic appendage,
    thigh, 76; formula leg, 78;
    references, 102

Nape, 11
Nasal canthus, 10
Nasal fossa, 15
Natal down, 188
Navigation: examples, 297
Nearctic avifauna, 203
Neck, 11
Neossoptile, 42;
    development, 45
Neotropical avifauna, 203
Nervous system, 91
Nest-building: relation of song, 320;
    "symbolic," 332; definition, 336;
    selection of site, 340; duration,
    stages in, 341; role of sexes,
    time, 342; in young birds, 343;
    study of, 345; references, 346
Nestling, 371
Nests: relation of song, 320;
    relief displays, 332; definition,
    development of, kinds of, 336;
    classification, 338;
    identification, selection
    of site, 340; false, "cock,"
    refuge, 343; fauna, parasites,
    protection of, re-use of, 344;
    study of 345; references, 346;
    sanitation, 391; helpers, 392;
    measuring height of, 431;
    marking site, 432
Niche: adaptation to, 5;
    definition, 219
Nictitating membrane, 10
Nidicolous young, 371

BRECKENRIDGE